Global Environments through
the Quaternary

Global Environments through the Quaternary

Exploring Environmental Change

SECOND EDITION

David E. Anderson

Andrew S. Goudie

Adrian G. Parker

OXFORD
UNIVERSITY PRESS

OXFORD
UNIVERSITY PRESS

Great Clarendon Street, Oxford, OX2 6DP,
United Kingdom

Oxford University Press is a department of the University of Oxford.
It furthers the University's objective of excellence in research, scholarship,
and education by publishing worldwide. Oxford is a registered trade mark of
Oxford University Press in the UK and in certain other countries

First Edition copyright 2007

Impression: 1

British Library Cataloguing in Publication Data
Data available

ISBN 978–0–19–969726–7

Printed in Great Britain by
Ashford Colour Press Ltd, Gosport, Hampshire

Contents

Preface

The first edition of *Environmental Change* was written in the hot, dry summer of 1976, and appeared in 1977. Two subsequent editions appeared and attempted to reflect the rapidly gathering pace and significance of research into the Quaternary – now taken to be the last 2.6 million years of geological time. After the third edition of 1992, there was a further huge expansion in our knowledge of the Quaternary, an increasing recognition of the relevance of studying this time period for understanding human-caused climatic change, and an increasing awareness of the importance of environment-related issues in general.

Taking into account the new ways in which Quaternary environments were being studied and understood, it seemed worthwhile in 2007 to produce a new title, *Global Environments through the Quaternary*, which updated and expanded upon the themes of the 3rd edition of *Environmental Change* while also introducing additional material. Its aim was to provide a broader and more comprehensive account of the Quaternary that also put more emphasis on the relationships between past environments and people. The expanded authorship not only made this task easier: it also benefited the book through collaboration and the sharing of ideas.

This second edition of *Global Environments through the Quaternary* has the same structure as the previous edition, though the order of the chapters has been modified. It is longer, attempts to incorporate the latest literature and ideas, and it goes into more depth on a series of topics related to research methods (including palaeoecological transfer functions, dating techniques, stable isotope geochemistry, and environmental modelling) and to climatic changes on various timescales (including the Cainozoic cooling trend, orbital forcing, millennial-scale changes involving ocean–ice–atmosphere interactions, and future predictions). Time spans and dates for changes and events are presented on a calibrated calendar year timescale as much as possible; but where dates must be given in uncalibrated radiocarbon years, for example when drawing from older research literature, this is made clear in the text. The sections about timescales and chronologies have been expanded in both Chapters 1 and 2 to explain the different systems and the notations used to distinguish them.

The book also contains many new figures and plates, improved guides to further reading, an expanded glossary (for terms emboldened in the text), and end of chapter review questions. The study of the Quaternary continues to be a burgeoning field of research and the literature is now huge. We hope this volume will provide a useful guide into this exciting area.

DEA
ASG
AGP
Oxford, *July 2012.*

New to this edition

- Topics surrounding climate change over various timescales, such as the Cainozoic cooling trend, and orbital forcing, have been expanded and updated, and more background is given on the different systems and notations used to distinguish timescales and chronologies.

- More depth is provided on research methods, including palaeoecological transfer functions, dating techniques, stable isotope geochemistry, and environmental modelling.

- Sections on short-term climate oscillations, future climate trends, and the relationship between human evolution and environmental change have been enhanced with an emphasis on highlighting the relevance of the Quaternary record to current issues.

- New review questions have been added to the end of chapters to aid learning and help readers to check their understanding.

Acknowledgements

We are very grateful to Ailsa Allen of the Oxford University Centre for the Environment for producing many of the diagrams. We are also grateful to Stephen Stokes for some ideas with respect to the 1st edition. Plates were kindly made available by NASA, Google Earth, Simon Underdown, Ian Cartwright, and Tony Wilkinson. We would also like to acknowledge and thank the reviewers of the 1st edition who provided many helpful comments and suggestions that have benefited this 2nd edition, as well as Caroline Davidson, Dewi Jackson, and Philippa Hendry who oversaw various stages of its production and Julian Thomas and Marion Moffatt for their quality copy-editing and proofreading of the final version of the manuscript.

We are grateful for permission from copyright holders to reproduce or modify various figures, and we apologize if any acknowledgement has inadvertently been missed. In cases where a request for permission to use a figure was sent but there was no reply, we assumed that there was no objection to its inclusion. We acknowledge the following sources of figures: **Fig. 1.1** from *The Holocene: An Environmental History, 2nd edition* (Roberts 1998) with permission by Blackwell Publishing Ltd; **Fig. 1.6** from *Science* (Zachos et al. 2001, figure 2) reprinted with permission of AAAS; **Fig. 1.7** reprinted from *Nature* (DeConto et al. 2008, figure 4) with permission by Macmillan Publishers Ltd; **Fig. 2.6** reprinted with permission from *Radiocarbon* (Reimer et al. 2009, figure 4); **Fig. 2.7** drawn from data held at www.radiocarbon.org with permission from *Radiocarbon*; **Fig. 2.8** from *Quaternary Dating Methods* (Walker 2005, figure 7.9) by permission of John Wiley & Sons Ltd; **Fig. 3.1** from *Science* (Zachos et al. 2001, figure 2) reprinted with permission from AAAS; **Fig. 3.3** drawn by K. Cohen as a summary from the full chronostratigraphical table (Cohen and Gibbard) in www.quaternary.stratigraphy.org reproduced with kind permission by K. Cohen; **Fig. 3.4** from *Journal of Geophysical Research* 100 (B2) (Verosub and Roberts 1995) with permission from the American Geophysical Union; **Fig. 3.6** from *Quaternary International* (Ehlers and Gibbard 2007, figures 6 and 8) reprinted with permission by Elsevier; **Fig. 3.7** from *The Physical Geography of North America* (Orme 2002) in chapter by Menzies, reprinted by permission of Oxford University Press; **Fig. 3.8** from *Quaternary Glaciations – Extent and Chronology – A Closer Look* (Ehlers et al. 2011) in Chapter 7 by Gibbard and Clark, figure 7.2, p. 77, reprinted with permission by Elsevier; **Fig. 3.10** from *The Ice Age World* (Andersen and Borns 1994, figure 2-3) reprinted with permission by Scandinavian University Press; **Fig. 3.13** from *Geomorphology and Soils* (Richards et al. 1985, figure 18.2 in section by Rose et al.) reprinted with permission of Allen & Unwin; **Fig. 3.15** from *Palaeogeography, Palaeoclimatology, Palaeoecology* (Soreghan et al. 2008) reprinted with permission of Elsevier; **Fig. 3.21** reprinted from *Nature* (Bond et al. 1993, figure 3) with permission by Macmillan Publishers Ltd; **Fig. 3.22** reproduced from the *Atlas of Palaeovegetation*, Quaternary Environments Network (edited by Adams and Faure 1997), by kind permission of J. Adams; **Fig. 3.25** from *Journal of Quaternary Science* (Reille et al. 2000, figure 7) reprinted with permission of John Wiley & Sons Ltd; **Fig. 3.26** reproduced from the *Atlas of Palaeovegetation*, Quaternary Environments Network (edited by Adams and Faure 1997), by kind permission of J. Adams; **Fig. 3.27** reprinted from *Nature* (Dansgaard et al. 1993, figure 1) with permission by Macmillan Publishers Ltd; **Fig. 3.29** from *Handbook of Holocene Palaeoecology and Palaeohydrology* (Berglund 1986), figure 1.2 in chapter by Birks, reprinted with permission of John Wiley & Sons Ltd; **Fig. 4.1** from *Quaternary Science Reviews* (deMenocal et al. 2000a, figure 3) reprinted with permission of Elsevier; **Fig. 4.2** from *Annals of Glaciology* (Thompson et al. 1990) with permission of the International Glaciological Society; **Fig. 4.3** from *Nature* (Lambert et al. 2008, figure 1) with permission by Macmillan Publishers Ltd; **Fig. 4.7** from *Quaternary Science Reviews* (Fitzsimmons et al. 2007) reprinted with permission of Elsevier; **Fig. 4.8** from *Great Warm Deserts of the World* (Goudie 2002) reprinted with permission of Oxford University Press; **Fig. 4.9** from *Quaternary Research* (Patrickson et al. 2010, figure 7) reproduced with permission by Elsevier; **Fig. 4.10** from *Quaternary Science Reviews* (Tchakerian and Lancaster 2001, figure 2) reprinted with permission of Elsevier; **Fig. 4.11** from *Quaternary Research* (Cross et al. 2001, figure 1) reprinted with permission of Elsevier; **Fig. 4.12** from *Great Warm Deserts of the World* (Goudie 2002) reprinted with permission of Oxford University Press; **Fig. 4.13** from *Palaeogeography, Palaeoclimatology, Palaeoecology* (Shanahan et al. 2006) reprinted with permission of Elsevier; **Fig. 4.14** from *Palaeogeography, Palaeoclimatology, Palaeoecology* (Scholz et al. 2011, figure 4) with permission of Elsevier; **Fig. 4.15** from *Journal of Quaternary Science* (Thomas et al. 2012, figure 1) reproduced with permission of John Wiley & Sons Ltd; **Fig. 4.16** from *Great Warm Deserts of the World* (Goudie 2002) reprinted with permission of Oxford University Press; **Fig. 4.17** from *Quaternary Science Reviews* (Robinson et al. 2006, figure 6) reproduced with permission of Elsevier; **Fig. 4.19** from *Quaternary International* (Kafri and Yechizli 2012, figure 8) reprinted with permission of Elsevier; **Fig. 4.21** from *Quaternary Research* (Ledru et al. 2005, figure 3) reprinted with permission of Elsevier; **Fig. 5.4** from *Nature* (Murton et al. 2010, figure 1) reprinted with permission by Macmillan Publishers Ltd; **Fig. 5.6** from *Quaternary Science Reviews* (Kutzbach et al. 1998, figure 1) reprinted with permission of Elsevier; **Fig. 5.9** from figures by H.J.B. Birks (1990, pp. 133–158) in *Landscape-Ecological Impact of Climatic Change* (Boer, M.M. and de Groot, R.S. eds) reprinted with permission of IOS Press; **Fig. 5.12** from *Global and Planetary Change* (Williams et al. 2009, figure 4) reprinted with permission from Elsevier; **Fig. 5.13** from *Vegetation History and Archaeobotany* (Ren 2007, figure 3) reprinted with permission of Springer-Verlag; **Fig. 5.17** from *Quaternary Science Reviews* (Wanner et al. 2008, figure 4) reprinted with permission from Elsevier; **Fig. 5.18** from *Geografiska Annaler* (Matthews and Briffa 2005, figure 2) reprinted with permission by Blackwell Publishing Ltd; **Fig. 5.20a,b** from *Nature* (Ritchie et al. 1985) reprinted with permission of Macmillan Publishers Ltd; **Fig. 5.21** from *Quaternary International* (Bubenzer et al. 2007, figure 2) reprinted with permission of Elsevier; **Fig. 6.4** from *Journal of Geophysical Research* (McNutt and Menard 1978) permission from the American Geophysical Union; **Fig. 6.11** from *Coral Reefs* (Toscano and Macintyre 2003, figure 6) reprinted with permission from Springer-Verlag; **Fig. 6.13** from *Quaternary Research* (Clark and Lingle 1979, figure 1) reprinted with permission of Elsevier; **Fig. 6.15** from *Quaternary Science Reviews* (Gehrels 2010b, figure 1) reprinted with permission from Elsevier; **Fig. 6.17** from *Boreas* (Behre 2007, figure 6) reprinted with permission from John Wiley & Sons Ltd; **Fig. 6.18a** from *Shorelines and Isostasy* (Smith and Dawson 1983), figure 8 in chapter by Shennan, reprinted with permission of Elsevier; **Fig. 6.18b** from *Boreas* (Behre 2007, figure 7) reprinted with permission from John Wiley & Sons Ltd; **Fig. 7.1** from *Geographical Journal* (Lamb and Morth 1978, figure 11) reprinted with permission of John Wiley & Sons Ltd; **Fig. 7.6** from *Journal of Climatology* (Wigley and Jones 1987, figure 2) reprinted with permission from John Wiley & Sons Ltd; **Fig. 7.7** from *Nature* (Wigley and Atkinson 1977) reprinted with permission of Macmillan Publishers Ltd; **Fig. 7.10** from *Geographical Journal* (Walsh et al. 1988, figures 4 and 5) reprinted with permission of John Wiley & Sons Ltd; **Fig. 7.19** from *Global Environmental Change* (Conway 2005, figure 3) reprinted with permission of Elsevier; **Fig. 7.22** from *Lakes and Reservoirs: Research and Management* (Micklin 2010, figure 5) reprinted with permission of John Wiley & Sons Ltd; **Fig. 7.27** from *Journal of Arid Environments* (Wang et al. 2004b, figure 6) reprinted with permission of Elsevier; **Fig. 8.1** reproduced by kind permission of Simon Underdown from his original; **Fig. 8.2** from *Earth and Planetary Science Letters* (deMenocal 2004, figure 11) reprinted with permission of Elsevier; **Fig. 8.3** from *Journal of Human Evolution* (Bobe et al. 2002, figure 5) reprinted with permission of Elsevier; **Fig. 8.4** from *Philosophical Transactions of the Royal Society, London, B* (Stringer 2002, figure 2) reprinted with permission of the Royal Society of London;

Fig. 8.5 from *Science* (Armitage et al. 2011, figure 3) reprinted with permission of AAAS; Fig. 8.6 from *Archaeology of Oceania*, edited by Lilley (2006), in chapter 2 by O'Connor and Veth, figure 2.1, reprinted with permission of Blackwell Publishing Ltd; Fig. 8.7 with permission from *Radiocarbon* (Gillespie 2002, figure 2); Fig. 8.9 from *Changing the Face of the Earth* (Simmons 1989, figure 4.1) with permission by Blackwell Publishing Ltd; Fig. 8.10 courtesy of Gareth W. Preston; Fig. 8.11 from *Geology* (Cullen et al. 2000, figure 4) reprinted with permission of The Geological Society of America; Fig. 8.13 from *Quaternary Research* (Mann et al. 2008, figure 5) reprinted with permission of Elsevier; Fig. 9.3 from *Solar Physics* (Hathaway and Wilson 2004, figure 9) with permission by Springer-Verlag; Fig. 9.4 from *Nature* (Lambert et al. 2008, figure 2) with permission by Macmillan Publishers Ltd; Fig. 9.7 from *Nature* (Molnar and England 1990, figure 4) with permission by Macmillan Publishers Ltd; Fig. 9.8 from *Nature* (Jouzel et al. 1993, figure 5) with permission by Macmillan Publishers Ltd; Fig. 9.9 from *Quaternary Science Reviews* (Broecker and Denton 1990) with permission by Elsevier; Fig. 9.10 from *Progress in Oceanography* (Richardson 2008, figure 16 from Gordon 1991) with permission by Elsevier and A.L. Gordon, who also kindly provided a high resolution version.

1

A Framework for Understanding Environmental Change

Chapter overview

The Quaternary, the last 2.6 million years, has been a time of great change in the world, and in this chapter we discuss how ideas about it have evolved over the last two hundred years. In particular we recount how the **Ice Age** was discovered, we outline how the degree of change that took place in the tropics came to be appreciated, and we show how our predecessors established the various climatic and sea level fluctuations that have taken place in both the Pleistocene and the Holocene. We also discuss the very different climatic conditions that preceded the Quaternary, including the way in which the Earth became cooler during the Tertiary, and we discuss different opinions about how to mark the Tertiary/Quaternary transition and the rationale behind the current scheme. We also outline some fundamental concepts relating to the way in which we reconstruct and interpret the environments of the past.

1.1 Introduction

This book is concerned with the time during which humans have inhabited the Earth, a period which more or less equates with what geologists call the Quaternary. The Russians, mindful that this period was unique because of human's presence, have sometimes called it the *Anthropogene*. Humans have only been evident on our planet for a few millions of years (see **Chapter 8**). Nonetheless, the environmental changes that have taken place during the time we have inhabited the Earth have not been slight. Changes in climate, sea levels, vegetation belts, animal populations, soils and landforms, have been both many and massive. Some changes are still taking place, and have been of such an order as to influence markedly both humans and the landscape. In the last 20 000 years alone, and humans, as already noted, have been in existence one hundred times longer than that, the area covered by glaciers has been reduced to one-third of what it was at the glacial maximum; the waters thereby released have raised ocean levels by over one hundred metres; the land, unburdened from the weight of overlying ice, has locally risen by several hundred metres; vegetation belts have swung through the equivalent of tens of degrees of latitude; permanently frozen ground has retreated from extensive areas of Europe; the rainforest has expanded; desert sand fields have advanced and retreated; inland lakes have flooded and shrunk; and many of the finest mammals have perished in the catastrophe called 'Pleistocene overkill'. Even at the present time, smaller fluctuations of climate are leading to changes in fish distributions in northern waters, marked fluctuations in valley glaciers, extensive flooding or desiccation of lakes, and difficulties for agricultural schemes in marginal areas.

During the past few million years humans have gradually diffused over the Earth and have now become powerful agents of environmental change in their own right (Goudie 2006). Some scientists, notably the atmospheric chemist Paul Crutzen, have argued that humans have already wrought profound enough changes on Earth's land, seas, and atmosphere since the industrial revolution to merit recognition of the beginning of a new geological epoch which he has called the *Anthropocene* (see Zalasiewicz et al. 2011 for a discussion of this idea).

The oldest records of human activity, crude stone tools which consist of a pebble with one end chipped into a rough cutting edge, have been found in conjunction with bone remains from various parts of East Africa. Elsewhere (see **Fig. 1.1**) the arrival of humans is considerably later, with the colonization of Australia approximately 50 000 years ago and widespread colonization of the Americas little more than 10 000 years ago: that of New Zealand, Madagascar and Oceania even later. The evolution and dispersal of humans,

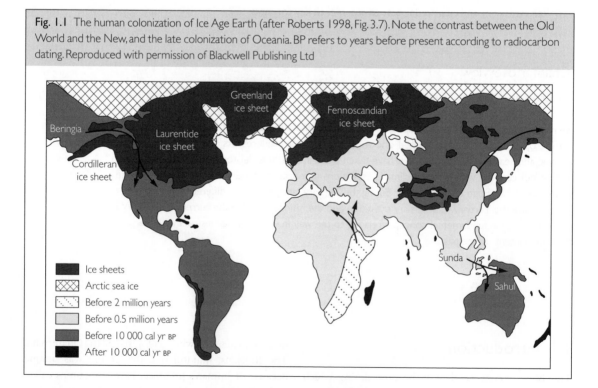

Fig. 1.1 The human colonization of Ice Age Earth (after Roberts 1998, Fig. 3.7). Note the contrast between the Old World and the New, and the late colonization of Oceania. BP refers to years before present according to radiocarbon dating. Reproduced with permission of Blackwell Publishing Ltd

and relationships with environmental change, are discussed in **Chapter 8**.

Subdividing geological time

The terminology used to describe some of the natural events and changes which provide the backcloth of human history and prehistory needs to be discussed here. As **Fig. 1.2** shows, the major geological divisions traditionally employed are as follows: the most recent **era** is called the Cainozoic ('recent life'). (The Cainozoic is also referred to as 'Cenozoic'.) This has been divided into two **periods**: the Tertiary and the Quaternary. The use of these two terms in a geological sense dates back to the eighteenth century as geologists for the first time attempted to classify rock formations and order them by relative age. In 1759, the Italian geologist Giovanni Arduino proposed a fourfold scheme in which he discussed geological time in terms of primitive, secondary, tertiary, and quaternary. The concept of the Quaternary as the most recent period was further established by Desnoyers in his 1829 study of Tertiary and Quaternary marine deposits in the Paris basin. As geological understanding increased through the nineteenth century, this basic four-part scheme was modified and elaborated; yet, while the terms 'primary' and 'secondary' dropped out of use, the terms 'tertiary' and 'quaternary' have remained commonly used as the names of the two most recent geological periods or **systems**. Although the terms 'period' and 'system' can seem to be used interchangeably in the literature, there is a difference in their meaning: the former refers to divisions of geological time; whereas the latter refers more specifically to the rocks and deposits themselves that formed during these time periods.

The Tertiary and Quaternary are in turn divided into various **epochs** or **series**, (the distinction being the

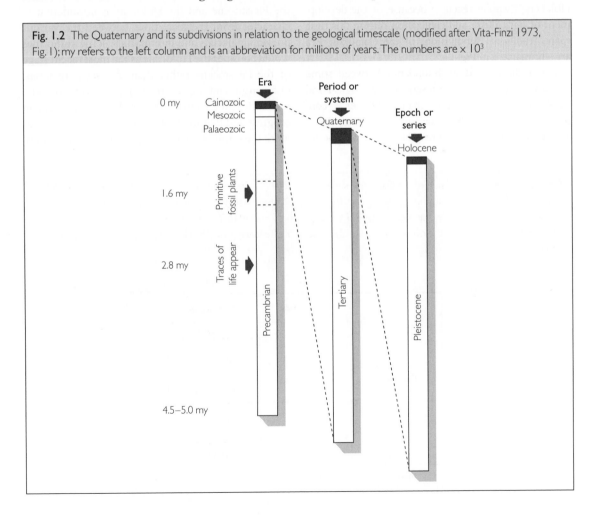

Fig. 1.2 The Quaternary and its subdivisions in relation to the geological timescale (modified after Vita-Finzi 1973, Fig. 1); my refers to the left column and is an abbreviation for millions of years. The numbers are × 10³

same as for periods and systems), with the Pliocene, for example, being the final series of the Tertiary. Lower down the hierarchy of the Geological Time Scale, epochs are divided into ages and series into stages. The Pleistocene and Holocene epochs/series together constitute the Quaternary Period/System – a mere 0.04 per cent of the 4600 million years or so since the formation of the Earth. Using the analogy of Earth's history being represented by a single day, this is equivalent to the last half minute of that day. Yet, as already mentioned, within this brief period there have been major climatic and environmental changes which have greatly influenced the world we inhabit, both in terms of the physical landscape and the evolution and distribution of organisms. The vast majority of Quaternary time is spanned by the Pleistocene ('most recent'), but a sliver of very recent geological time which extends from about 11700 years ago to the present is given status as a separate epoch, termed the Holocene ('wholly recent'), because of the development of agriculture and civilization during this time span. Traditional subdivisions of the Cainozoic Era are summarized in **Fig. 1.3**.

It is difficult to draw boundaries between some of these units for they represent terminological and classificatory devices rather than reality. As a consequence, there has been dispute over many years among authorities about how the Cainozoic should be subdivided. Recently the International Commission on Stratigraphy (ICS) has established usage of the *Palaeogene* and *Neogene* Periods to distinguish the early Cainozoic (prior to 23 million years) from the later Cainozoic extending up to the base of the Quaternary – and although the term 'Tertiary' Period/System remains in widespread use, it does not currently feature on the internationally recognized Geological Time Scale having been replaced by the Palaeogene and Neogene. There has also been debate recently over the formal status of the Quaternary in the Geological Time Scale because of some members of the ICS proposing to redefine the Quaternary as a sub-era within an extended Neogene Period. Some authorities have even advocated the abandonment of the term Quaternary in favour of the term 'Late Neogene', arguing that the Quaternary only represents an intensification of climatic trends associated with the onset of the Neogene.

Similarly, because the climatic transition between the Pleistocene and the Holocene is no different in character from the transitions between the various preceding **glacials** and **interglacials**, some authorities would consider the Holocene as the most recent stage of the Pleistocene rather than as a separate series. Other arguments against retaining the term Holocene for a series (or epoch) are that the Pleistocene is still in progress and that the Holocene has been too brief to merit the status of a series. In spite of these difficulties,

Fig. 1.3 Traditional subdivision of the Cainozoic showing the relationship between eras, periods and epochs; Myr refers to millions of years and kyr to thousands of years (not to scale). Note that in 2009 the International Union of Geological Sciences officially changed the base of the Quaternary Period to 2.6 Myr from its previous position at 1.8 Myr ago. Note also that on the official Geological Time Scale the Tertiary Period has been replaced by the Palaeogene and Neogene Periods

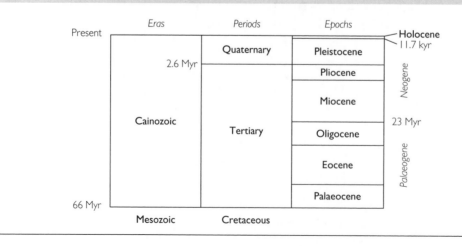

a decision in 2009 by the International Union of Geological Sciences sided with the traditional usage: the Quaternary retains its status as a Period/System, containing the Pleistocene and Holocene as its two epochs/series.

However we may divide up and name geological time periods, it is still clear that changes in environment have occurred over a wide range of timescales which range from the minor fluctuations within the period of instrumental record (with durations of the order of a few years or decades) to the major geological periods with durations of many millions of years. The shorter-term changes (described in **Chapter 7**) include such events as the period of warming that took place in the first decades of the twentieth century. The changes with durations of hundreds of years were characteristic of the Holocene (see **Chapter 5**) and include various phases of glacial advance and retreat such as the **Little Ice Age** between about AD 1500 and 1850. The fluctuations *within* the Pleistocene, consisting of major glacials and interglacials, lasted for the order of 10 000 to 100 000 years, while the Pleistocene itself consisted of a group of glacial and interglacial events that lasted in total about two and a half million years. Such major phases of ice age activity in Earth's geological history appear to have been separated by hundreds of millions of years.

Various curves which show the nature of these different orders of change are illustrated in **Fig. 1.4**. They serve to show the frequency and magnitude of changes during the Quaternary and to place such changes in the context of the longer span of Earth history.

1.2 The evolution of ideas

Appreciation of the fact that the world had such a history of environmental change emerged at much the same time that it was appreciated that the world had a history that was longer than the constricting 6000 years of Archbishop Ussher's biblical time span. This seventeenth century concept, which saw the Earth as having been created in 4004 BC, was relatively uncontested until the end of the eighteenth century; and it was accompanied by the belief that much of the **denudation** and deposition evident on the face of the Earth could be explained through the agency of Noah's Flood and other catastrophes. Gradually, however, these ideas were shown to be erroneous. In particular, Hutton and Playfair, the Edinburgh scientists, are often regarded as the men who most effectively propagated the new ideas, for in the geological record they saw 'no vestige of a beginning, no prospect of an end'. They realized that the complexity of the sedimentary record could be explained through the operation of processes akin to those of the present day extending over a long time span. A full discussion of the development of these ideas is provided by Chorley et al. (1964) and by Ehlers (1996).

The 'Ice Age' discovery

The idea that climate and other aspects of the environment had fluctuated or changed during this enlarged span of time resulted initially from the discovery that European glaciers had formerly extended further than their current limits. In 1787 de Saussure recognized **erratic** boulders of Alpine rocks on the slopes of the Jura Ranges and Hutton reasoned that such far-travelled boulders must have been glacier-borne to their anomalous positions. Playfair extended these ideas in 1802 but it was in the 1820s that the **Glacial Theory**, as it came to be known, really became widely postulated. In addition to evidence from the Alps put forward by Swiss scientists (such as Venetz and de Charpentier), much convincing evidence came from northern Europe, for example through the work of the Norwegian geologist J. Esmark and the German scientists Bernhardi and Schimper (Andersen and Borns 1994). However, the Swiss scientist Louis Agassiz, who was one of the originators of the term *Eiszeit* or Ice Age, did more than anyone else to promote the concept of past glaciation at the continental scale.

In spite of this convergence of opinion from numerous sources, these ideas were not easily assimilated into prevailing dogma, and for many years it was still believed that glacial **till**, called drift, and isolated boulders, called erratics, were the result of marine submergence, much of the debris, it was thought, having been carried on floating icebergs. Charles Lyell noted debris-laden icebergs on a sea-crossing to America, and found that such a source of drift was more in line with his belief in the power of current processes – the **uniformitarian** belief – than a direct glacial origin. For years even glacial or subglacial depositional features like eskers were thought to be of a marine origin, and were classified into fringe eskers, bar or barrier eskers, and shoal eskers. Moreover, in Britain, some drift deposits contained sea shells, a fact that gave prima-facie support to marine ideas. At last in 1840, Agassiz

Fig. 1.4 The different scales of climatic change: (a) Changes in the five-year average surface temperatures over the region 0–80° N; (b) Winter severity index for eastern Europe; (c) Generalized Northern Hemisphere air-temperature trends, based on fluctuations in alpine glaciers, changes in tree-lines, marginal fluctuations in continental glaciers, and shifts in vegetation patterns recorded in pollen spectra; (d) Generalized Northern Hemisphere air-temperature trends, based on mid-latitude sea surface temperature, pollen records, and on worldwide sea level records; (e) Fluctuations in global ice volume recorded by changes in isotopic composition of fossil plankton in a deep-sea core; (f) The spacing of ice ages in geological time

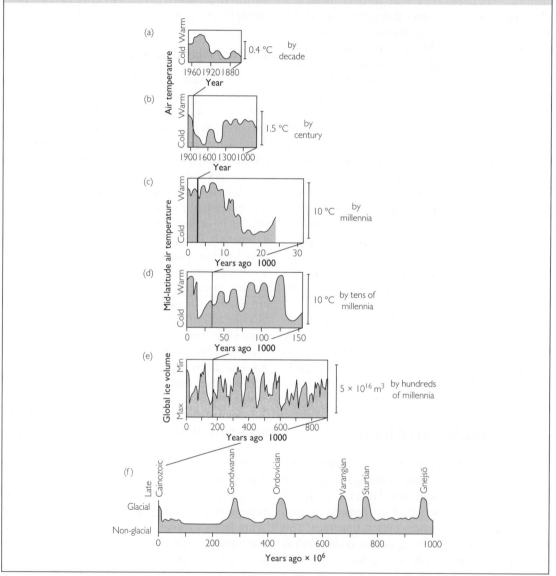

won over the initially sceptical Oxford geologist William Buckland to the evidence for former glaciers in the Scottish Highlands; and shortly after, Buckland persuaded Lyell, opening the way for widespread acceptance of the Glacial Theory in Britain.

The study of environmental change progressed somewhat further in the 1860s when Archibald Geikie, A.C. Ramsay, and T.F. Jamieson showed that the spread of the glaciers had not been a single event, and that there had been various stages of glaciation, represented by moraines, which had been separated by warmer phases, called interglacials.

Lower latitudes

In areas outside those which were subjected to Pleistocene glaciation, other types of fundamental change were recognized as having taken place, though Louis Agassiz, after a journey to Brazil, postulated somewhat wildly that even equatorial regions had undergone glaciation. He believed that the Amazon Basin had been overwhelmed, but he seems, like certain others at the time, to have allowed his enthusiasm to convince him that what was in reality deeply weathered soil and boulders produced by intense chemical activity under tropical conditions, was glacial boulder clay. A truer appreciation of the effects of climatic change in the tropics and subtropics came from Jamieson, Lartet, Russell, and G.K. Gilbert, all of whom studied the fluctuations of Pleistocene lakes in semi-arid areas, as represented by old **strandlines** and deltas. Their work established the supposed general relation between high-latitude glacials and mid- and low-latitude **pluvials**, and Russell was able to tie in the moraines of the former Sierra Nevada glaciers with old strandlines around Lake Mono in California. Gilbert demonstrated that Lake Bonneville (Utah), in much the same way as the glaciers, had fluctuated several times with alternations of high water and phases of desiccation. Other geologists investigating the west of the United States also believed that a diminution of rainfall might account for some of the anomalous drainage features they encountered on their travels, and the term pluvial was coined by Alfred Taylor in 1868. It means basically a period of comparatively abundant moisture in regions beyond the limits of the ice sheets. Whether such pluvial periods were in, or out, of phase with the high-latitude glacial periods is a controversial problem referred to in **Chapter 4**.

Thus in the south-west of the United States we have one of the germs of the idea that low-latitude pluvials were synchronous with high-latitude glacials. A corollary of this view, that post-glacial times will have been characterized by desiccation, spread through many parts of the world. A particularly notable focus in the early years of the twentieth century was Central Asia, particularly after the great explorations of the Tarim Basin, Lop Nor, and Tibet. In 1903 Andrew Carnegie financed an expedition to Turkestan and Persia, of which, among others, Raphael Pumpelly, W.M.

Davis and E. Huntington were members. In 1905–6 Huntington was able to go a second time to Asia. On these journeys, he noted the evidence for ruined cities and abandoned settlements, and recognized that lake terraces indicated a 'gradual desiccation of the country from early historical times down to the present'. After visiting the famed Tarim Basin, he wrote his *Pulse of Asia* (1907) and moved away from any simple idea of progressive desiccation to a view embracing multiple large fluctuations. He believed that the pulsations of climate had served as a driving force in the history of Eurasia, forcing nomadic invaders to overrun their more civilized neighbours whenever the climatic cycle reached a trough of aridity.

At much the same time as Huntington put forward his views, Prince Kropotkin put forward a simple view of progressive desiccation (1904, p. 722):

> Recent exploration in central Asia has yielded a considerable body of evidence, all tending to prove that the whole of that wide region is now, and has been since the beginning of the historic period, in a state of rapid desiccation ... It must have been the rapid desiccation of this region which compelled its inhabitants ... pushing before them the former inhabitants of the lowlands to produce those great migrations and invasions of Europe which took place during the first centuries of our era.

The concept of post-glacial desiccation was much favoured in Indian literature of the 1950s and 1960s. In particular, fears were expressed that the Thar Desert was on the move eastwards (National Institute of Sciences 1952). In East Africa, Nilsson (1940) proposed a possible correlation between African pluvials and Alpine glaciations, and, in his influential work on climatic geomorphology, Büdel (1982) maintained that the Sahara had been squeezed during glacial times as the westerly depression tracks were displaced from the north and rainforest and savannah encroached from the south.

Distrust of long distance correlations gradually arose, and was forcefully voiced by Zeuner (1945, p. 208):

> ... the suggested correlation of certain tropical pluvials with certain glaciations of Europe is not only unfounded, but in some cases definitely wrong. It was chiefly based on the mistaken, though most widely accepted, assumption of the strict contemporaneity of tropical glaciations and pluvials with European glaciations. This is a typical example of uncritical acceptance of an assumption born from the desire of simplifying, which is so current in scientific thought, though it often violates the facts.

In the 1950s we see the development of an increasing uneasiness about correlations between glacial events in high latitudes and pluvial events in low. This change of attitude was prompted by a combination of new dating technologies and by new types of evidence for environmental reconstruction and is well represented in the work of Zeuner (1959). Of especial importance was the recognition that some tropical areas, far from being highly desiccated at the present time, had experienced far more severe desiccation at some point, or points, in the past. Instrumental in this was the identification of ancient sand seas (**ergs**) in areas that are now relatively moist, most notably on the south side of the Sahara (Grove 1958).

Glacial and interglacial sequences

After the existence of *Die Eiszeit* had been established in the 1820s and 1830s, it gradually became evident that the Ice Age was a complex event in which there had been a number of glacial advances and retreats. Although monoglaciation had its adherents well into the twentieth century (see the review in Charlesworth 1957), the general view was of multiple glaciations. This was certainly the case for James Geikie (1874), who reviewed the evidence, particularly in the British Isles. In central Europe, between 1901 and 1909, A. Penck and E. Brückner produced a major work, *Die Alpen im Eiszeitalter*. For decades this formed the basis for much glacial chronology, and in particular they proposed a classic four-fold glaciation model of the Pleistocene (Table 1.1). By studying what was left of old moraines, they were convinced that there had been four great

glaciations of different intensities in the Alps. They also proved, from the preservation of plant remains at some sites, that some of the intervening periods were fairly mild. They further demonstrated the correspondence between these four glaciations and the successive gravel terraces of the Rhine and other rivers. They named the four glacial periods after valleys in which evidence of their existence occurred: the Günz, Mindel, Riss, and Würm (**Fig. 1.5**).

Penck and Brückner also provided an estimate of the relative duration of the glacial-interglacial cycles. Their estimate for the relative duration of each interglacial was:

Günz-Mindel	Mindel-Riss	Riss-Würm	post-Würm
3	12	3	1

Of particular note was their belief that the interglacial between the Mindel and the Riss lasted a longer time than any other. In the classic model this event was therefore termed *The Great Interglacial* (Bowen 1978). Their model was influential, widely accepted and almost attained the status of a law. Evidence for additional glaciations obtained by other workers was generally forced into their scheme, and assigned the secondary role of sub-stages or stadia within the four major glaciations (e.g. Eberl 1930). Traditional land-based chronological terminologies for northern Europe, the British Isles, and North America are shown in **Tables 1.2** and **1.3**.

The Holocene and the Late Glacial

'Holocene' was introduced by the French palaeontologist Paul Gervais in 1869 as a term for the last 10 000 years or so of Earth history. At the end of the nineteenth century it was recognized in Norway and Sweden that during the Holocene there had been a series of changes in vegetation. Axel Blytt, a Norwegian, argued that the remains of different plants (plant macrofossils) preserved at different levels within peat deposits indicated a pattern of species immigration into Norway following the last glacial, and that the sequence of these plant remains also reflected past climatic changes. He introduced the terms 'arctic', 'boreal', and 'atlantic' to characterize the different plant assemblages and the inferred climates (Blytt 1876, 1882). Sernander, a Swede, further developed the terminology and applied the concept more widely to characterizing the post-glacial vegetation development of Scandinavia (1894, 1908), and the 'Blytt–Sernander' scheme was born (Mangerud

Table 1.1 Penck and Brückner's scheme of glaciations (with related terraces) and interglacials in the Alpine region

	erosion
Würm Glaciation	Niedterrassen (Low Terrace)
Riss-Würm interglacial	*erosion*
Riss Glaciation	Hochterrassen (High Terrace)
Mindel-Riss interglacial	*erosion*
Mindel Glaciation	Younger Deckenschotter
Günz-Mindel interglacial	*erosion*
Günz Glaciation	Older Deckenschotter

(After Bowen 1978, table 2-1.)

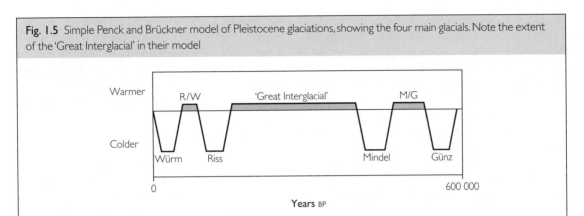

Fig. 1.5 Simple Penck and Brückner model of Pleistocene glaciations, showing the four main glacials. Note the extent of the 'Great Interglacial' in their model

Table 1.2 Traditional names for glacials and interglacials in different regions

Alps	Northern Europe	British Isles	North America
WÜRM	WEICHSEL	DEVENSIAN	WISCONSIN
R-W	Eemian	Ipswichian	Sangamon
RISS	SAALE	WOLSTONIAN	ILLINOIAN
M-R	Holstein	Hoxnian	Yarmouth
MINDEL	ELSTER	ANGLIAN	KANSAN
G-M	Cromerian	Cromerian	Aftonian
GÜNZ	MENAP	BEESTONIAN	NEBRASKAN

(Modified from Bowen 1978, table 1-4 and Andersen and Borns 1994, fig 4-7.)

Table 1.3 The glacials and interglacials of North America

Stage	Type-site/area
Wisconsin	State of Wisconsin: initial description based on end-moraines
Sangamon	Sangamon County, Illinois, where Sangamon soil lies between loess (Wisconsin) and till (Illinoian)
Illinoian	State of Illinois (Illinois Lobe of Laurentide ice-sheet): all deposits between the Yarmouth and Sangamon soils
Yarmouth	Yarmouth, Des Moines County, Iowa: soil separating Kansas and Illinoian glacial deposits
Kansan	Upper till at Afton Junction Pit, Union County, Iowa
Aftonian	Gravels at Afton Junction Pit, Union county, Iowa
Nebraskan	Lower till at Afton Junction Pit, Union county, Iowa

(Modified from Bowen 1978, table 2-9.)

et al. 1974; Mangerud 1982). As the sequence of peat layers, their degree of decomposition (humification) and associated plant remains appeared fairly uniform over large areas, many workers believed that the climatic subdivisions of the Blytt–Sernander scheme were broadly synchronous across north-west Europe. Hence, before the development of radiocarbon dating, the Blytt–Sernander scheme was widely used for correlating Holocene terrestrial deposits.

Shortly after the development of the Blytt–Sernander scheme, von Post introduced the method of pollen analysis (**palynology**) as a means of reconstructing past vegetation changes in an area (see **Chapter 2** for discussion of the technique). As the ideas of the Blytt–Sernander scheme were common currency among palaeoecologists in the first half of the twentieth century, pollen analysts (working with

lake or peat sediments) generally interpreted their Holocene pollen data within the Blytt–Sernander framework (Birks 1982). The first such detailed pollen-based vegetational reconstruction (**pollen zonation scheme**) was developed by Jessen (1935) for Denmark. Godwin (1940) published a similar sequence of Holocene pollen zones for England and Wales, and Jessen (1949) applied the system to Ireland. While each of these zonation schemes differed to some extent in plant species composition, they were all based on the assumption that the zonations were broadly synchronous over large parts of north-west Europe, and that the inferred changes in Holocene climate correlated with the Blytt–Sernander subdivisions. For example, pollen zone VIIa was seen to represent the height of the Atlantic period of the Blytt–Sernander scheme when it was inferred that temperature and precipitation were higher than at present and that north-west Europe experienced a 'climatic optimum'. **Table 1.4** summarizes the Blytt–Sernander scheme for north-west Europe and its relationship to the pollen zones.

It is worth noting however, that the idea of an 'Atlantic climatic optimum' in Europe contrasted with interpretations for the same time period in America. In the drylands of North America it was, from the human point of view, far from optimal. Evidence for a period of intense drought, with dune reactivation in the High Plains, the drying of lakes (Antevs 1936) and trenching of valley floors (Antevs 1955) led to the adoption of the term **altithermal** or **hypsithermal** (Deevey and Flint 1957).

Until the 1970s, most pollen analysts in Europe interpreted their data in relation to the Blytt–Sernander and Jessen–Godwin frameworks, and sometimes pollen-based vegetation reconstructions were thought to be flawed if they could not be fitted within this system. With the advent of radiocarbon dating, and its widespread application to terrestrial deposits, particularly in the 1970s, it has been shown that the palaeoecological and climatic subdivisions of

the Blytt–Sernander scheme do not represent contemporaneous changes across north-west Europe, and it is now known that the pattern of Holocene climatic change is much more complicated than the scheme suggests – both temporally and spatially (Lowe 1993). However, Mangerud et al. (1974) have argued that the scheme at least offers a set of terminology useful for subdividing Holocene time, and as shown in **Table 1.4** they have matched the Blytt–Sernander chronozones with the radiocarbon chronology (see **Section 2.5** for a discussion of **radiocarbon years** BP). Hence, the names of the Blytt–Sernander chronozones remain in use today primarily as chronological reference points.

Geomorphologists came to realize that, notwithstanding the views of people like Raikes (1967), who argued that the climate of the Holocene was rather stable, the climatic changes inferred from the study of plant remains and pollen had considerable geomorphological significance for fluvial systems (e.g. Starkel 1966) and, above all, for glaciers (Manley 1966). In 1939, Matthes introduced the term 'Little Ice Age' to define the period when glaciers were re-established in the Sierra Nevada of California following the postglacial climatic optimum. Moss (1951) and Nelson (1954) used the term **neoglaciation**, which was subsequently defined by Porter and Denton (1967) as a 'cool geologic-climate unit...indicating a probably world

---

Table 1.4 The classic chronology and pollen zonation of the Holocene for north-west Europe

Chronozone	Climatic characteristics	Jessen–Godwin pollen zones (Godwin 1975)	Associated vegetational characteristics	Chronology (^{14}C years BP)
SubAtlantic	period of generally deteriorating climate with cooler and wetter conditions	VIII	alder, birch, oak, beech	2.5 kyr BP to the present
SubBoreal	climatic optimum with warmer and drier conditions	VIIb	alder, oak, lime (elm decline)	5 to 2.5 kyr BP
Atlantic	climatic optimum with warmer and wetter conditions	VIIa	alder, oak, elm, lime	8 to 5 kyr BP
		VIc		
Boreal	climatic amelioration, warmer and drier	VIb	pine, hazel	9 to 8 kyr BP
		VIa		
PreBoreal	sub-arctic conditions	V	hazel, birch, pine	10 to 9 kyr BP
		IV	birch, pine	

(Dates of transitions between chronozones follow Mangerud et al. 1974 in thousands of radiocarbon years before present. See **Section 2.5** for a discussion of radiocarbon years.)

wide synchrony of glacier fluctuations in response to climatic change'. Radiocarbon dating revealed that glaciers had experienced a series of advances and retreats during the Holocene (see Grove 2004, for a review of both ideas and evidence).

Around the time that the Holocene climate-vegetation subdivisions were being established, evidence was also found pointing to a major climatic deterioration that had taken place near the transition between the Pleistocene and the Holocene. Towards the end of the last glacial phase (a period of time known as the **Late Glacial** when the climate was generally warming and ice sheets were receding) there was a short, but intense, cold phase known as the **Younger Dryas**, when glacial conditions were re-established in many parts of the world. For example, glaciers returned to the Scottish Highlands (the so-called Loch Lomond Readvance) (Sissons 1976) and intense valley excavation occurred in southern England (Kerney et al. 1964).

The Younger Dryas was first recognized in 1901, at Allerød in Denmark, where it was discovered that lake sediments showed a pattern of organic mud with birch remains sandwiched by lower and upper clay layers containing the remains of cold-adapted plants (Joosten 1995). This suggested that the warming following the end of the last glacial was interrupted by a shift back to cold conditions before the full interglacial conditions of the Holocene became established. Pollen analysis of sediments laid down immediately after full glaciation in other parts of north-west Europe showed a similar 'tripartite' sequence with upward progression from arctic/alpine pollen types, to dwarf-shrub and birch pollen, and back to arctic/alpine types. In the Jessen–Godwin

zonation scheme the initial rise in dwarf-shrubs following glacial retreat became classified as Zone I, the period of dwarf-shrubs and birch as Zone II, and the reversion back to arctic/alpine flora as Zone III. These pollen zones were also related to time periods termed the Bølling, Allerød, and Younger Dryas respectively. Some palynologists also interpreted a brief climatic deterioration from a decline in the pollen of dwarf-shrubs and other warm-loving taxa between the Bølling and the Allerød that became known as the Older Dryas. **Table 1.5** summarizes this traditional subdivision of the Late Glacial. Current views on the nature of the Late Glacial and the timing and severity of the Younger Dryas cold phase are discussed in detail in **Section 5.2**.

Sea level change

One of the great concerns of Quaternary scientists in the first half of the twentieth century was worldwide (**eustatic**) changes in sea level. Suess, the Austrian scholar, in Volume 2 of *Das Antlitz der Erde* (1888), believed that evidence of continental **transgressions** and **regressions** pointed to a remarkable synchrony of swings in sea level in widely spaced areas of the world. He later maintained that many continental areas had been absolutely stable during geological time, but that there had also been rapid subsidence of the ocean bottoms, which caused sea levels to fall in a punctuated manner, creating staircases of terraces.

Geomorphologists, finding flights of marine terraces, sought to correlate them over wide areas on the basis of altitude differences. It was argued that strandlines occurred at similar heights around much

---

Table 1.5 The classic chronology and pollen zonation of the Late Glacial for north-west Europe

Chronozone	Climatic characteristics	Jessen–Godwin pollen zones (Godwin 1975)	Associated vegetational Characteristics	Chronology (^{14}C years BP)
Younger Dryas	return of glacial conditions	III	open tundra, arctic/alpine taxa	11 to 10 kyr BP
Allerød	warming	II	park-tundra with birch	11.8 to 11 kyr BP
Older Dryas	phase of colder conditions	Ic	brief decline in dwarf-shrubs, including Juniper	12 to 11.8 kyr BP
Bølling	warming	Ib	open tundra, dwarf-shrubs	13 to 12 kyr BP
		Ia		

(Dates of transitions between chronozones follow Mangerud et al. 1974 in thousands of radiocarbon years before present. See **Section 2.5** for a discussion of radiocarbon years.)

of the Mediterranean (Depéret 1918–22; Lamothe 1918; Gignoux 1930). At a high level was the Calabrian (Pliocene or Pleistocene) bench with various others at lower levels (Sicilian, Milazzian, Tyrrhenian, Monastirian, etc.). Baulig (1928) was an exponent of the idea that erosion surfaces in the Massif Central were the result of eustatic changes, and influenced Wooldridge, who argued that a supposed bench in south-east England was a marine terrace of Calabrian (Plio-Pleistocene) age (Wooldridge 1951). There was also a view that a progressive decline in sea levels had taken place during the Pleistocene, with terraces forming in the main interglacials and with progressively lower low sea level stands during the four main glacials.

Perhaps one of the firmest advocates of eustatic sea level correlations was Zeuner (1945, p. 246) who produced a table, using the Mediterranean terminology, which purported to show the correspondence of shorelines by altitude in Algeria, the south of France, the Channel Islands, Northern France, southern England, South Africa, South Australia, the Sunda Archipelago, and North America! He produced a table of sea levels and approximate dates which he hoped would provide an absolute chronology of the Pleistocene that could be extended over the whole Earth (**Table 1.6**).

Such views were criticized, not least because of the large degree of tectonic instability found in an area like the Mediterranean basin, but also because there was limited valid chronological control on the ages of the terraces. The eustatic theory lingered on tenaciously,

but was plainly a great over-simplification of a complex problem (Beckinsale and Beckinsale 1989).

However, the debates over global eustatic correlations should not deflect us from appreciating the significance of **glacio-eustatic** changes during the Pleistocene and Holocene. As Daly (1934) pointed out, glacial-age sea levels stood at around 100 m below the sea level of today and had huge implications for the evolution of such features as coral reefs, deltas, and river valleys. Moreover, by the mid 1960s, radiocarbon dating had given a picture of the nature of sea level rise in post-glacial times, though there was some debate about whether Late Holocene sea level positions oscillated to points both above and below the present level.

Ideas of **glacio-isostasy**, whereby the addition and removal of ice caps caused crustal deformation, were developed initially in Scotland by Jamieson (1865). They were rapidly taken up in North America (Whittlesey 1868; Shaler 1874) where the term '**isostasy**' was introduced by Dutton (1889). Gilbert (1890) extended the idea to include the effects of water loading associated with the expansion of pluvial Lake Bonneville – hydro-isostasy. By the close of the 1880s, therefore, the hypotheses of glacio-isostasy and hydro-isostasy had been outlined (Smith and Dawson 1983). De Geer (1890) produced maps of coastal deformation for Scandinavia, and by the beginning of the twentieth century 'the pattern of crustal response to glacial unloading in Scandinavia was being reconstructed in impressive detail, providing a model for studies in other glacio-isostatically affected

Table 1.6 Zeuner's scheme of eustatic sea levels (heights in metres)

Shorelines	Average height	Algeria	South France	Jersey	North France	South England	South Africa	Sunda Archip.	South Australia	North America
Sicilian	100	103	90–100	–	103	c.96	–	–	(75)	81
									60	65
Milazzian	60	c.60	55–60	–	56–59	c.60 (36.5)	45–75	c.60?	(45)	49
Tyrrhenian	32	c.30	28–32	32–34	32–33	33.5	c. 32	–	27	29
Main Monastirian	18	18–20	18–20	18	18–19	15–18	18–20	c.15	19.5	20
Late Monastirian	7.5	–	7–8	7.5	8	5–8	6–7.5	–	7.5	8
Pre-Flandrian regression	−100	–	min. −92	–	min. −30	c.−100	−100	−70 to −100	−100	–

(From Zeuner 1945, p. 252.)

areas' (Smith and Dawson 1983, p. 371). Substantial work was also undertaken on glacio-isostatic warping of the Great Lakes of North America (Leverett and Taylor 1915, pp. 316–518). The deformation of shorelines in Scotland was discussed by Wright (1937) and was the subject of intensive research by Sissons and his pupils (see, for example, Sissons 1962). A good summary of views on glacio-isostasy in relation to sea level change was provided by Gutenberg (1941).

In areas that were glaciated the history of sea level change is usually the history of the race or struggle between glacio-eustatic and glacio-isostatic effects. Wright (1914 and 1937) argued that this was responsible for the complex form of raised beaches in areas like Scotland and the Baltic. The causes and effects of sea level change through time are discussed in detail in **Chapter 6**.

1.3 Early climate history of the Earth

Little is known about the very early climatic history of the Earth, for rocks deposited before about 2800 million years ago are generally so metamorphosed that it is extremely difficult to extract climatic information from them. In the early Proterozoic Eon (latter part of the Precambrian as shown in **Fig. 1.2**) evidence from Canada suggests that the first known glaciation occurred between 2400 and 2100 million years ago. It was probably a complex event with recurring glacial events rather than a continuous glaciation, and it was followed by a relatively warm phase that persisted until the Neoproterozoic Era (toward the end of the Precambrian). During the Cryogenian Period of the Neoproterozoic (between 850 and 630 million years ago) at least three major glacial episodes occurred, with the Sturtian (between 750 and 700 million years ago) and the Marinoan, or Varangian, (between about 660 and 635 million years ago) being the most significant (Macdonald et al. 2010).

These are also likely to have involved recurring glacial advances, with periods of extreme cold lasting millions of years followed by episodes of rapid warming. And as these glacial times were far more severe than anything seen during the Quaternary Period, (there is evidence for glaciers near sea level in the tropics!), the term **Snowball Earth** has been used to describe them; notwithstanding continued debate about just how extensive and continuous ice cover actually was. Earth could only come out of its 'snowball' state once enough

carbon dioxide had accumulated in the atmosphere from volcanism to strengthen the **greenhouse effect** to a point sufficient to initiate ice melting (Hoffman et al. 1998). This in turn led to positive feedback towards further warming as a reduction in ice covering the globe reduced reflection and increased absorption of solar energy: this is the **ice-albedo feedback** which is a key factor governing the Earth's energy balance and hence climate. (This of course works in both directions, and the general role of feedback mechanisms in Earth's climate is discussed in **Section 1.6**.)

At the end of the Proterozoic, the record of climate improves because the sedimentary record is better preserved and animals with shells that could be preserved as fossils had spread through the oceans. In general climatic conditions from the start of the Phanerozoic Eon (onset of the Palaeozoic on **Fig. 1.2**) until about 360 million years ago were relatively warm, though in the Sahara there is evidence of a brief glacial event in the late Ordovician around 440 million years ago. The Permo-Carboniferous was characterized by a long glaciation from about 330–250 million years ago, and ice sheets were extensive on the southern supercontinent of Gondwana.

During the succeeding Mesozoic Era the evidence of deep-sea cores becomes of greater utility. Temperatures ranged from 10 to 20 °C at the poles to 25–30 °C at the equator. There were, however, fluctuations in temperatures, with slight cooling taking place at the start of the Jurassic and with marked high-latitude warming taking place during the first half of the Cretaceous. Global cooling took place towards the end of the Cretaceous but there was a significant increase in temperatures at the onset of the Tertiary. However, a long-term cooling trend commenced shortly after the start of the Eocene Epoch some 50 million years ago following a major spike of warmth at the Palaeocene/Eocene boundary. The Cainozoic suffered various marked short-term cooling episodes: particularly in the mid Palaeocene (c. 58 million years ago), in the Middle Eocene (c. 45 million years ago), and at the Eocene–Oligocene boundary (c. 34 million years ago).

As described in more detail below, there is some evidence that the Antarctic glaciers may have come into existence in the Eocene and that about 10–15 million years ago mountain glaciers came into existence in the Northern Hemisphere. Earth's most recent transition from a relatively warm **greenhouse climate** (with no ice sheets on the planet) to a relatively cold **icehouse climate** (with ice sheets and glaciers) occurred as the Cainozoic climate cooled, and culminated as the Tertiary gave way to the Quaternary Period.

1.4 The Cainozoic climatic decline

Although global temperatures varied considerably through the Tertiary Period, on average the globe was significantly warmer during the Tertiary than during the Quaternary (**Fig. 1.6**); and particularly during the early Tertiary between 66 and 45 million years ago. The average global temperature of the Tertiary reached a peak in the late Palaeocene and early Eocene, a time referred to as the **Palaeocene–Eocene Thermal Maximum** (PETM) about 55 million years ago (Storey et al. 2007). At this time sea surface temperatures across the globe rose by 5 °C and more at the polar latitudes, with Arctic summer sea temperatures estimated to have reached over 23 °C! (Sluijs et al. 2006). Mean annual temperatures in many places were over 10 °C warmer than at present. Although the extreme warmth of the PETM itself lasted about 200 000 years (Röhl et al. 2000), a brief span compared with the duration of the Tertiary, the average global temperature continued to be much warmer than today for many more millions of years. Higher resolution studies of Late Palaeocene and Early Eocene climate, from sampling the sea-floor, have also shown other warm phases besides the PETM, such as an Eocene Thermal Maximum 2 at about 53.5 million years ago when coldest month mean temperatures over Arctic land masses were over 8 °C (Sluijs et al. 2009).

In the early to mid Eocene there was no ice at the poles, tropical climates extended 10 to 15° higher in latitude than today and subtropical plants survived as far as 60° N. Remarkably, fossils of alligators and other warmth-loving animals dating from this time have been found as far north as Ellesmere Island to the west of Greenland. With no polar ice, sea level was consequently much higher than today, and the atmosphere was more humid. By the later Eocene (45–40 million years ago) the climate had cooled enough for glaciers to form in Antarctica and for sea ice to form in the Arctic (Stoll 2006), and temperature dropped relatively sharply at the Eocene/Oligocene transition (34 million years ago). Despite later warming trends, temperatures never came close to reaching the peak attained during the Eocene, and by the end of the Tertiary (in the late Pliocene Epoch about 3 million years ago) ice sheets were also building up in the high northern latitudes.

Given the high degree of climatic variability during the Tertiary, it may be over-simplistic to talk of 'average Tertiary conditions' for comparison with the Quaternary. However, many workers have attempted to estimate the overall decline in temperatures through the entire Tertiary. On the basis of deep-sea sediment core studies, as early as 1961, Emiliani suggested that there was a broad temperature decrease amounting to about 8 °C for the middle latitudes during the Tertiary. A similar temperature depression is indicated in the Equatorial Pacific, but on the basis of land flora studies, somewhat greater changes than these have been proposed for the western United States between 40° and 50° N (Butzer 1972) and the picture for other areas is the same. Along with sea surface temperatures, the temperature of the deep ocean declined by over 12 °C from the early to late Tertiary (**Fig. 1.6**).

In Australia a comparable sequence of decline of Tertiary temperatures has been proposed, based on the former greater extent of the plants *Araucaria* and *Agathis*, including into Tasmania. At the present time they are limited to Queensland and the warmer parts of Australia. However, rainfall decline may have been equally important in such changes of floral distributions, and Gentilli (1961) has said that 'areas that now receive 12.5 cm of rain a year must then have received at least 125 cm with no rainy season. If there was even a short dry season the annual rainfall must have reached some 200 or 250 cm for these laurisilvae to grow as they did.' Large trees then existed in the Lake Eyre Basin and other stretches of the 'dead heart' of Australia (Gentilli 1961; Gill 1961).

Environments of the Tertiary

The North Atlantic region in the early Tertiary may have been characterized by a widespread, tropical moist forest type (Pennington 1969, Chapter 1). Indeed, the classic interpretation of the fossil plant evidence from the Palaeocene London Clay of southeast England, made by Reid and Chandler (1933), was that it represented a flora comparable to that of the Indo-Malayan region of today. Whether there really was a tropical rainforest environment in southern Britain in the Eocene is, however, a matter for debate. As Daley (1972, p. 187) has pointed out, 'Such a climate, with its markedly uniform character throughout the year, could not have existed at the latitude of southern Britain during the Eocene (40° N), since seasonal climatic variations would undoubtedly have occurred.'

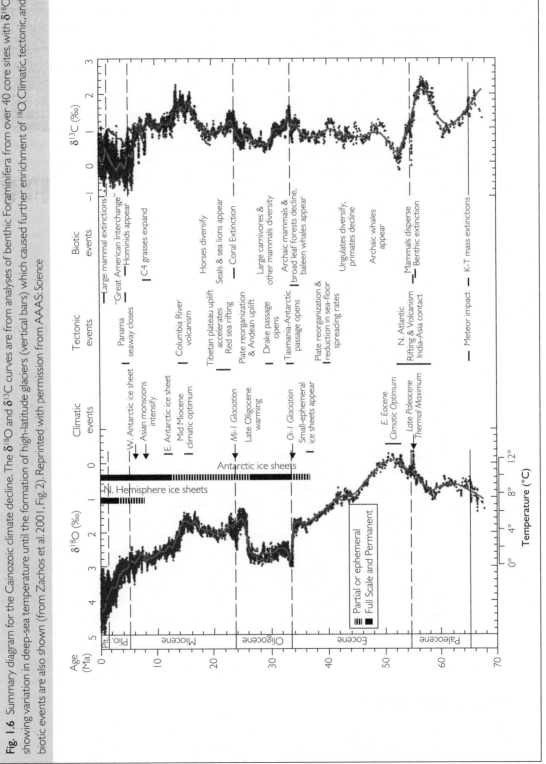

Fig. 1.6 Summary diagram for the Cainozoic climate decline. The $\delta^{18}O$ and $\delta^{13}C$ curves are from analyses of benthic Foraminifera from over 40 core sites, with $\delta^{18}O$ showing variation in deep-sea temperature until the formation of high-latitude glaciers (vertical bars) which caused further enrichment of ^{18}O. Climatic, tectonic, and biotic events are also shown (from Zachos et al. 2001, Fig. 2). Reprinted with permission from AAAS: *Science*

Nevertheless the picture of early and middle Eocene warmth is confirmed by many studies. For example, Collinson and Hooker (1987) report that beds of that age contain the highest proportion of potential tropical flora (up to 82 per cent) and the smallest proportion of potential temperate nearest living relatives (up to 42 per cent). Collinson and Hooker (1987, p. 267) consider that: 'A great many of the near living relatives of fossils co-exist today in paratropical rainforest in eastern Asia. Many are large forest trees, others lianas. The greatest diversity of these elements in the Early and Middle Eocene suggests the presence of a dense forest vegetation.' Tree rings were also particularly large, indicating rapid growth in response to high temperatures (Creber and Chaloner 1985).

The climatic deterioration at the end of the Eocene, which appears abrupt (Buchardt 1978) or gentle (Collinson et al. 1981), according to the type of evidence used, created a different climatic regime in the Oligocene. Then the climate of Britain was more comparable to that of a region like the southeastern United States, though there may have been very dry intervals, as indicated by the presence of gypsum evaporite crusts and ancient desert dunes in the Ludien and Stampian sediments from the Paris Basin. However, by Pliocene times, the 'tropical' vegetation of the North Atlantic region was being replaced by a largely deciduous warm temperate flora (Montford 1970). The last appearance of palms north of the Alps in Europe is in the Miocene flora at Lake Constance.

Immediately before the Ice Age, in a period which the Dutch call Reuverian B, the climate of the present temperate latitudes of the Northern Hemisphere seems to have favoured the development of woodland. At that time varied woodlands extended from the Atlantic coasts to the Sea of Japan, a picture that has not been repeated since (Frenzel 1973). Over wide reaches of the present warm and cool temperate latitudes, temperature and rainfall conditions were particularly favourable, and resembled those of a subtropical climate. In central and eastern Europe, temperatures were probably 3 to 5 °C higher than they are now, and precipitation levels were several hundred millimetres higher.

Explaining Tertiary climates

The causes of a warm Palaeogene (66 to 23 million years ago) in Britain are both local and global. At a local scale, Britain was at a lower latitude than today, being 10–12° further south in the Palaeocene. Movements of the continents may also have affected global climates by creating a unique configuration of the continental masses. Parrish (1987, p. 52–53) has suggested that:

By the beginning of the late Cretaceous, global palaeogeography had entered a mode of maximum continental dispersion, and the monsoonal circulation that had dominated palaeoclimatic history during the previous 250 million years had disappeared. From this time until the collision of India and Asia re-established strong monsoonal circulation during the Miocene, continental palaeoclimates were characterized by the zonal circulation that is the fundamental component of atmospheric circulation on Earth.

Global Palaeogene warmth is thought to have been caused largely by elevated levels of greenhouse gases in the atmosphere. Carbon dioxide levels were high because of frequent volcanic eruptions associated with high rates of sea-floor spreading and plate subduction (Barron 1985; Creber and Chaloner 1985). Superimposed on a long period of higher atmospheric carbon dioxide concentration than now were a number of distinct phases when greenhouse gases and temperature peaked.

The precise cause of the warming spike (the PETM) at around 55 million years ago continues to be an important subject of study, and various hypotheses have been put forward. Despite uncertainty over the trigger, there is no doubt that the PETM resulted from a massive release of greenhouse gases into the atmosphere over a very short period of time in geological terms (about 10 000 years). One view is that the event was mainly caused by the rapid release of a large quantity of methane into the atmosphere which intensified the greenhouse effect (methane, molecule for molecule, is over 20 times more effective as a greenhouse gas than carbon dioxide). The proposed source was the break down of over 1000 gigatonnes of gas hydrates – a combination of water and methane normally kept solid at high pressures and low temperatures deep beneath the oceans (Maslin 2004). Break down of methane hydrates could have occurred due to uplift of the sea-floor associated with the rifting of the Atlantic (Maclennan and Jones 2006). Another view puts less emphasis on methane hydrates and suggests a very large release of CO_2 in addition to methane. In this case, the enormous amount of carbon released to the atmosphere is attributed to flood basalt volcanism associated with the opening up of the North

Atlantic Ocean as rifting occurred between the continental masses of Greenland and north-west Europe (Storey et al. 2007). The great quantity of magma that up-welled during this episode of rifting effectively 'cooked' surrounding carbon-rich sediments, adding perhaps four to five thousand gigatonnes of carbon to the atmosphere in the 10 000-year span.

While the PETM was the most extreme (and is the best known) of these Tertiary warm spikes (or 'hyperthermals') during a relatively warm period of Earth's history, there were five other similar episodes of climate warming during the Palaeocene and Eocene, including the Eocene Thermal Maximum 2 referred to earlier. What all of these warm phases have in common is that they are believed to have been caused by phases of elevated greenhouse gas concentration in the atmosphere. However, a difference is in the source of the greenhouse gases: the PETM stands out as being driven by the release of carbon previously locked up in sediments; whereas the other phases of less extreme warming seem to relate to times of increased release of dissolved CO_2 from the world's oceans (Sexton et al. 2011). Each of these events ended as the efficacy of the oceans as a carbon 'sink' rather than source returned to previous levels.

The causes of the **Cainozoic climatic decline** following these hyperthermals that eventually led to the establishment of an icehouse climate towards the end of the Tertiary are still not fully understood, but the trend seems to be associated with the breakup of the ancient supercontinent of Pangaea and the reconfiguration of the continents as we know them today. Some 100 million years ago, well before the start of the Cainozoic, the crust of Australia and New Zealand began separating from Antarctica, with Antarctica gradually shifting into its present position mainly inside the Antarctic Circle. At the same time Eurasia and North America moved towards the North Pole. The positioning of more land at high latitudes was a prerequisite for the eventual development of ice caps and the initiation of ice-albedo feedback (which, through the high surface reflectivity of snow and ice, causes less absorption of solar energy leading to a cooler climate).

However, the change in the latitudinal position of the continents did not in itself account for the full Cainozoic cooling. Antarctica was already positioned over the South Pole during the Cretaceous Period, yet had a far warmer climate than it does today, once supporting polar forests. The former presence of forest in the interior of Antarctica has been known since fossils of ancient 'seed-fern' trees (*Glossopteris*) were collected from the Transantarctic Mountains by Captain Scott and his team during their ill-fated return journey from the South Pole in 1912. Until about 30 million years ago, forests of *Nothofagus* (southern beech) existed in Antarctica.

Important to the cooling of Antarctica during the Cainozoic was the opening up of a wide sea between Antarctica and Australia after *c.* 40 million years ago, followed by the separation of the Antarctic Peninsula from southern South America to create the Drake Passage. Some refer to this as a **gateway hypothesis** for Antarctic cooling, and there is some debate as to how much of a climate effect this had (Ruddiman 2008). These tectonic-caused changes in the positions of Southern Hemisphere land masses allowed the cold, west-to-east flowing Antarctic Circumpolar Current to develop in the Southern Ocean surrounding Antarctica, thereby isolating the continent from warmer ocean currents originating at lower latitudes. It was around this time when Antarctic glaciers began to form in highland locations.

Evidence for Eocene glaciation in Antarctica has been known for some time from deep-sea sediment cores taken from the Southern Ocean – a coexistence of microfossils of Eocene age, and of sediments of glacial type, was shown by Geitzenauer et al. (1968); and Drewry (1975) argued that eastern Antarctica ice-field initiation dates back at least to the lower Miocene. Confirmatory evidence of Eocene glaciation in Antarctica is provided by the nature of dated volcanic materials. Volcanoes that have erupted beneath an ice sheet display a suite of textural and structural characteristics that are especially distinctive in volcanoes composed of basaltic lava, and such materials have been found in Antarctica back to the Eocene. Similar evidence implies also that glaciation may well have been uninterrupted up to the present in parts of this South Polar region (LeMasurier 1972).

A key transition in Cainozoic cooling occurred at the start of the Oligocene about 34 million years ago. Not only did Antarctic glaciers expand significantly, but the climate of the Northern Hemisphere land masses also cooled at the same time (Bowen 2007). A major extinction event ensued in which tropical and subtropical Eocene flora and fauna gave way to types adapted to cooler and drier conditions in many parts of the world. Synchronous global cooling at this time suggests that a global factor was at work in addition to the important regional tectonic changes affecting the distribution of land and sea. In their reconstruction of Palaeogene atmospheric carbon dioxide concentrations, Pagani et al. (2005) inferred that CO_2 dropped sharply from levels between 1000 and 1500 ppm during the Eocene to around 500 ppm in the early Oligocene. As shown in **Fig. 1.7**, levels have remained below 500 ppm throughout the Neogene up to the present day.

Major changes in the operation of the global carbon cycle must have occurred to account for such large changes in atmospheric CO_2 levels. The longer-term trend of decreasing levels during the Eocene may relate to a reduction in the rate of sea-floor spreading and subduction, causing a *reduced average input* of CO_2 to the atmosphere from volcanoes worldwide. Since the Eocene, the uplift of mountain ranges and plateaux, notably the Himalayas and the Tibetan Plateau, has caused an *increase in the average output* of CO_2 from the atmosphere through the

process of hydrolysis (chemical weathering of silicate minerals in rock).

This effect works in the following way: with the uplift of mountains comes an increase in precipitation as incoming air parcels are forced to rise over upland terrain, cooling and losing their moisture. Higher precipitation speeds up the delivery of carbon dioxide to the ground surface in rainwater, where, in the form of carbonic acid it reacts with silicate minerals in rock. For a discussion of the many factors involved in rates of chemical weathering, see Goudie and Viles (2012). As the rock is broken down, clay minerals are formed and carbon is carried away in streams in the form of dissolved ions to eventually reach the ocean, where it may become part of a marine organism's shell before eventually being deposited on the sea floor to form new sedimentary rock. Rates of hydrolysis in steep, mountainous terrain are also enhanced by the high rates of physical weathering and erosion that occur in such environments, continually breaking up and

Fig. 1.7 Reconstructed atmospheric CO_2 concentration spanning the Eocene, Oligocene, and Miocene compared with climate modelled thresholds for the critical level of CO_2 below which Antarctic and Northern Hemisphere glaciation can occur (horizontal lines). The arrow refers to a CO_2 drop thought to be linked with the expansion of Antarctic ice at the Eocene/Oligocene boundary (from DeConto et al. 2008, Fig. 4). Reprinted by permission from Macmillan Publishers Ltd: *Nature*

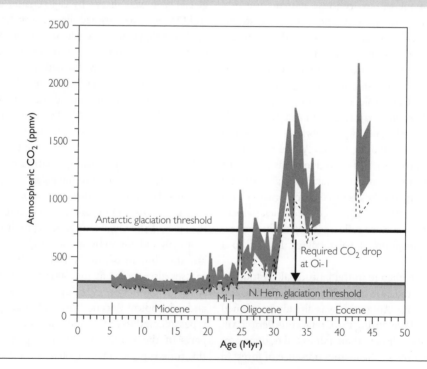

exposing fresh bedrock to chemical weathering. In this way, much of the drop in CO_2 levels during the Oligocene shown in **Fig. 1.7** can probably be accounted for by the collision of the Indian subcontinent with Asia and the resultant uplift of continental crust and enhancement of chemical weathering.

Following a late Oligocene shift back to relatively warm conditions, climate cooled again at the start of the Miocene. Interestingly, this cooling correlates both with an acceleration of Tibetan Plateau uplift as well as a phase of low **eccentricity** and low **obliquity** (axial tilt) in Earth's orbit (Zachos et al. 2001) (See **Section 9.4** for a discussion of orbital variation and effects on climate). It's therefore likely that cooling was caused *both* by an increase in the removal of atmospheric CO_2 through chemical weathering and an orbital configuration conducive to cooler summers.

Climate warmed again in the middle Miocene; but then cooled significantly about 14 million years ago. This cooling saw a major growth phase of ice in Antarctica, with the East Antarctic ice sheet reaching a size similar to today about 13.8 million years ago; while in the Arctic, sea ice became much more extensive. Although it has been suggested that an increase in organic carbon burial in the oceans reduced atmospheric CO_2 causing the climate to cool – the 'Monterey Hypothesis', owing to the carbon-rich sediments dating to this time along the Monterey coast of California (Vincent and Berger 1985) – a reduction in atmospheric CO_2 is unlikely to have played as important a role as in earlier Cainozoic cooling episodes. More recent reconstructions of past CO_2 suggest that levels varied relatively little (and were already low) across this transition (**Fig. 1.7**). Instead, the causes were probably both orbital and tectonic: the cycles of eccentricity and obliquity of Earth's orbit again reached a low point (Abels et al. 2005), and the cooler summers enabled glaciers to expand in Antarctica and more sea ice to develop in the Arctic. And, as ice spread at the poles, the ice-albedo feedback referred to earlier would have helped amplify the orbitally-induced cooling. It's also been suggested that constriction of the ancient Eastern Tethys Sea (as India continued to move into Eurasia) caused a reduction in the transport of warm waters towards the Antarctic continent. (Flower and Kennett 1995; Shevenell et al. 2004).

From the period of middle Miocene cooling onwards, global climate continued to cool fairly steadily through the late Miocene and Pliocene. Glaciers began forming about 10 million years ago in Alaska,

the West Antarctic ice sheet was fully formed about 6 million years ago, and a Greenland ice sheet was forming about 3 million years ago. With lower levels of atmospheric CO_2 than during the Eocene and Oligocene, the Earth's climate became more responsive to orbital cycles. At certain times these cycles coincided to push climate across thresholds whereby ice at the poles expanded and climate cooled yet further. Important changes in ocean currents and wind systems also occurred, fundamentally altering the transport of heat and moisture across the globe. Some of the tectonic changes that influenced these patterns during the Pliocene are discussed in relation to the Tertiary / Quaternary transition in **Section 1.5**.

Late Cainozoic cooling also promoted the expansion and development of some of the world's great deserts (**see Chapter 4**). For instance, in late Cainozoic times (van Zinderen Bakker 1984) aridity became a prominent feature of the Saharan environment, probably because of the occurrence of several independent but roughly synchronous geological events (Williams 1985):

i   As the African plate moved northwards there was a migration of northern Africa from wet equatorial latitudes (where the Sahara had been at the end of the Jurassic) into drier subtropical latitudes.

ii During the late Tertiary and Quaternary, uplift of the Tibetan Plateau had a dramatic effect on world climates, helping to create the easterly jet stream which now brings dry subsiding air to the Ethiopian and Somali deserts.

iii The progressive build-up of polar ice caps during the Cainozoic climatic decline created a steeper temperature gradient between the equator and the poles, and this in turn led to an increase in trade wind velocities and their ability to mobilize sand into dunes.

iv Cooling of the ocean surface may have reduced the amount of evaporation and convection in low latitudes, thus reducing the amount of tropical and subtropical precipitation.

1.5 The Tertiary/Quaternary transition

The Quaternary has long been distinguished from the preceding Tertiary by global cooling associated with the expansion of glaciers, particularly across high- to mid-latitude land masses in the Northern Hemisphere. However, the precise time of transition from the Tertiary to the Quaternary (which equates with the Pliocene/Pleistocene series boundary) has been, and continues to be, a matter of interpretation and debate. It was originally thought, from the traditional Alpine glaciation model of Penck and Brückner (1909), that the Pleistocene began about 600 000 years ago; however, over succeeding decades it became clear that the Pleistocene needed a much longer chronology than that.

Furthermore, the traditional view of glaciation being confined to the Quaternary must be discarded. Previously it had been widely believed that the period from the Triassic through the Tertiary was a lengthy span when ice sheets and glaciers were absent, and when climatic fluctuations were less frequent and less severe than they were to be in the Pleistocene. However, this idea became increasingly untenable as more was learnt about Tertiary environments, particularly from studies of ocean sediments (see **Section 2.3** for an explanation of methods). For example by studying the *Globigerina* type of **Foraminifera** preserved in ocean-sediment cores, Bandy (1968) found evidence for a major expansion of polar faunas no less than between 10 and 11 million years ago. This expansion was followed, he suggested, by a Pliocene expansion around 5 million years ago, and the classic Pleistocene expansion after some 3 million years ago. Bandy also argued that the magnitudes of planktonic faunal changes in the Tertiary 'are almost as great as those of the classic Quaternary'.

As early as the 1970s Flint remarked that 'perhaps the most stirring impression produced by recent great advances in the study of the Quaternary Period is that the Quaternary itself is losing its classical identity' (Flint 1971, p. 2). Indeed, Antarctic glaciation (as reviewed in **Section 1.4**) stretches much further back, into the late Eocene, perhaps starting more than 40 million years ago.

Given the variable nature of Tertiary climates and the early onset of Cainozoic cooling, it is difficult to make any division which is both logical and rigid between the Pleistocene and the Pliocene. Indeed the Villafranchian (until recently the earliest unit of the European Pleistocene), together with its marine equivalent, the Calabrian, have only been assigned to the Pleistocene rather than to the Pliocene since 1948. And, as explained later in this section, a new age for

the Pliocene/Pleistocene boundary (and hence also the Tertiary/Quaternary boundary) was agreed by the International Union of Geological Sciences as recently as 2009.

On faunal grounds the Pleistocene came to be regarded as the time when many of the modern genera, including elephant, camel, horse, and wild cattle, first appeared. Attempts were made to place the boundary between the Upper Pliocene and the Villafranchian of the Lower Pleistocene by using tectonic breaks in the stratigraphic succession, though this provides a generally inadequate and unusable correlation basis, except on a very local scale. In Britain the division between the Pliocene and the Pleistocene was traditionally placed at the boundary between the Coralline Crag and the Red Crag of East Anglia. At this point there is a relatively clear stratigraphic break, a marked increase in the proportion of modern forms of marine Mollusca, and of Mollusca of northern aspect, and the first arrival (in the Red Crag) of elephant and horse.

In Europe the base of the Pleistocene has traditionally been identified by the appearance in Late Cainozoic sediments occurring in various parts of Italy of a cold-water marine fauna (the Calabrian) differing from the underlying Pliocene fauna. The newer fauna are characterized by the appearance of a dozen species of North Atlantic molluscs, and by certain Foraminifera. In northern Italy the marine strata of the Calabrian grade into Villafranchian and Upper Villafranchian continental sediments containing a distinctive mammal fauna (Emiliani and Flint 1963).

At a meeting of the International Geological Congress (IGC) in Moscow in 1984 it was decided that an exposure of marine sediments spanning the Pliocene/Pleistocene transition located at Vrica in Calabria, Italy, should serve as the internationally recognized reference section for the Pliocene/Pleistocene boundary (and base of the Quaternary), with the first appearance of the cold-loving marine **ostracod** *Cytheropteron testudo* being used to mark the onset of the Pleistocene. This point of transition lies conveniently close in time to the end of the Olduvai **geomagnetic event** when the Earth's magnetic field switched from normal polarity to reversed polarity; and, as described in **Section 2.5**, geomagnetic events recorded in sedimentary sequences can be used as chronological markers. Thus, the Olduvai geomagnetic event was formerly used as a marker to identify the beginning of the Pleistocene series (and

Quaternary Period) throughout the world. At the time that the Vrica **Global Stratotype Section and Point (GSSP)** was established, it was thought that the point of transition in the section dated to 1.64 million years ago; but since then, improvements in the **palaeomagnetic** timescale have pushed the date back to 1.81 million years ago. Those who supported this way of marking the boundary were said to favour the 'short Quaternary chronology'.

However, many other researchers favoured a longer Quaternary in which the Pliocene / Pleistocene boundary is placed at a point between 2.7 and 2.4 million years ago – a period when various forms of evidence from around the world indicate the onset of significant cooling. Such evidence includes the appearance of ice rafted debris in deep-sea sediment cores from the North Atlantic and North Pacific Oceans, the onset of **loess-palaeosol** accumulation in China, and the appearance of sub-antarctic molluscs in New Zealand (Gibbard et al. 2005, 2010). Although there have been mountain glaciers in the Northern Hemisphere for at least 10 million years (this is shown for instance in the White River Valley area of Alaska where lithified glacial deposits are inter-bedded with volcanic lavas dated to between 9 and 10 million years old), ice sheets only began forming between about 3 and 2.5 million years ago.

Several explanations have been put forward to explain the cooling trend and expansion of Northern Hemisphere ice sheets at this time. Through a study of deep-sea sediments from the sub-arctic Pacific Ocean, Haug et al. (2005) suggested that major Northern Hemisphere glaciation may have been triggered by a shift to an increased seasonal range of sea surface temperatures (SSTs) in the Northern Pacific dated to 2.7 million years ago. In particular, warmer late summer and autumn SSTs in the sub-arctic Pacific had the effect of increasing the moisture content of westerly airflow, thereby significantly increasing annual snowfall over the northern parts of North America.

Important tectonic events influencing climate in the late Pliocene to early Pleistocene included the formation of the Isthmus of Panama, and the convergence of Australia and Asia with the consequent emergence of Indonesian islands. The former event severed the Atlantic–Pacific connection between the Americas, and the latter event restricted the passage of waters between the Pacific and Indian Oceans. Both events were crucial for establishing the modern pattern of ocean currents, and the variation in these currents, that has played such an important role in climate change over the Quaternary Period (see **Section 9.7** for a discussion of ocean circulation and climate). From about 4.6 million years ago onwards, the uplift of seabed and gradual emergence of the Isthmus of Panama was a precondition for the formation of the Gulf Stream and North Atlantic Drift Currents, these being crucial for delivering heat and moisture to high latitudes in the Northern Hemisphere. This may also have been important for increasing precipitation and snowfall in the North Atlantic region and northern Eurasia (Haug and Tiedemann 1998). As discussed later in this book (**Section 3.6**), variation in these currents is closely connected with the large oscillations in climate that are so much a feature of the Quaternary – especially in the North Atlantic region.

It has been suggested that the northward movement of New Guinea and emergence of Indonesian islands (especially Halmahera) between 4 and 3 million years ago was critical to the aridification of east Africa as well as for global cooling, as very warm Pacific equatorial waters could no longer penetrate westward into the Indian Ocean (Cane and Molnar 2001). This change in Pacific Ocean circulation was also probably a precursor to the changes in North Pacific SSTs referred to above that increased moisture transport and snowfall over North America, enabling ice sheets to build up.

Although the precise workings and influence of these 'oceanic gateway' changes continue to be debated, a general way of looking at the tectonic background to the establishment of the cold and variable climate characterizing the Quaternary Period is to see that these changes led to a global pattern of ocean circulation that had less equatorial transfer of water between oceans and more isolation of **oceanic gyres** than has typically been the case when Earth has experienced warmer climates (Smith and Pickering 2003). Furthermore, as described in the previous section, a background of relatively low levels of atmospheric CO_2 and an already cold, ice-covered Antarctica set the stage for the initiation of Northern Hemisphere glaciation and its pacing by cycles in Earth's orbital parameters. The effects of orbital variation on climate are treated in more detail in **Chapter 9**.

To return to the issue of when to mark the beginning of the Quaternary, it has become increasingly evident through recent research that a more important global climate threshold was crossed between 3 and 2.5 million years ago than at the start of the Calabrian, 1.8 million years ago, represented at Vrica. Therefore, the majority of workers currently advocate a longer Quaternary chronology. Hence the Subcommission on Quaternary Stratigraphy (a subgroup of the International Commission on Stratigraphy) made the proposal that the start of the Quaternary Period (and Pleistocene Epoch) should be pushed back to 2.6 million years ago, close to an earlier and more significant geomagnetic transition: the Gauss–Matuyama boundary. The proposal was ratified by the International Union of Geological Sciences (IUGS) in 2009, and therefore the base of the Quaternary is now officially recognized as 2.6 million years ago, rather than 1.8 million, in the Geological Time Scale.

The GSSP for the new base of the Quaternary is located in the Monte San Nicola section in southern Sicily. The precise point in the cliff section of exposed sediments actually lies 1 m above the Gauss–Matuyama reversal and has been dated by orbital tuning to 2.588 million years ago (Gibbard et al. 2010). (See **Section 2.5** for a discussion of orbital tuning.) The point also marks the beginning of the Gelasian Stage which used to be assigned to the Pliocene but is now the first stage of the Pleistocene. The Vrica GSSP referred to earlier still marks the start of the Calabrian Stage which is now the second stage of the Pleistocene. Fortunately, the acceptance by the IUGS of the proposal by the Subcommission on Quaternary Stratigraphy has safeguarded the status of the Quaternary as a geological period, preventing it from being downgraded to a sub-era of an extended Neogene as some in the geological community argued for. This maintains the traditional view of the distinctiveness and importance of the Quaternary, as well as the long-running convention of equating the start of the Quaternary with the start of the Pleistocene. A GSSP has recently also been established for the base of the Holocene Epoch at 11.7 thousand years 'b2k' (before AD 2000) as described in **Section 5.3**.

1.6 Reconstructing and interpreting past environments: key concepts

A remarkable array of evidence is used to infer, or reconstruct, past environments, but similar approaches and assumptions underlie most of the methods. Most of the evidence is stratigraphic in nature, meaning that deposits of sediment (or other material) laid down during the time span of interest are sampled and investigated for clues as to the characteristics of the environment at the time of deposition. If deposition

occurred continuously over a fairly long period of time, then the result is a **stratigraphic sequence** in which changes in the characteristics of the deposit through the sequence are used to infer changes in the environment through time. In addition to the sampling and analysis of deposits, mapping deposits and landforms of similar type and age is another key aspect of palaeoenvironmental study, being necessary for estimating the geographical expression of the past environmental conditions and changes inferred from stratigraphic records.

The types of material studied are diverse, including hard rock, glacial till, fine lake or marine sediments, soil, peat, tree rings and ice. Samples of stratigraphic sequences are either taken as sections from exposed layers of sediment, for instance in a quarry or along a cliff face, or in the form of cylindrical cores using manual or machine operated coring devices that penetrate deposits from the surface downwards (or from the outside inwards in the case of tree rings, corals and cave deposits). With most types of deposits, it is possible to make inferences about past environments (that is to conduct palaeoenvironmental reconstruction) both from the physical characteristics of the deposit itself and from the chemical and biological constituents within the deposit. When used in this way, deposits provide an indirect measure of past environmental or climatic conditions and are often referred to as sources of **proxy data** because they substitute for the lack of direct measurements during the time period of interest.

The activity of reconstructing past environments is also becoming increasingly linked with efforts to understand the causes of environmental change and to model the environment and climate for predictive purposes. To appreciate the contribution that such studies make towards understanding contemporary environmental change, it is necessary to have some knowledge of the 'systems approach' and to be familiar with the main concepts underlying environmental and climate modelling. Therefore, in addition to discussing the important concepts and assumptions behind palaeoenvironmental reconstruction, the following sections also provide background to environmental systems and modelling.

The law of superposition

When examining a stratigraphic sequence and analysing the proxy data that it contains for palaeoenvironmental reconstruction it must be assumed that material at the bottom of the sequence was deposited first and material towards the top was deposited later. Although obvious, the importance of this assumption for palaeoenvironmental research cannot be over-stated. This is known as the 'law of superposition' stating that underlying rock must be older than over-lying rock unless the layers of rock have been folded or faulted in such a way that the original sequence becomes inverted. With this simple yet crucial insight the pioneering geologist William Smith was one of the first to systematically describe relationships between different rock layers (strata), and this made it possible for him to construct the first modern geologi-cal map in the world: his geological map of England and Wales (1815).

As with rock, this assumption can be safely applied to other materials such as lake sediments, in which sed-iment carried into a lake by streams settles on top of older sediment on the lake bottom; or ice on a glacier, in which snow that eventually compacts into glacier ice is continuously added above previously formed ice. However, post-depositional reworking can sometimes disturb or invert a stratigraphic sequence causing problems for interpretation, and this must be borne in mind when choosing deposits to sample. Using the example of lake sediments, the influence of strong currents, lake overturning and high biological activ-ity on a lake bed all contribute to disturbing the sedi-mentary sequence, and better stratigraphic sequences can be obtained from lakes where these processes are less important. Considering ice, the margins and base of ice sheets are less suitable for sampling because the ice layers in these places, away from the summit, are more likely to be deformed due to tensional or com-pressional stresses.

Diagenesis, equifinality, provenance, and taphonomy

Another key assumption for using stratigraphic sequences for palaeoenvironmental reconstruction is that the physical, chemical, and biological properties of the deposit are the product of the past environment *when* the deposit was laid down. However, this may not always be the case due to **diagenesis** in which post-depositional processes of compaction and chemical alteration change the physical nature of the material from what it was at the time of deposition. This tends to be more of a problem with older deposits that have had more time to be chemically altered, although the

degree to which diagenesis interferes with interpreta-tion of the past environment depends very much on the type of material being analysed. In addition to the problems of diagenesis, it is important to be aware of the problem that **equifinality** can pose for interpreting deposits and landforms. This applies when the distin-guishing features of a deposit or landform could arise from more than one set of processes and environmen-tal conditions; though like diagenesis, this too varies in importance depending on what is being investigated.

When reconstructing past environments it is also often assumed that the deposit under investigation is representative of, and provides information about, the environment *where* the deposit is located. While this is generally the case, for example organic-rich sedi-ments on a lake bed can reveal much about past con-ditions and productivity of the lake's ecosystem, this type of assumption is not always as straightforward as it might first appear. For instance, fine, wind-blown dust (**loess**) can be transported thousands of kilome-tres from its point of origin before being deposited; hence such deposits owe as much to processes that occurred far away as to local conditions. In fact, it is sometimes very difficult to determine the source area (**provenance**) of a deposit accurately and to make cor-rect inferences about whether the deposit provides past environmental information at the local, regional, continental, or global scale. There can be uncertainty even when comparing different deposits of very simi-lar nature, as for example the spatial scale represented by evidence preserved in lake sediments may vary with the size and characteristics of the lake, with what part of the lake is sampled, and with atmospheric, hydrological and ecological conditions prevailing at the time of deposition.

Taphonomy is the study of how a group, or 'assem-blage', of fossils found in a particular place came to be there, and environmental reconstruction based on fossils found within a deposit can be complicated by taphonomic problems. Strictly speaking, fossils are chemically altered remains or impressions of organ-isms left behind in rock; although the principles of taphonomy also apply to the remains of former organisms not yet petrified (for instance plant or ani-mal remains buried in peat or lake sediments) as is often the case in Quaternary deposits. The term 'sub-fossils' is sometimes used to distinguish such remains from remains that are fully petrified.

Taphonomy relates to the problem of provenance described previously in that it is not always easy to determine whether a fossil organism lived where it is found (being *in situ*) or if it was transported from elsewhere before burial. The taphonomy of a fossil assemblage also refers to the conditions which led to preservation, and it is important to be aware that different environmental conditions will influence the extent to which different kinds of organisms are preserved and hence their representation within the assemblage. This means that the proportions of various organisms within an assemblage are unlikely to reflect the actual proportions in the ecosystem when the organisms were alive. For example, shelled organisms such as snails will be better preserved in alkaline deposits, such as marl, than in acidic deposits, such as peat, where calcium carbonate shells are dissolved. On the other hand, cellulose plant material is well preserved in peat because the acidic and anoxic conditions hinder decomposition. By taking into account and correcting for differential transport and preservation, taphonomic studies improve the reconstruction of past plant and animal communities from fossil evidence.

Uniformitarianism and neo-catastrophism

Another geological concept important for reconstructing past environments is **uniformitarianism**; often phrased as 'the present is the key to the past'. Introduced by Hutton in the late eighteenth century, and further developed in the nineteenth century by Lyell, this is the idea that geological processes observed at present have been occurring in the same fundamental ways for great stretches of time in the past. For example, observations of the deposition of mud in an estuary tell us something about how mudstone is formed and how long it takes to form a layer of mudstone of a certain thickness. This insight completely altered earlier views about the age of the Earth (as described in **Section 1.2**) as it became apparent that great spans of time were necessary to explain the formation of the sedimentary rock strata exposed at the surface – not to mention what lies beneath.

While the nineteenth-century concept of uniformitarianism overlooked the sometimes rapid and catastrophic, and at times very different, geological processes that have occurred in the past, it made an essential contribution to our understanding of geological time as well as our understanding of the development of life. Indeed, Darwin's theory of evolution by natural selection rested on the recognition (made possible by uniformitarianism) of enormous spans of

geological time during which natural selection could work. Relatively recent discoveries, from research in the late twentieth century, of rapid changes and major extinction events in the geological record do contrast with the slow and continuous geological changes originally envisioned by Lyell; and new ideas about the environment experiencing long periods of relatively little change interrupted by brief phases of major change are sometimes grouped under the heading of **neo-catastrophism**.

The term neo-catastrophism distinguishes modern views about the causes of catastrophic change from the earlier (late eighteenth century) concept of 'catastrophism' in which a much shorter geological timescale was assumed, and major changes were explained by drawing on events described in the Bible such as the Great Flood. There is now much evidence pointing towards rare, but highly devastating, asteroid impacts and volcanic 'super-eruptions' as the most important agents of catastrophic geological change, and the asteroid impact theory for the extinction of the dinosaurs at the end of the Cretaceous Period (66 million years ago), first proposed by Luis and Walter Alvarez in 1980 (Alvarez et al. 1980), is perhaps the best known example of such an event. The fundamental ideas inherent in uniformitarianism and neo-catastrophism are not, however, mutually exclusive. The Earth's geological history should be understood as being shaped by slow processes acting over long periods of time as well as by infrequent, but high magnitude, events.

In the reconstruction of Quaternary environments, uniformitarian principles are often applied when observations of present-day processes are used to surmise how past processes must have worked to create the deposits or landforms under investigation. For instance, such logic is used when modern glaciers are studied in order to infer the behaviour of extinct Pleistocene glaciers in areas that no longer have glaciers today. Observation of the rates of glacial processes acting today also enables inferences to be made about how long it took for various glacial deposits and landforms to form in the past. Similarly, present-day measurement of the accumulation rate of sediment on a lake bed can be used to approximate the number of years represented per unit depth downwards along a core sample of the lake sediment so that the time spanned by the sedimentary sequence can be estimated (although this method is not without its problems as described in **Chapter 2**). These examples illustrate the **analogue approach** in which observable processes serve as analogues, or substitutes, for understanding past processes that cannot be observed.

This approach is also widely used for interpreting Quaternary environments from the organic remains contained in sediments. Pollen grains, for example, are often abundant and well preserved in terrestrial deposits, and by identifying and counting the different types of pollen and spores in a deposit it is possible to infer past vegetation cover in an area: the Jessen–Godwin pollen zonation scheme for the Holocene and Late Glacial described earlier offers an early example of this approach. By looking at relationships between modern plant communities and climate, it then becomes possible to infer climate conditions at times in the past where evidence for similar plant communities can be derived from the pollen record. This generally works well for the study of the Quaternary Period because most modern plants have existed through the duration of the Quaternary, and, in most cases, there has not been enough time for evolution to significantly change their climatic tolerances. Of course, using modern plant communities as analogues for past vegetation becomes less valid the further removed in time the past vegetation is from the present. Another potential problem with this type of analogue approach is that even in very recent geological time (just thousands of years ago) there are cases when climatic and environmental conditions were so different from at present that specially adapted plant communities existed for which there are no equivalent modern analogues.

Despite these difficulties, a great deal of what has been inferred about Quaternary changes in climate (and changes in related variables such as sea surface temperature) has been through the study of fossil and sub-fossil remains preserved in a wide range of terrestrial and marine depositional contexts. A number of approaches to environmental reconstruction from biological evidence have been devised, and several examples are included in **Chapter 2**. These either involve the study of one or more **indicator species** or the relation of a fossil species assemblage to climate parameters through **palaeoecological transfer functions**. The use of indicator species is straightforward in that if the present climatic tolerance range of a species is known, then its presence in a deposit indicates that the past climate when the organism lived must have been within this climate range. (Of course this assumes that the organism's tolerances are the same today as they were in the past.) This works best for

organisms that can be identified to species level, that have fairly narrow tolerances, and that do not have much 'ecotype' variation. The type of information gained by the presence of indicator species usually enables maxima or minima to be established (e.g. minimum mean January temperatures) for the location being studied. Furthermore, changes in the percentage abundance of indicator species through a sedimentary sequence enables inference of climatic trends, such as of temperature or precipitation, that affected the location.

A more complicated approach, but one that can deliver a more comprehensive and precise picture of past climate, is the use of palaeoecological transfer functions in which a collection of different species in a fossil assemblage and their relative abundances in relation to each other are used to infer past climate through a mathematical model. Statistically, this involves developing a modern 'training set' from which multiple regression models are derived relating the differing components of the modern assemblages to differing climate conditions. Once the statistical function has been calculated, it is then 'transferred' from the modern training set to the fossil assemblage in order to make a quantitative estimate of the past climate. Transfer functions based on pollen data have been used to estimate past temperature and precipitation, and those based on **diatoms** and other aquatic organisms have been used, among other things, to estimate past water temperature.

While such transfer functions are widely used in Quaternary research, and are being continually improved, it is important to be aware of their limitations: as suggested already, the procedure assumes that the assemblage–climate relationship established from the training set is applicable to the past, and this may not always be the case. The method also assumes that the fossil assemblage was *in equilibrium* with the prevailing climate of the time. As discussed later in **Chapters 3** and **4**, many past climatic changes occurred so quickly that there were significant time lags between the establishment of a different climate and the adjustment of vegetation to the new climate. This is especially a problem with pollen data relying upon tree genera as the migration of trees (leading to change in forest composition) is a relatively slow process (see **Section 5.3**). Therefore, the estimates yielded from pollen-based transfer functions may be more valid for certain times in the past than for others. Detailed treatment of the principles and methods behind transfer functions are contained in Birks and Birks (1980) and Birks and Gordon (1985).

In addition to uniformitarianism (and derived ideas such as the analogue approach discussed above) insights of neo-catastrophism also contribute to understanding and reconstructing past environments. Super-eruptions such as those of Yellowstone at 2.1, 1.3, and 0.65 million years ago and of Toba on Sumatra 73500 years ago caused major environmental and climatic changes on a scale well beyond what has been observed from volcanic eruptions in historic times. In addition, the operation of 'non-linear processes', by which minor climate perturbations trigger disproportionately large environmental changes, is being increasingly recognized. As explained in detail in **Section 3.6**, the discovery of rapid and high magnitude climatic changes during the Pleistocene from **ice core** studies points to the existence of thresholds in the climate system. Rather than changing slowly and gradually as uniformitarian thinking would imply, environmental conditions have often changed in abrupt and 'catastrophic' ways on the crossing of a critical threshold in some climate factor (analogous to the 'straw that broke the camel's back'). When this happens, a minor change to the system that manages to push it over the threshold is amplified by positive feedback mechanisms (discussed further later in this section) until conditions settle into a new state of stability in which the climate and environment is significantly different from what it was before. Different types of threshold processes relevant for understanding Quaternary environmental change are dealt with later in this book; suffice it to say that such processes also account for abrupt changes revealed in the geological record in addition to infrequent, high magnitude events like super-eruptions.

Environmental systems

An important paradigm for understanding the Earth's climate and environment is the idea of environmental systems, and within this conceptual framework, uniformitarian and catastrophic-type changes can be seen as different modes of operation within the system. Seeing the Earth as a set of interconnecting 'environmental systems' derives from general systems theory which explains phenomena in terms of flows and transformations of energy and matter. There is a seemingly infinite variety of natural systems ranging widely in type and scale: from the biological system of a single cell to the physical system governing circulation of the entire atmosphere. Yet, the operation of all systems can be understood according to a set of fundamental principles. A throughput of energy is of course essential for driving the system, and the state (or characteristics) of the system at any one time is a result of interactions between its various component parts. Over time the system may change in a variety of ways depending on changes in the balance between inputs and outputs of energy and matter, as well as on changes in the internal functioning of the parts that make up the system.

In systems terminology, a closed-system is one in which all the matter is circulated inside the system with only inputs and outputs of energy; whereas an open-system has inputs and outputs of both matter and energy. The environmental systems paradigm sees the Earth's total environment as being controlled by the interaction of numerous open-systems, such as atmospheric, hydrological, geological, and biological systems, and changes in the Earth's environment cannot be explained without an understanding of how these component systems interact with each other. One example of the way such open-systems interact is in the transfer of H_2O. H_2O existing as liquid on the Earth's surface is transferred to the atmosphere because the input of solar energy causes evaporation. Later, when vapour condenses in the atmosphere to form cloud droplets, the energy originally used to evaporate the water is released as latent heat (warming surrounding air) due to the phase change from vapour back to liquid. Provided that droplets grow large enough, the H_2O will then leave the atmospheric system as precipitation and, if over land, enter a drainage basin system. From there it could follow any number of pathways, perhaps being taken up by vegetation or returning to the ocean via runoff where it may eventually be evaporated again to re-enter the atmosphere. The hydrological cycle is just one of many different ways by which environmental systems interact in terms of material and energy transfer. What makes understanding environmental change so complex is that the component systems themselves are governed by the interactions of smaller scale sub-systems, which in turn, contain their own sub-systems and so on.

Systems analysis was brought into the earth sciences in the mid twentieth century; enabling a more integrated view of the environment and a better understanding of how it changes. An important offshoot of this was the recognition of feedback mechanisms in environmental change. The idea of positive feedback has already been introduced in the context of threshold changes in which an initial change in the system leads to further changes that reinforce the initial change – causing a 'snowballing' effect. On the other hand, negative feedback mechanisms are processes that dampen the initial change; thereby working to bring the system back to its original state. Negative feedbacks are equivalent to homeostasis: the processes by which an organism regulates internal conditions (such as body temperature) in response to external stimuli. The environmental systems paradigm recognizes that both positive and negative feedback mechanisms operate when systems interact with each other, and that fluctuations in environmental conditions can be viewed as the result of the interplay between different feedbacks.

Because of the complexity of interactions within environmental systems, environmental and climatic conditions can respond to changes in a particular variable (or forcing mechanism) in a number of different ways. As summarized in **Fig. 1.8**, environmental responses can be proportional, insensitive or sensitive to the forcing mechanism, as well as abrupt in the case of the threshold-type response. In proportional change, environmental conditions alter to an extent in line with the change in the forcing mechanism.

The insensitive condition describes a situation where a variable changes, but the system is largely immune to the applied forcing. In the sensitive condition environmental conditions respond strongly to the applied forcing, and in a threshold response there is a rapid shift in the system from one state to another.

Within the context of Quaternary studies, the application of systems theory and an understanding of feedback mechanisms and different system responses are essential for going beyond description to explanation of palaeoenvironmental changes inferred from the evidence. Discovering relationships between past changes in a forcing variable and past changes in the environment is of great importance for understanding how the environment may respond to similar changes in the future, and it is through this type of systems analysis that palaeoenvironmental studies are most relevant to contemporary debates about human-caused environmental changes such as global warming.

Timescales, stratigraphic correlation, and dating

If we are to explore the workings of environmental systems in the past, a major focus of Quaternary

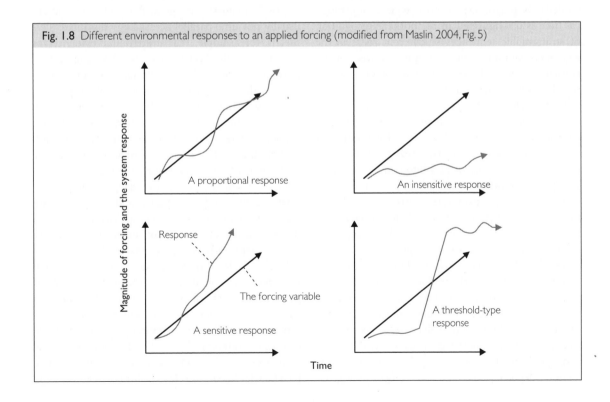

Fig. 1.8 Different environmental responses to an applied forcing (modified from Maslin 2004, Fig. 5)

research, then placing past events within a universal chronological framework that enables comparison and correlation of the timing of events from widely spread localities is essential. Subdivisions of the Geological Time Scale have already been introduced in this chapter, as have various climatic and environmental changes through ancient and more recent geological time. However, the actual methods of establishing chronologies, common conventions in talking about geological time, and approaches and issues involved in correlating stratigraphic sequences from different places also require some introduction.

Because the Geological Time Scale and its subdivisions are the product of work done by scientists over more than 200 years, a variety of different criteria have been used to segment geological time and to mark the boundaries between these segments. Fundamentally, layers of rock (and the geological time over which the rock layers formed) can be classified on the basis of the fossil organisms contained in the rock (**biostratigraphy**) or by the properties of the rock itself (**lithostratigraphy**). (Both approaches are used in describing and classifying Quaternary-aged rocks and unconsolidated sediments.) Where important biostratigraphic or lithostratigraphic changes have been identified in rock sequences, these have tended to be the markers where one phase of geological time is distinguished from another. Recognition of stratigraphic sequences of different rocks in the eighteenth and nineteenth centuries allowed for the construction of relative local chronologies, where rock layers exposed in one locality could be classed as relatively older or younger than rock layers exposed elsewhere using the law of superposition and by comparison with the universal geological column as it was known at the time (**relative dating**). In the twentieth century methods of **absolute dating**, especially using radiometric methods such as potassium-argon and uranium series dating (see **Section 2.5**), allowed for actual ages in terms of years before present (BP) to be assigned to the major bio- and lithostratigraphic changes observed in rock stratigraphy.

This paved the way for the development of **chronostratigraphy** and **geochronology**: the former referring to the rocks or deposits laid down over a specific time interval, and the latter referring to the interval of time itself (Lowe and Walker 1997). As referred to earlier in this chapter, the terms 'system', 'series', and 'stage' are chronostratigraphic, while the terms 'period', 'epoch', and 'age' are geochronological.

Intervals of geological time are pinned down, as it is said, by a 'golden spike' – a **Global Stratotype Section and Point (GSSP)** – placed in a specific stratigraphic sequence in a specific locality. Once established a GSSP becomes an internationally agreed reference point for chronostratigraphic correlation and for the assignment of past events to geochronological intervals. And, such formally recognized divisions of the geological column are fixed and therefore 'time-parallel' across the globe.

Another way of subdividing geological time is on the basis of past climatic changes inferred from stratigraphic sequences. This approach has a long tradition in Quaternary research (because recognition of the Quaternary and its subdivisions is very much climate-based); but it is important to be aware that climatostratigraphic units suffer from the problem of being **time-transgressive**. Many climatostratigraphic terms have already been introduced in this chapter: the named glacial and interglacial phases described in **Section 1.2** (as well as **stadials** and **interstadials**, and the pluvial phases experienced in arid lands) are climatostratigraphic subdivisions. They are time-transgressive in nature because the various lithological, geomorphological, geochemical, and biological changes in the palaeo-record used to identify them respond to climate change at different rates and are usually not time-parallel across large distances. This means, for example, that stratigraphic evidence for the onset of the Weichselian glacial in one part of northern Europe will not necessarily be time equivalent with similar evidence from another part, or indeed, with similar evidence from another part of the world. Another problem is that different workers have applied different criteria to separate climatostratigraphic units – for instance, should a boundary be drawn when climate has begun to change, at the midpoint of the change, or once the change is complete? A further confusion is that glacials and interglacials are often referred to as 'cold stages' or 'warm stages' in the literature – yet they are not equivalent to stages in the strict chronostratigraphic sense explained above.

The example of the Mangerud chronozones shown in **Table 1.4** was an attempt to address these kinds of problems by establishing time-parallel divisions in a chronological framework that was originally climatostratigraphic; thereby promoting consistency in the classification and correlation of Holocene stratigraphic sequences. Another example is the recognition of a **Last Glacial Maximum (LGM)** chronozone. Mix et al.

(2001) suggested that it be set between 23 and 19 thousand calendar years ago to ensure consistency between workers comparing evidence for what the climate and environment was like at the height of the last glacial phase (Weichsel-Wisconsin). However, more recently its onset is recognized as occurring earlier, *c*. 26 thousand years ago, on the basis of when maximum global ice volume was reached (Clark et al. 2009).

Another approach to dealing with the time-transgressive nature of climatostratigraphic units is to agree on a common climate record against which climatic changes inferred from other stratigraphic sequences can be compared. The climatostratigraphic subdivisions that are most applicable globally are the stable oxygen isotope variations preserved in marine sediments which reflect global fluctuations in the volume of glacier ice (see **Sections 2.3** and **3.2**). As the detail, resolution and age control of these isotopic variations has improved, so too has their utility as a 'master record' of Quaternary climate change, in which transitions between glacials and interglacials are drawn at the mid-points between major isotope excursions. The numbered system of cold and warm **marine isotope stages (MIS)** is explained in **Section 2.3** and shown in **Fig. 3.3**.

To return to the correlation of different stratigraphic sequences, the aim is to compare the timing of events. In other words, to investigate leads and lags between environmental and climatic changes and the various effects, be they geomorphological, biological, or related to events in human prehistory and history. This allows for hypothesized causes and effect relationships to be tested, ultimately leading to a better understanding of environmental systems. Given the relatively short span of geological time encompassed by the Quaternary, and the relatively high time resolution examined compared with studies of other times in Earth's geological history, it is even more important that the distinction between chronostratigraphic and climatostratigraphic schemes is borne in mind when correlating sequences. If working with million-year timescales, changes in the biological and lithological characteristics of rocks and the related climate changes seem to have occurred instantaneously; but this is not the case when scrutinizing time periods measured over millennia and centuries, or less, as is often the case in Quaternary research. Hence, there has been a great deal of effort put into advancing dating methods (**Section 2.5**) so that 'independent chronologies' can be constructed for individual stratigraphic records and the events contained therein. The better dated stratigraphic sequences are, the more precise correlations can be.

Another issue that mainly applies when correlating recent stratigraphic records at a high time resolution is that there is more than one system of nomenclature for giving age estimates. For reasons that are discussed in **Chapter 2**, radiocarbon-based chronologies that give the date of an event (or the time spanned by a sequence) in terms of years before present (BP) always do so in *years before AD 1950*. Therefore, such a date is not time equivalent to a date given in AD or BC. Furthermore, it is important to check whether the chronology for the sequence is given in uncalibrated or calibrated years. An uncalibrated date means that the age estimate is in **radiocarbon years** and is not directly equivalent to calendar years. Such a date is commonly given the notation ^{14}C yr BP. The difference can be very significant as shown in **Fig. 2.6**. If calibrated, then the radiocarbon-based age quoted has already been converted into calendar years BP (but still giving years before 1950). The common notation is **cal yr** BP. Another system adopted especially by the ice core community is to quote age in terms of years before AD 2000 (b2k). It is crucial that such differences in age reporting are taken into account before attempting to correlate stratigraphic sequences.

Contributions of stable isotope analyses

The contributions of **stable isotope** analyses of stratigraphic sequences to our understanding of Quaternary environments and climates cannot be overstated. Isotopes are variations in the number of neutrons within the atomic nucleus, causing atoms of the same element to vary in atomic mass. For instance oxygen exists as three different isotopes (^{16}O, ^{17}O, ^{18}O) depending on whether it has eight, nine or ten neutrons in addition to the eight protons. Oxygen in the form of ^{16}O (by far the most abundant at nearly 99.8% of all oxygen) is said to be 'light' and ^{18}O 'heavy'. Most carbon is in the form of ^{12}C (about 99%) with nearly 1% ^{13}C and a trace amount of ^{14}C. Isotopes are referred to as 'stable' when they do not decay like radioactive isotopes such as ^{14}C. Of course the study of radioactive isotopes has also made a major contribution to Quaternary research, particularly for methods of dating described in the next chapter. The ratio of heavy to light isotopes of a particular element in a material is measured using **mass**

spectrometry. This technique, developed in the 1950s and 1960s, involves the vaporisation of a sample to produce ions which are sent through an electro-magnetic field. On passing through the field, isotopes of different masses are more or less deflected, allowing for the proportion of different isotopes of an element contained in the original sample to be detected and compared.

The key concept behind stable isotope studies is that a process called fractionation occurs when substances are changed due to chemical or biological processes, leading to the enrichment of one isotope in relation to another for a given element; thereby changing the isotopic ratio for that element. The amount of fractionation that occurs can also often be directly related to the type or strength of the chemical or biological process. Therefore, in stable isotope analyses, it is changes in the ratio that are of interest. The most widely studied isotope ratios for palaeoenvironmental and palaeoclimatic reconstruction are the ratios of oxygen ($^{18}O/^{16}O$), carbon ($^{13}C/^{12}C$) and hydrogen ($^{2}H/^{1}H$); and so that analyses are comparable worldwide, ratios are calculated in relation to internationally agreed standards and are reported on a per mille basis. The main standards (agreed in Vienna where the International Atomic Energy Agency is based) are:

- Vienna Standard Mean Ocean Water (VSMOW) for oxygen and hydrogen isotope ratio analyses of water and ice

- Vienna Pee Dee Belemnite (VPDB) for carbon isotope ratio analyses and for oxygen isotope analyses from carbonates (the name originates from the Cretaceous belemnite fossil from the Pee Dee formation of South Carolina that was the original standard for the $^{13}C/^{12}C$ ratio)

The isotope ratios are reported as deviations in parts per mille (‰) from the standard in the following way:

$$\delta^{18}O = \frac{(^{18}O/^{16}O)\text{sample} - (^{18}O/^{16}O)\text{standard}}{(^{18}O/^{16}O)\text{standard}} \times 1000$$

Note that this same equation applies to carbon ratios ($\delta^{13}C$) and hydrogen ratios ($\delta^{2}H$ or 'δD' as ^{2}H is also known as 'deuterium'). Values for $\delta^{18}O$ range between about −55 and 5 and for $\delta^{13}C$ between about −35 and 2.

In relation to environmental systems, fractionation of stable isotopes as substances undergo change and transfer between different 'stores' (e.g. oceans, atmosphere, ice, vegetation) allows reconstruction of past processes and changes assuming the isotope ratios in substances used for analysis have not changed since they were deposited (i.e. have not undergone diagenesis).

The applications of stable isotope analysis are numerous and will be encountered many times in this book. The hydrological cycle referred to already offers a good example: as H_2O is evaporated from the seas, fractionation occurs as $H_2^{16}O$ is more readily taken up causing sea water to become enriched in $H_2^{18}O$. This is why precipitation has a more negative $\delta^{18}O$ than sea water and why an increase in the amount of snow and ice on the planet will increase the $\delta^{18}O$ of sea water as more ^{16}O is taken out of the ocean store. Because marine organisms (such as Foraminifera) incorporate oxygen from surrounding sea water as they build their shells, after extraction from a marine sediment sequence, $\delta^{18}O$ analysis of fossil forams enables reconstruction of past changes in ^{18}O enrichment of the sea and hence past changes in exchange of H_2O between ocean and **cryosphere** (the world's ice). This process (as well as a temperature dependent effect on $\delta^{18}O$) is discussed further in **Section 2.3**.

Fractionation also occurs as vapour condenses in the atmosphere with $H_2^{18}O$ more readily condensed. As air cools, it progressively loses water vapour through condensation causing progressive enrichment of ^{16}O in remaining water vapour. This explains why snow with the most negative $\delta^{18}O$ is found at the poles, and this concept underpins the reconstruction of past temperature using $\delta^{18}O$ (and δD) of layers of ice preserved in ice cores (see **Section 2.4**).

Analysis of $\delta^{18}O$ can also be used as a tracer for changes in source water. For instance, local to regional scale changes can reveal past episodes of meltwater discharge into areas of ocean (e.g. through influx of ^{18}O depleted freshwater), changes in surface and deep ocean circulation, and changes in connection or isolation of bodies of water from the sea. As low-latitude precipitation is more enriched in ^{18}O than higher-latitude precipitation, measures of $\delta^{18}O$ in lake sediments and cave carbonate deposits (**speleothems**) have allowed reconstruction of past strength and latitudinal distribution of **monsoon** rainfall.

Stable carbon isotopes are fractionated during the process of photosynthesis with $^{12}CO_2$ more readily taken up than $^{13}CO_2$ giving organic carbon a negative $\delta^{13}C$. In a marine context, more biological productivity means that surrounding sea water becomes enriched in ^{13}C, and therefore $\delta^{13}C$ can be used as a tracer for source water and ocean currents. Marine carbon isotope ratios are also related to fluctuations in exchange with atmospheric carbon (mainly CO_2 but also CH_4 as during the PETM). In a terrestrial context, $\delta^{13}C$ has been used to study past changes in the abundance of plants using the C_3 or C_4 photosynthetic pathways. C_3 plants (most plants and 95% of plant biomass) have tissue with $\delta^{13}C$ ranging between −33 and −24; whereas C_4 plants (some shrubs and grasses adapted to heat, drought, and lower CO_2 concentration) have less fractionation resulting in values between −16 and −10. Particularly in tropical environments, $\delta^{13}C$ analyses of sediment sequences have allowed reconstruction of changing abundances of C_3 and C_4 plants and therefore also of the climatic effects related to these changes.

Another important application of stable isotope analysis is in prehistorical and archaeological studies. As marine and terrestrial organisms have contrasting $^{13}C/^{12}C$ ratios (and contrasting $^{15}N/^{14}N$ ratios), isotopic analyses of human remains can be used to infer past diets, which in turn can shed light on past food resources and cultural practices. Examples of this type of application are included in **Chapter 8**.

Environmental modelling

Predictions of environmental change require the construction of models, and reconstructions of past environments are playing an increasingly important part in the design and testing of such models.

In this sense, the term 'model' refers to a simplified representation of a natural system that is used to investigate the changes to some aspect of the environment that result when certain variables (environmental factors) are changed. When designing such a model, modellers seek to represent only the interactions within the natural system that are thought relevant to the variable of interest while ignoring those processes that are believed to be negligible. Because of the enormous complexity of the environment, and our imperfect knowledge of all of the processes operating within it, it is impossible to design a model that represents all interactions perfectly; and this is why all models, even the most complicated, are simplifications of reality resting on certain assumptions about the workings of the system. Assumptions about what processes should be included or excluded, and about the way in which the system will respond to forcing, lay

models open to criticism; although as understanding of natural systems improves through experiment and observation, so too does the accuracy of the assumptions used in model design.

Models can be qualitative or quantitative, with the former useful for visualizing connections between factors and for drawing general conclusions about cause and effect, and the latter necessary for precise predictions of the magnitude of change. The quantitative models can be classified according to the spatial scale investigated by the model (meso-scale, regional scale, and global scale) as well as by their complexity and spatial resolution. At the global scale, General Circulation Models (GCMs) used for predicting the amount of atmospheric warming that will occur in response to human-caused emissions of greenhouse gases are the best known and most complicated of the quantitative type. The outputs of different GCMs from various universities and institutions around the world form the basis for the Intergovernmental Panel on Climate Change (IPCC) predictions of climate change over the next century (see **Section 7.9**); and they have been important for examining the causes for changes in climate over the past century that are described in **Chapter 7**.

These state-of-the-art mathematical models are run on the world's fastest super-computers, yet a single GCM simulation of atmospheric behaviour over a hundred-year span takes over a month to compute. GCMs represent the atmosphere as a number of interacting grid boxes across the surface of the globe, and the current generation of models can represent the whole atmosphere with boxes measuring less than 200 by 200 km across and 1 km high. To start the model, 'boundary conditions' need to be defined (such as the solar energy input, land topography, and the areas covered by sea, land, or ice) and each box has to be assigned initial characteristics (for example temperature, humidity, cloudiness, windspeed, etc.). In addition to this 'initialization', a process of 'parameterization' is required so that weather phenomena, such as convective cloud formation, that happen on a finer scale than represented by the grid boxes can be converted to values representative of a whole box.

When the model is set up, it can be used to explore the effect on climate at different timescales resulting from changes in different variables. For instance, carbon dioxide levels in each grid box can be changed (altering the absorption of long wave energy), and when the model is run, first principle physical equations are used to calculate the effect on other variables in each box as well as how each box will influence adjacent boxes. Typically the model will recalculate every variable in every grid box on a simulated timescale ('time step') of once every hour or so.

It is hard to imagine the enormous number of calculations required to compute such a model for even a one-year simulation at this spatial and temporal resolution. GCMs have become increasingly sophisticated in recent years as the atmospheric models (AGCMs) have been 'coupled' with oceanic models (OGCMs) to create Atmosphere–Ocean General Circulation Models (or AOGCMs); and some GCMs also factor in processes involving variation in the land surface, the biosphere and the cryosphere. However, limits to computing power mean that when the aim is to model a longer term process operating over centuries or millennia, e.g. the effects of orbital changes on ice sheet expansion and retreat, a model with a lower spatial and temporal resolution is required: these are called 'Earth Models of Intermediate Complexity' (EMICs).

Palaeoenvironmental reconstruction contributes to modelling efforts in two main ways. First, proxy data from stratigraphic evidence sometimes reveal environmental changes in the past of a type or magnitude previously unimaginable within the prevailing orthodoxy about how climate and environmental change works. This throws up new challenges to theoreticians because assumptions about how environmental systems interact must be revised in light of the new evidence. The challenge then passes to modellers who must find ways to incorporate previously unknown processes into their models so that they more closely represent the real world.

An excellent example of this type of contribution concerns the discovery of abrupt and high magnitude climatic changes during the transition from the Late Glacial to the Holocene. As discussed further in both **Chapters 3** and **9**, it became apparent in the 1990s from various forms of proxy data derived from terrestrial, marine and ice sheet contexts that Atlantic Ocean circulation can fluctuate dramatically over short timescales in response to freshwater input causing major climatic changes of a speed and scale that might have been thought impossible if it were not for the palaeoenvironmental evidence. In response, climate modellers have been redesigning AOGCMs to take into account this new knowledge of how ocean, ice and atmosphere can interact, and these developments were at the root of widely publicized concerns about global warming actually causing north-west Europe to cool if increased

meltwater from polar snow and ice disrupts the North Atlantic's warm currents (see **Section 9.7** for discussion of recent views on the sensitivity of this system and implications for climate). Other modelling challenges thrown up by study of past climates include simulating the processes associated with the dramatic cooling to Snowball Earth during the Neoproterozoic and the rapid warming of the PETM during the Tertiary.

The other way in which palaeoenvironmental reconstructions contribute to modelling is by providing snapshots of the climate and environment at certain times in the past that model simulations can try to replicate. Sometimes termed 'hind-casting', this is important in the process of 'validating' GCMs. If a climate model can adequately reproduce climatic conditions at a specific time in the past when the boundary conditions of the model are altered to reflect the past background characteristics (such as different distribution of land, sea, ice sheets, and concentrations of greenhouse gases and dust in the atmosphere) then more confidence can be placed in the model's ability to predict change in the future. This is an important reason for some of the large-scale, coordinated efforts in palaeoenvironmental reconstruction undertaken in recent decades.

Two such collaborations that pioneered this approach, and involved researchers from many institutions, were the CLIMAP project (1981) and COHMAP (Cooperative Holocene Mapping Project) (1988). The CLIMAP project pooled temperature proxy data from marine sediment cores to make maps of sea surface temperatures (SSTs) across the world as they were during the Last Glacial Maximum. This provided an essential boundary condition (the SSTs) for incorporation into GCMs that try to simulate the glacial climate. The COHMAP project was concerned with modelling the changes in global climate that occurred from the LGM up to the present day in 3000-year 'time slices'. With this type of exercise, the ability of the model to simulate the climate at any one time can be tested against the climate reconstructions based on Quaternary proxy data.

Since CLIMAP and COHMAP, efforts have continued to focus on improving our knowledge of past climate for the testing and validation of climate models. Through the EPILOG collaboration (Mix et al. 2001) the original CLIMAP reconstructions of the LGM have been updated and improved (see **Section 3.6**). The Paleoclimate Modeling Intercomparison Project (PMIP) compares different model simulations of climate for the LGM and for the mid Holocene to identify areas of agreement and disagreement between models and palaeo-data as well as between different models. In this way, modellers are able to identify those processes and phenomena that are modelled well and those less well (Crucifix 2008).

In summary, collecting proxy data and constructing models are both important activities for understanding environmental change, and each has much to contribute to the other. While proxy data may reveal processes previously unaccounted for in models, modelling studies may in turn draw attention to processes and factors initially overlooked in the palaeoenvironmental record; thereby stimulating further development and new directions in palaeoenvironmental research. However, environmental modelling is a highly technical and rapidly developing field, more concerned with contemporary and future change than with the past, and detailed description of modelling methods is outside the scope of this book. For understanding the environmental changes discussed in this book it is more important to survey the various types of evidence underpinning environmental reconstructions, and this is the subject of the next chapter.

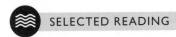

SELECTED READING

On the development of ideas:

A very good book for its overview of the development of ideas about the Ice Age and its concentration on interpretation of glacial landforms and deposits: B.G. Andersen and H.W. Borns Jr. (1994) *The Ice Age World*.

A brief and accessible popular account of the discovery and characteristics of the Ice Age: J. Gribbin and M. Gribbin (2001) *Ice Age*.

Although published some time ago, still one of the best accounts of how climate cycles during the Ice Age were recognized and how evidence from deep-sea sediments led to confirmation of the orbital theory in pacing these cycles: J. Imbrie and K.P. Imbrie (1979) *Ice Ages: solving the mystery*.

On subdividing geological time:

The Subcommission on Quaternary Stratigraphy website (www.quaternary.stratigraphy.org.uk) provides an accessible overview of the procedures and rationale behind the establishment of the Monte San Nicola GSSP for the base of the Quaternary and includes downloadable chronostratigraphical correlation tables. There are also links to articles on the idea of the post-industrial 'Anthropocene'.

On Cainozoic climate trends:

An excellent book for comprehending the way plants have influenced, and been influenced by, the atmosphere through geological time: D. Beerling (2007) *The Emerald Planet: How Plants Changed Earth's History*. Chapter 6 discusses the polar forests in Antarctica that once extended as far as 85° S, and Chapters 7 and 8 include discussion of plants and levels of greenhouse gases in the evolution of climate through the Tertiary.

B. Lovell (2010) *Challenged by Carbon* provides a fascinating account in Chapter 2 of the discovery of the Paleocene-Eocene Thermal Maximum and its relevance as an analogue for present-day greenhouse gas emissions.

A full treatment of the factors governing climate change on different timescales, including hypotheses that have been put forward to explain global cooling over the past 50 million years: W.F. Ruddiman (2008) *Earth's Climate: Past and Future* (2nd edition).

On environmental change and the Quaternary in general:

An important major work on the Quaternary was: D.Q. Bowen (1978) *Quaternary Geology*.

More recent general reviews of long-term environmental change include: R.J. Huggett (1997) *Environmental Change: The Evolving Ecosphere*, A. Mannion (1998) *Natural Environmental Change*, and P.D. Moore, B. Chaloner, and P. Stott (1996) *Global Environmental Change*.

Other textbooks more specifically on the Quaternary include: M.A.J. Williams et al. (1998) *Quaternary Environments* (2nd edition), R.C.L. Wilson, S.A. Drury, and J.L. Chapman (2000) *The Great Ice Age: Climate Change and Life*, and M. Bell and M.J.C. Walker (2005) *Late Quaternary Environmental Change, Physical and Human Perspectives* (2nd edition), and a European perspective is provided in: J. Ehlers (1996) *Quaternary and Glacial Geology*.

An accessible introduction to climate modelling: K. McGuffie and A. Henderson-Sellers (2005) *A Climate Modelling Primer* (3rd edition).

? REVIEW QUESTIONS

1. Summarize the main forms of evidence behind the development of the Glacial Theory and the eventual recognition that glaciation during the Quaternary was not a single event.

2. In what ways was it proposed that past environmental change in lower latitudes were related to glacials and interglacials recognized in higher latitudes?

3. Consider some arguments as to why long-term changes in the position of the continents alone cannot explain the transition from 'greenhouse' to 'icehouse' climate conditions during the Cainozoic.

4. Despite debate surrounding the division and classification of recent geological time, consider the reasons why it is important that a common framework is adopted.

5. In what ways is study of the PETM relevant to current studies and prediction of human-caused climate change (global warming)?

6. Using examples, discuss the extent to which studies of past environmental and climatic change are relevant for climate modelling, and thereby also for predicting future climates.

2

Sources of Evidence for Reconstructing Past Environments

Chapter contents

Chapter overview

To gain a picture of the changes that have taken place over the Quaternary, we need first to be able to reconstruct past environments and second to date the major events that have taken place. In this chapter we discuss the evidence that can be gained from on land, from the ocean floors and from cores that have been drilled through ice caps. In recent decades the range of evidence that has become available has been greatly increased, though traditional techniques such as pollen analysis still have great validity and utility. No less important revolutions have taken place in dating techniques. By combining detailed environmental reconstruction with accurate and plentiful dates, we have now gained a much more impressive picture of how and when global environments have changed.

2.1 Introduction

There are numerous sources of evidence that provide information about past environments, and great strides have been made recently in both the variety of evidence used and in the precision and accuracy of the analyses and interpretations based upon that evidence. Because a single line of evidence from a single place offers only a specific and limited range of information, scientists are increasingly making use of multiple lines of evidence when trying to piece together a picture of past environmental and climatic conditions. These various sources of evidence can be grouped in different ways, but to provide a broad overview they are presented in this chapter as either terrestrial, marine or ice-based. The aim of analysing various forms of evidence falls into one or both of two categories: either to date or to give environmental information. A separate section (**Section 2.5**) at the end of this chapter focuses specifically on methods of dating and establishing chronologies.

The sections that follow concentrate on the most important sources of evidence and techniques relevant to reconstructing Quaternary environments, although many of these have also been applied in similar ways for the study of earlier periods. Since the Second World War the number and type of techniques have vastly expanded; but the new techniques have tended to supplement rather than to replace the more traditional ones. It is beyond the scope of this chapter to cover all the traditional and modern methods of environmental reconstruction in detail: the selected reading at the end of the chapter refers to fuller treatments of the topics discussed here.

2.2 Terrestrial evidence

The study of terrestrial deposits and landforms has the longest history within Quaternary research, dating well back into the nineteenth century. Owing to their relatively recent origin in geological time, Quaternary deposits are often not lithified, and therefore tend to be relatively loose and unconsolidated (although hard igneous rocks of Quaternary age are abundant in areas of the world with active or recent volcanism). Such unconsolidated deposits include glacial **till**, **periglacial** deposits, ancient soils (**palaeosols**), deposits of wind-blown material (**loess**), and accumulations of partially decomposed plant matter (peat). Although deposited in water, lake and stream sediments are categorized here as terrestrial-based because such deposits may be found on land long after the lake or stream where the sediments were laid down has disappeared. To this list can also be added cave sediments which may contain material from outside that was washed into the cave, as well as chemicals of external **provenance** incorporated into **speleothem** (secondary mineral deposits) formed on the walls of a cavern. While the interpretation of past environments is often based on analyses of various kinds of deposits and depositional landforms, the study of erosional landforms also yields important

information. Tree rings have also proven particularly important.

Glacial evidence on land

The presence of moraine indicates the past existence of a glacier (**Plate 2.1**), and the characteristics of the till making up the moraine provide information about the characteristics of the glacier at the time of deposition and the way in which it deposited the material. Mapping till deposits (moraines) enables reconstruction of former ice limits and flow directions, and such information can enable palaeoclimatic inferences in relation to the glaciated region. Till is a 'diamicton', meaning that the deposit is unsorted and non-stratified (**Plate 2.2**). In other words, there is a mix of different particle sizes (hence the synonymous term 'boulder-clay') and there is no evidence of layering as would be expected if the deposit were laid down by water or wind. Before the concept of continental glaciation was established (see **Section 1.2**), deposits now known to be glacial till were thought to derive from the biblical flood. The related term 'drift' was often applied to till deposits because of the nineteenth century idea that much of the material was carried in drifting icebergs during a great flood and then deposited as the icebergs melted. Even today drift is still sometimes used as a general term for all kinds of glacial sediments.

Tills are analysed for a range of attributes including particle size distribution, particle shape, type and provenance, degree of compaction, and fabric (meaning the orientation of stones within the till), all of which help to distinguish between different types of till (see for example, Sugden and John 1976). Two main categories are 'ablation till' and 'lodgement till'. The former refers to debris being deposited because of the melting away of ice around it, whereas the latter refers to debris carried beneath a glacier being stuck, or 'lodged', into the ground as the glacier moves over it. In terms of attributes, the particles making up ablation till tend to be larger, more angular, less compact and show little flow orientation compared with lodgement till because the ablation till has not been affected by crushing and transport along the ice/ground interface.

For the purposes of Quaternary reconstructions, recognizing different types of till is essential for correctly mapping past distributions of glaciers and ice sheets. For instance, it is important to distinguish till that was deposited along the margins of ice in the form of end or lateral moraines from other types. Indeed, moraines take on many different forms depending on the characteristics of the glacier at the time of deposition and where the deposition of till occurred in relation to the glacier. For example, the process of debris lodgement beneath a glacier creates ground moraine that in some cases might be streamlined into mounds called **drumlins** that are aligned with the past flow direction of the ice. Ground moraine can also be sculpted into linear mounds called 'fluted moraine' which, like drumlins, are oriented parallel to past ice flow but have a much greater length to width ratio. Where the terminus of a glacier remains relatively stationary for hundreds or thousands of years, large end moraines develop as rock debris carried along within the glacier flow reaches the terminus and is dumped out. Seasonal retreat and advance of the glacier builds up the moraine ridge further as debris deposited along the margin during summer melting (ablation) is later bulldozed forward during the period of winter advance. End moraines are called 'terminal moraines' when they mark the furthest extent of a former glacier and 'recessional moraines' when they mark stages during glacier retreat.

A summary of the main types of moraine is given in **Table 2.1**. In general, the identification and mapping of terminal and recessional moraines is important for determining the maximum extent reached by glaciers in the past and the pattern of their retreat, while types of ground moraine, such as drumlins and fluted moraine, help to interpret past flow directions. When mapping moraines, it is important to distinguish them from depositional features created by glacial meltwater in the form of ice-contact **glaciofluvial** deposits and outwash. The size and shape of the deposits can sometimes give an indication of their origin, but often such landforms need to be viewed in section to determine whether they are glacial or glaciofluvial.

Interpreting the past dimensions of a glacier or ice sheet is less straightforward than it might at first seem because end and lateral moraines are not always well-developed, well-preserved, or easily identified. Depending upon the manner in which a glacier retreated, sometimes a recessional moraine can be larger and more prominent than the glacier's terminal moraine and can be wrongly interpreted as the latter. In areas affected by multiple glaciations it can be difficult to assign moraines and till to the correct glacial episode, and the more ancient the phase of glaciation, the more fragmentary the depositional record because of post-depositional reworking and erosion. It is therefore important that glaciologists use all available evidence when reconstructing past glaciation, including glacially transported **erratics** and features of glacial erosion (**Plate 2.3**). For instance, striations and chatter marks on hard rock caused by glacial abrasion provide another source of evidence for interpreting past directions of glacier movement, and 'trimlines' marking the highest elevation of glacial erosion in a valley help in interpreting the thickness of a former glacier ice mass and its surface contours. A reconstruction of the former altitude of Loch Lomond Stadial (**Younger Dryas**) ice in Wester Ross, north-west Scotland, provides a good example of the identification and interpretation of trimlines (Ballantyne et al. 1997).

Table 2.1 Summary classification of morainic forms

Linear features parallel to ice flow	Linear features transverse to ice flow	Non-linear features lacking orientation
Subglacial streamlined forms:	*Subglacial forms:*	*Subglacial forms:*
– Fluted and drumlinized ground moraine	– Rogen or ribbed moraine	– Low relief ground moraine (till plain)
– Drumlins	– De Geer or washboard moraine	– Hummocky ground moraine
– Crag and tail ridges	– Subglacial thrust moraine	
Ice-marginal forms:	*Ice front forms:* (seasonal/terminal/recessional)	*Ice surface forms:*
– Lateral moraine	– End moraine (dump moraine)	– Disintegration moraine (hummocky moraine)
– Medial moraine	– Push moraine	
	– Ice thrust/shear moraine	

(Modified from Sugden and John 1976.)

Once the dimensions of a former glacier or ice sheet are estimated, then many other inferences about past environmental and climatic conditions are possible. For example, past temperature and precipitation regimes can be estimated from reconstructions of the equilibrium line altitude (ELA) of the former glacier or ice sheet. The ELA refers to the altitudinal line along a glacier that separates the upper area of net annual snow accumulation (the accumulation zone) from the lower area of net annual ablation (the ablation zone). When a glacier is in a steady state (neither advancing nor retreating) total accumulation and ablation are balanced on either side of the ELA and glacier ice that forms from compacted snow in the accumulation zone flows down slope to replace ice that has been melted near the glacier's snout. In modern glaciers, the ratio between the area of net accumulation and the total area of a glacier in a steady state is usually between 0.6 and 0.65. Assuming that this relationship also applied in the past, the ELA can be approximated once the total area and surface contours of a former glacier are known. A modern **analogue approach** (as introduced in **Section 1.6**) can then be taken by measuring temperature and precipitation at the ELA of existing glaciers and using these values as past climate

parameters for the area where the ELA was positioned in the past (**Fig. 2.1**). This type of analysis has yielded estimates of mean annual temperature depressions on the order of 5 to as much as 15 °C lower than at present during the height of the last glacial phase in mountainous areas such as the Rockies and the Andes (Rodbell 1992). Carr et al. (2010) discuss methods for reconstructing the dimensions of former glaciers and the different ways of estimating the ELA.

In addition, the presence of glacial **cirques** can be used to infer the positions of former snowlines, which are themselves climatically controlled. The median level of a cirque floor tends to be at, or just above, the local snowline, so that the lowest cirque floor of a group of contemporaneous cirques will give a close approximation of the local snowline. Thus the height of Pleistocene snowlines can be compared with present-day snowlines, and some estimate of temperature change can be obtained. However, the main difficulty in palaeoclimatic inferences based on reconstructions of the past snowline or ELA lies in isolating temperature and precipitation effects. It is the case, for instance, that in mountainous areas with high precipitation, such as the Cascade and Coast Ranges in North America's Pacific Northwest, glaciers reach

Fig. 2.1 Scatter plot showing the relationship between summer temperature and precipitation (snow accumulation shown as millimetres water equivalent) at the equilibrium line altitudes (ELAs) of 70 glaciers from around the world. Data sourced from Ohmura et al. (1992). If either temperature or precipitation can be constrained using other forms of proxy data, then the ELA-climate relationship can be applied to estimate the other variable at the location of a former glacier a lower altitude than glaciers further inland, such as those in the Rockies. It must be borne in mind when estimating temperature variations from a past change in the ELA, that such a change could also be related to changes in precipitation (or changes in the input of wind-blown snow) that affected the glacier's accumulation rate.

Reconstructed temperatures for a glaciated region can be adjusted for altitude above or below the ELA by using the average lapse rate (temperature change with height) through the atmosphere of 0.6 °C per 100 m. For example, on the basis of glacial reconstructions for the Scottish Highlands it is estimated that mean summer temperature at sea level was about 6 °C (9 degrees colder than at present) during the Younger Dryas cold phase (Sissons and Sutherland 1976; Sissons 1980b).

Evidence of periglaciation

Periglaciation refers to the action of non-glacial cold-climate processes on the landscape. Periglacial environments are characterized by long, cold winters, during which temperatures rarely exceed 0 °C and the average temperature of the coldest month is below −3 °C, and by relatively low mean annual precipitation (French 1996). Unlike glacial environments, ground must be seasonally free of snow and ice, and there must be fluctuation around 0 °C during the summer months because repeated freezing and thawing near the ground surface is necessary for the operation of most periglacial processes. Periglacial environments are most extensive in the Northern Hemisphere where there is a lot of land at high latitude that is not covered in glacier ice, for example across much of Alaska, Canada, and Siberia. However, periglacial environments are also found in some mid- to low-latitude locations where high altitude creates similar cold climates. Be they at high latitudes or high altitudes, such environments are typically open and treeless, with only low growing tundra vegetation and are often underlain by **permafrost**.

Periglacial processes give rise to many distinctive landscape features, and when such features are identified in areas outside present zones of periglacial activity, they provide evidence for colder climates in the past. By studying contemporary formation of landscape features in periglacial areas, it has been possible to narrow down the associated climatic conditions so that the presence of relict features in an area can be used as a fairly accurate proxy of the past climate

when the features were formed. A good example concerns the use of **ice-wedge casts** as palaeoclimate indicators. Ice wedges form in areas of permafrost affected by frost cracking in winter and thawing of the ground surface (the active layer) in summer. In areas of extreme cold when temperature drops well below 0 °C in winter, the ground contracts forming a polygonal network of vertical cracks across the surface. When ice near the surface of the ground thaws in summer, meltwater seeps into the permafrost cracks to eventually freeze later in the season. This process is repeated each year and the crack widens due to further contraction of the ground and further accumulation of ice within each vertical crack. Over hundreds of years the cracks (which look like wedges in cross-section) can become as much as 3 m wide and 10 m deep. If the climate warms and the ice melts, then the crack becomes filled in with sediment forming a 'cast' that preserves the shape and size of the original ice wedge.

Ice-wedge casts have been found in many lowland, mid-latitude regions of Eurasia and North America, attesting to the former presence of permafrost during the height of the last glacial phase. For example, maps of their distribution suggest that permafrost extended across most of the British Isles with the exception of parts of south-west England and Cornwall (see **Fig. 3.13**). Where ice-wedge casts are found in fine-grained sediment, the mean annual air temperature must have been at least as low as −3 °C when the ice wedge was forming, and in coarser sediment probably as low as −6 °C (Ballantyne and Harris 1994). Annual precipitation must also have been low, usually less than 250 mm, because frost cracking is limited by the insulating effect of snow on the ground. In addition to ice-wedge casts, **involutions** also indicate the former existence of permafrost.

There are several other relict features indicative of colder climates in the past. A common effect of periglaciation is the production of patterned ground which includes such features as stone circles, polygons, steps, and stripes (**Plate 2.4**). These various types of patterned ground are formed from repeated processes of frost heave and sorting on flat or sloping ground. Seasonal thawing of ice in the active layer leads to the process of **gelifluction** in which soil and regolith (saturated with pore water due to the impermeable frozen ground below preventing downward percolation into the rock) flows slowly downhill. Over time, gelifluted material builds up along the flanks of hills to form gelifluction lobes and benches and in valleys to form head deposits. Deposits resulting from gelifluction can be several metres thick and are identifiable as a mixture of fine sediment, sand and angular, frost-shattered stones that have long axes oriented down-slope in the former flow direction. Dry valleys are yet another relict landscape feature of periglacial climates. By preventing downward percolation of precipitation and meltwater in areas where the bedrock would otherwise be permeable, permafrost causes excess water to flow over the surface causing fluvial erosion and creating valleys. As the climate warms and the permafrost disappears, the local water table falls and the valleys produced during periglacial conditions become dry remnants of former streams.

Palaeosols and wind-blown deposits

Palaeosols are soils that formed in the past and that have since been buried beneath more recent sediments. Like the glacial and periglacial deposits, they are another type of relict feature reflecting past climatic and environmental conditions. However, the interpretation of palaeosols is often less straightforward because of problems related to **diagenesis** and difficulties in distinguishing true palaeosols from ancient soils that are still developing. Even distinguishing palaeosols from other deposits of weathered material can sometimes be difficult. Soils contain a mixture of mineral and organic matter, and they develop through physical, chemical, and biological processes. These processes give rise to horizontal layers in soils, and the characteristics of these layers strongly reflect the prevailing climate and vegetation during the time of formation, as well as the relief, the bedrock, and the amount of time over which the soil developed.

Palaeosols are normally identified and sampled from exposed sedimentary sections, and are commonly analysed for variations in colour, texture (particle size) and mineral constituents across the different horizons. These characteristics can reveal much about the past environment. For instance, the colour can indicate the extent to which soil-forming conditions were oxygenated (reddish) or anoxic (blue-grey) serving as a proxy for past moisture levels. In addition, high precipitation levels in the past should result in a palaeosol with a leached profile, having a relatively low pH and more minerals stripped from the A-horizon and deposited in the B-horizon, whereas more arid conditions would leave behind a more alkaline palaeosol

having more minerals (particularly calcium carbonate) in the A-horizon. Palaeosols can also be useful for the organic material they contain, and if preservation is good, plant remains and pollen give an indication of the past vegetation cover when the soil developed.

Deposits of wind-blown material are another important source of palaeoenvironmental information, sometimes found inter-bedded with palaeosols in stratigraphic sections (**Plate 2.5**). Unconsolidated, wind-blown deposits can consist of relatively coarse material (sand-sized) or fine material (silt-sized), the former type forming sand sheets and dunes and the latter forming loess. In a stratigraphic section, the presence of wind-blown sediment indicates a time when the landscape upwind of the sediment was more susceptible to wind erosion usually because of relatively arid conditions and sparse vegetation cover. In some mid-latitude regions thin soils are underlain by extensive sand deposits indicating that the surrounding landscape was drier in the recent past, an example being the grass covered sand hills of western Nebraska in the American Great Plains.

In many parts of the world fossil sand dunes can be a valuable source of palaeoclimatic information. For example, as we shall see in **Chapter 4**, large continental dunes only develop over wide areas where precipitation levels are below about 100–300 mm annually. Above that figure sand movement is drastically reduced by the development of an extensive vegetation cover. Thus if fossil dunes are currently found in areas of high rainfall, it tends to suggest that rainfall levels have increased since they were formed. Conversely, some low-latitude areas of active and extensive dunes today, known as sand seas or **ergs**, overlie palaeosols and stream and lake deposits testifying to wetter conditions in the past (see **Section 5.10**).

As distinct from sand sheets and sand dunes, 'coversand' refers more specifically to flat, sandy wind-blown deposits formed near the margins of ice sheets during phases of continental glaciation. Till and glaciofluvial deposits and outwash plains along glacial margins are exposed to strong polar winds, and owing to little vegetation cover the sand-sized particles are easily picked up by the wind, transported and eventually deposited downwind of their glacial source. As the climate warms and glaciers retreat, coversand becomes stabilized by vegetation and soil begins to form on top. Coversands can be several metres thick, and are found extensively beneath present-day soils near the former margins of the North American and European ice sheets. Silt-sized particles of glacial origin can be carried much further by the wind, and hence thick deposits of loess are found further away from the former ice sheet margins than coversands. Loess is found in many parts of the world and aspects of the formation and distribution of loess are discussed in more detail in **Section 3.5**.

The Loess Plateau in north-central China contains the thickest loess in the world, reaching as much as 300 metres thickness and spanning the whole Quaternary. These loess deposits are interspersed with darker coloured palaeosols, and studies of the alternating succession of loess and palaeosols have yielded important information about climate changes since 2.5 million years ago (Kukla and An 1989). The area's geography makes it uniquely sensitive to variations in the strength of the Asian **monsoon**, and the thickness and continuity of the loess sequences are unique among sources of terrestrial evidence in enabling a reconstruction of climate change that is long and detailed enough to be correlated with the reconstructions based on marine sediment cores (see **Section 3.2**). The palaeosol layers represent times of a strengthened monsoon associated with higher summer precipitation and enhanced soil development whereas the layers of loess indicate more arid periods associated with continental glaciation when strong northerly winds brought in great quantities of dust from the Siberian plain. In addition to the colour differences, palaeosol and loess units are also distinguished and classified by their degree of mineral magnetic susceptibility (very fine grained particles contained in the palaeosols show a higher magnetic susceptibility) and by their chemical constituents (see **Fig. 3.4**).

Lake, peat, and fluvial deposits

Lake, peat, and fluvial deposits have long been used for reconstructing past environments: the peatland-based Blytt–Sernander scheme of post-glacial climate change being a case in point (see **Section 1.2**). However, recent years have seen many advances and developments in the techniques used for both analysing and dating these sediments. In contrast with most other types of deposits, lake sediments and peats often provide continuous stratigraphic records which can yield detailed time series of depositional changes for a variety of different time and space scales. Continuous deposits are especially important for estimating the ages of past environmental changes and for calculating the

deposition rates which may provide insights into the rate and mode of sedimentation and landscape **denudation** at different time periods. Interpretation of fluvial sediments can, however, be more complicated than lake sediments because of changes in sedimentation due to channel migration over relatively short periods of time – not to mention problems of post-depositional reworking.

Lake sediments and peat represent two extremes along a continuum of depositional environments related to water level. Lake (limnic) deposition is subaqueous, while the plant matter that accumulates to form peat is mostly derived from above the local water table. Intermediate between lake and peat sedimentation is 'telmatic' sedimentation occurring in marshes or swamps that are seasonally inundated. Depending on the size of a lake, the climate and sedimentation rates, infilling of a lake can occur taking it through the telmatic phase to eventually become a fen or bog; or alternatively, an increase in the height of the local water table could cause a peatland to be flooded eventually becoming a lake. For these reasons the **stratigraphic sequences** obtained from lake beds and from peatlands can contain a combination of limnic, telmatic, and terrestrial deposits, although many of the analytical methods applied to these three types are similar.

The composition and characteristics of lake sediments are influenced by a variety of factors related to the lake itself and to the surrounding landscape. Lake sediments include both 'autochthonous' and 'allochthonous' material, the former originating from organisms within the lake and the latter being washed into the lake from the surrounding drainage basin (or catchment) (Moore et al. 1991). The balance of these depends on the productivity of the lake ecosystem in relation to rates of weathering and erosion in the catchment. The characteristics of the lake sediment are also affected by the conditions for preservation on the lake bed and by the types of organisms dominating the lake ecosystem. As climate and environmental changes alter a lake's physical, chemical, and biological properties, changes in sedimentation occur leaving behind a record of change that can be interpreted through analysis of the lake bed stratigraphy. **Table 2.2** summarizes characteristics of lake sediments on the basis of nutrient status.

Under special conditions, lake sediments may develop laminations (alternating bands of sediment with distinctive composition, texture, and appearance)

Table 2.2 Characteristics of lake sediments based on nutrient status

Nutrient status	Sediment characteristics
Eutrophic (high productivity)	Characterized by a green-brown, organic-rich sediment (termed 'gyttja' or neckron mud) with a high proportion of autochthonous material. In shallow lakes gyttja often includes coarse plant detritus. In calcareous regions precipitation of calcium carbonate forms marl.
Oligotrophic (low productivity, scarcity of nutrients)	If erosion rates in surrounding catchment are high sediments will be predominantly inorganic (clastic), if low the sediments may be dominated by in washed humic substances derived from the surrounding catchment – such conditions are termed 'dystrophic' producing a dark brown sediment ('dy' or gel-mud). Sediments generally have a high proportion of allochthonous material, but some oligotrophic lakes support large diatom floras which form a whitish diatomite sediment.

(After Moore et al. 1991 and Lowe and Walker 1997.)

which result from rhythmic variations in sedimentation. Development of laminations is inhibited by bioturbation caused by bottom-dwelling (benthic) organisms and/or by shifting of sediment across the lake bed due to currents. Laminae are most likely to be preserved when mixing of the water column is strongly seasonal (or absent) and when water at depth (the hypolimnion) remains anoxic, due to little or no mixing with the oxygenated surface waters (the epilimnion) over the course of the year. Laminations often occur on an annual basis and have proven valuable as an incremental dating method (Saarnisto 1986) (see **Section 2.5**). At mid to high latitudes such laminations, also known as **varves**, comprise pairs or couplets attributable to a seasonal rhythm. Such layers were produced by the great summer inputs of meltwater from glaciers into lake basins. The coarser material was deposited in the summer, but the fine suspended clays in the meltwater did not settle out until the autumn or following winter, and it is this which produces the banding. Some years would tend to produce particularly thick bands, and varve chronology

depends on the correlation of such distinctive bands in different localities, and on the assumption that the couplets are annual. Analysis and interpretation of varve chronologies was pioneered with great effect by de Geer in Sweden.

Provided that conditions are sufficiently wet on a land surface, plant matter will accumulate more quickly than it is decomposed, forming peat. Peat can be defined specifically as an organic rich soil containing at least 30 per cent organic material by dry weight. If the organic component exceeds 75 per cent of dry weight, it is classed as an 'organic peat' rather than a 'mineral peat' (Heathwaite et al. 1993). Peat forming systems can be classified in a variety of different ways depending on climatic, geological, ecological and geomorphological conditions (Steiner 1997). **Tables 2.3** and **2.4** summarize various peatlands on the basis of ecology, form, and source of water influx.

In general, sediments derived from lakes, peatlands, and streams are analysed for the physical and chemical characteristics of the sediment itself and/or for the biological remains that they contain. Sometimes the sediments are found as exposures along a natural or man-made free-face and can be viewed and sampled easily once the face is cleaned. In most cases, however, the sediment sequences are sampled from the top down using a coring device. After sediment cores have been extruded and their stratigraphy examined (**Plate 2.6**), they are typically divided along their length into subsamples which are then analysed for a variety of **proxy data**.

As a starting point, most sediment cores are described in terms of colour changes and physical appearance and then analysed along their length for changes in bulk density, moisture content and organic content (through loss on ignition). Sub-samples are also extracted for radiocarbon dating (**Section 2.5**) so that sedimentation rates can be determined. The inorganic material is sometimes studied for its particle size distribution because changes in the proportions of clay, silt, sand, and gravel can provide information about the source of the material, and crucially for fluvial sediments, the transportation capacity and competence of the flow which carried the sediment prior to deposition.

When applied to fluvial sediments, particle size information can be used to infer past levels of discharge in stream systems which may reflect past climatic changes that affected the hydrological cycle, and/or past changes in the drainage basin system.

Table 2.3 Characteristics of peatlands

Type	Description
Ecological characteristics	
Marsh	Peatland subjected to periodic or regular flooding. Abundant sedges, grasses, and rushes.
Swamp	Wetland or peatland subjected to periodic or regular flooding with trees in addition to herbs.
Fen	Meadow-type peatland dominated by sedges, grasses, reeds, and shrubs. Sometimes with trees.
Bog	A peatland that is acidic and nutrient poor. Dominated by *Sphagnum* mosses, ericaceous shrubs, sedges, and grasses that can tolerate oligotrophic conditions.
Physiognomy	
Basin and valley mires	Peatland development in natural depressions which collect water.
Blanket peats	Peatland development across sloping landscapes in very wet climates.
Flood plain mires	Usually marshes or swamps.
Upland soligenous mire	Following upland springs and streams.
Raised mire	Peat accumulation above the local water table – dome shaped. Found in large, open valleys in regions with cool and wet climatic conditions.

(Modified from Shotyk 1992 and Heathwaite et al. 1993.)

Table 2.4 Characteristics of peatlands based on source of water influx

Source of water influx	Description
Soligenous	Controlled by catchment drainage patterns which cause groundwater and runoff to flow through the peatland. Generally nutrient rich due to rheotrophic (minerotrophic) conditions.
Topogenous	Water collects within a natural depression causing peat accumulation. Peatland receives water via groundwater and precipitation, although the relative proportions of these sources can vary. Hence topogenous mires can be minerotrophic, mesotrophic, or oligotrophic.
Ombrogenous	Peatland receives all water only from precipitation and receives no groundwater and runoff inputs.

Furthermore, the identification and dating of alluvial deposits in river valleys can indicate times of high or low levels of sediment delivery, which in turn, can relate to past climates. For instance, by compiling the frequency of Holocene alluvial deposits of different ages in Britain, Coulthard and Macklin (2001) suggested a relationship between shifts to wetter climate interpreted from peatlands (see below) and increased fluvial sedimentation.

It is remarkable how much can be inferred about the past environment through relatively simple physical analyses of sedimentary sequences. For example, by studying changes in percentage and type of inorganic material and sedimentation rates represented in core samples from Lake Pátzcuaro, central Mexico, O'Hara et al. (1993) were able to identify previously unrecognized phases of heightened soil erosion in the surrounding uplands, which on the basis of radiocarbon dating, led to the conclusion that there was overintensive agriculture and land-use in the area prior to Spanish colonization. Historical rates of upland soil erosion have also been based on physical and depositional studies of floodplain sediments, for instance by Trimble (1981) in south-west Wisconsin. Layers of peat with relatively high amounts of mineral matter have been interpreted as indicating phases of heightened erosion in south-west Scotland (Edwards et al. 1991).

One of the most important physical characteristics of peat is its state of humification which refers to the degree of decomposition of the plant material making up the peat. Highly humified peat is dark in colour and is made up of plant material that is mostly decomposed and barely recognizable; whereas less humified peat has good preservation, enabling the identification, sometimes to species level, of many of the plant components. Changes in the degree of humification through a peat sequence primarily relate to past changes in the wetness of the peatland surface which altered the effectiveness of aerobic bacterial decomposition. During times of prolonged waterlogging aerobic decomposition slows down, and as growth continues on the bog surface, over time plant matter becomes incorporated into the lower, permanently saturated layers of the bog without significant alteration. On the other hand, if the bog surface dries out and the water table falls, then plant matter on the surface undergoes more decomposition before transfer to the lower layers. Because the rate of decomposition is over 100 times slower in the lower saturated zone of a peatland (the 'catotelm') compared with the oxygenated surface layer (the 'acrotelm') (Clymo 1984; Malmer 1992), the state of humification acquired by plant matter while near the surface undergoes little change for thousands of years as it becomes buried beneath younger peat. Hence, the humification of a sub-sample taken from a peat core is mainly a function of the amount of time that plant matter was exposed to aerobic decomposition

in the acrotelm before being subsumed within the anaerobic catotelm (**Fig. 2.2**).

It has long been recognized that layers of less humified peat may represent times of wetter and/or cooler climate which increased effective precipitation and enhanced waterlogging, while more humified layers may indicate times of drier and/or warmer climate which aerated peat surfaces. Such interpretations contributed to the development of the Blytt–Sernander scheme of post-glacial climate change described in **Section 1.2**. Around the turn of the twentieth century attempts were made to correlate peat layers of similar humification across north-west Europe leading to the identification of various low humification 'recurrence surfaces' which were thought to represent European-wide climatic shifts to wetter conditions during the Holocene. The most well-known of these was termed the 'Grenzhorizont' (boundary horizon) by the German scientist C.A. Weber to mark a major

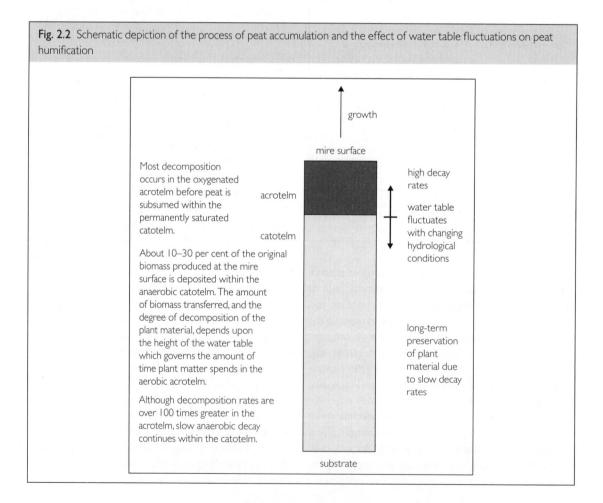

Fig. 2.2 Schematic depiction of the process of peat accumulation and the effect of water table fluctuations on peat humification

growth

mire surface

Most decomposition occurs in the oxygenated acrotelm before peat is subsumed within the permanently saturated catotelm.

acrotelm

catotelm

high decay rates

water table fluctuates with changing hydrological conditions

About 10–30 per cent of the original biomass produced at the mire surface is deposited within the anaerobic catotelm. The amount of biomass transferred, and the degree of decomposition of the plant material, depends upon the height of the water table which governs the amount of time plant matter spends in the aerobic acrotelm.

Although decomposition rates are over 100 times greater in the acrotelm, slow anaerobic decay continues within the catotelm.

long-term preservation of plant material due to slow decay rates

substrate

transition to less humified peat believed to be associated with a shift to a wetter climate at the end of the Bronze Age (SubBoreal/SubAtlantic transition) between 800 and 500 BC (Barber 1981). Broad geographical correlations were also made between layers of highly humified peat containing tree stumps (buried 'forest beds') believed to represent phases of drier climate (**Plate 2.7**).

However, these early attempts at palaeoclimatic reconstruction from peatlands were later shown to be over-simplifications of Holocene climatic change – both temporally and spatially. With the advent of radiocarbon dating, inferred temporal correlations between peat layers could be tested, and in many cases such correlations were falsified (see for example the radiocarbon dating of pine stumps from Scottish blanket peat by H.H. Birks 1975). The importance of autogenic (internally driven) changes in bog wetness had also been under-appreciated, and there was debate during the mid twentieth century about whether externally driven climatic signals could in fact be detected from peat stratigraphy. Through a detailed study of the stratigraphy of Bolton Fell Moss in Cumbria, Barber (1981) showed that peat bogs, particularly of the ombrogenous type (see **Table 2.4**), can contain sensitive records of climatic change, and from the 1980s onwards, renewed confidence in peat-based palaeoclimatic reconstructions led to many more peatland studies and further development of the techniques.

Modern analyses of peat stratigraphy differ from the early twentieth century interpretations in several respects, the most important being the way in which chronologies are constructed for peat sequences. Constructing age-depth models using radiocarbon dating and other forms of chronological control applies to a whole range of sediment types (e.g. lake and marine sediments) besides peat and is discussed in **Section 2.5**. However, the important point here is that temporal correlations between the climatic shifts interpreted from different peatlands can now be made with more confidence than in the past because of advances in dating. The techniques used to detect past hydrological changes (which may be climatically driven) from peat sequences have also developed far in the past couple of decades. Methods of measuring the degree of peat humification through a sequence have increased in precision (e.g. Aaby 1986; Blackford and Chambers 1993) and subtle changes in the height of the water table can be reconstructed through the analysis of 'plant macrofossils' contained in the peat

(particularly changes in the abundance of species of *Sphagnum*) and various microfossils, such as rhizopods (testate amoebae) (Charman et al. 2007). (Plant macrofossils are plant remains that are visible without magnification.) A good summary of recent developments in peat-based palaeoclimatic reconstruction is contained in Barber and Charman (2005), and a study of Scottish peat sequences by Langdon and Barber (2005) provides an example of a multi-proxy approach, using the full range of available methods.

The previous discussion has focused primarily on the specific nature of lake sediments and peat and on what inferences can be made about the past environment and climate on the basis of their physical make-up. Another important source of palaeoenvironmental evidence comes from the biological and chemical materials incorporated within the sediments. Perhaps the most widely studied of the biological constituents are pollen and spores.

Pollen analysis

Pollen analysis, or **palynology**, was pioneered by Von Post, a Swedish botanist, who published the first quantitative pollen diagram in 1916. (See **Fig. 3.25** for an example of a pollen diagram.) The technique makes use of the fact that sediments often contain pollen grains and spores which mostly come from the air by fallout (pollen-rain), and they are thus derived from the regional and local vegetation. The pollen and spores of most temperate plants are produced in abundance, widely dispersed and have an 'exine' (outer wall) resistant to decay; and as a result, pollen is often found in abundance in a variety of sediment types (particularly lake sediments and peat) when other microfossil types are rare or absent. Vegetational changes, which may be caused by climatic, edaphic, biotic, or human factors, can be recorded by the preservation of pollen in a sequence. Pollen grains may be counted and recorded by dispersing sediments with appropriate agents and then looking at them under a high-power binocular microscope (pollen grains and spores vary in size but are typically 20–40 µm in diameter). Results of such analysis give a picture of the vegetation at a given point in time, and also allow the sequence of vegetational change to be examined over a period. Because of their abundance, pollen and spores are particularly well suited for quantitative and statistical analyses (Birks and Gordon 1985). However, the interpretation of pollen data does have its problems, especially involving taxonomic precision, taxon over- or under-representation, and **taphonomy**.

In recent years the length of record obtained from pollen analysis has been increased by the taking of long cores from lakes and swamps. Among the longest known pollen records are those from Tenaghi Philippon in north-east Greece extending back over a million years (Tzedakis et al. 2006), and from Bogotá, Colombia where a core 357 m long yielded a record extending back 3.5 million years (Hooghiemstra 1989). Horowitz (1989) reported continuous pollen records from the Hula and Dead Sea lakes in Israel also going back 3.5 million years, and a 1400 m long pollen core, going back well over 3 million years, has been retrieved from Lake Biwa, Japan (Fuji 1988).

Thus, palynology has contributed to our understanding of both Holocene and Pleistocene vegetation changes in many parts of the world and has many applications for long-term ecological studies (Seppä and Bennett 2003). Palynology has also proven an important means for palaeoclimatic reconstruction. Quantitative estimates of past climate (such as past temperature and precipitation) based on pollen can be achieved in three main ways: the 'indicator species approach', the 'modern analogue technique', and through the development of 'transfer functions' that relate specific pollen assemblages to climatic parameters (Birks 2005). The indicator species approach (introduced in **Section 1.6**) has a long history, see for example Iversen's classic study of the pollen of *Viscum album*, *Hedera helix*, and *Ilex aquifolium* (mistletoe, ivy, holly) in Denmark (Iversen 1944). As these three plants have well defined climatic ranges, the mutual presence of their pollen in the same fossil pollen samples allowed Iversen to quantify the past climate range and show that compared with today mid Holocene temperatures in Denmark were 2–3 °C warmer in summer and 1–2 °C warmer in winter.

The modern analogue approach can be summarized as using statistics to identify a present-day pollen assemblage that is most similar to the fossil pollen assemblage (the statistical methods are in fact designed to yield and compare 'dissimilarity measures' between fossil and modern assemblages). The present-day climatic conditions associated with the least dissimilar modern pollen assemblage are then assumed to be the climatic conditions that prevailed at the time and location of the fossil assemblage. However, the approach can suffer from the 'no modern analogue' problem where the fossil assemblage bears little resemblance to

any known modern assemblage (and therefore to no known climate regime); or from the problem of multiple analogues when a similar modern assemblage can occur under different climatic regimes. Account must also be taken of taphonomic issues concerning differential representation and preservation of pollen types in the fossil assemblage. Notwithstanding these problems, the technique has been successfully applied: one of the most important such studies in Europe, based on pollen analysis of sediments from La Grand Pile, France, has yielded a robust quantitative climatic reconstruction extending back through the most recent glacial/interglacial cycle (e.g. Guiot et al. 1993). (See **Fig. 3.24** for the Grand Pile pollen sequence.)

In contrast with the modern analogue technique which uses statistical comparisons, the transfer function approach involves calculation of a multivariate mathematical model to directly relate changes in the abundance of various pollen types in the sample to changes in a controlling climatic variable. As it is used on many types of biological data besides pollen, the principles are outlined further at the end of this section.

Other forms of terrestrially-based biological evidence

In addition to pollen, there are numerous other types of biological remains that may be found in lake and peat sequences and analysed to give palaeoenvironmental information.

Plant macrofossils have already been mentioned in connection with interpreting peat stratigraphy; but they are found in a variety of depositional environments besides peat, including lake sediments and palaeosols. Because plant macrofossils (seeds, twigs, leaves, wood fragments, etc.) are so large compared with pollen, they tend not be transported far from their source. In peat, plant macrofossils are generally *in situ*, whereas in lake sediments they often represent flora from around the lake margin. While plant macrofossils do not give an indication of the regional vegetation in the way that pollen can, plant macrofossil analysis, in conjunction with palynology, provides important site-specific information, often giving proof of the local presence of a specific species at a particular time. Those seeds or woody fragments that are *in situ* also provide valuable material for radiocarbon dating. Plant macrofossils may also be preserved in the form of charcoal. Such charcoal can, through microscopic analysis, give an indication of both past vegetation conditions and the incidence of fires (Tolonen 1986).

A highly valuable depositional context for plant remains in drylands is the packrat midden. In the south-west USA, for example, the middens of *Neotoma* are found in many caves, and these contain plant material foraged by the rats from the local area. Cemented by dried rat urine, these middens may remain preserved for tens of thousands of years, and provide an inventory of past vegetation conditions around a site (Wells 1976; Betancourt et al. 1990).

An equally laborious technique to pollen analysis, but one which has produced good results, is that which utilizes non-marine molluscs, remains of which are found very commonly in Pleistocene deposits. Molluscan assemblages have been found to be indicative of particular types of climate. Cold faunas show a dominance of a few species in great numbers, and temperate faunas a larger number of species, many of which occur frequently.

Similarly techniques have been developed to use remains of beetles (Coleoptera) (Coope et al. 1971). Their wing cases are found in suitable sediments, and as distributions of living species are known quite well, especially in Scandinavia, it has proved possible to interpret palaeoenvironments on the basis of insect faunas. With knowledge of the climatic tolerance ranges of various beetle species, it is possible to infer from an assemblage of beetle remains the prevailing climate at the time that the beetles lived. This involves finding the 'climatic overlap' between the beetle species in a particular assemblage – the 'mutual climatic range' approach illustrated in **Fig. 2.3**. Results have tended to correlate well with results obtained by pollen analysis, and by the study of non-marine Mollusca, though certain discrepancies have been noted. At Lea Marston, Warwickshire, for example, a deposit dated to about 9500 **radiocarbon years** before present shows a relatively warm beetle fauna but a cool vegetation assemblage dominated by birch, dwarf willow and pine. This appears to be because the very mobile insect fauna was able to react to the very rapid climatic amelioration around that time in advance of the more slowly migrating trees (Osborne 1974). As explained in **Chapter 5** (see **Sections 5.2** and **5.3**), temperature reconstructions based on beetle remains are an important source of terrestrial evidence for rapid warming events in the Late Glacial and early Holocene that can be correlated with evidence from **ice cores** (**Section 2.4**).

Fig. 2.3 Illustration of the 'mutual climatic range' approach. The axes represent climatic variables, for example, mean January and mean July temperatures. The area of overlap indicates the climatic range within which all of the species represented in the assemblage could have lived. The technique works well with beetle remains but can be applied to other biological data, e.g. pollen and diatoms

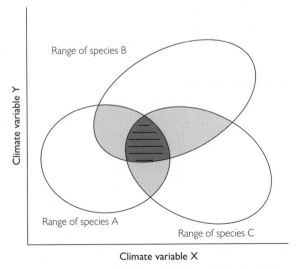

Assemblages of **diatoms** (microscopic unicellular algae) which are often preserved in lake sediments, can give indications of past lake-water quality, salinity, pH, and nutrient status. Diatom studies can also be used to study the interface of fresh and saline environments, thereby enabling statements to be made about such phenomena as the locations of past shorelines (Battarbee 1986). There has also been much progress in the use of diatoms to generate quantitative reconstructions of various climate variables, such as levels of ice-cover, surface water temperatures, and prevailing air temperatures (Mackay et al. 2005b). Microscopic animals may also be preserved, including Cladocera (freshwater invertebrates) (Frey 1986) and **ostracods** (Loffler 1986), which also give an indication of past hydrological conditions. Ostracods (small, bivalved crustaceans) are also useful for geochemical analysis of their shells which can enable reconstruction of the past water temperature and salinity on the basis of both the concentration of certain trace elements and the **stable isotope** ratios of the shell material (Holmes and Engstrom 2005). Chironomid remains (non-biting midges) are being increasingly used as another source of quantitative temperature reconstructions because of the abundance of chironomid larvae in lake sediments, and the narrow temperature ranges of many chironomid taxa (Walker 2001; Brooks et al. 2007).

The remains of larger animals, both living and extinct, are also important sources of evidence for past climate and environment; and as with plant remains, especially so for organisms that are more specialist in their environmental and climate requirements. For example, remains of hippopotamus and pond tortoise in England attest to a warmer climate than today during the Eemian interglacial (**Section 3.6**).

Geochemical and other sediment analyses

Geochemical analyses of sediments provide yet another important source of palaeoenvironmental evidence. The distinction between biological and geochemical forms of evidence can sometimes be blurred, for instance in the case of interpreting the geochemistry of shell fragments – as already noted with respect to ostracods. Generally speaking however, biological evidence could be seen to refer more to information gained through the presence and abundance of various taxa within the sediment whereas geochemical evidence involves studying the chemical

make-up of both organic and inorganic constituents of the sediment.

A good early example of the use of both biological and chemical analyses of lake sediments for interpreting past environments is the work of W. Pennington and colleagues, for instance in northern Scotland (Pennington et al. 1972). In addition to diatom and pollen analyses, they undertook elemental analyses and were able to identify phases of enhanced soil erosion where there were higher concentrations of potassium, sodium, and magnesium within the sediments. It has long been recognized that increases in these, and some other elements, in lake sediments result from increased delivery of unweathered minerals into lake basins (Mackereth 1966). Such episodes could reflect either natural (climate and/or vegetation related) or human-caused landscape changes affecting a drainage basin. Trace elements (such as lead, zinc, and copper) have also been analysed in recent lake sediments and in ombrotrophic peat (e.g. Shotyk 1996) to infer changing levels of pollution. In tropical settings, shifts between wetter and drier climates can be detected in lake cores. For instance, evaporite layers may be identified, and regarded as being the product of dry conditions (e.g. Kendall 1969). In Lake Kivu, East Africa, for instance, Fe-Ni sulphide contents were thought to indicate strongly reducing conditions, and thus high water stands, while high Al and Mn contents were thought to indicate lowstands (Degens and Hecky 1974). Many of the most widely utilized chemical analyses and their interpretation are detailed by Engstrom and Wright (1984) and Bengtsson and Enell (1986).

Mineral magnetic analyses, which have already been mentioned with respect to loess and palaeosol sequences, can be applied to all manner of sediments. Measures of magnetic susceptibility indicate how easily a material can be magnetized; whereas measures of isothermal remanent magnetization show how a material acquires and retains magnetization following exposure to different magnetic field strengths. Magnetic analyses help in identifying the provenance of material making up a sediment, and they can indicate changes in catchment erosion, fire frequency, and anthropogenic activity (Thompson and Oldfield 1986; Walden et al. 1999).

Analyses of stable isotopes in sediments by **mass spectrometry** have a very wide range of applications. The underlying principles behind stable isotopes and their interpretation were introduced in **Section 1.6** and more specific applications are discussed further in the context of marine sediments and ice cores in **Sections**

2.3 and 2.4). As the oxygen and hydrogen ratios in precipitation are partly dependent on temperature, the aim of many isotope analyses of lake sediments and peat is to reconstruct past temperatures by seeing how isotope ratios vary through the sequence. In lake sediments this is often achieved by measuring the isotope ratio of shell material, e.g. from ostracods or molluscs, as the isotopic composition of a shell generally reflects the isotopic composition of the lake water at the time that the shell was formed. In peat it can be achieved by measuring the ratio in the plant cellulose of an abundant peat former, such as *Sphagnum*. Of course, the link back to air temperature requires the study of lakes and peatlands that can safely be assumed to be primarily rain-fed and free from significant groundwater input. They should also be 'open system' lakes so that the lake water is representative of the isotopic characteristics of the regional precipitation, because in closed lake systems (especially in drylands), isotopic changes are more influenced by the precipitation/evaporation ratio (Leng and Marshall 2004). There are other factors influencing isotope fractionation in lakes and peatlands, but under the right conditions, convincing palaeotemperature estimates can be gained from $\delta^{18}O$ and δD analyses (e.g. Brenninkmeijer et al. 1982; Dupont 1986; Goslar et al. 1993).

Studies of $^{13}C/^{12}C$ ratios can be used for investigating past changes in the carbon dioxide content of the atmosphere and for reconstructing changes in carbon cycling. A particularly good example is the interpretation of changes in the $^{13}C/^{12}C$ ratio of lake sediments and peat in tropical/subtropical regions where shifts have occurred over time in the abundance of plants utilizing either the C_3 or C_4 photosynthetic pathways. Terrestrial C_3 plants have lower ratios while C_4 plants, adapted to moisture stress and low levels of CO_2, have higher ratios; hence in some tropical peats (e.g. Sukumar et al. 1993) and lake sediments (e.g. Street-Perrott et al. 1997) increases in the $^{13}C/^{12}C$ ratio of organic matter mark phases of C_4 plant dominance which may have occurred in response to enhanced aridity and/or reduced atmospheric CO_2. Leng (2005) provides a useful summary of the variety of methods and applications of stable isotope analysis of lake sediments.

Cave deposits

Caves are an important source of palaeoenvironmental information because they are natural places for material to collect over long periods of time. Many types of material can be found in caves, including mineral matter that has been washed or blown inside and organic matter from animals that occupied the cave. Among the more important organic materials include skeletal remains of animals that lived in (or brought their prey to) the cave, and dung (coprolites) that may help in deducing what plants and animals lived in the vicinity of the cave. As caves were often used in prehistory for shelter, for handling animals, and for burials, they are also an important source of archaeological and anthropological information.

In limestone cavern systems the study of speleothems has proved particularly valuable for reconstructing past changes at a high temporal resolution. Speleothem refers collectively to forms of secondary mineral deposits, such as flowstones, stalactites, and stalagmites, that develop by the precipitation of calcium carbonate (and sometimes other minerals) from supersaturated water as it flows or drips through a cave. Under the right conditions, speleothem growth can be continuous over long periods of time and a chronology for the time span of growth can be constructed by dating speleothem using uranium-series methods (Lauritzen 2005). Speleothems provide palaeoenvironmental information in a variety of ways. Through analysis and dating of a speleothem in cross-section it is possible to identify periods of relatively slow or rapid growth that, especially in the mid to high latitudes, are closely related to interglacial/glacial and **interstadial/stadial** oscillations in climate. This is because during phases of warmer climate there is typically more groundwater percolation and more CO_2 dissolved in the groundwater due to increased respiration of soil fauna.

A more subtle palaeoclimatic record can be gained through analysing changes in the stable isotope ratios of speleothem growth layers. The $^{18}O/^{16}O$ ratio of speleothem carbonate is controlled by both the temperature in the cave at the time of calcite precipitation and by the $^{18}O/^{16}O$ ratio of the source water; and, through various means, a temperature and/or palaeoprecipitation signal can be isolated (Lauritzen 2005). The long Devils Hole record from Nevada shown in **Fig. 3.2** is such an example. Analyses of $\delta^{18}O$ on speleothem from caves in China have enabled detailed reconstruction of changing strength of the East Asian monsoon going back over 200 000 years. Phases when the monsoon was strong are indicated by more negative $\delta^{18}O$ values due to the large amount of $H_2^{18}O$ depleted precipitation associated with intense rainfall

(Wang et al. 2008). Speleothems also tend to contain impurities which themselves have value for palaeoenvironmental reconstruction. For example, varying concentrations of humic material incorporated within the speleothem can be indicative of past changes in local vegetation cover and productivity. Variations in humic content can be assessed by measuring the luminescence of the speleothem, and this can also aid in identifying growth layers which are sometimes annual (Baker et al. 1993).

Tree rings

Tree rings, in contrast with most types of deposits (varved lake sediments and some speleothems being exceptions) have the great advantage of providing annually resolved palaeoenvironmental information. Most trees produce visible annual rings because wood formed in the spring tends to have larger and more thin-walled cells than wood formed later in the growing season. The sequence of annual rings for an individual tree can be easily sampled with a coring device that extracts a core of wood extending from the outer bark to the centre of the tree; and after sanding and polishing, the rings can be counted to determine the age of the tree. Of fundamental importance is the fact that annual rings vary in their widths depending on the conditions for growth in any particular year; and if many trees of the same species are sampled in a given area, it is possible to identify specific narrow ring-width years shared by a whole population that must therefore relate to some external factor that impaired growth in those years. In fact, the ring patterns found in individual trees can be likened to 'bar codes' that have matches in other nearby trees, be they younger or older, provided that the trees overlap for part of their life spans. It is the matching ('crossdating') of ring patterns from trees of different, but overlapping, ages that makes dating by tree rings (**dendrochronology**) possible over spans far longer than the life span of a single tree. While simple in concept, the actual practice of crossdating is complicated by differences in growth rates related to the age of trees and to local factors, as well as the fact that annual rings are not always clear and can sometimes be missing for particularly poor growth years. Large numbers of trees need to be sampled in an area to overcome these problems when developing a chronology.

Much of the pioneering work in dendrochronology was undertaken in the south-west USA on various species of pine, and the longest continuous tree ring record from this region is based on the extremely long-lived bristlecone pine (*Pinus longaeva*). Individual bristlecone pines can live for over 4000 years, and by matching the ring patterns of living and sub-fossil bristlecone pinewood dating to different periods it has been possible to extend the record back 8700 years. The other tree most widely used for dendrochronology is oak. The longest continuous record in the world is the Hohenheim oak–pine chronology, based on German oaks extending through the Holocene and German and Swiss pines taking the record back to almost 12 600 calendar years ago (Friedrich et al. 1999, 2004; Schaub et al. 2008). Work is continuing on this record to extend it back through the Late Glacial (Kaiser et al. 2010). Long chronologies extending back over 7000 years have also been developed in Sweden and Finland using pine, in Siberia using larch, and in Ireland and England using oak.

Dendrochronology has many applications. As discussed further in **Section 2.5**, the method is classed as a form of 'incremental dating', and among its many uses is the dating of wood preserved in ancient structures thereby providing age estimates useful in archaeological research. For example, dendrochronology in the eastern Mediterranean has made important contributions to our knowledge of the timing of prehistoric events in the region (Kuniholm 1996). Also as discussed further in **Section 2.5**, dendrochronology has been used to calibrate the radiocarbon timescale: this has been achieved by extensive radiocarbon dating of wood of known age.

However, tree ring analysis is not limited to purely chronological applications. Because of the relationship of annual ring widths (and wood density) to conditions for growth in specific years, it is possible to make inferences about past climatic change. This aspect of tree ring research (dendroclimatology) is less straightforward than dendrochronology because of the numerous environmental factors that can affect the growth of trees. Nonetheless, where certain types of trees are growing near the margins of their climatic tolerance range it is sometimes possible to identify the dominant climatic factor in influencing growth rates. For instance, in northern Scandinavia ring widths of Scots pine are thought to relate primarily to changes in mean summer temperature (Briffa 1994) while in south-west USA ring widths have been linked to winter precipitation (e.g. D'Arrigo and Jacoby 1991). Confidence in palaeoclimatic inferences based on tree rings is enhanced when recent ring data from the population of trees being sampled is robustly calibrated against recent instrumental climate records (see Salzer et al. 2009 for an example of comparison with meteorological data). It is then assumed that the climate variable found to be most influential on the recent ring patterns was also the dominant factor affecting ring widths during times prior to the instrumental record. Complicating factors include the 'carbon fertilization effect' of recent times (more rapid tree growth due to the post-industrial increase in atmospheric carbon dioxide) and statistical methods of 'detrending' the tree ring width records (to take into account the reduction in ring width that occurs in an individual tree as it ages) which can cause longer, multi-centennial climate signals to be masked (Esper et al. 2002).

Where a climatic shift can be identified in the tree ring record, there is the added advantage of knowing precisely when the shift occurred (assuming little or no time lag between the shift and the response of trees). This has proven particularly useful in dating brief, but major, climatic changes caused by large volcanic eruptions. For example, dendroclimatic studies of Irish oak offer some firm evidence for correlation between large volcanic eruptions and climatic change, particularly at the time of the eruption of Santorini in the Aegean Sea in 1626 BC (Baillie and Munro 1988), as do studies of bristlecone pine in western USA (Salzer and Hughes 2006). The key principles and applications of dendroclimatology are well surveyed by Fritts (1976) and by Hughes et al. (1982).

Geomorphic evidence

Although the data provided by the examination of various types of deposits and tree rings are central to reconstructing past environments and climates, the importance of fossil landforms should not be forgotten. It is not possible here to go into detail about the relations between landforms and climate, and fossil landforms and fossil climates, but there are certain landforms which can give relatively precise information about past environments. Some such landforms have already been discussed in the context of glacial and periglacial evidence, for example cirques indicating the positions of former snowlines and patterned ground indicating the former presence of permafrost; but there are many others. For instance, one important contribution of geomorphology to palaeoenvironmental research is the identification and mapping

of fossil lake shorelines which indicate hydrological changes from wet to dry; and attempts have been made on the basis of old lake volumes to estimate former precipitation levels. On the other hand, fossil sand dunes are indicative of former dry periods. Another example of geomorphic evidence is the identification of debris cones in upland areas, particularly where these bury soil or peat, as they are indicative of destabilization of the landscape – sometimes due to a combination of climatic change and human-caused vegetation clearance (e.g. Wild et al. 2001). **Table 2.5** summarizes these and other forms of geomorphic evidence.

However, it needs to be appreciated that the formation of most landforms results from a wide range of factors, of which climate forms but one group. Climate itself is also extremely complex. The problems of interpretation that these considerations present can be examined through a study of river terraces. They may sometimes form as a result of non-climatic causes, such as tectonic change, sea level change, glacial invasion of catchment areas, and so forth. However, even if one can eliminate non-climatic causes, it is difficult to draw precise inferences as to climate from alluvial stratigraphy within terraces because of the variety of possible climatic influences: the amount of precipitation, distribution of precipitation throughout the year, mean and seasonal temperature, and other climatic variables. Moreover, the response of a stream, in terms of load and discharge, to changes in such climatic variables, will be influenced by vegetation cover, slope angle, the range of the altitude of the basin, and other circumstances. Thus a change in any one climatic factor might, within one area, lead to different responses in different streams, and even in different segments of a single stream. Consequently, extreme care needs to be exercised in the utilization of landforms such as terraces for the reconstruction of past climates and environments.

Palaeoecological transfer functions

Transfer functions were introduced in **Section 1.6** and referred to earlier in this chapter in the context of palynology. Since the development of the technique on forams in marine sediments to reconstruct past sea temperature (see **Section 2.3**) by Imbrie and Kipp (1971), transfer functions have been applied to many types of biological data to reconstruct a wide range of different environmental variables. For

Table 2.5 Some quantitative and semi-quantitative geomorphic indicators of environmental change

Landform	Indicative of	Example
Pingos, palsen, ice-wedge casts, giant polygons	Permafrost, and through that, negative mean annual temperatures	Williams (1975), Great Britain
Cirques	Temperatures through their relationship to snowlines	Kaiser (1969), European mountains
Closed lake basins	Precipitation levels associated with ancient shoreline formation	Dury (1973), New South Wales
Fossil dunes of continental interiors	Former wind directions and precipitation levels	Grove and Warren (1968), Saharan margins
Tufa mounds	Higher groundwater levels/wetter conditions	Butzer and Hansen (1968), Kurkur Oasis
Caves	Alternations of solution (humidity) and aeolian deposition, etc. (aridity)	Cooke (1975) NW Botswana
Angular screes	Frost action with the presence of some moisture	McBurney and Hey (1955), Libya
Misfit valley meanders	Higher discharge levels which can be determined from meander geometry	(Dury, 1965) Worldwide
Aeolian-fluted bedrock	Aridity and wind direction	Flint (1971), Massachusetts, Rhode Island, Wyoming
Oriented and shaped deflation basins	Deflation under a limited vegetation cover	De Ploey (1965), Congo Basin
Dune breaching by rivers	Increased humidity	Daveau (1965), Mauritania
Lunette type	Hydrological status of lake basin	Bowler (1976), Australia
Old drainage lines	Higher humidity	Graaf et al. (1977), Australia
Fluvial siltation	Desiccation	Mabbutt (1977), Australia
Colluvial deposition	Reduced vegetation	Price-Williams et al. (1982), Swaziland

example, transfer functions based on forams, diatoms, Cladocera, chironomids, and ostracods have variously been used to reconstruct past water temperature, pH, salinity, and nutrient load; and those based on pollen, chironomids, and Coleoptera have been used for past air temperature (e.g. annual mean, seasonal maxima and minima) as well as past precipitation. Transfer functions based on testate amoebae have been used to reconstruct past water table levels of ombrotrophic peatlands (e.g. Charman et al. 2007), and functions based on ostracods have been used to reconstruct palaeolake levels (e.g. Wrozyna et al. 2009).

Regardless of the type of organism used or the environmental variable of interest, all transfer functions rely on a modern 'training set' that relates modern assemblages of the organism (or organic materials such as pollen grains) to variations in the environmental variable. A mathematical relationship is computed between variations in the modern assemblages and

variations in the chosen environmental variable, and this 'function' is then applied to the fossil assemblage to arrive at a quantitative estimate for the environmental variable at the time represented by the fossil assemblage.

There are a number of different numerical techniques for deriving the transfer function. These are designed to relate assemblages containing many tens or hundreds of different taxa to just one or a few different environmental variables. The taxa within assemblages are usually represented by their percentage abundance in the sample, and 'weighted averaging' can be used to relate the percentages of taxa represented in multiple samples to their environmental range and optima. Commonly used techniques include 'two-way weighted averaging' (WA) and 'weighted averaging partial least squares regression' (WA-PLS). In his review of these numerical methods, H.J.B. Birks (2005) notes that transfer functions are only reliable provided that certain assumptions are valid. His list is summarized below:

- The taxa in the modern data training set are related to the environment in which they live (i.e. not 'out of equilibrium' with the prevailing environment or climate conditions).

- The environmental variable of interest is an important determinant of the assemblage of taxa.

- The taxa in the training set are the same 'biological entities' as in the fossil assemblage, and their ecological tolerances to the environmental variable of interest have not changed since the fossil assemblage was formed.

- The numerical method chosen correctly represents the taxa–environment relationship (and the error range inherent in the method should be significantly less than the range in the environmental variable that is being reconstructed).

- Where other environmental factors have an influence on the assemblage, the influence should be small compared with the variable under investigation, and its effects should not have changed since the fossil assemblage was formed.

In order for these assumptions to be valid, Birks goes on to set out a number of requirements. Important ones being:

- That transfer functions are based on systems with 'abundant and preservable fossils'.

- That training sets are high quality and representative of 'the likely range of past environmental variables'.

- That fossil assemblages and modern training sets are consistent in terms of taxonomy, counting and preparation procedures, and stratigraphic context.

The extent to which these requirements are met, as well as the quality of the chronology for the sequence containing the fossil assemblages, determines the accuracy and precision of the transfer function-based reconstruction.

2.3 Evidence from oceans

Over recent decades there has been a rapid development in our knowledge of the oceans, and in particular

of our knowledge of what has been happening on the sea-floor. Many of these developments have depended on the recovery of a continuous core of deep-sea sediment, a possibility that was created with the development of the Kullenberg piston corer in 1947. The importance of the oceans is that their record is less disrupted than the record of continental sediments. As Ewing expressed it (1971, p. 572):

…the problem of estimating the number of glaciations from evidence on continents is contrasted with the problem of estimating the number of glaciations from evidence in deep-sea sediments. The first may be compared in complexity to estimating the number of times a blackboard has been cleaned; while the second may be compared to finding the number of times the wall has been painted.

Although by no means completely stable (burrowing organisms, solution and currents can create problems for interpretation), the sea-floor does offer a more continuous and lengthy stratigraphic record than do most terrestrial sections; and therefore, ironically, it is to the oceans that we must look for a complete picture of the pattern of glaciation on land through the Quaternary Period. As described in **Section 1.5**, deep-sea sediment analyses have also been integral to debates about the timing of the Tertiary/Quaternary transition. Moreover, recent years have seen an increasing appreciation of the linkages between the oceans and the atmosphere that influence Earth's climate, and reconstructions of past sea surface temperature (SST) and ocean currents from evidence contained in marine sediment are of much value for understanding both past and future climatic change.

Several different types of proxy data can be gained from deep-sea sediment cores, and the most widely utilized are summarized in **Table 2.6**. As with terrestrial sediment, these proxies can be divided broadly into physical, biological and geochemical types; but first it is worth looking briefly at the various depositional processes that form marine sediments.

Marine depositional environments

Sea-floor sediments are made up of both mineral matter and biological remains: the relative proportions of the two depending on the conditions of the marine environment at the time of deposition. The mineral matter can comprise the fall-out of fine

Table 2.6 Some palaeoenvironmental information from deep-sea cores

Indicator	Environmental information
$^{18}O/^{16}O$ ratios	Ice volume, temperature
$^{13}C/^{12}C$ ratios	Nutrient content
Cadmium/Calcium ratios	Nutrient content
Coarse debris	Iceberg rafting
Fluvial sediments	River inputs
Aeolian dust	Aridity
Aeolian sand turbidites	Aridity
Clay minerals	Weathering regimes on land
Sensitive species Foraminifera	Temperature
Total species Foraminifera	Temperature
Coccolith carbonates	Temperature
Alkenone ratio (U_{37}^{K})	Temperature

airborne material (dust and volcanic ash), material brought from the land by erosion and stream transport, precipitation of compounds out of solution from surrounding seawater, and material dropped to the sea-floor from melting ice. The biological component consists primarily of the shells of microorganisms that were either surface-dwelling (planktonic) or bottom-dwelling (benthic). Deep-sea sediments rich in biological remains are known as marine oozes; and such sediments tend to be fine grained with mineral matter mostly in the form of clays, because only the finest particle sizes (clays, dust, ash) are able to be carried in suspension great distances from shore, by wind or water, before settling out. In shallower, near-shore settings there tends to be a higher proportion of terrestrially-derived ('terrigenous') material, and the particle size displays a greater range. Such terrigenous material can be transported down the slopes of continental shelves by density currents and transported along the sea-floor by deepwater currents following sea-floor contours (contour currents). At high latitudes material dropped to the sea-floor from melting icebergs is an important constituent of marine sediment and is known as 'ice rafted debris' (IRD). IRD ranges in particle size from clay to the size of boulders because it represents the whole range of material carried in glacier ice – where glaciers meet the sea, 'calving' occurs and chunks of ice break off of the glacier's snout forming icebergs that can float a long

way before becoming completely melted. In this way, coarse material derived from the interior of a glaciated continent, such as Greenland, can find its way to areas of the deep ocean that would not normally receive much terrigenous material.

It has already been noted that marine sediments often contain longer and more continuous stratigraphic records than can be found on land; however, there is a trade-off when investigating marine sediments between length of record and resolution. Marine oozes on the deep-sea floor tend to accumulate slowly (on the order of a centimetre or so every thousand years), and bioturbation by benthic organisms can affect the seabed to a depth of about 5 centimetres. This means that the signal of any environmental changes inferred from the sediment will be smoothed out over a period of thousands of years; and therefore marine oozes are best suited to looking at broad environmental trends over hundreds of thousands of years. Indeed, some such records cover many millions of years spanning the full Quaternary and extending into the Tertiary (e.g. Zachos et al. 2001, see **Fig. 3.1**). Resolving environmental changes on sub-millennial timescales requires more rapidly accumulating marine sediment, often in near-shore environments; but these sediments are generally less continuous and span a shorter length of time. Hence choosing where to sample marine sediments depends on the time span of interest; and much care needs to be taken to select locations of the sea-floor where preservation of microorganisms is good, where bioturbation is minimal, and where sedimentation is relatively undisturbed by bottom currents. Moreover, different regions of the sea-floor may be more or less suitable depending on the focus of investigation, for example whether the aim is reconstruct SSTs, glacial/interglacial cycles, sediment delivery from land or past ocean circulation.

Physical characteristics

Changes in the physical make-up of marine sediment can be an important indicator of large-scale changes in the global environment and climate. Through the Quaternary, increases in the proportion of mineral matter in deep-sea sediment cores generally correlate with glacial phases for two main reasons: lower sea level during glacials means stream-derived material is transported further into the ocean basins, and expanded periglacial areas and more arid continental interiors during glacials means increased input of wind-blown material to the oceans. For example, **aeolian** debris on the ocean floors (e.g. Parmenter and Folger 1974) together with the presence of large quantities of unweathered minerals, including feldspars, has been used to assess whether tropical climates were dominantly arid and semi-arid, or whether they were dominantly humid during particular phases. During arid phases rivers would tend to carry unweathered feldspars, while under more humid conditions the quantity of feldspars relative to the more stable quartz would be less. Similarly, in oceanic sediments off West Africa, opal **phytoliths** and freshwater diatoms are common in sediments deposited when waters were warm, but less common in sediments deposited when waters were cold.

Of particular importance for understanding the dynamics of large continental ice sheets is the presence of ice rafted debris. Identification of IRD layers in deep-sea sediment cores reveals periods of calving activity at the seaward margins of glaciers, and more extensive ice sheets (and colder SSTs) during glacial phases causes icebergs to reach lower latitudes than during interglacials. Therefore the presence of IRD in sediment cores can be used both to indicate glaciation and to track differences in how far icebergs travelled and how much glacier ice was discharged into the oceans at different times. The presence of IRD in high latitude marine sediments is one of the forms of evidence used for establishing the timing of the onset of glaciation in the Antarctic and Arctic (**Sections 1.4** and **1.5**). Over the most recent glacial phase several distinct IRD layers have been found in deep-sea sediments from mid-latitude areas of the North Atlantic Ocean. These layers, named 'Heinrich layers' after their discoverer, were deposited during episodes of massive iceberg discharge, primarily from the margins of the **Laurentide ice sheet** that was centred over the north-eastern part of North America (Heinrich 1988). The recognition and analysis of these layers have led to major advances in our understanding of the behaviour of ice sheets in the North Atlantic region, and in our understanding of rapid climatic change (see **Section 3.6** for a full discussion of **Heinrich events**).

Studies of particle size distribution can also be of use in studies of deep-sea sediments, particularly for reconstructing past changes in the strength of deep-water currents (coarser material reflecting stronger flow). For example, McCave et al. (1995) used grain size data in sediment cores from the north-east Atlantic to reconstruct variations in the strength of North Atlantic Deep Water flow during the Late Glacial.

Biological data

Most marine oozes are dominated by the calcareous remains of **Foraminifera** and **Coccolithophora**. Foraminifera (forams) are a group of single-celled organisms that consist of planktonic and benthic types. They range widely in size, from tenths of a millimetre to several centimetres in diameter, and they have calcareous shells (tests) that are often coiled. Coccolithophora on the other hand are single-celled, photosynthetic algae. They are much smaller than forams, usually being less than 100 μm in diameter, and are surrounded by calcareous platelets (called coccoliths). As different species of forams and coccolithophores have different ecological preferences (such as water temperature and salinity), by investigating changes in the abundance of various foram tests and coccoliths preserved in marine sediment it is possible to reconstruct past changes in the marine environment. Palaeoenvironmental information can also be gained from studying the siliceous remains of diatoms and Radiolaria. The former have been discussed above in the context of lake sediment analysis. The Radiolaria are marine protozoans, up to 2000 μm in diameter, with silica-based skeletons. Marine oozes rich in Radiolaria and diatom remains are found in deep areas of the oceans, usually in excess of 4 km (where calcareous material is rare because of dissolution beyond the **carbonate compensation depth – CCD**) and at high latitudes.

One of the most productive ways of examining the cores has been the study of changes in the frequency of particular 'sensitive species' of Foraminifera. These are thought to reflect changes in the temperature of ocean waters (Kennett 1970). *Globorotalia menardii* tests, for example, may be counted, and the ratio of their number to the total weight (mg) of foram tests can be worked out. The ratio can range from as high as 10 or 12, to as low as nearly zero. A high ratio appears to be associated with the warm water of interglacial conditions, while a low ratio appears to be associated with the colder water of glacial stages. Thus, by taking samples along the length of a core, the alternations of warmth and cold can be established. Similarly, another of the Foraminifera, *Globorotalia truncatulinoides*, can be used to the same end. In any portion of a core-sample some of its tests will show a left-hand direction of coiling, and others will show a right-hand direction of coiling. It has been found that right-coiling tests are associated with warmer conditions, and left-coiling tests with cooler. Thus ratios of left and right coiling may enable an assessment to be made of palaeoclimates.

Some workers have attempted to use more sophisticated methods of utilizing Foraminifera, and instead of approaching the problem through the study of 'sensitive species', they have tried to establish climatic sequences based on 'total fauna' (see Shackleton 1975 for a discussion). Applying logic similar to the quantitative climate reconstructions from pollen assemblages discussed previously, statistical methods can be used to match fossil foram assemblages to the most similar 'modern analogue' assemblage. Using training sets of water temperatures characterizing modern assemblages, quantitative temperature estimates can be made from fossil assemblages using transfer functions (see **Section 2.2**). This technique, introduced by Imbrie and Kipp (1971), has since played an important role in the reconstruction of Quaternary SSTs, notably through the CLIMAP project (1981).

A more recently developed technique for SST reconstruction involves the measurement of long-chain carbon compounds called 'alkenones' (a type of 'ketone') in coccolithophore remains. The ratio of two different types of the unsaturated alkenone C37 produced in living coccolithophores is temperature dependent, and as these molecules resist decay over millions of years, their ratio in deep-sea sediments reflects the sea temperature at the time the coccolithophore was alive. Alkenones are analysed using gas chromatography. The notation for the 'unsaturated index' of C37 is $U^{K'}_{37}$ and change in this index is linearly related to change in SST (Prahl and Wakeham 1987). This method can provide a useful corollary and check to SST estimates based on foram assemblages (e.g. Peeters et al. 2004). The results from CLIMAP and from more recent temperature reconstructions from marine sediment cores are discussed in **Chapter 3**.

Another type of marine-based biological evidence involves sessile organisms which are restricted to certain water depths or the intertidal zone. The remains of such organisms, such as various types of corals, oysters and barnacles, can be used as palaeo-sea level

indicators. These and other methods for reconstructing past sea levels are discussed in **Section 6.2**.

Geochemical data

Another way in which the Foraminifera can be utilized is by the measurement of the $^{18}O/^{16}O$ ratios in the calcitic tests. Stable isotopes were introduced in **Section 1.6** and have been discussed in the context of lake and peat sediments (**Section 2.2**), and the basic principle is the same as described for freshwater ostracods and molluscs in that the isotopic composition of the foram test reflects the isotopic composition of the water when the foram lived. (Interpretation can be complicated, however, by taxonomic differences in the way forams fractionate oxygen when extracting sea water for building tests.) In marine sediments lacking forams, $^{18}O/^{16}O$ analyses can be conducted on diatoms (see Swann and Leng 2009 for a review).

Oxygen isotope analyses on marine sediments were pioneered in the 1950s by Emiliani and others (Emiliani 1961). He originally supposed that the $^{18}O/^{16}O$ ratio depended substantially on the temperature of the water in which the Foraminifera lived. While water temperature does have an effect (colder water relates to higher ^{18}O relative to ^{16}O), it was later discovered that most of the down-core variability in foram $^{18}O/^{16}O$ ratios through the Quaternary is due to changes in the global volume of ice rather than to changing sea temperature. (The relative influence of ice volume versus temperature on isotope ratios can be estimated through use of independent temperature proxies in the same core such as $U^{K'}_{37}$ described previously.) To understand the ice volume effect, it is important to review the isotopic properties of oxygen.

As explained in **Section 1.6**, ^{16}O and ^{18}O differ in atomic weight, with the former being 'light' and the latter 'heavy'. The ratio of the two in a sample is measured in relation to a standard ratio, in the case of forams the VPDB standard. The ratio is then expressed as a per thousand deviation from the standard (symbolized as $\delta^{18}O$). Of fundamental importance is the effect of evaporation on the $^{18}O/^{16}O$ ratio of water. $H_2^{16}O$ is evaporated more readily than $H_2^{18}O$, and, as shown in **Fig. 2.4**, this causes enrichment of the heavier isotope in the seawater relative to the vapour yielded from evaporation. The build up of large ice sheets during a glacial phase requires a lot of seawater to be evaporated and then to condense and be deposited as snow over the continents, and this results in an

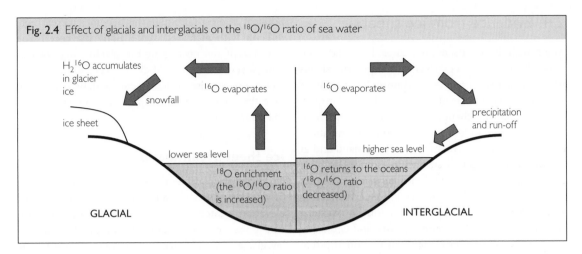

Fig. 2.4 Effect of glacials and interglacials on the $^{18}O/^{16}O$ ratio of sea water

$H_2{}^{16}O$ accumulates in glacier ice

snowfall

ice sheet

^{16}O evaporates

^{16}O evaporates

precipitation and run-off

lower sea level

higher sea level

^{18}O enrichment (the $^{18}O/^{16}O$ ratio is increased)

^{16}O returns to the oceans ($^{18}O/^{16}O$ ratio decreased)

GLACIAL

INTERGLACIAL

enrichment of ^{18}O in the oceans which reaches a peak when global ice volume is at its maximum and sea level at its minimum. During interglacials the melting of glacier ice returns 'lighter' H_2O to the oceans causing the $\delta^{18}O$ of seawater (and hence the $\delta^{18}O$ of foram tests) to decrease. The difference in the measured $\delta^{18}O$ in forams between glacials and interglacials ranges between about 1 and 2 ‰, and a change of 0.1 ‰ is thought to reflect about a 10 m change in sea level. While the Foraminifera-based $\delta^{18}O$ records from different deep-sea sediment cores vary in the absolute $\delta^{18}O$ values, for example the ratios are considerably higher when measured from benthic as opposed to planktonic forams, there is remarkable similarity in the pattern of $\delta^{18}O$ change through time preserved in marine sediments throughout the world. This shows that the oxygen isotope data are reflecting global-scale changes and that the data can be used for correlating marine sediment sequences worldwide.

Phases of higher or lower $\delta^{18}O$ spanning deep-sea sediment cores have been assigned to different **marine isotope stages** (**MIS**) – the periods of higher $\delta^{18}O$ (colder / more ice) being given even numbers and periods of lower $\delta^{18}O$ (warmer / less ice) being given odd numbers. An important core from the west equatorial Pacific (V28-238) showed 23 such stages extending back over 800 000 years (Shackleton and Opdyke 1973), and further study of long marine sediment sequences has enabled the record to be extended back through the entire Quaternary Period. Data from the Ocean Drilling Program (Site 677 in the tropical Pacific) has enabled identification of over 100 oxygen isotope stages over the past 2.6 million years (Shackleton

et al. 1990); and, depending on how the data are interpreted, there appear to have been between 30 and 50 cold / temperate cycles since the Quaternary began (see **Fig. 3.2**). More than any other form of proxy data, $\delta^{18}O$ measurements have revolutionized our understanding of the waxing and waning of ice sheets through Quaternary time. Oxygen isotope stratigraphy has shown that there were far more periods of ice sheet advance and retreat than originally envisioned (see for example the classic four-fold division described in **Chapter 1**) and that these cycles have a periodicity related to orbitally-induced variations in the distribution of solar energy over the globe as originally put forward by the Serb astronomer Milankovitch. The character and timing of glacials and interglacials as interpreted from deep-sea sediment $\delta^{18}O$ data are discussed in **Chapter 3**, and the use of isotope stages for dating sediments is discussed in **Section 2.5**.

Geochemical analyses of marine sediments have also focused on reconstructing ocean circulation. It has been found, for instance, that Cd/Ca ratios of benthic foram tests provide a proxy of past nutrient content (primarily phosphorus) of the seawater, with higher ratios indicating more nutrient-rich waters. On the other hand, relatively high $^{13}C/^{12}C$ ratios are indicative of nutrient-depleted waters. These proxies were used to great effect by Boyle and Keigwin (1987) in reconstructing past changes in North Atlantic Deep Water flow based on cores from the Bermuda Rise area of the Atlantic floor: times of vigorous deepwater flow brought nutrient-depleted waters across the sea bed, decreasing Cd/Ca ratios and increasing $\delta^{13}C$.

2.4 Evidence from ice sheets

Along with deep-sea sediment cores, from the 1960s onwards, analyses of ice cores from the polar ice sheets have contributed a great deal to our understanding of Quaternary environments, particularly with respect to abrupt climatic change. Long ice cores, representing tens of thousands of years of snow accumulation, have been drilled in Antarctica, in Greenland and at Devon Island in Arctic Canada. Cores have also been recovered from high altitude ice caps in the tropics (see, for example, Thompson et al. 1986). Pioneering studies were based at Byrd Station in western Antarctica, and at the Camp Century and Crete stations in Greenland. The Camp Century core (Epstein et al. 1970), put down in north-west Greenland, was remarkable in its day, attaining a length of no less than 1390 m. Since then, improvements in site selection and drilling have enabled the extraction of much longer cores of ice. The Vostok ice core (eastern Antarctica), started in 1980, achieved its maximum length of 3623 m in 1998, and it provides a climate record extending back some 420 000 years. At Dome C, also in eastern Antarctica, the EPICA (European Project for Ice Coring in Antarctica) team succeeded in extracting an ice core containing the longest temporal record, extending from the present to about 800 000 years ago. In Greenland a new record for the Arctic was set in 2004 with the attainment of a 3085 m core from the NorthGRIP station. This ice core (and the GRIP and GISP2 cores drilled in the summit area of Greenland in the early 1990s) reaches back through the most recent glacial phase into the previous interglacial (the Eemian) over 110 000 years ago. GRIP refers to the European Greenland Icecore Project and GISP2 refers to the North American Greenland Ice Sheet Project 2; both are of tremendous importance for understanding climatic change in the North Atlantic region as discussed in **Section 3.6**. The NEEM (North Greenland Eemian Ice Drilling) project, which hit bedrock in 2010, is focused on taking the Greenland record back further into the Eemian interglacial.

Ice cores are extremely useful archives of palaeoenvironmental change because several types of proxy data are preserved within the ice, and ice layers can sometimes be resolved annually. The fact that ice cores contain annual layers means that, if the layers are visible, environmental changes revealed in the ice stratigraphy can be dated simply by counting the number of layers from the top of the core downwards to the point of interest (although this becomes increasingly difficult with increased depth along the core). The duration (if not necessarily the absolute age) of certain palaeoenvironmental events recorded deeper in an ice core can also be determined with high precision by estimating the number of ice accumulation years spanning the event.

Ice sheet accumulation

Ice sheets are continental-sized glaciers, with an area of at least 50 000 km² and with the highest surface elevation located around the centre of the ice mass. This gives ice sheets a domed shape; and unlike valley glaciers in which ice flow is down a slope in one direction, in ice sheets the ice flow radiates outwards from the centre. Today, about 85 per cent of the world area covered by glacier ice is accounted for by the west and east Antarctic ice sheets covering nearly 13.8 million km². The Greenland ice sheet accounts for about 11 per cent, covering some 1.8 million km². As with any glacier, an ice sheet has a high elevation accumulation zone and an ablation zone around the margins – this accounts for the continual movement of glacier ice outwards from the central area of net accumulation to the marginal areas of net ablation where ice is lost through melting, and especially by calving of icebergs from ice shelves along the margins of Antarctica. However, a glacier is a dynamic system, constantly adjusting to changes in climate that affect the balance between accumulation and ablation and such changes leave their mark in the ice stratigraphy.

Bearing in mind the morphology of an ice sheet, the deepest ice cores containing the longest and least disturbed records of ice accumulation will be obtained by drilling in the vicinity of the central accumulation zone. Because of seasonal changes in accumulation, annual layering can be identified along some ice cores. For example on a glacier where there is some melting in summer, mineral matter becomes more concentrated in the snow giving it a darker colour, whereas the winter snow is lighter in colour. In addition to visual inspection, annual cycles of variation in certain proxies (such as dust content, electrical conductivity, and stable isotope ratios) can also aid in identifying annual layers. Annual layer counting from the surface downwards based on Greenland ice cores extends back to about 40 000 years ago; but with increasing depth in an ice sheet, layers become increasingly difficult to identify because of compaction and flow-induced distortion. Thus certain theoretical assumptions have to be made about ice sheet accumulation and processes at depth in the interpretation of the lower parts of ice cores. Identification of annual ice layers is more difficult in Antarctica because accumulation rates are slow and seasons less marked in comparison with Greenland.

Annual layering in ice sheets is not only valuable for chronological purposes: changes in the thickness of layers give an insight into past changes in the accumulation rate. For instance, from analyses of ice layering in the GISP2 ice core Alley et al. (1993) showed that the accumulation rate doubled in just three years at the Younger Dryas/Holocene transition reflecting a rapid and large-scale climate reorganization over central Greenland that greatly increased annual snowfall.

Stable isotope evidence

Stable isotope ratios are one of the most important and widely studied forms of proxy evidence obtainable from ice cores. Dansgaard and co-workers showed in the 1950s from studies of Greenland ice that a linear relationship exists between the $^{18}O/^{16}O$ ratio preserved in ice layers and the air temperature at the time the precipitation fell to form the layer. As illustrated in **Fig. 2.4**, precipitation contains a lower $^{18}O/^{16}O$ ratio than seawater; however, in addition to this, when it is colder precipitation is even more depleted in ^{18}O relative to ^{16}O. This is because as air cools and vapour condenses, 'heavy water' ($H_2^{18}O$) condenses out more readily and with continued cooling of air, water vapour becomes increasingly enriched in the lighter isotope. Thus in colder areas, the vapour source for precipitation is isotopically lighter than in warmer areas. The oxygen isotope ratios of samples of ice are measured in relation to the standard known as VSMOW (Vienna Standard Mean Ocean Water) and expressed as per thousand deviations from the standard ($\delta^{18}O$) (**Section 1.6**). The $\delta^{18}O$ of ice is always negative in relation to VSMOW, but becomes more negative as temperatures decrease. For every $1\,°C$ change there is a change in $\delta^{18}O$ of around 0.7 ‰. In the GISP2 ice core for example, values range between about -43 and -35 ‰, and this translates, therefore, into an amplitude of mean annual temperature variability of the order of $10\,°C$ in central Greenland over the past 100 000 years (see **Fig. 3.17** for Greenland $\delta^{18}O$ data).

Hydrogen isotope ratios ($^2H/^1H$) of ice layers also record mean annual temperature changes according to the same fundamental principles as described for $^{18}O/^{16}O$ ratios: as temperature declines, precipitation becomes more depleted in ^2H (deuterium) and the δD relative to the standard becomes increasingly negative. Analysis of δD captures more variability in Antarctica and is the primary source for reconstructing past temperatures from the long Antarctic ice cores.

Other ice core proxies

There are many other ways in which ice cores can be used besides isotope-based temperature reconstructions, and **Table 2.7** summarizes the variety of palaeoenvironmental information that can be gained. In addition to aiding in the identification and counting of ice layers, dust concentration reflects the past loading of dust in the atmosphere and therefore indicates large-scale changes in aridity and wind erosion. Far more dust is deposited on ice sheets during glacials than during interglacials. Much can also be gained from analysing the chemistry of the ice. By measuring the electrical conductivity it is possible to identify periods of high sulphuric and nitric acid deposition relating to past volcanic activity (higher conductivity reflects higher acid content) (e.g. Taylor et al. 1993). If tephra (volcanic ash) is also present in an ice layer, then geochemical analysis combined with ice layer counting enables specific prehistoric eruptions to be identified and dated. Analyses of sea salt (e.g. forms of Na,

Table 2.7 Some types of palaeoenvironmental information from ice cores

Indicator	Environmental information
$^{18}O/^{16}O$ and $^2H/^1H$ ratios	Past temperatures
Other stable and radioactive isotopes	Solar activity (e.g. ^{10}Be), geomagnetic field strength, bomb testing
Ice layer thickness	Past accumulation rates (link to past snowfall and/or temperature)
Electrical conductivity	Acidity/alkalinity, past volcanic activity
Dust content	Past aeolian activity, aridity
Other aerosols	Tephra fall-out, pollution
Chemical composition	Past atmospheric circulation, pollution
Composition of trapped air	Past atmospheric concentration of CO_2, CH_4, N_2O, etc.

Cl, Mg, K, and Ca) within the ice enable reconstruction of past changes in the source area of moisture. For example, O'Brien et al. (1995) interpreted phases of higher concentrations of sea salt in the GISP2 core as representing times when atmospheric circulation was more meridional (less zonal) in the North Atlantic region. Historic atmospheric pollution can also be detected from ice cores, for instance, elevated lead concentrations in Greenland ice have been linked to ancient Greek and Roman silver mining and smelting activities (Hong et al. 1994).

Air bubbles within ice cores effectively make it possible to sample the past atmosphere. (The air is not completely trapped until the transformation from snow to glacier ice is complete, and therefore the air in tiny ice cavities is younger than the surrounding ice by an amount that must be estimated with respect to inferred ice accumulation rates.) Measurements of the amount of carbon dioxide and methane within the trapped air have provided important information about natural variability in the atmospheric concentration of these greenhouse gases against which human-caused increases can be compared (e.g. Petit et al. 1999). (CO_2 measurements from Greenland ice are contaminated by wind-blown carbonates from Canada, and therefore Antarctic ice cores are used for long-term reconstruction of atmospheric CO_2.) Even more importantly, such analyses enable relationships between greenhouse gas concentrations and climatic change over the long term to be examined – a theme that is explored in **Chapter 9**. Furthermore, changes in the concentration of methane in ice cores can be used to infer past changes in the global quantity of wetlands (wetlands being a major source of the gas) and, because the signal is global, can be used to correlate the Arctic and Antarctic ice core temperature reconstructions (e.g. Blunier et al. 1998).

2.5 Chronologies and correlation

Constructing chronologies is essential for understanding the sequence and timing of past changes inferred from the proxy evidence as well as for correlating different proxy records so that ideas about linkages and causation in environmental change can be tested. Of course the better the chronology, the more rigorous such testing can be; particularly with respect to the study of rates of change and identifying leads, lags, and feedbacks within environmental systems. Early

palaeoenvironmental research depended largely on examination of the position of changes within stratigraphic sequences for working out the relative age of events according to the law of superposition (**Section 1.6**). Applying **uniformitarian** principles, it is also possible to make an estimate of the actual age of an event in a sedimentary sequence by measuring the present-day accumulation rate of the sediment. A problem with this approach is that accumulation rates are rarely constant through time; but with careful examination of a sequence it is sometimes possible to identify where sedimentation sped up, slowed or halted and to make the appropriate adjustments to the chronological interpretation. The classic schemes of Pleistocene and Holocene subdivision discussed in **Chapter 1** were mainly based on this kind of interpretation. While general patterns of change were revealed in these schemes, age estimates were imprecise and it was difficult to correlate different sequences with each other. Often faulty correlations were made by assuming that similar looking stratigraphic layers in different places were formed at the same time when this was not actually the case (e.g. the correlation of peat recurrence layers described in **Section 2.2**); an error referred to as 'homotaxial'. The distinction between climatostratigraphy and **chronostratigraphy**, and the **time-transgressive** nature of divisions based on the former, is explained in **Section 1.6**.

Human artefacts, which may occur abundantly in some Quaternary deposits, have also long been used for dating and correlating deposits. While pottery, coins, and other metal objects may give relatively precise dates for deposits laid down in the last few thousands of years, the use of stone tools for dating older deposits offers much less precision. Such tools may often be derived from even earlier deposits, their characteristics tend not to have changed very rapidly through time, and changes have rarely been concurrent in different areas. In general it is normally preferable to date stone artefacts using other dating methods than to expect the artefacts to provide precise dates themselves.

All this changed with the development of radiocarbon dating and other **radiometric dating** methods from 1949 onwards. The classic schemes could at last be tested using dating methods independent of prior assumptions about the stratigraphic relation of events, and global-scale correlations could begin to be made on a much firmer basis. In addition, great strides have been made in recent decades in the use of incremental dating; that is in using regularly accumulating records such as annually resolved tree rings, varves, and ice layers for dating. The application of radiometric and incremental methods has made it possible to achieve much more accurate and precise age estimates than in the past, though as described below, these methods are not without their various problems. There have also been important developments in the use of 'age-equivalent' markers that are contemporaneous regionally or globally. Such markers include **palaeomagnetic** events, tephra layers, and oxygen isotope shifts in deep-sea sediments. If the age of the marker can be established in one location, for example by using radiometric or incremental methods, then the marker can be used to fix a date on the position where it appears in a sequence somewhere else. Even relative methods of dating have seen many advances recently, for example, through the study of degrees of weathering, degrees of soil formation (pedogenesis) and rates of break-down of organic compounds.

There is now a wealth of dating techniques available for Quaternary-aged stratigraphic sequences and landforms, many of which overlap in their time ranges and can be used in conjunction with each other in many situations to cross-check and increase confidence in the constructed chronology. However, along with these advances come certain pitfalls: different methods have different error margins associated with them, and often different conventions for citing the resulting age estimates which can lead to confusion when comparing independently dated proxy records and past events. As detailed in **Section 1.6**, when age estimates are given in years before present (BP), this refers to years *before AD 1950*, and must not be confused with the BC/AD system usually used for reporting the dates of archaeological materials and historical events. As described below, it is also important to be aware

that **radiocarbon years** BP (^{14}C yr BP) are not the same as calibrated calendar years BP (cal yr BP); and some age estimates, notably from ice cores, are given as calendar years before AD 2000 (b2k). It is important to be aware of these differences to avoid erroneous correlations, and so that the necessary conversions are made before comparing proxy data sets that have been dated in different ways.

It is beyond the scope of this book to explain all of the various dating methods in detail; instead, the rest of this section provides an introduction to the most commonly used techniques.

Radiocarbon dating and other radioactive isotopes

Of especial importance are isotopic dating techniques, especially radiocarbon, uranium series, and potassium-argon. Some of the features of these methods are shown in Table 2.8. These three isotopic dating techniques all depend on the measurement of amounts of elements which through time are either formed by, or are subject to, radioactive decay. The rate of decay being known for a particular element, the time-interval may be assessed between the present, and the time when the particular parent material was fixed and its decay began. Thus, for example, a growing organism incorporates radiocarbon, and on its death the radiocarbon is trapped and then begins to decay. As half the radioactivity will be lost after an interval calculated to be about 5730 years, by measuring the radioactivity of fossil material containing carbon, the date at which death took place can be determined.

Radiocarbon or ^{14}C dating looks at the carbon in a sample in terms of the relative quantities of ^{14}C and the stable isotopes ^{12}C (98.9% of all carbon) and ^{13}C: the ^{14}C

Table 2.8 Some isotopic methods of dating Quaternary deposits

Name	Isotope	Half-life (years)	Range (years)	Materials
Radiocarbon	^{14}C	5730 ± 40	0–75 000	Peat, wood, shell, charcoal, organic muds, algae, tufa, soil carbonate, coral
Uranium series	^{234}U	250 000	50 kyr–1 Myr	Marine carbonate, coral, molluscs, cave carbonate
	^{230}Th	75 000	0–400 000	Deep-sea cores, coral and molluscs, cave carbonate
	^{231}Pa	32 000	5 kyr–120 kyr	Coral and molluscs
Potassium-Argon	^{40}K	1.3×10^9	over 100 kyr	Volcanic rocks, granites, and other igneous rocks

is always in trace quantities (about one part per trillion), but becomes increasingly scarce in a sample over time as ^{14}C decays. Radiocarbon is continually produced in the atmosphere as cosmic rays interact with ^{14}N to produce ^{14}C, and due to this replenishment, the proportion of $^{14}CO_2$ in the atmosphere stays relatively constant over long periods of time. Through photosynthesis and the building of calcareous shells, radiocarbon becomes incorporated into living matter and food chains of ecosystems so that all organisms are in 'isotopic equilibrium' with the atmosphere. This is why it is only on the death of an organism when it stops taking in new ^{14}C from its environment that the radiocarbon clock begins. In order to calculate a date for when the organism was alive from the half-life of ^{14}C (which in radiocarbon labs is taken as 5568 to maintain consistency with ^{14}C dating done before the revised and now accepted actual half-life of 5730 was known), a starting ^{14}C concentration must be assumed (taken as the atmospheric concentration in 1950). The fact that the ^{14}C concentration of the atmosphere (and therefore in living organisms) *has not* been completely constant over time is why radiocarbon dates require calibration to be turned into calendar year dates as described further below.

Radiocarbon dating, formerly used mainly for organic carbon, in the form of peat and wood, has been extended to a wider range of Holocene and Late Pleistocene materials, especially soil carbonates and Mollusca. The technique has evolved steadily since its first application by Libby and co-workers (Libby et al. 1949), and it provides a chronology for approximately the last 50 000 years, though practical problems become severe beyond about 45 000 years ago. Some laboratories can bring material 75 000 years old within dating range (Grootes 1978). The standard method cannot be used to date very recent samples (within the last 200 years) because of major human-caused fluctuations in the proportion of $^{14}CO_2$ in the atmosphere. On the one hand, fossil fuel use since the Industrial Revolution has been adding old (^{14}C depleted) carbon to the atmosphere leading to erroneously old dates on recent samples (this is called the Suess effect); on the other, atomic bomb testing greatly elevated ^{14}C levels in the 1950s and 1960s.

For sediment sequences within the range of ^{14}C dating, a series of radiocarbon dates can be obtained from organic material sampled at different depths in a core, and a chronostratigraphy can be constructed simply by plotting the radiocarbon ages against depth in the sequence. An age-depth curve for the sequence can then be derived by linear interpolation between the radiocarbon dated horizons, as illustrated in **Fig. 2.5**, or by fitting a higher order polynomial function through the series of radiocarbon dates. A long-standing problem is obtaining sufficient organic carbon for dating the horizons of interest. For conventional dating, which measures the radioactivity of a sample, radiocarbon laboratories require large bulk samples (sometimes spanning several centimetres of a core) to provide sufficient carbon after pretreatment, and this will reduce dating precision. An important innovation in this regard has been the **accelerator mass spectrometry** (**AMS**) method of radiocarbon dating, which directly measures the $^{14}C/^{12}C$ ratio of a sample, thereby overcoming the problem of sample size. Through the AMS method it is possible to date, for example, seeds, leaves, and even pollen grains (e.g. Brown et al. 1992).

There are further problems which need to be considered in assessing the reliability of the very large number of ^{14}C dates which are now available. In particular, contamination of samples may take place. Humic acids, organic decay products, and fresh calcium carbonate may be carried downward to contaminate underlying sediments. In the case of inorganic carbonates, 'young' carbonate may be precipitated in, or replace, the carbonate which one is interested in dating, and removal of the contaminant from pore spaces and fissures is almost impossible. In the case of biogenic marine carbonates, radiocarbon dates are typically 'too old' because when the organisms were alive they were incorporating radiocarbon from seawater rather than from the atmosphere. Especially in deep areas of the oceans, there may be a long time interval between carbon entering the sea from the atmosphere and subsequently being incorporated into an organism, by which time it will have lost some of its radioactivity. This is referred to as a 'reservoir effect' and different radiocarbon laboratories may have used different assumptions in correcting for it. It must also be remembered that there is a degree of error in the laboratory methods for measuring the radioactivity (or $^{14}C/^{12}C$ ratio) of a sample. This is why all radiocarbon dates are given with a 1 sigma error range, for example ±50 years to indicate that there is a 68% probability that the actual radiocarbon age lies anywhere between 50 years older or younger than the radiocarbon date (and a 95% probability that it lies between 100 years older or younger).

An important distinction needs to be explained here about the difference between 'radiocarbon years' and 'calendar years'. As shown by radiocarbon dating individual tree rings, the age calculated by radiocarbon

Fig. 2.5 Illustration of an age-depth curve based on linear interpolation through a sequence of radiocarbon dates on a sediment core from a Scottish peat bog (from Anderson 1996). (a) The stratigraphy (using Troels-Smith symbols) and the positions of uncalibrated radiocarbon dates. (b) The age-depth curve after calibrating ^{14}C dates into calendar year age estimates, and estimates of sediment accumulation rates. Note that early Holocene sedimentation was rapid as the basin infilled, and accumulation rates slowed through the transition from lake sediments to peat. Also note the increasing discrepancy in the mid to early Holocene between radiocarbon and calendar ages

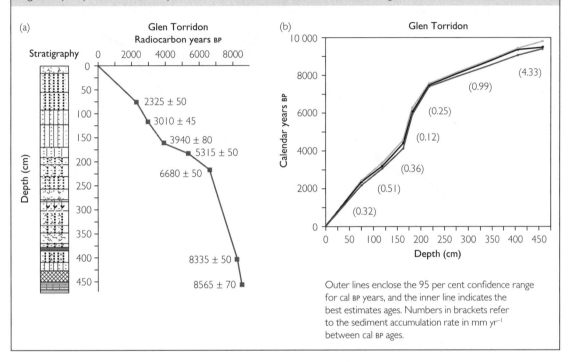

Outer lines enclose the 95 per cent confidence range for cal BP years, and the inner line indicates the best estimates ages. Numbers in brackets refer to the sediment accumulation rate in mm yr^{-1} between cal BP ages.

dating usually does not correspond with the real (or calendar) age derived from dendrochronology (**Section 2.2**). This is because of past changes in the production rate of ^{14}C in the atmosphere. As already referred to, this means that the radiocarbon clock does not necessarily start with the same amount of radiocarbon for organisms that lived at different times. Many factors affect atmospheric ^{14}C over different timescales, including changes in solar activity and the geomagnetic field, which changes the cosmic ray flux, and deep ocean **upwelling** which changes the exchange of carbon between oceans and the atmosphere (see Stuiver and Braziunas 1993 for a detailed discussion). A period of gradually declining ^{14}C in the atmosphere will result in a **radiocarbon plateau**: a lengthy period over which radiocarbon dates obtained from organic material are approximately the same. During these times, such as during the Late Glacial and early Holocene, it is difficult to obtain a chronologically consistent set of radiocarbon dates on a sediment sequence.

It has also been found that radiocarbon ages become increasingly too young in relation to the true calendar age with increased age of the sample. For instance, radiocarbon dates on the major mid-Holocene decline in elm pollen (see **Section 5.8**) at numerous sites in the British Isles average out between 5000 and 5100 years ago (Parker et al. 2002), which in calendar years is around 6000 years ago. In the early Holocene the discrepancy is even larger: from radiocarbon dating, the start of the Holocene has been placed at *c.* 10 000 years ago whereas incremental dating methods place the boundary at 11 700 years ago. As shown in **Fig. 2.6**, this relates to the higher proportions of ^{14}C that existed in the atmosphere compared with more recent times. Problems with radiocarbon dating events in the Late Glacial and early Holocene are discussed further in **Sections 5.2** and **5.3**.

Through comparison with the tree ring record (dendrocalibration) it has been possible to calculate the calendar year probability ranges for specific radiocarbon ages, and this makes it possible to convert radiocarbon

Fig. 2.6 Relationship between radiocarbon years BP and calibrated calendar years (cal yr BP) extending back 50 000 calendar years, and per mille variations in the concentration of atmospheric ^{14}C (from Reimer et al. 2009, Fig. 4). Note that the IntCal09 calibration curve becomes thicker with age reflecting an increased error range. 0 = AD 1950. The error range also widens for the ^{14}C variations and is depicted by the parallel lines. Reprinted with permission of *Radiocarbon*

year chronologies into calendar year chronologies. By comparing radiocarbon dates with annually laminated marine sediments and uranium-series dates on fossil corals and speleothems it has been possible to extend the calibration curve beyond the 12 500 cal yr BP or so reach of dendrochronology to 50 000 calendar years ago (**Fig. 2.6**). At the time of writing, the 'International Calibration' curve in current use is INTCAL09, and this curve underpins the BCal, CALIB, and OxCal software calibration programmes (Reimer et al. 2009). This curve replaced INTCAL04 which stretched back 26 000 years, and it too will eventually be revised and replaced as new data becomes available.

In summary, it is important to take into account the problems caused by atmospheric variability in ^{14}C when interpreting radiocarbon dates, and to realize that a single radiocarbon date will, in fact, represent a range of possible calendar dates sometimes spanning centuries. In some cases, however, increased precision can be gained through a technique known as

'wiggle matching'. In this case, a precise calendar year estimate based on radiocarbon dating is achieved by comparing the age-depth curve derived from a closely spaced sequence of radiocarbon dates with the closest matching wiggles in the dendrocalibration curve (van Geel and Mook 1989 provide a detailed discussion). As previously noted, dates given in years before present (BP) need to be interpreted as representing raw, uncalibrated 'radiocarbon years'; whereas dates expressed as 'cal yr BP' reflect radiocarbon dates that have been calibrated into calendar years. And as BP means before AD 1950, 1000 cal yr BP, for example, equates to AD 950.) If it is not clear whether a study has calibrated its radiocarbon dates, it is best to assume that the age is being given in radiocarbon years – this is especially the case for older research literature (e.g. pre 1990s) before calibration became common place. **Figs 2.6 and 2.7** show the relationship between radiocarbon dates and calendar years BP on longer and shorter timescales and can be used to approximate the conversions.

Fig. 2.7 Relationship between radiocarbon years BP and calibrated calendar years (cal yr BP) extending back 12 000 calendar years drawn from IntCal04 raw data stored at www.radiocarbon.org. Error margins are not shown. The IntCal09 curve is unchanged from IntCal04 between 0 and 12 000 calendar years. 0 = AD 1950. Data sourced with permission of *Radiocarbon*

Since the early 1960s potassium-argon (^{40}K/^{40}Ar) dating has been applied to Pleistocene and Pliocene chronology and has greatly changed our views on the length of the Pleistocene, and on the time when glaciation was initiated. Whilst radiocarbon dating utilizes organic and inorganic carbonates, potassium-argon dating, which can cover a theoretically unlimited time span, utilizes unaltered, potassium-rich minerals of volcanic origin in basalts, obsidians, and the like. It is however, only usable in practice for materials older than about 100 000 years.

Also since the 1960s the thorium-uranium and other uranium series dating methods have been applied to such materials as molluscs and coral. Although still subject to certain deficiencies, particularly for molluscs, these techniques are extremely valuable when applied to coral in bridging the gap between radiocarbon and potassium-argon techniques. These methods have been used for materials up to about 200 000 years in age with some success, and uranium series dates obtained for coral terraces have caused a change in ideas on the fluctuations of sea level before the last glaciation (see **Chapter 6**).

Recent sediments, of the order of decades old, can be dated radiometrically using isotopes with very short half-lives, such as using lead-210 (^{210}Pb) and caesium-137 (^{137}Cs), the latter isotope having been produced by nuclear testing since 1945.

Luminescence dating

In contrast with measuring ratios between parent and daughter nuclides to arrive at an age estimate, radioactivity also enables dating through the measurement of its effects on the crystal structure of minerals. Background radiation in the environment causes electrons to accumulate within structural defects ('traps') inside the crystal lattice of minerals, and the quantity of trapped electrons is proportional to the amount of time that the mineral has been exposed to radiation. Electrons are released from these traps, and return to their original positions within the crystal lattice, when a material is either heated or exposed to sunlight, thereby 'resetting' the 'luminescence clock'. **Luminescence dating** involves artificially releasing and measuring the trapped electrons in the laboratory to arrive at an estimate of the time since the sample was last 'reset'; for example, when it was last exposed to sunlight before becoming buried to accumulate another dose of radiation. The method also requires that the material being dated is 'calibrated' (to see how it reacts to known doses of radiation), and that the background radiation to which the material was exposed is estimated (in terms of a radiation dose per year).

The technique known as thermoluminescence dating (TL) can be employed on a wide range of materials including burnt artefacts (such as ceramics and flint), volcanic material, cave carbonate, and quartz grains in loess, dune sand, and glacial deposits. The material being dated by TL is heated, and emitted light caused by the release of trapped electrons is measured. This is called a 'glow curve', the intensity of which depends in part on the time over which the sample has accumulated its radiation dose (Aitken 1989). The glow curve is analysed to estimate the radiation dose required to produce it: this is assumed to be the 'palaeodose' that, when divided by the 'dose rate' (dose per year estimated for the background radioactivity of the sample and its original location) allows for an age estimate in years. The age range of the method depends on the material being analysed, but is generally applied to time spans of tens of thousands to hundreds of thousands of years.

Closely related to TL dating is optically stimulated luminescence dating (OSL), an approach that was first demonstrated by Huntley et al. (1985). This has great potential for the dating of quartz grains of wind-borne and water-borne origin. The luminescence emission caused by optical excitation has the advantage that it is only derived from the light-sensitive sources of luminescence from within the grains, i.e. those sources that are most likely to undergo resetting – 'bleaching' – in the process of being transported and deposited, as compared to the spectrum of both light-sensitive and light-stable traps that are sampled during a TL analysis (Rhodes 1988). The wavelength of light used depends on the mineral analysed: the luminescence of a sample of quartz is stimulated using blue or green light whereas for feldspar it is infrared. As with TL, OSL yields an estimate of an 'equivalent dose' or 'palaeodose' required to produce the luminescence that, when divided by the dose rate, gives an estimate in years since the sample was last exposed to light. Compared with TL, the OSL method is preferred for

dating sedimentary deposits, and its range spans as little as a few hundred years up to around 100 000 years. It has proven especially useful for dating deposits, such as sand dunes, that lack organic carbon for radiocarbon dating (Stokes 1999). The method is explained in detail in Aitken (1998).

Fission track, ESR, and CN dating

Another technique, which like K/Ar dating can be applied to volcanic materials, is fission-track dating. This is based on the principle that traces of an isotope, ^{238}U, occur in minerals and glasses of volcanic rocks, and that this isotope decays by spontaneous fission over time causing intense trails of damage, called tracks. These narrow tracks, between 5 and 20 microns in length, vary in their number according to the age of the sample. Thus by measuring the numbers of tracks an estimate can be gained of the age of the volcanic minerals. The method has been applied to the dating of **tephrochronological** events and provides cross-checks on other methods such as K/Ar dating.

The application of fission-track dating to relatively young geological events is summarized by Naeser and Naeser (1988). At the moment the time frame over which the technique can be used is limited by the time required to develop a statistically significant number of tracks. This is generally in excess of 100 000 years.

Electron spin resonance (ESR dating) has emerged as an important dating technique in recent decades (Ikeya 1985). It is a method of measuring the paramagnetic defects in minerals or the skeletal hard parts of organisms. These defects are created by penetrating radiation from radioactive elements within, or in the sediment surrounding, the sample. It shares many fundamental principles with luminescence dating, though has the advantage of being applicable to a wider range of materials and over a longer timescale. It has been applied successfully to the dating of materials such as corals, molluscan fossils, and tooth enamel, and can be used to date carbonate fossils back to c. 400 000 years, with an uncertainty of ±10–20 per cent.

Cosmogenic nuclide (CN) dating is used for dating the time since exposure of surface rock (a method of 'surface exposure dating'). Once a rock surface is exposed it becomes bombarded by cosmic rays that pass through the atmosphere, and these cosmic rays cause the formation of certain stable and radioactive isotopes within rock minerals. Over time these cosmogenic nuclides (for example isotopes of helium, beryllium, neon, aluminium, and chlorine) accumulate in the surface rock at known rates, and by accelerator mass spectrometry their amount can be measured in surface samples and a calculation of the time since exposure can be made. The technique spans a wide time range, from 1000 years to 10 million years, depending on which nuclides are analysed, and it has proven useful for geomorphological and archaeological studies (see Gosse and Phillips 2001, for a review).

Incremental methods

Much incremental dating has been focused on annually laminated lake sediments (varves), tree rings, and annual layers of glacier ice as discussed previously in **Sections 2.2** and **2.4**. Varves have proven particularly useful for dating glacial retreat in high-latitude areas, such as Scandinavia, because varves are often a feature of proglacial lakes and the number of varves can give an indication of the age of the lake. As explained in **Section 2.2**, dendrochronology has a wide range of environmental and archaeological applications. It has been especially important for calibrating the radiocarbon timescale as described above. The counting of annual layers of glacier ice has been extended back furthest on ice cores from Greenland (to about 40 000 years ago), and this has meant that many palaeoenvironmental changes inferred from the last glacial and Holocene portions of the cores can be dated with annual precision. Combined with electrical conductivity and tephra analysis, this has proven especially useful for the dating of prehistoric volcanic eruptions (precise dating of past eruptions can also be achieved with varved sediments).

One characteristic of these forms of incremental evidence that requires mention is the existence of 'floating' chronologies. This is a case in which a continuous set of varves, tree rings, or glacier ice layers are identified but, because of gaps in the record, cannot be 'anchored' with complete confidence to a year-by-year master chronology extending back from the present. The actual time spanned by floating chronologies can be estimated by other means (such as radiocarbon dating) and such chronologies can be useful for determining the duration and relative timing of environmental changes, even if the absolute age is less certain.

Other sources for incremental dating are corals and certain types of long-lived molluscs ('sclerochronology'). Some species of coral produce seasonal banding as they grow; and coupled with geochemical analyses of coral carbonate it is sometimes possible to reconstruct various environmental changes, such as water temperature, dissolved CO_2, and runoff from land, with annual precision. Long-lived molluscs, such as the clam *Arctica islandica* which can live for as long as 500 years, show annual bandings in the growth of their shells which like tree rings vary depending on conditions for growth. Using cross-dating of *A. islandica* shell samples (e.g. Wanamaker et al. 2008), an Iceland area marine sclerochronology has been constructed extending back over a thousand years. Under certain conditions, speleothems can also be annually banded (see **Section 2.2**).

One other incremental method needs mentioning: **lichenometry**. This has become increasingly important since it was developed in the 1950s, and is especially useful for dating glacial events over the last 5000 or so years. It is believed that most glacial deposits are largely free of lichens when they are formed, but that once they become stable, lichens colonize their surfaces. The lichens become progressively larger through time. Thus by measurement of the largest lichen thallus of one or more common species, such as *Rhizocarpon geographicum*, an indication of the date when the deposit became stable can be attained. Good reviews of lichenometry, which stress the inherent problems in the method, are provided by Innes (1985) and Worsley (1990).

Age-equivalent markers

One of the most important forms of 'age-equivalent' dating evidence is **palaeomagnetism**. Currently the Earth has what is termed a 'normal' magnetic field so that at the north magnetic pole a compass dips vertically towards the Earth's surface. However, the magnetic field, for reasons not fully understood, can switch to become 'reverse'. As ferromagnetic minerals in some rocks and sediments may preserve the characteristic signal of the magnetic field during the time the unit was deposited, it has been possible to produce a calendar of magnetic events marked by switches from 'normal' to 'reverse'. As many of these switches have now been dated by independent means (such as K/Ar dating for example), in a conformable sequence of sediments these magnetic switches enable a particular section to be dated against a master system (Glass et al. 1967). Thus, sediments from deep-sea cores can be given an age-scale to considerable length on the basis of the sequence of palaeomagnetic events revealed in the stratigraphy.

A two-level system of names has been introduced to describe the observed sequence of polarity reversals. At the lower end are polarity events – short intervals of normal or reversed polarity lasting in the order of 150 000 years or less. At the higher level are the polarity epochs – longer intervals during which the magnetic field was predominantly of one polarity, and which may contain one or more events (Cox et al. 1968). The sequence of these epochs and events is shown in **Fig. 2.8**.

It has also proved possible to date recent sediments (Holocene age) on the basis of secular changes in the Earth's magnetic field. Over centuries and millennia the location of the magnetic pole changes with respect to the axis of Earth's rotation. This results in changes in the 'magnetic declination' (the angle between magnetic north and true north) as well as 'inclination' (the angle of dip of the magnetic field relative to the horizontal at a particular latitude). However, unlike polarity reversals which, wherever identified in rock or sedimentary sequences, can be related to a global master calendar, reconstructions of past secular changes need to be compared to regionally specific master calendars. Master sequences of secular variation have been constructed from varved or radiocarbon dated lake sediments in various places, such as in northern Europe (e.g. Saarinen 1998).

Volcanic eruptions can also provide important stratigraphic markers for the Quaternary. Different ash falls may be recognizable on the basis of petrology and chemical composition. The falls of ash can be placed in chronostratigraphic position by radiocarbon dating of associated sediments, or by K/Ar dating of the source of the volcanic unit. Ash falls from some Late Glacial and Holocene aged volcanic eruptions have been identified in Greenland ice cores and have been given more precise age estimates by ice layer counting; and similar precision can also be achieved where ashes are identified in varved lake sediments. Once the age has been established, an ash layer can be used as a marker horizon in otherwise undated sections. This technique is termed **tephrochronology**, and its importance as a dating method has increased greatly in recent years as the database of dated ash layers in various parts of the world has continued to expand (see, for example, Davies et al. 2002, 2008), and the techniques for 'fingerprinting' specific tephras have improved.

As the oxygen isotope record of deep-sea sediments (**Section 2.3**) primarily reflects changes in the Earth's

Fig. 2.8 The most recent 3.5 million years of the geomagnetic polarity timescale. Black indicates a time of mainly normal polarity, white mainly reversed polarity. Potassium-argon dates for transitions are shown on the left, orbitally-tuned ages on the right. Polarity 'epochs' are named on the left, and polarity 'events' are named on the right. The orbitally-tuned ages of boundaries are the ages commonly used (from Walker 2005, Fig. 7.9). Reprinted with permission from Wiley

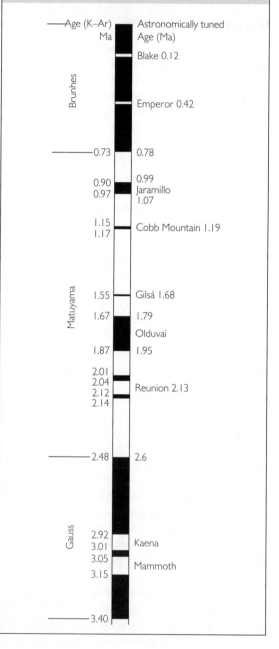

total volume of glacier ice, shifts in the $\delta^{18}O$ signal represent globally synchronous time horizons that can be used for correlating deep-sea sediment cores worldwide. As described previously, the record of $\delta^{18}O$ variability has been divided up into a system of stages with odd numbers referring to phases of lower $\delta^{18}O$ (warmer/less ice) and even numbers referring to phases of higher $\delta^{18}O$ (colder/more ice) (see **Fig. 3.2**). The global sequence of marine isotope stages (MIS) continues to be refined and extended as more $\delta^{18}O$ records from individual deep-sea sediment cores are studied. The process of constructing a master $\delta^{18}O$ record involves averaging the signal interpreted from a set (or stack) of individual deep-sea core records so as to maximize the 'signal-to-noise' ratio. Once a master curve is constructed, spectral analysis of the time series data enables the identification of periodicities. In the 1970s it was discovered that periodicities in $\delta^{18}O$ curves correlate with predicted periodicities of ice sheet expansion and contraction on the basis of orbit-ally-induced variations in distribution of solar energy across the planet (the **Milankovitch Theory** described in **Chapter 9**). Because the timescales of these orbital variations are known and can be computed, it is possible to assign ages to the corresponding cycles of variation in the $\delta^{18}O$ curves – an approach known as 'orbital tuning'. Thus, transitions in the global system of MIS can be dated and then used as age-equivalent markers where located in any individual deep-sea sediment core. Construction of a timescale for the system of MIS has also been aided by comparison with the palaeomagnetic record.

An important step forward in the development of an orbitally-tuned master sequence of isotope stages was the SPECMAP work (Imbrie et al. 1984) which made use of five different marine $\delta^{18}O$ records and extended back about 750 000 years. The timescale covered by SPECMAP has since been refined and extended. For example, making use of 57 different benthic foram-based records of $\delta^{18}O$ derived from deep-sea sediment cores from the Pacific, Atlantic and Indian Oceans, Lisiecki and Raymo (2005) improved and extended the MIS timescale back to 5.3 million years ago.

Relative methods

These methods involve placing things or events in chronological order. As discussed previously, the use of relative methods has a long history in palaeoenvironmental research, beginning with the study of rock strata in the nineteenth century (**Section 1.6**). Even though such methods do not yield exact age estimates, they remain important both as a cross-check on **absolute dating** methods and in cases when radiometric or incremental dating is impractical or impossible.

One example is the use of rock weathering data to establish the relative time since various rocks were first exposed to weathering. The products of rock weathering can be measured in a variety of ways, and one such way is through measuring the thickness of the weathering rind (a layer of altered rock that extends from the outer surface of the rock inwards). Another way is by testing rock hardness, for example by using a Schmidt hammer (e.g. McCarroll 1994). The Schmidt hammer is calibrated so that a stronger rebound of the hammer indicates harder (and therefore less weathered) rock.

A technique that has been widely used by archaeologists is obsidian hydration dating because obsidian, a type of volcanic glass, was often used to make cutting tools in the Stone Age. When a fresh obsidian surface is exposed, it absorbs water to form a hydrated rind that increases in thickness with time; and although many factors affect the rate of obsidian hydration, thickness of the hydration layer can be measured and used to develop relative chronostratigraphies (e.g. Trembour and Friedman 1984).

Another important means of **relative dating** is based on the fact that protein preserved in the skeletal remains of animals undergoes a series of chemical reactions, many of which are time-dependent. After death of the organism the protein slowly degrades, as does the nature of the amino acids which form the basis of the protein. Thus an examination of the amino acid composition of bone and of carbonate fossils has potential for dating purposes (Miller et al. 1979). The method has the advantage that only very small amounts of sample are required. It also covers a wide time range. However, the degree of change in amino acid composition depends on factors other than time, and therefore involves making certain assumptions, particularly about temperature conditions, that may create substantial errors. The technique seems to have most potential for ascertaining the relative age of mollusc shells (both marine and non-marine) because with this material the rate of amino acid diagenesis is less affected by external factors than is the case with bone and other protein residues. Prospects and applications of **amino acid geochronology** are surveyed in Wehmiller and Miller (2000).

Table 2.9 Categories of dating methods

Category	Technique
Radiometric and related methods	Radiocarbon dating
	Potassium-Argon dating
	Uranium-series dating
	Lead-210 dating (and other short-lived isotopes)
	Thermoluminescence dating
	Optically stimulated luminescence dating
	Fission track dating
	Electron spin resonance
	Cosmogenic nuclide dating
Incremental methods	Dendrochronology
	Varves
	Glacier ice layers
	Sclerochronology
	Speleothem accumulation layers
	Lichenometry
Methods that establish age equivalence	Palaeomagnetism
	Tephrochronology
	Marine oxygen isotope stages
Relative age methods	Rock hardness and weathering rinds
	Soil development studies
	Obsidian hydration dating
	Amino acid geochronology

Summary

A wide range of dating techniques is now available which can be broadly categorized as follows: radiometric methods (which rely on radioactive decay and related changes), incremental methods (based on the accumulation through time in a regular manner of biological or lithological materials), methods that establish age equivalence (chronostratigraphic markers), and relative methods that establish the relative order of antiquity of materials. Various techniques in each of these categories are summarized in **Table 2.9**.

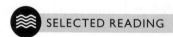 SELECTED READING

The book *Reconstructing Quaternary Environments* (2nd Edition) by J.J. Lowe and M.J.C. Walker (1997) is strong on techniques for environmental reconstruction, as is N. Branch et al. (2005) *Environmental Archaeology: Theoretical and Practical Approaches.*

A general and detailed reference book is R.S. Bradley's (1999) *Paleoclimatology: Reconstructing Climates of the Quaternary* (2nd Edition).

The book edited by B.E. Berglund (1986) *Handbook of Holocene Palaeoecology and Palaeohydrology* remains an important reference for many methods and sources of evidence related to reconstructing Holocene environments, and more recent developments in this area are surveyed in A. Mackay et al. (2005) *Global change in the Holocene*.

Techniques for analysing glacial sediments are reviewed by D.J.A. Evans and D. Benn (2005) *A practical guide to the study of glacial sediments*, and the interpretation of periglacial features and landforms is detailed by H.M. French (1996) *The Periglacial Environment* (2nd Edition).

Much important information on the use of lake sediments to reconstruct changing environments in presented in A.S. Cohen (2003) *Paleolimnology: The History and Evolution of Lake Systems*, while the use of peatlands for palaeoenvironmental reconstruction is surveyed by D.J. Charman (2002) *Peatlands and Environmental Change*.

Detailed treatments of pollen analysis and its many applications can be found in Birks and Birks (1980), Faegri and Iversen (1989), Delcourt and Delcourt (1991), and Moore et al. (1991).

R.B. Alley's (2002) *The Two-Mile Time Machine* is an excellent popular introduction to the methods and results pertaining to ice cores.

A wide ranging and detailed account of dating methods is contained in Walker (2005) *Quaternary Dating Methods*.

The book *Bones, Rocks and Stars: the Science of When Things Happened* by Chris Turney (2008) is an excellent popular introduction to the science of age determination.

The NOAA Palaeoclimatology webpages (http://www.ncdc.noaa.gov/paleo/paleo.html) contain links to most forms of evidence relevant for the Quaternary, and the webpage for the journal Radiocarbon (www.radiocarbon.org) contains information and links about radiocarbon dating.

The outreach pages of the Quaternary Research Association (www.qra.org.uk/outreach) also contain many links to information about Quaternary palaeoenvironmental reconstruction.

? REVIEW QUESTIONS

1. What insights can be gained from the study of relict features of glaciation and periglaciation, and what difficulties exist in the interpretation of such features?

2. Consider what microfossil sources of evidence from terrestrial and marine stratigraphic sequences have in common in terms of how they are sampled, analysed, and interpreted.

3. For what reasons can it be said that a full understanding of environmental changes on land during the Quaternary requires study of marine sediments?

4. Compare and contrast stable isotope evidence from deep-sea sediments and ice cores.

5. Discuss the ways in which radiocarbon dating, incremental dating methods, and age-equivalent markers can be used in conjunction with each other to improve our chronologies of environmental change.

3

Pleistocene Climatic Change and Environments of Mid to High Latitudes

Chapter overview

The evidence discussed in **Chapter 2** is the basis for the reconstruction of the Pleistocene record that we provide in **Chapter 3**. We also describe the results of the frequent, severe and often abrupt climate changes that occurred, including the waxing and waning of ice caps and glaciers, the spread and retreat of permanently frozen ground (permafrost), and the deposition of wind-blown dust (loess). Above all we discuss the nature of glacials and interglacials, and the changes in the biosphere that accompanied them.

3.1 Introduction

The Pleistocene, as discussed in **Chapter 1**, was composed of alternations of great cold (**glacials, stadials**), with stages of relatively greater warmth (**interglacials, interstadials**). There is still a marked degree

of controversy over the number of these events. This exists partly because of the problems of definition, a matter discussed further in **Section 3.6**. There is also a lack of agreement with regard to correlations of events between different areas. Nevertheless, the continued development and application of dating techniques and, in particular, advances in the analysis of deep-sea sediment cores, have enabled some statements about Pleistocene climatic change to be made with a greater degree of confidence than hitherto. Indeed, there are those who would claim that Pleistocene studies were transformed by new techniques in the latter part of the twentieth century. Bowen (1978, p. 193), for example has argued:

> There is no doubt whatsoever that Quaternary systematics have been subject to a revolution comparable to that of plate tectonic theory on geology as a whole.

In particular, he points to the fact that whereas traditionally the 'Quaternary Ice Age' was regarded as having comprised four, five, or at most six major glacials (see **Section 1.2**), now, depending on how the data are interpreted, there are indications from the ocean cores that there have been between 30 and 50 cold stages spanning the Quaternary, with at least 20 of these resulting in extensive glaciation (Ehlers and Gibbard 2007).

3.2 The Pleistocene record

The various terrestrial sequences that have been compiled by classic stratigraphic techniques have demonstrated beyond doubt that major glacial and interglacial fluctuations have taken place and that the Pleistocene was a period of remarkable geomorphological and ecological instability. However, because of the incomplete nature of the evidence it is not possible, using this information, to construct a correct, long-term model of environmental changes in the Pleistocene. For this to be achieved it is necessary to look at the more complete record of deposition preserved in the deep-sea sediment cores; and, to a lesser extent, in the **loess** sequences of the Old World, and the long **ice core** records recently obtained from Antarctica.

Deep-sea sediments

Many sediment cores have now been extracted from the oceans, and as recounted in **Chapter 2**, many techniques have been used to extract palaeoenvironmental information and to obtain a reliable dating framework. The most important information has been gained from the oxygen isotope composition ($\delta^{18}O$) of **foraminiferal** tests. (The principles underlying interpretation of $\delta^{18}O$ data are discussed in **Section 2.3**.) By

compiling benthic foram $\delta^{18}O$ records from over 40 Deep Sea Drilling Project (DSDP) and Ocean Drilling Program (ODP) sites, Zachos et al. (2001) have been able to extend the marine oxygen isotope record back through the entire Cainozoic Era. As shown in **Fig. 3.1**, the $\delta^{18}O$ data show a long-term (though not uninterrupted) cooling trend from the 'Eocene Climatic Optimum' about 50 million years ago to the Pleistocene – the **Cainozoic climatic decline** as described in **Section 1.4**. Prior to the formation of Antarctic ice sheets, the gradual increase in $\delta^{18}O$ is the result of deep-sea cooling: following the Eocene Climatic Optimum deep-sea temperature cooled by over 7 °C from its peak of around 12 °C. The sharp increase in $\delta^{18}O$ at the Eocene/Oligocene boundary probably reflects a geologically swift ^{18}O enrichment of the oceans as isotopically light $H_2^{16}O$ became locked up in growing Antarctic glaciers. From this time onwards the $\delta^{18}O$ record is mainly reflecting past changes in global ice volume. Hence, $\delta^{18}O$ values increase further once Northern Hemisphere glaciation is underway, culminating in peak values of around 5 ‰ during periods of maximum glaciation in the Pleistocene. Besides simply increasing in value, $\delta^{18}O$ also exhibits a trend of increasing variability through the Pliocene and into the Pleistocene, reflecting a gradual increase in the amplitude of global climate variation between glacials and interglacials.

An important milestone in understanding the pattern and timing of Pleistocene glacial/interglacial cycles came with the analysis of deep-sea sediment core V28-238 which has long been used as a type sequence for the $\delta^{18}O$ record of the last 900 000 years (Shackleton and Opdyke 1973). It was taken from the Pacific near the Equator (01° 01′ N 160° 29′ E), and was raised from a water depth of 3120 m (Kukla 1977). The upper 14 metres of the core were subdivided into 22 **marine isotope stages** (MIS) numbered in order of increasing age. As already described in **Section 2.3**, odd-numbered units are relatively deficient in ^{18}O (and thus represent phases when temperatures were relatively high and the ice sheets small), while the even-numbered units are relatively rich in ^{18}O (and represent the colder phases when the ice sheets were more extensive). Boundaries separating especially pronounced isotopic maxima from exceptionally pronounced minima have been called **terminations**. They are in effect rapid deglaciations, and are conventionally numbered by Roman numerals in order of increasing age. The segments bounded by two

Fig. 3.1 Compilation of benthic foram $\delta^{18}O$ data from over 40 DSDP and ODP sites (simplified after Zachos et al. 2001, Fig. 2). The 0 to 12 °C range refers to deep-sea temperature – until the expansion of glaciers, increasing $\delta^{18}O$ was due to deep-sea cooling alone. The solid vertical bars indicate times of extensive glaciation, and the striped bars indicate times of partial or ephemeral glaciation (reproduced with permission from AAAS: *Science*)

terminations are called glacial cycles. The youngest glacial cycle, which started with isotopic stage 1, and followed Termination I, is not yet completed. The previous glacial cycle extends from stage 2 (which includes the **Last Glacial Maximum** or **LGM**) back to Termination II, which marks the boundary between stage 6 and stage 5. Because the amplitude of variation between MIS 4 and 3 is not as great as that between stages 5 and 2, the MIS 4 and 3 represent stadial and

interstadial fluctuations contained *within* the most recent glacial cycle. Peaks and troughs in $\delta^{18}O$ within the isotopic stages themselves are denoted by lower case letters; or, as is increasingly the case, by decimals. For instance MIS 5 contains substages 5a, 5b, 5c, 5d, and 5e (or alternatively 5.1 to 5.5, with '5.0' denoting the transition between MIS 4 and 5). As explained further in **Section 3.6**, MIS 5e represents the previous interglacial.

The timing and character of these glacial cycles spanning the past 2.6 million years is well represented in core ODP-677 (**Fig. 3.2**). This core from the eastern tropical Pacific (01° 12′ N 83° 44′ W) was recovered in two parts from depths of 3472 and 3447 m (Shackleton et al. 1990). It extends the MIS record of V28-238 back in time, and through orbital tuning (see **Section 2.5**) it has improved the chronology of the isotope stages and terminations.

An extraordinarily long **speleothem** record derived from a fissure in Nevada, known as Devils Hole,

displays a similar pattern of isotope variation to the deep-sea sediment cores. The Devils Hole record extends from 60 000 to 568 000 years BP and contains major shifts in δ¹⁸O that can be related to the glacial cycles and terminations of the MIS record (Winograd et al. 1992). However, the dates for Terminations II and III (marked on **Fig. 3.2**) are around 10 000 years older than corresponding dates derived from orbital tuning of stacked marine records. The reasons for this are still not fully understood; but the Devils Hole record is likely to be less representative of broad climate

Fig. 3.2 Quaternary glacial cycles as revealed in the isotopic composition of a tropical Pacific core (ODP-677) and the vein calcite record from Devils Hole, Nevada. Roman numerals indicate glacial terminations. Some marine oxygen isotope stages are shown, with even numbers referring to cold stages and odd numbers to warmer stages. In the palaeomagnetic record black indicates times of normal polarity and white indicates times of reversed polarity. Terminations II and III in the Devils Hole record occur earlier than in the marine record – see the text for an explanation. Sources: World Data Center for Paleoclimatology and NOAA Paleoclimatology Program (Shackleton and Crowhurst 1996) and USGS (Landwehr et al. 1997)

patterns averaged for the whole globe compared with stacked marine records. Analysis of another speleothem record from Nevada, some 400 km northeast of Devils Hole, gives a later date for Termination II that agrees more closely with the marine record (Shakun et al. 2011).

The results of the analysis of core ODP-677, illustrated in **Fig. 3.2**, can be summarized thus. There have been ten completed glacial cycles and eleven terminations in the last one million years. Oscillations of global ice sheet volume within each glacial cycle seem to follow a sawtooth pattern, progressing from an early minimum to a late maximum. The number of secondary fluctuations in isotopic composition within each glacial cycle varies. Commonly it varies between four and six. Over the last million years the general shape and amplitude of the glacial cycles do not show great variation, although some of the cold stages are marked by deeper and/or longer lasting isotopic highs than others. For instance, MIS 12 was severe while cold stage 14 seems to be one with an exceptionally low ice volume. The warm peaks (excepting stage 3) all seem to approach a similar level, indicating that global ice volume and sea surface temperatures in peak interglacials were similar to those of the present day. During the last million years there have been ten episodes with global climate comparable to, and sometimes warmer than, today's. In other words, there have been ten interglacials. These only constitute about 10 per cent of the time, and seem to have had a duration of the order of 10^4 years, while a full glacial cycle seems to have lasted on the order of 10^5 years. Conditions such as those we experience today have thus been relatively short-lived and atypical of the Pleistocene as a whole.

Since Shackleton and Opdyke's pioneering work on core V28-238, analysis of ODP-677 and many other cores has enabled the system of oxygen isotope stages to be extended far beyond 1 million years (e.g. Lisiecki and Raymo 2005). As shown in **Fig. 3.3**, 104 isotope stages have been identified in the deep-sea sediment record spanning the last 2.6 million years (Gibbard et al. 2005; Cohen and Gibbard 2011). Of the 50 or so warm/cold oscillations during the Quaternary Period, it seems at least 20 resulted in the growth and retreat of ice sheets to an extent that could merit the term 'glacial cycle'. As seen in both **Fig. 3.2** and **Fig. 3.3**, before 1 million years ago fluctuations in $\delta^{18}O$ are less marked and have a different periodicity. Spectral analysis of $\delta^{18}O$ data shows that from the early Oligocene until the middle Pleistocene, expansions and contractions of global ice volume occurred with a periodicity of approximately 41 000 years – on the same period as variations in the Earth's tilt on its axis of rotation (the **obliquity** cycle as explained in **Chapter 9**). Around 900 000 years ago, the amplitude between cold and warm stages became more extreme, glaciation became more extensive, and the dominant pacing of glacial cycles shifted to a 100 000-year periodicity matching the **eccentricity** cycle of Earth's orbit. This important change in climatic variability has been referred to as the 'mid-Pleistocene revolution'.

There is much debate as to why this change in amplitude and pacing occurred, and some of the theories are discussed in **Chapter 9**; however, it can be said that some new combination of climate feedbacks must have taken hold to amplify changes in the eccentricity cycle because variation in orbital eccentricity by itself has a very small effect on varying the solar energy input across the Earth.

Loess sequences

The frequency of glaciations indicated by the deep-sea core record has been confirmed in the loessic record from various parts of Central Europe, the former USSR and China. During glacial phases **aeolian** silt was deposited as loess, while in warmer phases of soil stability and denser vegetation cover **palaeosols** developed. Kukla (1975) has, from his work in Eastern Europe, found eight cycles of glacials and interglacials in the 800 000 years or so of the Brunhes epoch. Work in Central Asia, employing both **palaeomagnetic** and **luminescence dating** of 200 m thick loess profiles has indicated that there may have been as many as 45 phases when palaeosols formed over the past 2.5 million years. Of these, eight cycles have taken place since the **Brunhes–Matuyama boundary**. A similar picture has emerged from the classic loess terrains of China.

Although there may be slight differences in the record from these three loessic areas, one is none the less impressed by the way in which the frequency of change broadly mirrors that found in the oceanic record (Goudie et al. 1983). For example, **Fig. 3.4** shows a close correspondence between SPECMAP orbitally-tuned marine oxygen isotope stages and Chinese loess/palaeosol horizons identified by analysis of mineral magnetic susceptibility. (See **Section 2.5** for an explanation of the SPECMAP.) Peaks in magnetic susceptibility indicate palaeosols which match phases of reduced $\delta^{18}O$ in the composite deep-sea record

Fig. 3.3 Global chronostratigraphical correlation table – including palaeomagnetism, isotope stages, and composite Chinese loess. (Courtesy of K. Cohen, modified from Cohen and Gibbard 2011. See www.quaternary.stratigraphy.org. uk/charts/ for the full version.)

(indicating interglacial phases). As with the oxygen isotope record, fluctuations in magnetic susceptibility become more marked from about 900 000 years ago onwards; however, correlations between peaks in susceptibility and δ¹⁸O minima can be found throughout the record extending back to MIS 103 approximately 2.6 million years ago (**Fig. 3.3**).

Ice cores

It used to be that ice cores did not extend far enough back in time to examine long-term Pleistocene climatic trends. This situation changed with the completion of drilling at Vostok, eastern Antarctica in 1998, which provided the first ice core record spanning four glacial cycles (Petit et al. 1999); and since then, the European Project for Ice Coring in Antarctica (EPICA) has recovered an ice core from Dome C, about 550 km distant from Vostok, that pushes the ice record back to some 800 000 years BP. This is equivalent to the last 20 isotope stages revealed in deep-sea sediment cores; and, in contrast with deep-sea sediments, the data obtained from ice cores are at a high temporal resolution and provide a wide range of climate-related information – including temperature, precipitation, atmospheric dust levels, and concentrations of greenhouse gases.

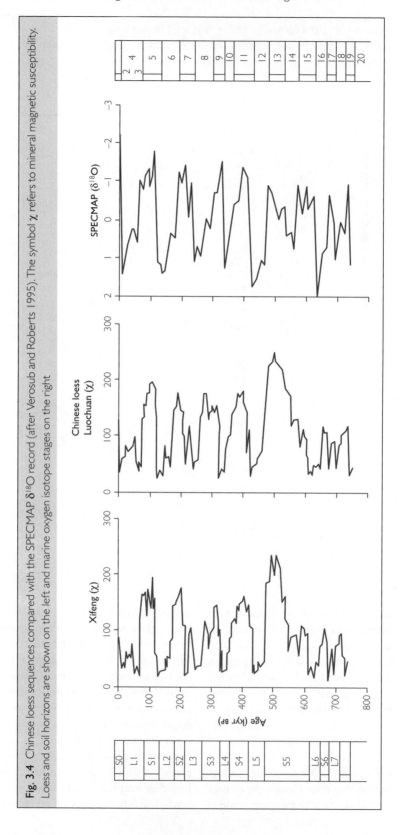

Fig. 3.4 Chinese loess sequences compared with the SPECMAP δ¹⁸O record (after Verosub and Roberts 1995). The symbol χ refers to mineral magnetic susceptibility. Loess and soil horizons are shown on the left and marine oxygen isotope stages on the right

The climate record of the past 740 000 years reconstructed from the EPICA Dome C (EDC) core, has revealed important new insights about the Middle Pleistocene (EPICA Community Members 2004). The deuterium/hydrogen ratio (δD) data show that *c.* 425 000 years ago at Termination V (onset of MIS 11) there was a transition from a long period of relatively cool interglacials and lower amplitude glacial/interglacial temperature variation, to warmer interglacials with higher amplitude temperature variation (**Fig. 3.5**). The causes for this 'mid-Brunhes event' are not fully understood, but must have involved an intensification of the climate feedback mechanisms that began pacing glacial/interglacial cycles at a 100 000-year periodicity around 900 000 years ago. Significantly, carbon dioxide and methane concentrations in the Vostok and EDC cores rise sharply at each glacial termination suggesting a close link between glacial/interglacial temperature changes and atmospheric greenhouse gas concentrations. Concentrations of CO_2 are typically 180–200 ppmv during glacials and 280–300 ppmv during interglacials. The values for CH_4 are typically 320–350 ppbv during glacials and 650–770 ppbv during interglacials (Petit et al. 1999). In general, CO_2 and CH_4 both begin rising gradually with the increasing temperatures at each termination, and each initial increase is followed by a particularly sharp and rapid rise in CH_4 at the transition to the interglacial climate. There are also major shifts in the quantity of aeolian dust within the Antarctic cores. Dust concentrations in the ice are usually more than 20 times higher during times of peak glaciation in comparison with interglacials. In addition to greenhouse gases, levels of dust in the atmosphere affect the Earth's energy balance (primarily through reflection of sunlight) and could also be implicated in glacial/interglacial cycles (**Chapter 9**). Over the past 420 000 years covered by both the Vostok and EDC cores, the two records display very similar patterns of greenhouse gas and dust variability; and back to 340 000 years ago trends are also corroborated by the East Antarctic Dome Fuji ice core.

Another important insight from the EDC core is that the interglacial within MIS 11 was surprisingly long compared to other interglacials in the record: it lasted for about 30 000 years, and over much of this time period δD values show that temperatures were, on average, as warm as the Holocene. MIS 11 is thought to be a good analogue for our present interglacial because of similar orbital parameters (relatively low eccentricity in Earth's orbit), and this has led some to speculate that the Holocene (which has lasted 11 700 years so far) could also be a relatively long interglacial, regardless of any human modification of climate trends due to greenhouse gas emissions. However, as explored further in **Section 9.6**, it has also been argued that, notwithstanding a long MIS 11, the trend from Holocene warmth towards the next glacial cycle should have begun two to three thousand years ago were it not for the human impact (Ruddiman 2005a, b).

The long ice core records from Antarctica add to the information gleaned from the even longer deep-sea sediment cores in one other important way – they confirm that rapid climatic changes first seen in the Greenland ice cores spanning the most recent glacial cycle are also a feature of earlier cycles extending back into the Middle Pleistocene. The timing and character of such rapid climate change events are discussed in Section 3.6.

Relation between land-based subdivisions and MIS

Since the pioneering work of Penk and Brückner on Alpine glacial chronology, several land-based schemes for subdividing the Pleistocene have been developed in various parts of the world. For example, **Tables 1.2** and **1.3** (**Chapter 1**) show the traditional schemes of subdivision pertaining to the Alps, Northern Europe, the British Isles and North America. The traditional schemes of these and other areas have been variously extended and modified as new glacial and interglacial deposits have been identified and analysed; and application of new dating methods has improved correlations between the land-based glacial chronologies of different regions and continents.

It is, however, important to recognize that such schemes do not have the global applicability of the system of marine isotope stages. For reasons already discussed, the oxygen isotope evidence from deep-sea sediments reflects changes in the Earth's total ice volume; whereas the land-based chronologies can be thought of as representing variations on the global theme, specific to the regions for which they were devised. Hence, the cart should not be put before the horse when relating the land-based and MIS systems. When referring to various phases within the Pleistocene, workers increasingly use the isotope stage or

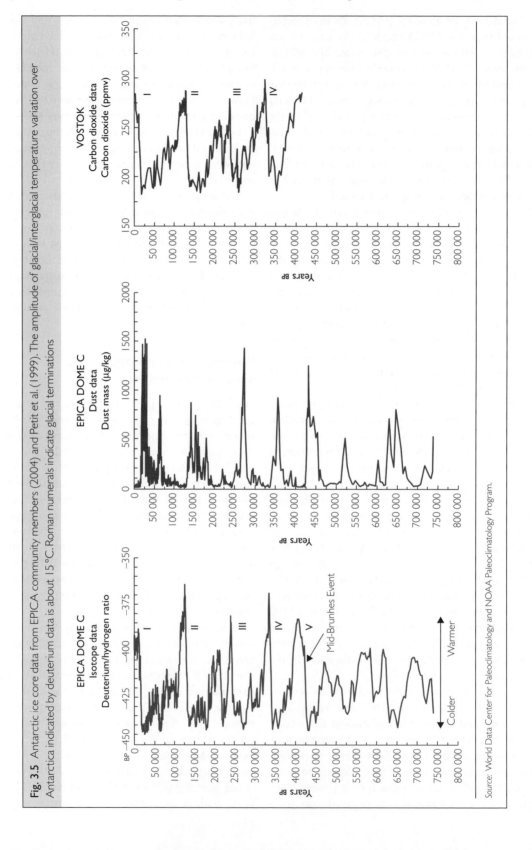

Fig. 3.5 Antarctic ice core data from EPICA community members (2004) and Petit et al. (1999). The amplitude of glacial/interglacial temperature variation over Antarctica indicated by deuterium data is about 15 °C. Roman numerals indicate glacial terminations

Source: World Data Center for Paleoclimatology and NOAA Paleoclimatology Program.

substage, rather than the regional name of a glacial or interglacial to avoid potential confusion and ambiguity. For example, MIS 5e is globally recognized as representing the most recent interglacial prior to the Holocene; and from deep-sea sediments, it has been dated to between approximately 130 000 to 115 000 years ago (see **Section 3.6** for a fuller discussion). However, in northern Europe, the British Isles, Russia, and North America it is variously named the Eemian, Ipswichian, Mikulinian, and Sangamonian respectively; and in each region the character of, and dates on, the associated deposits vary. (See **Section 1.6** for a discussion of the problems correlating 'climatostratigraphic' schemes which are by nature **time-transgressive** over large areas.) The problems with dating and correlating land-based glacial chronologies become even more acute earlier in the Pleistocene.

While the system of isotope stages provides the best global template for subdividing Pleistocene time, it is still important to be familiar with the terminology associated with land-based chronologies and with how these chronologies tie in with the MIS 'master curve'. Many of the names contained in traditional schemes continue to be used – particularly in the context of regionally-based glacial histories. **Figure 3.3** provides a summary of the relationship between land-based Pleistocene stages recognized in various parts of the world and the global system of isotope stages.

3.3 The changing extent of glaciers and ice caps

The Pleistocene glaciers did not show precisely the same spread in each of the cold stages, and it wasn't until MIS 22 (*c.* 880 000 years ago) that ice sheets became extensive across the Northern Hemisphere mid latitudes – previously they were mainly restricted to polar and mountain regions (Ehlers and Gibbard 2007). Quaternary glaciations reached their greatest extents during the Middle and Late Pleistocene. In Europe, the Saalian glaciation is regarded as representing the maximum spread of ice, with the Illinoian probably representing the maximum spread of ice in North America. At their maximum extent, glaciers worldwide probably covered up to *c.* 45 million km², which is rather more than their extent during the last glaciation some 35 to 40 million km². For CLIMAP modelling of the LGM (see **Section 1.6**) a global glacier ice cover of 35 million km² was used (Ruddiman 2008). For comparison, the present extent of glacier ice cover is *c.* 15.8 million km² (**Table 3.1**). In other words during maximum Pleistocene glaciation,

Table 3.1 The former and present extent of glaciated areas

	Maximum Late Pleistocene extent(m km²)	(% of total)	Present day extent (m km²)	(% of total)
Antarctica	14.50	32.4	13.5	86.0
Greenland	2.35	5.2	1.8	11.5
Laurentia	13.40	30.0	(Arctic 0.24)	–
North America Cordillera	2.60	5.8	(Alaska 0.05) (Rockies 0.03)	– –
Andes	0.88	2.0	0.03	–
European Alps	0.04	–	0.004	–
Scandinavia	6.60	14.8	0.004	–
Asia	3.90	8.7	0.12	–
Africa	0.0003	–	0.0001	–
Australasia	0.07	–	0.001	–
British ice sheet	0.34	–	–	–
Total	44.68		15.78	

(Modified from data in Smithson et al. 2002.)

Fig. 3.6 Comparison of the possible maximum extent of Pleistocene glaciation in the Northern Hemisphere (left) with the extent during the Last Glacial Maximum (right). C = Cordilleran ice; L = Laurentide ice; S = Scandinavian ice. Notice the slightly less extensive ice cover across North America during the LGM compared with a much reduced south-eastern margin of Eurasian ice. An area of uncertainty concerns the limits across Arctic continental shelves: glacier ice may have been more extensive in the Arctic region than shown here. There also remains some uncertainty about the extent and timing of confluence of British and Scandinavian ice across the North Sea basin during the last glacial cycle, and it has been proposed that confluence occurred before, but not during, the LGM (from Ehlers and Gibbard 2007, Figs. 6 and 8)

ice covered around three times as great an area as it does today (**Fig. 3.6**). However, both the Antarctic and Greenland ice sheets, because of the restrictive effects of iceberg-calving into deep ocean water, differed little in area from their present sizes, though they were much thicker. The greatest reductions in the glacial area have taken place through the melting of the North American **Laurentide** and the Scandinavian ice sheets, which have now lost 99 per cent of their former maximum bulk. In terms of volume, although precise assessment is difficult because of the problems of reconstructing former ice sheet profiles, the Riss/Saale/Illinoian glaciation had a volume of 84–99 million km³ compared to an estimated 32 million km³ at the present time.

The Last Glacial Maximum (**Fig. 3.6**) saw a total glacier ice volume estimated between about 80 and 90 million km³ (Mix et al. 2001), with sea level data suggesting c. 84 million km³ (Yokoyama et al. 2000). Most ice sheets had reached their maximum size by 26 500 years ago, as indicated by the time when the global sea level minimum was reached (Peltier and Fairbanks 2006), and significant deglaciation did not begin until c. 19 000 years ago (Clark et al. 2009). However, there were differences in the exact timing

of glacial expansion and the onset of deglaciation in different parts of the world. Details of regional histories of glaciation are provided in Ehlers (1996), Ehlers and Gibbard (2004), Ehlers et al. (2011), and for the last glacial in Clapperton (1997). Detailed summary tables of the chronology of Cainozoic glaciation for North America, Europe, and other regions of the world are contained in Ehlers and Gibbard (2007).

North America

During the maximum glacial phases of the Pleistocene, including the late Wisconsin maximum around 20 000 years ago (during MIS 2), ice was continuous or nearly so across North America from Atlantic to Pacific, and was composed of two main bodies, the Cordilleran glaciers associated with the Coastal Ranges and the Rockies, and the great **Laurentide ice sheet** (Menzies 2002) (**Fig. 3.7**). The former were most extensive in the mountains of British Columbia, and diminished both northward into Alaska and Yukon, and southward through the western United States. The southern limit of continuous Cordilleran ice was south of the Canada–USA border, down to the Columbia River and the Columbia Plateau. The Okanogan lobe of

Fig. 3.7 Outline map of ice limits in North America. The solid lines indicate maximum Pleistocene extent, and the dashed line indicates the Last Glacial Maximum. Dotted lines indicate uncertain marine ice margins

Source: Menzies 2002.

the Cordilleran ice sheet diverted the Columbia River east of its current course, causing it to erode into the basalt of the Columbia Plateau and develop the canyon scenery of the Grand Coulee. The rest of the scoured and eroded 'channeled scabland' scenery of the Columbia Plateau was formed over successive cycles of formation and outburst flooding of Glacial Lake Missoula in the Montana Rockies during deglacial times. The enormous 'Bretz floods' (after J Harlen Bretz) occurred each time the ice dam of the Purcell lobe failed, causing water volumes of *c.* 2000 km³ from Glacial Lake Missoula to discharge rapidly westwards across Washington State (see Anderson 2006 for a review).

South of the Cordilleran ice sheet there were numerous localized ice caps and glaciers, notably in the Yellowstone region and in the Sierra Nevada

Range. In the Sierra Nevadas the thickest ice, in the lee of the Coastal Ranges, was as much as 2300 m deep. Whether or not there was an 'ice free corridor' between the Cordilleran and Laurentide ice sheets during the late Wisconsin glaciation has been a matter of some debate (Mandryk and Rutter 1996).

The Laurentide ice sheet, which, as already mentioned, was with the Scandinavian ice sheet largely responsible for the great difference in world glacial areas between glacial phases and interglacials, reached its extreme extent in the Ohio-Mississippi basin at latitude 39° in Wisconsin times, and 36°40′ in Illinoian times. It extended to approximately the present positions of St. Louis and Kansas City in pre-Wisconsin times but west of this area the southern ice margin trended north-westwards, leaving western Nebraska and western South Dakota largely ice-free. On the basis of post-glacial **isostatic** readjustment it seems likely that the thickest ice occurred over Hudson Bay and attained a thickness of around 3300 m. Its volume may have been about 33 million km³. In some regions the maximum ice advance post-dated the LGM. For example, maximum southerly ice limits of the Cordilleran ice sheet in Washington State were reached thousands of years later than the limits of the easterly margins of the Laurentide ice sheet. The extent of pre-Illinoian glacier ice is less well known, but various studies have extended the glacial chronology of North America back into the late Pliocene.

The British Isles

The British Isles ice sheets merged with those of Scandinavia in some glacial phases, but also possessed large local centres of ice dispersal, including the Scottish Highlands, the Lake District, the Southern Uplands, the Pennines, the Welsh Mountains, and various mountains in Ireland. The extent of the various glaciations (**Fig. 3.8**),

notably in South Wales (Bowen 1973) and in Wessex, has been the subject of some dispute; but it is likely that the maximum extent of glaciation ran through northern parts of the Isles of Scilly (Mitchell and Orme 1967, Coque-Delhuille and Veyret 1989), the north coast of the south-west peninsula, the area just to the south of Bristol, the Oxford region, and Essex. The limits of glaciation in the Devensian are better known, and they appear to have been mostly further north than during the earlier glaciations, so that in East Anglia, for example, the ice only just touched the Norfolk coast at Hunstanton, and in the south-west ice extended close to the north coast of Devon (Hubbard et al. 2009). It is proposed that a separate, thin plateau icefield developed on the northern part of Dartmoor, and similar small ice caps might also have formed on Exmoor and Bodmin Moor (Evans et al. 2012). In previous glaciations ice reached down just to the north of London, with the Anglian Glaciation (*c.* 430 000 years ago, MIS 12) seeing the British ice sheet reach its greatest extent (Lee et al. 2011).

Details of glacial sequences in the British Isles are given in Jones and Keen (1993), and **Fig. 3.9** summarizes

Fig. 3.8 Maximum glacial limits in southern Britain. Note that the North Sea basin and English Channel were above sea level and that during each maximum the British–Irish ice sheet was confluent with the Scandinavian ice sheet in the North Sea basin (from Gibbard and Clark, Fig. 7.2 in Ehlers et al. 2011). Reprinted with permission from Elsevier

Fig. 3.9 The English Pleistocene sequence (from Bowen et al. 1986b, Table 2). The four Lp numbers represent pollen stages in the Ludham Borehole. Note the frequency of glacial events in this interpretation

	Stage Names			Glacial Episodes
LATE PLEISTOCENE	FLANDRIAN			
	DEVENSIAN	Late		Loch Lomond Glaciation
				Dimlington Glaciation
		Middle Early		
	IPSWICHIAN			
MIDDLE PLEISTOCENE	WOLSTONIAN			Glaciation in North-East England
	HOXNIAN			
	ANGLIAN			Lowestoft Glaciation/ North Sea Drift Glaciation
	CROMERIAN			
EARLY PLEISTOCENE	BEESTONIAN			Glaciation in West Midlands and North Wales
	PASTONIAN			
	PRE-PASTONIAN			Glaciation in West Midlands and North Wales
				Glaciation in West Midlands and North Wales
				Glaciation in West Midlands and North Wales
	BAVENTIAN	Lp 4		Glaciation in North Sea Region
	BRAMERTONIAN			
	ANTIAN	Lp 3		
	THURNIAN	Lp 2		
	LUDHAMIAN	Lp 1		
	PRE-LUDHAMIAN			

Pleistocene glacial chronology as it has been interpreted for England. Reviews of the British–Irish ice sheet during the Devensian and earlier glacials are provided by Bowen et al. (2002) and Gibbard and Clark (2011), and the results of recent work modelling its cycles of growth and retreat between 38 000 and 10 000 years ago are reported in Hubbard et al. (2009). More recent reconstructions for the ice sheet's maximum extent during the Devensian suggest it was larger than previously thought – covering *c.* 750 million km², becoming confluent with the Scandinavian ice sheet (Carr et al. 2006), reaching a thickness of over 1500 m in its central areas, and attaining a volume about a third of today's West Antarctic ice sheet.

Europe and Asia

In mainland Europe and Asia there were three main centres of ice, the Alpine, the Siberian, and the Scandinavian.

The Alpine glaciers reached down to altitudes of 500 m on the north, and 100 m on the south side. The ice may have been over 1500 m thick in places. There was an ice-free corridor between this ice mass and the Scandinavian ice mass to the north.

The Siberian Plains did not experience widespread glaciation until the Middle Pleistocene. At its maximum, the eastern margin of the ice sheet that stretched from Scandinavia across Russia reached about 110° E longitude. Unlike the maximum extent of North American ice, which was only slightly greater than the area covered during the Last Glacial Maximum, the spread of Eurasian ice south-eastwards across western Siberia during the LGM was much less than the maximum reached during the Middle Pleistocene. In the mountains of north-eastern and southern Siberia, the ice sheets were less developed and alpine glaciers prevailed. During the Late Pleistocene the extent of Siberian ice (Zyryanka glaciation) in Russia has been the subject of debate and controversy. One group believes that in the Last Glacial Maximum this ice sheet was coalescent with a large ice sheet and ice shelves in the Barents Sea, Kara Sea, and Arctic Ocean, and that it also merged with the Scandinavian ice sheet to the west. Others have argued that it had a much more limited extent. The competing arguments are reviewed in Dawson (1992) and Velichko and Spasskaya (2003).

The Scandinavian ice sheet (**Fig. 3.10**) at its greatest known maximum, may have been coalescent with ice spreading from the Ural Mountains of Russia, and,

in the south-west, with glaciers of British origin. It extended an unknown distance into the Atlantic off Norway, and may have merged with ice over Spitzbergen. In the south, the Elster and Saale glacial borders trend along the northern bases of the central European Highlands, and the Saale ice sheet penetrated far down the basins of the Dniepr and Don Rivers. Although the thickness of this ice sheet is not known with precision, it may well have exceeded 3000 m both over the Sognefjord region of western Norway, and at the head of the Gulf of Bothnia, though the average thickness was probably about 1900 m. The ice also extended across the present North Sea basin, which in times of maximum glaciation was largely above sea level. It has long been thought that Dogger Bank, which rises some 20 m higher than the surrounding sea-floor, is possibly the remnant of a great moraine some 250 km long, and 100 km wide (Stride 1959), and further study has confirmed that this feature represents the southern margin of confluent British and Scandinavian ice moving through the North Sea basin (Carr et al. 2006).

There is a controversy as to whether large portions of northern Eurasia, and particularly the Arctic continental shelf, were covered by an integrated ice sheet. One theory is that sea ice thickened during cold phases until it became grounded on continental shelves. Once formed these shelves might have continued to grow into domes by the addition of snow and the freezing of any meltwater draining off the land. This would be particularly the case in sheltered seas where iceberg calving was restricted, such as the Kara and Barents Seas, and in the bays and inter-island channels of the Queen Elizabeth Islands. Under this theory, sometimes called the 'maximum reconstruction' (Denton and Hughes 1981), huge ice masses would have existed over areas of the high latitudes in the Northern Hemisphere that are now covered by sea. By contrast the older and more firmly established 'minimum reconstruction' envisages the existence of several disconnected ice caps and ice sheets with continental shelves, aside from those fringing the Scandinavian ice sheet, remaining unglaciated, as were the Arctic Seas.

Further south, the mountains around the Mediterranean basin were also subject to glaciation (Hughes et al. 2006). Glaciers were present in the High Atlas of North Africa, in Lebanon and Turkey, Greece and the Balkans, Corsica, the Italian Appenines, the Pyrenees, and other mountain ranges in Spain.

The extent and sequence of glaciation in the high mountains and plateaux of central Asia are still the

Fig. 3.10 Maximum glacial limits in Europe. (from Andersen and Borns 1994, Fig. 2-3): 1. The Weichselian (MIS 2) 2. The Warthe phase of the Saalian (MIS 6) 3. The Saalian (MIS 10 and 8) 4. The Elsterian (MIS 12). B (Berlin), W (Warsaw), M (Moscow), P (Prague), x (erratics at Belchatow), xx (erratics north-west of Celle), y (erratics at Horni Rasnice, Czech Republic). Reprinted with permission of Scandinavian University Press

subject of considerable controversy. Some workers from China and elsewhere have maintained that large ice caps developed over wide areas (e.g. Kuhle 1987), while others have suggested that the extent of glaciation was much more modest (e.g. Holmes and Street-Perrott 1989). Much of the controversy has arisen because of problems of identification of **tills**, and their superficial similarity to other deposits produced by other processes (including debris flows, deep weathering, etc.). The maximalist view, as expressed by Kuhle, is that a large ice sheet enveloped the Tibetan Plateau and surroundings, extending over 2.0 to 2.4 million km², an area considerably larger than the present Greenland ice (1.8 million km²). The extent of ice in this area remains uncertain, although it is thought that cover was more extensive during the Saalian than during the LGM (Ehlers and Gibbard 2007).

The southern continents

In the southern continents it is clear that glaciation was markedly less extensive than in the Northern Hemisphere. A good review is provided by Clapperton (1990),

who suggested that there is a clear general synchroneity of events in the two hemispheres with, for example, similar dates for the LGM. However, there was an anti-phase relationship between the two hemispheres, most closely seen at mid to high latitudes, during the **Late Glacial** (the time after the LGM but before the start of the Holocene) (see **Section 5.2**). Clapperton also tentatively suggested that earlier Pleistocene glaciations in some of the southern continents (most notably in Patagonia and Tasmania) may have been more extensive than later ones.

In South America the ice sheets developed from the Andes were greatly expanded (**Fig. 3.11**), and in the far south the glaciated zone was over 200 km wide and the ice may have been over 1200 m thick. A more or less continuous zone of glaciation extended to about 30° S, though north of 38° S the ice tended not to expand very far from the Cordillera, either to the Pacific in the west or into the plainlands in the east. The Patagonian ice cap was especially extensive at about 1.2 million years ago, when it covered an area of 300 000 km² (compared with 100 000 km² in the last glaciation) (Clapperton 1990). The furthest north ice body was that which

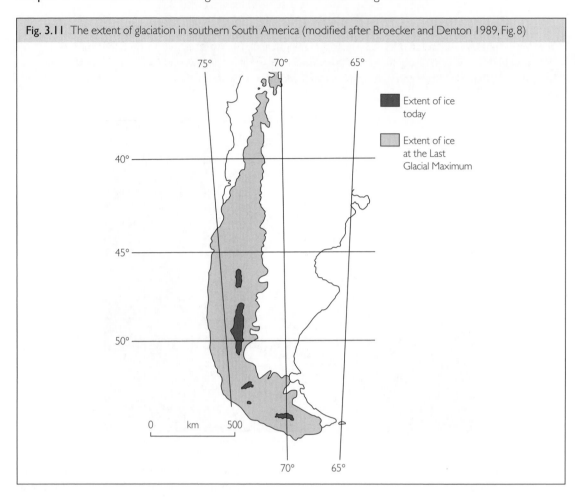

Fig. 3.11 The extent of glaciation in southern South America (modified after Broecker and Denton 1989, Fig. 8)

capped the Sierra Nevada de Santa Marta in Colombia. Other large glaciers developed in the Sierra Nevada de Merida in Venezuela (Schubert 1984).

Africa is more noted for its coral strands than for its icy mountains and under present conditions glaciation is limited. However, in the Pleistocene some of the higher areas were glaciated, though the chronology is insecure and fragmentary, and there remains controversy in some areas as to whether or not the evidence for past glaciers is reliable. The greatest development of glaciation was in the East African mountains (Mahaney 1989). The presently glaciated mountains, Kilimanjaro, Mt. Kenya, and Ruwenzori, have a total ice cover of around 10 km², but the same mountains, together with the currently unglaciated Aberdares and Elgon, had a maximum combined extent in the Pleistocene of 800 km² of glacier ice (Hastenrath 1984). These mountains probably experienced multiple phases of glaciation, and Mahaney

et al. (1991) have identified five Pleistocene and two Holocene glacial advance phases on Mt. Kenya.

Farther north there was extensive Pleistocene glaciation in Ethiopia, with ice covering 10 km² in the Simen Mountains, possibly over 600 km² in the Bale Mountains, and 140 km² on Mount Badda (Messerli et al. 1980). It is also possible that there was limited glaciation in the Saharan mountains (e.g. Tibesti), but the extent of glaciation in Southern Africa is less well known. There is evidence for the past existence of small, isolated glaciers in the Drakensberg range. For instance, Mills et al. (2009) identified moraines marking the dimensions of a glacier in eastern Lesotho of approximately 0.24 km² in area with an equilibrium line altitude of about 3100 m elevation. Geomorphic evidence of past glaciation in the region is reviewed by Lewis (2008); and, despite few radiocarbon or surface exposure dates from glacial deposits to provide absolute ages, the relatively fresh, unweathered character of those features interpreted to be glacial suggests a recent age, probably correlating with the LGM.

It was long thought that climate conditions in the Eastern and Western Cape during the LGM (or Bottelnek Stadial as known locally) were too dry to support glaciation; but, now that the evidence for past glaciers is more secure, this view is being questioned. Despite higher winter precipitation than today now postulated for the high Drakensberg, the environment was marginal for glaciation, with glaciers only forming in shady, high elevation locations where microclimate and topography were favourable for the accumulation of snow – both from direct precipitation and from a varying contribution of snow blown in from surrounding upland areas (Gellert 1991, Lewis 2008, Mills et al. 2009).

In Australia, because of the low relief and arid interior, there was limited glacier development. Former glaciation of the mainland was confined to a single zone in the Snowy Mountains with an area of barely 52 km². In Tasmania, however, glaciation was more extensive, and a large ice cap developed in the Central Plateau, where at least four glaciations have been identified (Colhoun 1988).

New Zealand, with its greater relief and oceanic conditions, has some major glaciers at the present day and in the Pleistocene the New Zealand Alps in the South Island were intensively and widely glaciated, though the North Island was largely unaffected. The great ice caps and glaciers (**Fig. 3.12**) helped to form the extensive fjords and lakes that are such a feature of the landscape of South Island (**Plate 3.1**).

The glaciation of the Antarctic is as yet very imperfectly known, though it is clear that both the margins and the thickness of the ice sheet varied

Fig. 3.12 Glacial limits in the South Island (redrawn from New Zealand Geological Survey 1973). The Otiran is the name given to the last glaciation. The other limits are shown in order of increasing age

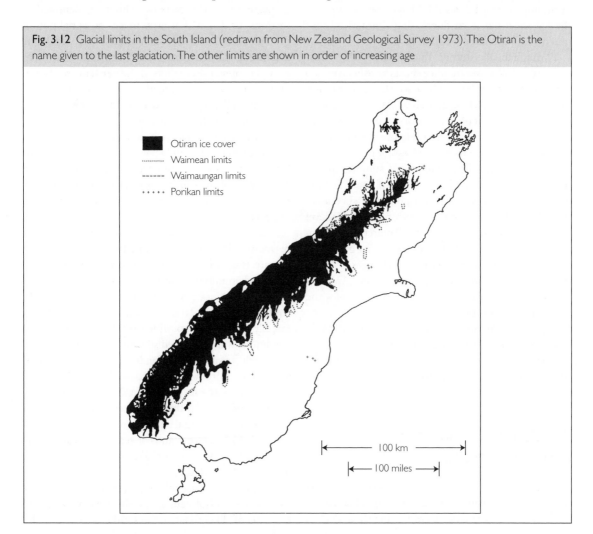

during the course of the Pleistocene and Late Tertiary. Glaciers in Antarctica may have formed as early as the Eocene (see **Section 1.4**), and oxygen isotope studies of sediments from deep-sea cores in the Southern Ocean indicate a major cooling at 34 million years ago and a further cooling at 15–13 million years ago. The East Antarctic ice sheet may first have reached something near its present size with the second of those cooling events (Barrett 1991). The extent to which it has waxed and waned since that time is the subject of considerable ongoing research. In particular, controversy surrounds the issue as to whether Antarctica experienced massive deglaciation during the globally warmer climate of the Pliocene around 3 million years ago. One group, the 'stabilists' argue that a polar ice sheet has persisted in the East Antarctic for at least 14 million years whereas the 'dynamists' argue that the ice sheet has been fluctuating dramatically for much of its existence (Sugden 1996).

Through the Pleistocene, fluctuations in the size of the East and West Antarctic ice sheets have been in step with the Northern Hemisphere ice sheets. For example, during the LGM the dates bracketing the episode of maximum extent of Antarctic ice agree closely with those for maximum Northern Hemisphere ice cover (Weber et al. 2011), which is interesting given that the orbital effect controlling amounts of summer insolation in the two hemispheres is anti-phase. Changes in the size of Northern Hemisphere ice sheets are thought to affect Antarctic ice both through sea level (e.g. a lowering causes areas of the Antarctic continental shelf to emerge above sea level enabling grounded ice to advance further), and through changes in **Atlantic Meridional Overturning Circulation (AMOC)**. AMOC (described further in **Section 3.6**) varies the heat flux between the two hemispheres. During the LGM, maximum cold conditions in the north were linked with *relatively* warmer conditions in the south, probably allowing more snowfall to occur over Antarctica contributing to ice advance there.

3.4 Permafrost and its extent in the Pleistocene

Beyond the limits of the great Pleistocene ice sheets there were, particularly in Europe, great areas of open tundra. These areas were frequently underlain

by **permafrost**. Permafrost is a frozen condition in soil, alluvium, or rock, and is currently concentrated in high northern latitudes, reaching thicknesses of as much as 1000 m.

The current southern boundary of continuous permafrost coincides, approximately, with the −5 or −6 °C mean annual isotherm. The limits of discontinuous and sporadic permafrost are rather higher, but mean annual air temperatures have to be negative. Thus in Europe continuous permafrost is restricted to Novaya Zemlya, and the northern parts of Siberia, whilst discontinuous permafrost extends into northern Lapland. However, there is very strong evidence for the former extension of such permanently frozen subsoil conditions to wide areas of Europe during cold phases. The evidence consists of the casts of ice wedges which form polygonal patterns in areas of permafrost. The casts can either be identified in sections, or detected as crop marks on aerial photographs. These have been encountered very widely, for instance in southern and eastern England. Elsewhere are to be found other relics of the permafrost, including the remains of miscellaneous mound features ('palsas' and 'pingos') and ground ice depressions. Ballantyne and Harris (1994) provide a full analysis of the effects of **periglacial** and permafrost conditions in Great Britain. The only part of mainland Britain to have been unaffected by permafrost is probably the extreme tip of the south-western peninsula (Williams 1975). The distribution of such features is shown in **Fig. 3.13**.

With regard to the former southern extension of permafrost in mainland Europe there is considerable dispute (**Fig. 3.14**), though even the most northerly proposed boundary indicates that only the central and southern Balkans, peninsular Italy, Iberia, and southwest France were largely unaffected. This indicates forcibly the degree to which tundra and periglacial conditions were displaced southwards, and the extent to which temperature conditions were depressed in much of Europe. On the basis of a −5 °C limit for permafrost it seems likely that in eastern England, where conditions in the Pleistocene may have been made more continental by the drying out of the North Sea during glacial lowstands of sea level, mean annual temperatures were depressed by 15 °C (or more) during the Last Glacial.

In North America relatively less is known about the southern displacement of the permafrost zone; but it seems likely that as the southern extent of the

Fig. 3.13 The distribution of Late Devensian periglacial soil structures in the British Isles (after Rose et al. Fig. 18.2, in Richards et al. 1985). Reprinted with permission of Allen & Unwin

Wisconsin ice sheets was further south than that of the Würm–Weichselian in Europe, the zone of more severe periglacial conditions was probably more restricted and was mostly confined to a narrow belt 80–200 km south of the Laurentide ice sheet. None the less, Johnson (1990), on the basis of the analysis of patterned ground, suggests that in the late Wisconsin period permafrost extended as far south as 38° 30′ N.

Fig. 3.14 Various dashed lines indicate different reconstructions of the permafrost distribution in Europe during the LGM. Numbers are the inferred mean annual temperature depressions in °C (from French 1996, Fig. 13.5, with modifications derived from Bell and Walker 2005, Fig. 4.5)

There was also widespread growth of alpine permafrost in parts of Idaho, North Dakota, Wyoming, and Colorado.

Although the presence of permafrost suggests that mean annual temperatures were at least as low as those now experienced in tundra areas, it is probable that the periglacial climates of glacial Europe and America were different in character from any now found on Earth. Because of latitude, especially in America, days were longer in winter and shorter in summer than in any high-latitude periglacial area at present. Also, the Sun would have risen higher in the sky, giving both higher midday temperatures, and more marked diurnal changes. Evaporation rates would have been higher. Wind action may have been of greater importance than it is in high latitudes today because of the intense high pressure systems that developed over the great Pleistocene ice sheets and the compression and concentration of the major climatic zones. The nature of these differences has been well reviewed by French (1996).

3.5 Loess formation, distribution, and accumulation rates

Loess has been the subject of an enormous literature, ever since Lyell (1834) drew attention to the **loamy** deposits of the Rhine Valley. Many theories have been advanced to explain its formation. It was, however, von Richthofen (1882, pp. 297–298) who cogently argued that these intriguing deposits probably had an aeolian origin and that they were produced by **dust storms** transporting silts from deserts and depositing them on desert margins.

Loess is largely non-stratified and non-consolidated silt, containing some clay, sand, and carbonate (Smalley and Vita-Finzi 1968). It is markedly finer than aeolian sand. It consists chiefly of quartz, feldspar, mica, clay minerals, and carbonate grains in varying proportions. Many parts of the world possess long sequences of loess and palaeosols (Rutter et al. 2003), and these provide a major source of palaeoenvironmental information that can be correlated with that obtained from ocean cores (**see** Sections 2.2 and 3.2).

Loess distribution

Quaternary loess accumulations cover as much as 10% of the Earth's land surface (Muhs et al. 2004) (**Fig. 3.15**). Over vast areas (at least $1.6 \times 10^6 \, km^2$ in North America and $1.8 \times 10^6 \, km^2$ in Europe) these blanket pre-existing relief, and in Tajikistan have been recorded as reaching a thickness of up to 200 m (Frechen and Dodonov

1998). In the Missouri Valley of Kansas the loess may be 30 m thick, in European Russia sustained thicknesses, often 10 to 30 m, reach over 100 m in places, while in New Zealand, on the plains of the South Island, thicknesses reach 18 m. Loess profiles thicker than 50 m are known from the Pampas of Argentina (Kröhling 2003).

Loess is known from some high-latitude regions, including Greenland, Alaska (Muhs et al. 2004), Spitzbergen, Siberia (Chlachula 2003), and Antarctica (Seppälä 2004). Loess has also been recorded from various desert regions. In Arabia, Australia, and Africa, where glaciation was relatively slight, loess is much less well developed, though an increasing number of deposits in these regions is now becoming evident. Of all the world's loess deposits, those of China are undoubtedly the most impressive for their extent and thickness, which near Lanzhou is 300–500 m.

The main areas of loess in the USA include southern Idaho, eastern Washington, north-eastern Oregon, and even more importantly, a great belt from the Rocky Mountains across the Great Plains and the Central Lowland into western Pennsylvania. Loess is less prominent in the eastern USA as relief and climatic conditions for deflation, and the nature of outwash materials, seem to have been less favourable than in the Missouri–Mississippi region. There are at least four middle-to-late Quaternary loess units

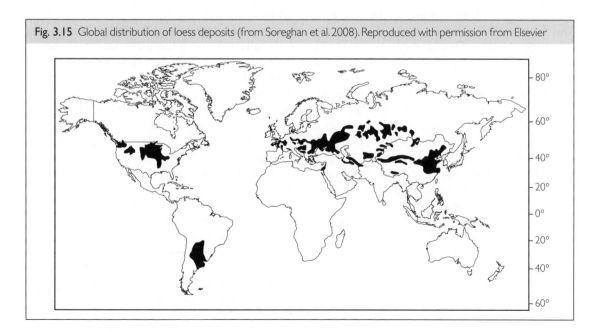

Fig. 3.15 Global distribution of loess deposits (from Soreghan et al. 2008). Reproduced with permission from Elsevier

in the High Plains, which from oldest to youngest are the Loveland Loess (Illinoian glacial), the Gilman Canyon Formation (mid to late Wisconsinian), the Peoria Loess (late Wisconsinian) and the Bignell Loess (Holocene) (Pye et al. 1995; Muhs et al. 1999). The loess deposits of the USA have been reviewed by Bettis et al. (2003), who suggest that the Last Glacial (Peoria) loess is probably the thickest LGM loess deposit in the world, being more than 48 m thick in parts of Nebraska, and 41 m thick in western Iowa. Some of the Peoria loess, including that in Nebraska, may not be glaciogenic, having been transported by westerly to northerly winds from parts of the Great Plains not directly influenced by the Laurentide ice sheet or alpine glaciers (Mason 2001). However, this has been a matter of some controversy, for Winspear and Pye (1995) favoured a more glacial explanation for the Peoria Loess in Nebraska. Some of the loess in the Great Plains (the Bignell Loess) is of Holocene age (Mason and Kuzila 2000; Mason et al. 2003; Jacobs and Mason 2005). Miao et al. (2005) believe that much of the Holocene loess, most of which dates from approximately 10 000 to 6500 years ago, was produced in dry phases as a result of the winnowing of dune fields.

In South America, where the Pampas of Argentina and Uruguay has thick deposits, a combination of semi-arid and arid conditions in the Andes rain shadow, combined with glacial outwash from those mountains, created near-ideal conditions (Zarate 2003). The Argentinian loess region is the most extensive one in the Southern Hemisphere, covering 1.1×10^6 km^2 between 20° S and 40° S. Zinck and Sayago (2001) have described a 42 m thick loess-palaeosol sequence of Late Pleistocene age from north-west Argentina, though generally thicknesses are less than this.

New Zealand has the other major loess deposits of the Southern Hemisphere. They cover extensive areas, especially in eastern South Island and southern North Island (Eden and Hammond 2003). The loess has been derived mainly from dust deflated by westerly winds from the many broad, braided river floodplains. Some of the loess may have been derived from the continental shelf at times of low sea levels. Parts of the New Zealand loess is of considerable antiquity, and in the Wanganui region of North Island there is a 500 kyr record of 11 loess layers and associated palaeosols (Palmer and Pillans 1996). On South Island the Roma-hapa loess/palaeosol sequence is at least 350 kyr old (Berger et al. 2002).

In Europe the loess is most extensive in the east, where, as in the case of America, there were plains and steppe conditions. The German loess shows a close association with outwash, and in France the same situation is observed along the Rhône and Garonne. These rivers carried outwash from glaciers in the Alps and Pyrenees, respectively. The Danube was another major source of silt for loess in eastern Europe. Britain has relatively little loess. Indeed, in Britain wind-lain sediments of periglacial times are conspicuous only for their rarity and 'loess is more of a contaminant of other deposits than one in its own right' (Williams 1975). The maximum depth of loess in Britain is only about two or three metres. In southern Europe, Late Pleistocene loess, up to 10 m thick, occurs in the Granada Basin of south-east Spain (Günster et al. 2001).

Loess is probably more widespread in South Asia than has often been realized. Given the size of the Thar Desert and the large amounts of sediment that are transported to huge alluvial plains by rivers draining from the mountains of High Asia, this is scarcely surprising. In Pakistan there are loess deposits in the Potwar Plateau (Rendell 1989) and Kashmir (Dilli and Pant 1994), while in India loess has been identified from the Delhi Ridge of Rajasthan (Jayant et al. 1999), various tributary valleys of the Ganges plain (Williams and Clarke 1995), and the central Himalayas (Pant et al. 2005). It has also been found in Gujarat (Malik et al. 1999).

One of the most striking features of Central Asia, and one it shares with China (Bronger et al. 1998), is the development of very thick (some more than 200 m thick) and complex loess deposits dating back to the Pliocene (Ding et al. 2002). They are well displayed in both the Tajik (Mestdagh et al. 1999) and Uzbek Republics (Zhou et al. 1995), where rates of deposition were very high in Late Pleistocene times (Lazarenko 1984). The nature of the soils and pollen grains preserved in the loess profiles suggest a progressive trend towards greater aridity through the Quaternary, and this may be related to progressive uplift of the Ghissar and Tien Shan Mountains (see Davis et al. 1980). A thermoluminescence (TL) chronology for the loess deposits of Tajikistan was provided by Frechen and Dodonov (1998), though some of the earlier TL dates for the deposits are believed to be unreliable (Dodonov and Baiguzina 1995; Zhou et al. 1995). Nonetheless, as in China, the loess profiles contain a large number of palaeosols that formed during periods of relatively moist and warm climate. Rates of loess deposition have been very modest in the Holocene, whereas in the Last Glacial rates of accumulation were as high as 1.20 m per 1000 years (Frechen and Dodonov 1998). Ding et al. (2002) believe that the alternations of loess and soil horizons in Central Asia can be well correlated with the Chinese loess and deep-sea isotope records.

Loess in China and east Asia

Loess reaches its supreme development in China, most notably in the Loess Plateau, a 450 000 km² area in the middle reaches of the Yellow River. North-west of Lanzhou, the loess attains a thickness of 334 m, while in Gansu Province, a thickness of 505 m has been reported (Huang et al. 2000), but over most of the plateau 150 m is more typical. The classic study is that of Liu (1988). Loess deposits occur in locations other than the Loess Plateau, including the mountainous regions (Rost 1997; Lehmkuhl 1997; Sun 2002), the Tibetan Plateau (Lehmkuhl et al. 2000), parts of northern Mongolia (Feng 2001) and Korea (Yatagai et al. 2002).

In some areas loess *sensu stricto* overlies the Pliocene Red Clay Formation (PRCF) which is also in part a product of aeolian dust accumulation (Liu et al. 2003; Yang and Ding 2004). The boundary between the loess and the PRCF has been palaeomagnetically dated at 2.5 million years ago. The abrupt commencement of loess deposition on a large scale at that time implies a major change in atmospheric conditions and the ongoing uplift of the Tibetan Plateau may have contributed to this (Ding et al. 1992). The appearance of loess beds alternating with numerous palaeosols indicates a cyclical climatic regime, with dry cold conditions being dominated by the north-westerly **monsoon** and humid warm conditions being dominated by the south-easterly monsoon. This contrasts with the more continuous warm climate that prevailed in the preceding 3 million years during the Pliocene. It appears that the accumulation of dust then accelerated rapidly from about 1.2 million years ago and that the front of loess deposition was pushed 600 km further southeastwards from 0.6 million years ago (Huang et al. 2000). At the Jiaxian section (Qiang et al. 2001) rates of sedimentation were about 6 m per million years between 5 and 3.5 million years ago, rising to 16 m per million years between 3.5 and 2.58 million years ago, and reaching 20–30 m per million years thereafter.

Immediately above the PRCF is the Wucheng Loess. Above that in turn are the Lower Lishi Loess,

the Upper Lishi Loess, and the youngest unit, the Malan Loess (late Pleistocene). There may also have been some relatively limited Holocene loess deposition, but average rates of loess accumulation in the Loess Plateau were higher, possibly by a factor of two, in the later part of the Last Glacial period than during the Holocene (Pye and Zhou 1989). The Last Glacial appears to have been a time when soil moisture contents were low, dunes became destabilized, and the desert margin shifted southwards towards the Loess Plateau (Rokosh et al. 2003).

The loess units contain large numbers of palaeosols with as many as 32 soils present above the PRCF. Differences in the nature of these soils and of the loess in between have been used to establish the history of climate over the last 2.5 million years (Liu and Ding 1998). The loess can furnish a high resolution record of change so that sub-millennial scale variations have been picked up (Heslop et al. 1999). Porter (2001) has argued that high-frequency fluctuations in dust influx during the period of Malan dust deposition may be correlated with North Atlantic **Heinrich events** (see **Section 3.6**). At longer timescales various periodicities have been identified in Chinese loess-palaeosol sequences, associated with orbital fluctuations, including 100 kyr and 400 kyr cycles (Lu et al. 2004).

In general terms, periods of loess deposition are associated with cold phases (which by implication were dry) while the palaeosols are associated with warmer phases (An et al. 1990; Sartori et al. 2005), indicating their origin as products of deflation and subsequent transport and deposition by dust storms. During the last glacial cycle, it was westerly and north-westerly winds that were the most important agents for the transport of dust to the Loess Plateau (Lu and Sun 2000). A comparison of the magnetic signatures of the loess with sands from the Taklamakan, suggest that some of the loess was derived from that source region (Torii et al. 2001), while the presence of calcareous nannofossils in the Malan Loess suggests transport by westerly winds from the Tarim basin (Zhong et al. 2003).

The dust that formed the Chinese loess appears to have been trapped downwind by an *Artemisia*-dominated grassland vegetation through the last 130–170 thousand years (Jiang and Ding 2005; Zhang et al. 2006), with but sparse evidence that over the same period there was a widespread forest cover (Liu et al. 2005). Unexpectedly, C4 plant abundance declined during glacials, but increased during palaeosol formation in interglacials (Vidic and Montañez 2004).

Loess accumulation rates

The data in **Table 3.2** show a range of values between 22 and 4000 mm per 1000 years. Pye (1987, p.265) believes that at the maximum of the last glaciation loess was probably accumulating at a rate of between 500 and 3000 mm per 1000 years, and suggests that 'Dust-blowing on this scale was possibly unparalleled in previous Earth History'. By contrast, he suggests that 'During the Holocene, dust deposition rates in most parts of the world have been too low for significant thicknesses of loess to accumulate, although aeolian additions to soils and ocean sediments have been significant'. Pye also hypothesizes that rates of loess accumulation showed a tendency to increase during the course of the Quaternary. Average loess accumulation rates in China, Central Asia and Europe were of the order of 20–60 mm per 1000 years during Matuyama time (early Pleistocene), and of the order of 90–260 mm per 1000 years during the Brunhes epoch (post 0.78 million years ago). He also points out that these long-term average rates disguise the fact that rates of loess deposition were one to two orders of magnitude higher during Pleistocene cold phases, and one or two orders of magnitude lower during the warmer interglacial phases when pedogenesis predominated.

An analysis of loess accumulation rates in China is provided by Kohfeld and Harrison (2003). They indicate that in the glacial phases (e.g. MIS 2) aeolian mass accumulation rates were c. 310 g m^{-2} yr^{-1} compared to 65 g m^{-2} yr^{-1} for an interglacial stage (e.g. MIS 5) – a 4.8× increase. A comparable exercise was carried out for Europe by Frechen et al. (2003). They found large regional differences in accumulation rates, but suggested that along the Rhine and in Eastern Europe, rates were from 800 to 3200 g m^{-2} yr^{-1} in MIS 2. Loess accumulation rates over much of the USA during the LGM were also high, being around 3000 g m^{-2} yr^{-1} for mid-continental North America (Bettis et al. 2003). From 18 000 to 14 000 years ago, rates of accumulation in Nebraska were remarkable, ranging from 11 500 g m^{-2} yr^{-1} to 3500 g m^{-2} yr^{-1} (Roberts et al. 2003).

3.6 The nature of glacial/interglacial cycles

The degree of climatic change in glacials

Although the presence of greatly expanded ice sheets and permafrost conditions gives a broad indication of the extent to which temperatures changed during glacial intervals, it is possible, through a variety of techniques, to gain some more precise and quantitative measures of the degree of change that has occurred. Using various assumptions of glacial boundary conditions derived from palaeo-data (see **Section 1.6**) the Paleoclimate Modeling Intercomparison Project II (PMIP2) generated a range of simulations of average mean annual global cooling during the LGM of between 3.6 and 5.7 °C compared with the pre-industrial average (Braconnot et al. 2007). Ruddiman (2006) suggested a mean global temperature depression of c. 4.5 °C. A significant drying of global climate was also simulated. However, global averages hide a great deal of regional variation: in general, the high and especially the mid latitudes experienced much greater temperature depressions than the lower latitudes.

Empirical evidence of glacial/interglacial temperature changes has been obtained through five main sources: isotopic measurements, the levels of **cirques**, the extent of permafrost, the limits of frost-affected sediments, and the nature of floral and faunal remains. These methods are all subject to certain difficulties and pitfalls in that temperature may be only one of the controls which influence, say, the position of the tree- and snowlines. Similarly, the interpretation of the palaeoclimatic significance of snowline levels, represented by cirque floor heights, depends very much on the estimation of probable local lapse rates.

Table 3.2 Loess accumulation rates for the Late Pleistocene

Location	Rate (mm per 1000 years)
Negev (Israel)	70 – 150
Mississippi Valley (USA)	700 – 4000
Uzbekistan	50 – 450
Tajikistan	60 – 290
Lanzhou (China)	250 – 260
Luochaun (China)	50 – 70
Czech Republic and Slovakia	90
Austria	22
Poland	750
New Zealand	2000

(From various sources in Pye 1987, Table 9.6, and Gerson and Amit 1987.)

Lapse rates are the mean rates at which temperatures change with altitude (generally 0.6 °C/100 m), but they are subject to local fluctuations.

The isotopic methods have proved fruitful, particularly with respect to the $\delta^{18}O$ and δD measurements on ice core records (**Section 2.4**). In central Greenland the $\delta^{18}O$ data indicate that air temperature during the Last Glacial Maximum was 15 °C colder or more than at present (between −50 and −45 °C compared with today's −30 °C) (Johnsen et al. 2001; Clark et al. 2012). Modelling studies indicate over 30 °C of surface cooling compared with the present over the former Laurentide and Scandinavian ice sheets (Braconnot et al. 2007). Reconstructions from Antarctica, such as those from the Vostok and Byrd ice cores, indicate that glacial/interglacial surface air temperature differences were less extreme than in the Arctic – the amplitude being between 7 and 15 °C (Blunier et al. 1998).

The downward movement of snowlines in glacials indicated lowering of temperatures, especially summer temperatures, though it has to be remembered that precipitation and cloudiness could also affect the level, as does the local lapse rate (**Section 2.2**). A knowledge of local lapse rates is required to relate the altitudinal shift of the snowline to temperature change. The position of the Pleistocene snowline is also subject to some error in its assessment, in that it is determined by a study of the position of cirque floors. These tend to cluster at or just below the 0 °C summer isotherm. In general the cirque floor measurement is only valid in areas where the former glaciers never grew beyond the corrie type. Values that have been determined by this method suggest a mean temperature depression during glacials of c. 5 °C. The varying degree of snowline depression from region to region gives a spread in temperature depression values of from only about 2 °C to over 10 °C. This reflects the fact that snowline depression values ranged from as little as 600–700 m in the northern Urals, the Middle Atlas, and the Caucasus, to as much as 1300–1500 m in the northern Pyrenees, Kilimanjaro, the Apennines, and the Tell Atlas.

The former extent of permafrost has already been discussed in relation to Europe and North America. In that the current boundary of permafrost in Siberia, Scandinavia, and North America can be related to mean annual temperatures, it is possible to infer Pleistocene temperatures from this source. The permafrost data tend to give somewhat higher values for the amount of temperature depression than do the snowline data, with values of around 15 °C being recorded

for the Midlands of England and for East Anglia, values of over 15 °C throughout much of France and Germany (**Fig. 3.14**) and values of 10–15 °C for parts of central North America.

In many parts of the world, where conditions are now both too warm and too dry for frost activity to be important in rock disintegration, there are screes of angular debris, which have been widely interpreted as being the product of frost weathering. These have, for example, been described from Libya (Hey 1963). Periglacial deposits of this type have been used to suggest a 11 °C depression of glacial temperatures in the south-western USA (Galloway 1970), a greater than 9 °C depression in south-eastern Australia (Barrows et al. 2004), and a depression of at least 6 °C in the Cape Province of South Africa (Lewis 2008).

The data provided by organisms and plants can be difficult to interpret because there are a variety of factors that influence their distribution and abundance besides temperature. However, the analysis of specific **indicator species**, and/or the application of 'transfer functions' to fossil biotic assemblages (**Section 2.2**), has enabled Pleistocene temperature reconstructions from land and sea in many parts of the world. Many species of beetle, for example, have narrow temperature ranges; and the presence of certain species in dated terrestrial deposits therefore enables fairly precise temperature reconstructions. For instance, beetle evidence averaged across the British Isles suggests mean annual temperatures of around −10 °C during the LGM, with mean July temperatures of only 10 °C (Atkinson et al. 1987). Beetle-derived temperature reconstructions are similar elsewhere in northern Europe; but with LGM mean July temperatures of less than 10 °C in places further from the Atlantic such as southern Sweden and central Poland (Coope and Lemdahl 1995). In Beringia (an area incorporating Alaska and the presently submerged Bering Land Bridge) beetle assemblages show that summer temperatures were 2 to 4 °C cooler during the LGM and winter temperatures about 8 °C cooler (Elias 2000).

Beetle remains have also been used alongside pollen data for reconstructing the temperature of the Last Glacial. This has been put to good use on the long pollen sequence from La Grande Pile, eastern France, where the mean annual temperature was just under 0 °C during the LGM – a temperature depression of over 15 °C in relation to today (Guiot et al. 1993). Similarly, evidence from the southern Appalachians of south-east USA indicates a mean annual temperature of around 0 °C during the LGM (Brook and Nickmann 1996).

The mean July temperature of the LGM at La Grande Pile was about 10 °C, only slightly warmer than mean July temperatures across northern Europe. Such studies tie in well with simulations of the glacial climate across Europe. Compared with the present, climate modelling suggests that LGM winter temperatures were cooler by 10–20 °C in northern Europe, by 7–10 °C in parts of southern and central Europe, and by 2–4 °C in Spain and near the Black Sea (Barron et al. 2003).

Foraminifera from the glacial segments of deep-sea cores, when compared with those of the present in the same locations indicate significant sea surface temperature (SST) depressions during glacials. The CLIMAP Project (1981) used an array of cores collected throughout the world's oceans to estimate SSTs for the LGM. This work showed a severe SST depression, on the order of 10 to 14 °C colder, relative to the present in the mid- to high-latitude North Atlantic Ocean. During the LGM the polar front (the boundary between cold, polar water and warmer water) in the north-east North Atlantic was at the latitude of Spain; whereas today it is at the latitude of Iceland (see **Fig. 5.3**). CLIMAP suggested that on a world basis the average anomaly between present and glacial surface-water temperatures was on the order of 2.3 °C, with most of the cooling at higher latitudes and relatively little cooling at low latitudes. The original CLIMAP data and SST reconstructions can be viewed and compared with more recent reconstructions from the National Climatic Data Center website (http://www.ncdc.noaa.gov/cgi-bin/paleo/hsstplot.pl).

Since the CLIMAP data were published, it has been found that CLIMAP underestimated SST cooling in the tropics and subtropics (Mix et al. 2001). For example, analysis of corals shows a glacial SST depression of 5 °C or more in the tropical South Pacific (Beck et al. 1997), and analysis of marine sediments in the Indian Ocean suggests SST depression of between 3 and 4 °C (Bard et al. 1997; Barrows and Juggins 2005). In the subtropical North Atlantic average SSTs appear to exhibit a range of at least 8 °C between glacials and interglacials (Lehman et al. 2002). Averaged across the world's tropical oceans, Ballantyne et al. (2005) have estimated SST cooling of 2.7 ± 0.5 °C during the LGM compared with the present. From their study of subtropical Indian Ocean SSTs over multiple glacial cycles, Bard and Rickaby (2009) show a cooling of 4 °C during the LGM, and an even greater cooling of 6 °C compared with the present during MIS 10 and 12.

The fall in the temperature of the oceans strongly affected the distribution of marine life. This can be illustrated from a study of coral reefs (**Fig. 3.16**). The present effective limit of reef growth is approximately that of the 20 °C ocean water isotherm (Stoddart 1973). By subtracting the glacial falls in temperature for each major ocean derived from miscellaneous palaeo-temperature observations a map of probable Pleistocene coral reef growth can be constructed. It shows the considerable degree of contraction which must have taken place in reef distribution. Over large areas reef corals would have died because of the relatively cool conditions. This effect would have been heightened further by the lowstands of sea level during glacial phases.

Various climate model simulations (e.g. COHMAP 1988; Webb et al. 1997) have suggested that averaged surface temperature over land areas and seas not covered by ice at mid to high latitudes may have been about 8 °C cooler during the LGM relative to the present, with cooling in the tropics averaging around 5 °C (Ballantyne et al. 2005). Glacial cooling in some high-latitude locations, e.g. the ice core evidence from Greenland already referred to, was much more severe.

The calculation of former precipitation levels is even more beset with difficulties than the calculation of temperature changes, in that most of the methods attempt in fact to measure not precipitation, but evaporation/precipitation ratios. They are thus partially dependent on temperature estimation. Decreases in temperature of the type discussed above would in many areas be sufficient by themselves to account for certain **pluvial** (see **Section 1.2**) or **lacustral** phenomena which have been interpreted in the past as being the result of increased precipitation. Phenomena that can be used to assess changed precipitation/evaporation ratios include the volumes and stratigraphy of lakes, the nature of cave fillings, the distribution of dune fields, the characteristics of fossil soils (palaeosols), and the nature of former stream regimes as deduced from sedimentological and morphological evidence. It is not easy to obtain any quantitative data from these sources though various attempts have been made.

In general, the world climate was drier, windier, and dustier during glacials compared with interglacials. However, major shifts in the pattern of

Fig. 3.16 Possible extent of the contraction of the coral-reef seas during the Pleistocene. The isotherm of 20 °C is taken as the effective limit of reef formation, and the map is constructed by subtracting the glacial falls in temperature for each major ocean derived from published palaeotemperature analysis from the present-day sea surface temperatures of the coldest month (from Stoddart 1973, Fig. 3)

Pleistocene reef growth (>20 °C)

Marginal belt
(present reef growth >20 °C)

0   km   4000

atmospheric circulation, such as weakening of subtropical monsoons and southward migration of mid-latitude storm tracks in the Northern Hemisphere, meant that during the height of glacials the global distribution of precipitation was very different from today. While some regions dried out, in other places lakes had much higher volumes at various stages of the Pleistocene, and examples from low latitudes are discussed in **Chapter 4**.

In America, using snowline-derived temperature data and related measures, geologists and hydrologists have appraised the water budgets of pluvial lakes in the Basin and Range Province during their maximum Late Pleistocene levels. Estimates of the increase in mean annual precipitation (from present average values over the drainage areas, compared with those during the lake maximum) ranged from 180 to 230 mm, and for the decrease in mean annual temperature, from 2.7 to 5.0 °C. Thus, for example, at Spring Valley in Nevada, Snyder and Langbein (1962) proposed a pluvial rainfall of 510 mm compared with about 300 mm at the present. It needs, however, to be stressed that these estimates are based on low temperature depression values. By contrast, on the basis of temperatures implied from

periglacial features, Galloway (1970) proposed that in the south-west United States temperatures were depressed by 11 °C. He calculated on this premise that far from precipitation levels being higher during the pluvial phases, precipitation levels were only about 80 to 90 per cent of current levels. On the other hand a general survey of the evidence, notably in Australia and the United States led Dury (1967) to propose that shrunken lake levels and misfit streams (streams too small for their valleys) indicate an increase in mean annual precipitation during pluvials of the order of 1.5 to 2 over present totals, even when allowance is made for temperature reductions. Moreover, climate modelling by COHMAP (1988) suggests that the presence of large ice sheets over North America caused the westerly jet stream to split, creating a southern branch that tracked precipitation bearing depressions across areas of the American west and south-west that are currently semi-arid.

In some parts of the Southern Hemisphere, such as in South America and southern Africa, there is also evidence for an equatorward shift in mid-latitude westerly winds, which brought more precipitation to areas that are now relatively dry. For example, the formation of small glaciers in the high Drakensberg of southern Africa suggests formerly higher winter precipitation (as well as colder temperatures) during the LGM than exists in the region today (Mills et al. 2009). Chase and Meadows (2007) review the evidence for higher rainfall during the LGM in presently dry areas of south-west Africa.

Rapid climatic change events (RCCEs)

The previous discussion highlights remarkable differences between the climate as it was at the height of glacials and the interglacial climate of today. What is even more remarkable is that the transitions between glacial and interglacial-type climates (and transitions between stadials and interstadials) often occurred extremely rapidly – over decades to centuries rather than over thousands of years as previously thought. From $\delta^{18}O$ studies of deep-sea sediments in the 1970s, it became clear that glacial/interglacial cycles were indeed paced by cyclic variations in the Earth's orbit: a mechanism first proposed by James Croll in the nineteenth century and refined and computed by Milutin Milankovitch in the early twentieth century. This seemed to lead logically to the conclusion that the large amplitude climatic changes associated with glacial/interglacial transitions also occurred over timescales similar to orbital variation and in a gradual manner. It was a great surprise, therefore, when analyses of Greenland ice cores (which have a much higher temporal resolution than marine sediments) showed evidence for abrupt, high magnitude temperature changes both during the last glacial cycle and at the Pleistocene/Holocene transition. The discovery of these rapid climatic change events (RCCEs) fundamentally altered earlier views about climatic and environmental change during the Quaternary. It is now believed that the climate of glacial periods (and to a lesser extent interglacials) is far less stable than used to be assumed, and that various environmental feedback processes must be implicated in climatic change to amplify and speed up the changes that are initiated by variations in orbital parameters. The precise ways in which environmental processes interact to produce 'non-linear' and 'threshold' type climatic changes (see **Section 1.6**) is an area of intensive research with as much relevance for predicting future climate as for understanding the changes of the past. The study of RCCEs has thrown much light on such processes; but before considering how these processes work it is necessary to survey the evidence.

The 'sawtooth' pattern evident in $\delta^{18}O$ data from deep-sea sediments suggests that glacial terminations occurred more rapidly than glacial onsets; but the rapidity of warming was not fully appreciated until the 1990s, by which time enough ice cores had been analysed to be confident that polar temperatures increased extremely quickly at the onset of the Holocene. For example, from $\delta^{18}O$ analysis of the Dye 3 ice core from Greenland, Dansgaard et al. (1989) suggested that the local mean annual temperature rose by at least 7 °C in less than 50 years. The evidence for RCCEs seen in earlier Greenland ice cores was corroborated and extended by the European GRIP (Greenland Ice Core Project) and North American GISP2 (Greenland Ice Sheet Project 2) cores which were completed in the early 1990s. From the GISP2 core Alley et al. (1993) showed that the rate of ice accumulation doubled in just three years at the onset of the Holocene, representing an incredibly rapid climatic shift over Greenland. The rapidity (within years) and high magnitude of climate shifts spanning the last Termination are also shown in changes in deuterium excess data and other proxies in the North Greenland Ice Core Project (NGRIP) core (Steffensen et al. 2008).

In addition to showing rapid warming at the Pleistocene/Holocene transition, the GRIP, GISP2, and NGRIP ice cores provide evidence for rapid and severe warming and cooling events throughout the last glacial cycle. These abrupt temperature oscillations are named **Dansgaard–Oeschger cycles** (D–O cycles) after their discoverers; and they contrast starkly with the gradual changes in high latitude insolation receipt as illustrated in **Fig. 3.17**. D–O cycles are the Greenland expression of stadial and interstadial fluctuations spanning the Last Glacial: there were at least 24 of these cycles, and they represent mean annual air temperature shifts of around 15 °C occurring on, at most, a centennial timescale (Johnsen et al. 2001). Besides Greenland, the whole of the North Atlantic region experienced these rapid and high magnitude temperature oscillations, as shown in various other proxies such as beetle evidence from the British Isles. The most recent D–O cycle was the oscillation from warmer conditions during the Allerød interstadial to colder conditions during the **Younger Dryas** stadial. The nature and timing of this major climatic oscillation, which occurred during the Late Glacial, is discussed in depth in **Chapter 5**.

The time period preceding the glacial D–O cycles in the Greenland ice cores represents the interglacial

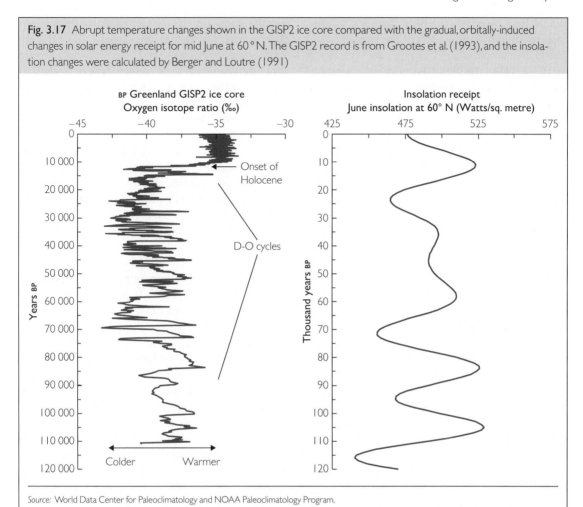

Fig. 3.17 Abrupt temperature changes shown in the GISP2 ice core compared with the gradual, orbitally-induced changes in solar energy receipt for mid June at 60° N. The GISP2 record is from Grootes et al. (1993), and the insolation changes were calculated by Berger and Loutre (1991)

Source: World Data Center for Paleoclimatology and NOAA Paleoclimatology Program.

of MIS 5e (the Eemian of Europe). In this portion of the record the δ¹⁸O data from the GRIP and GISP2 cores diverge, with the former core showing dramatic fluctuations in contrast with the latter. The NGRIP core already referred to was begun in 1995, partly to resolve this discrepancy and to determine whether or not rapid temperature changes were also a characteristic of the last interglacial (Dahl-Jensen et al. 2002). NGRIP was completed in 2004. It does not extend through the entire Eemian; but what record there is provides support for a less variable Eemian, and it appears that the fluctuations seen in the GRIP core are an artefact of processes acting at depth within the ice sheet. Interestingly, the ice cores concur in showing that Greenland temperature during the Eemian was 3–5 °C warmer than the present day. (And the Greenland ice sheet itself was considerably smaller than now.) The evidence from NGRIP has helped to refine

understanding of RCCEs during the Pleistocene while also contributing further insights into the nature of Holocene climatic change. The more recent NEEM (North Greenland Eemian Ice Drilling) project will provide further insights into the nature of Eemian climate and fluctuations over the last glacial cycle.

An important question concerns the extent to which the abrupt climatic changes inferred from the Greenland ice cores were manifested globally. This is not an easy question to answer because **stratigraphic sequences** with high enough resolution and good enough chronological control for comparison with D–O cycles are relatively rare. Nonetheless, there is now enough **proxy data** (for example from high resolution marine sediments, lake sediments and speleothems) to be confident that the D–O cycles do represent global-scale climatic shifts. For example, similar short-term fluctuations in SSTs during the Last

Glacial have been inferred from deep-sea sediment cores taken from various parts of the world's oceans. The highest amplitude SST changes are seen in sediment cores from the North Atlantic, but corresponding SST and other marine changes have been found in areas far from Greenland, e.g. the Santa Barbara Basin in the north-east Pacific (Behl and Kennett 1996) and the Cariaco Basin off the coast of Venezuela (Hughen et al. 1996). In the Mediterranean Sea there were declines in SSTs on the order of 5 to 8 °C during the coldest phases of D–O cycles (Rohling et al. 1998b). Loess sequences in China show layers of coarser aeolian material (indicating more wind and aridity) correlating with cold phases of D–O cycles (Porter and An 1995). Leuschner and Sirocko (2000) reviewed global correlatives of D–O events, and presented evidence for D–O cycle related changes in the fraction of aeolian dust present in cores from the Arabian Sea that relate to changes in strength of the summer and winter monsoons in that region.

Of particular importance is the close correlation between D–O cycles and levels of methane in the atmosphere. The methane signal in ice cores is believed to represent the global extent of wetlands and their productivity because wetlands (primarily in the tropics) are the main source for the gas. In the GRIP core, D–O cycles are associated with large fluctuations in methane on the order of between 400 and 600 ppbv, showing that the Greenland temperature changes are related to large-scale hydrological changes in the tropics. For example, the southern China Hulu Cave speleothem record, used to reconstruct changes in the East Asian monsoon, shows a close correlation between phases of strengthened monsoon and the warm phases of the D–O cycles (Rohling et al. 2009).

As the methane signal is a global one, methane has also been the primary means for putting the Arctic and Antarctic ice core records on a common time-scale (annual layer counting has only been possible on Greenland cores back to around 40 000 years ago). This is, however, less straightforward than it may at first seem because (as described in **Section 2.4**) the age of gases trapped in ice is younger than the surrounding ice by a factor that needs to be estimated. Another problem with comparing ice cores from the two hemispheres is the considerable difference in accumulation rates: ice in Greenland accumulates faster than ice in the Antarctic. However, by comparing the relatively high resolution Byrd ice core from western Antarctica with the GRIP core, Blunier et al. (1998) were able to

identify stadial/interstadial temperature fluctuations that correspond with the larger Greenland D–O cycles. As illustrated in **Fig. 3.18**, the Antarctic fluctuations are more subdued than their counterparts in Greenland and are anti-phase: that is, declining temperatures in Greenland correlate with increasing temperatures in Antarctica. More recent comparison between the NGRIP record and a relatively high resolution Antarctic ice core from Dronning Maud Land near the Atlantic sector of the Southern Ocean confirms this anti-phase relationship, showing Antarctic warming equivalents for each Greenland cooling phase (EPICA Community Members 2006).

It is currently believed that the asynchrony between Antarctic and Greenland temperature records spanning the Last Glacial is the result of variations in ocean heat transport; particularly in relation to the strength of different deepwater currents in the Atlantic basin (the AMOC already referred to and described later in this section). A general survey of the various ways in which changes in ocean circulation cause climatic change is contained in **Section 9.7**. However, some discussion of the ocean–atmosphere links operating in the Atlantic is necessary to understand the D–O cycles and their anti-phase relationship with Antarctic temperatures.

The climate of north-west Europe (and the northern North Atlantic region generally) is presently mild for its latitude because of the transfer of heat to the atmosphere effected by the warm, northward flowing Gulf Stream and North Atlantic Drift currents and the prevailing westerly surface winds. It has been known for many years from studies of North Atlantic deep-sea sediment cores that there have been frequent and large-scale shifts in the latitudinal position of the oceanic polar front, and hence in the northward extent of the warm surface currents (**Fig. 3.19**). During the LGM the boundary between polar and subpolar waters was over 20° farther south compared with its position today.

The present extension of the North Atlantic Drift to high latitudes is driven largely by the formation of North Atlantic Deep Water (NADW) in the Greenland-Iceland-Norwegian Sea: the descent and subsequent southward flow of water along the bottom of the North Atlantic perpetuates the northward flow of warm waters to the regions of NADW formation (**Fig. 3.20**). This is a part of a global-scale ocean circulation system known as the 'Great Ocean Conveyor Belt' (Broecker 1991). The process in the North Atlantic relies upon northward flowing water becoming denser than surrounding water so that it is able to sink at high

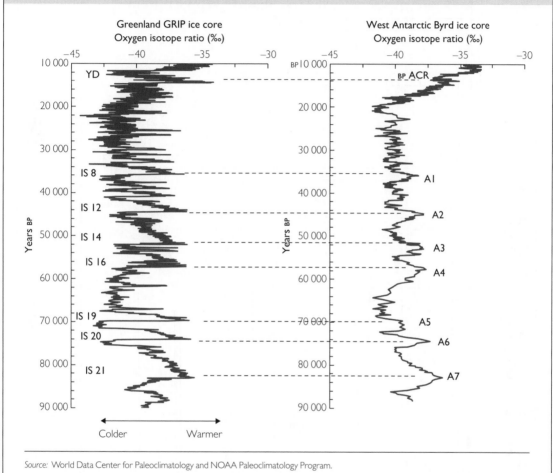

Fig. 3.18 Oxygen isotope data ($\delta^{18}O$) for the GRIP and Byrd ice cores (from Blunier and Brook 2001) showing the anti-phase relationship between Greenland and Antarctic temperatures through the last glacial cycle. A1–A7 refer to Antarctic warm phases that precede Greenland interstadials (IS 8–21). ACR refers to the Antarctic Cold Reversal and YD to the Younger Dryas event

Source: World Data Center for Paleoclimatology and NOAA Paleoclimatology Program.

latitudes, and this happens because during northward transport the water becomes colder and saltier (the salinity increases as surface water is evaporated). It has been possible to reconstruct past variations in the strength of this temperature/salinity driven system (known as the **thermohaline circulation** system, or THC) from deep-sea sediment cores, and it has been found that times of weakened THC correlate with reduced North Atlantic SSTs and the cold phases of D–O cycles (e.g. Boyle and Keigwin 1987; Keigwin et al. 1994; McManus et al. 2004). Such correlations are particularly clear from the high accumulating marine sediments of the Bermuda Rise area (see **Fig. 3.20**).

Various modelling studies have also shown that the formation of NADW (and hence the strength of the THC) is highly sensitive to freshwater forcing: in other words, an increased input of freshwater to the high-latitude North Atlantic via increased glacial melt can reduce salinity (thereby reducing water density) resulting in either a weakening or a rapid 'shut-down' of the THC (e.g. Manabe and Stouffer 1995; Rahmstorf 1995).

In more recent research literature, the term Atlantic Meridional Overturning Circulation (AMOC) is often used to refer to the entire Atlantic surface and deep ocean circulation system connecting the north and south high-latitude regions; and it is increasingly

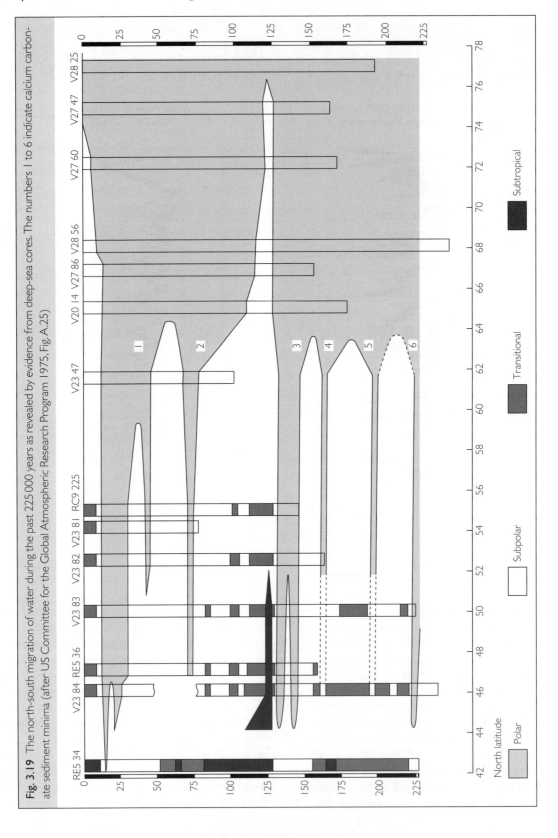

Fig. 3.19 The north–south migration of water during the past 225 000 years as revealed by evidence from deep-sea cores. The numbers 1 to 6 indicate calcium carbonate sediment minima (after US Committee for the Global Atmospheric Research Program 1975, Fig. A.25)

Fig. 3.20 The North Atlantic region and a generalized illustration of the thermohaline circulation system. Note that NADW consists of an upper branch (UNADW) and a lower branch (LNADW). Formation of NADW (and resultant northward transport of warm waters) is highly sensitive to freshwater forcing (from Anderson 1997)

recognized that wind systems also play a key role in the creation of poleward currents and the surface salinity changes that influence the formation of deep water (O'Hare 2011). When the AMOC is in the glacial mode, the THC in the North Atlantic is in the 'shut down' state: NADW is drastically reduced and Antarctic Bottom Water (AABW) replaces NADW at depth through much of the Atlantic basin. On the other hand, during interglacials and interstadials the AMOC is characterized by strong overturning in the North Atlantic (a strong THC): warm surface currents move north, and density contrasts cause water to sink and form NADW which moves south towards Antarctica. An intermediate mode of AMOC is also recognized in which THC operates in a weakened state resulting in less vigorous NADW formation and a more limited Gulf Stream and North Atlantic Drift.

Whatever the precise mechanisms, it is believed that phases of weakened overturning in the North

Atlantic, with the consequent cessation of northward ocean heat transport, are directly related to the rapid temperature declines detected in the Greenland ice cores. Broecker has also argued that these phases of THC shut down led to dramatic advances in sea ice across the North Atlantic: this in turn helps to explain the large declines in air temperature in the North Atlantic region, particularly in winter (Broecker 2010). The rapid warmings of D–O cycles are the result of sudden re-establishment of the strong mode of the THC when, following a freshwater pulse, salinity recovers to a critical level re-initiating NADW formation. The threshold-type behaviour of the THC system in response to changing salinity (the 'salt oscillator' model) was described by Broecker et al. (1990), and Clark et al. (2002b) have reviewed the ways in which THC variability results in abrupt climatic change.

Alley (2004) used the analogy of 'tipping a canoe' to understand the threshold behaviour of the THC and related abrupt climate change events. It is difficult to determine exactly how far the canoe can be tipped before it will flip over; likewise the precise amount of freshwater forcing required to shut down the THC is unknown. Yet there is a point of no return, and when crossed, the climate rapidly flips to a different state.

In addition to reconstructing AMOC variability, deep-sea sediment cores have also provided evidence for periods of large-scale iceberg discharge in the form of layers of ice rafted debris (IRD as described in **Section 2.3**). These **Heinrich events** (after Heinrich 1988) involved the episodic release of tremendous quantities of icebergs into the North Atlantic Ocean with consequent sea level rises of 10–35 m. Mineralogical analyses of the IRD from the sea-floor show that the icebergs were mainly from the Laurentide ice sheet via Hudson Strait. It therefore appears that at

various times during the Last Glacial the Laurentide ice sheet was prone to large-scale surges, and with each surge there was a catastrophic release of icebergs that, on melting, caused large-scale freshening of the North Atlantic. Dating of IRD layers by Bond et al. (1993) showed that Heinrich events correlate with the coldest phases of the D–O cycles and immediately precede abrupt interstadial warming. The relationship between the Heinrich events and the D–O cycles is shown in **Fig. 3.21**.

It is important to note that not all of the stadial/interstadial fluctuations revealed in Greenland ice cores are associated with Heinrich events. Instead, the Heinrich events are associated with the largest amplitude stadial/interstadial temperature swings, and between each of these there is a period of smaller-scale variability. The term '**Bond cycles**' is used to

Fig. 3.21 The relationship between sea surface temperatures, Heinrich events and D–O cycles. The top two curves show changes in percentage of left coiling *N. pachyderma* (high percentage indicates cooler sea surface temperatures) from different deep-sea sediment cores. These changes correlate well with the temperature oscillations reconstructed from the GRIP ice core. Heinrich events are numbered H1 to H6, each correlating with a major shift from cold to warm. Heinrich events separate 'Bond cycles' during which there are D–O cycles superimposed on longer-term cooling trends as shown in the bottom curve (from Bond et al. 1993, Fig. 3). Reproduced with permission of Macmillan Publishers Ltd: *Nature*

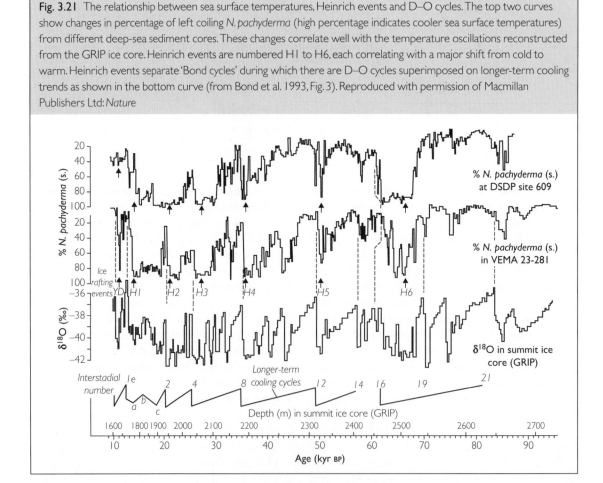

indicate the groups of D–O cycles (the sequence of stadials and interstadials) that occur between each Heinrich event. In general, there is a trend towards cooler interstadials and more severe stadials over the duration of each Bond cycle. It is also important to notice that Heinrich events do not precede the cold phases. This means that Heinrich events do not trigger stadials; instead, when Heinrich events occur they *amplify* stadial cooling after it is already underway. In addition to Heinrich layers, North Atlantic sediment cores also contain more frequent, smaller-scale IRD layers that were deposited at intervals of 2000 to 3000 years – corresponding closely with the D–O cycles (Bond and Lotti 1995). For example, IRD layers in a high resolution sediment core taken from near the Faeroe Islands (Rasmussen et al. 1997) indicate times of increased iceberg calving from the Fennoscandian ice sheet. These IRD layers correlate with reduced SSTs, as indicated by increases in the abundance of the left coiling polar foram *Neogloboquadrina pachyderma*, and periods of freshening of the North Atlantic, as indicated by shifts towards lighter $^{18}O/^{16}O$ ratios in planktonic foram tests due to the presence of ^{18}O deficient meltwater.

Hence the stadial events recorded in Greenland ice cores are closely tied to phases of increased iceberg discharge from ice sheets bordering the high-latitude North Atlantic. The resultant freshening of North Atlantic surface water would have caused the THC to shift from a strong mode to a weak mode thereby reducing northward transport of warm water and causing regional air temperatures to decline. For reasons that are not fully understood, every 7000 to 10 000 years these stadial events led to much larger-scale iceberg discharges (the Heinrich events) from the margins of the Laurentide ice sheet causing more severe freshening of the North Atlantic and wholesale collapse of the THC from an already weakened mode. It is at these times that Greenland ice core records show their coldest temperatures. Once the meltwater pulse associated with a Heinrich event had run its course, the THC 'switched back on' leading to the rapid interstadial warming seen in both the SST records and the ice cores. Various hypotheses have been advanced to explain how Heinrich events are triggered, and these focus on both internal ice sheet dynamics and external forcing. Given the similarity in timing of fluctuations in the Laurentide and Fennoscandian ice sheets throughout the Last Glacial (Fronval et al. 1995), it is likely that external forcing is important, with sea level

fluctuation playing a key role. Some possible causes are discussed in **Section 9.7**.

The above discussion highlights the complex and closely coupled behaviour of ice, ocean, and atmosphere in the North Atlantic region during the Last Glacial. Less is known about RCCEs during earlier glacials because Greenland ice cores do not extend back to the penultimate glacial cycle (although an aim of the NEEM project is to recover ice extending this far). However, temperature records from the long Antarctic ice cores suggest that rapid changes were not unique to the last glacial cycle. Because the AMOC is interconnected with the global system of ocean currents, past changes in the AMOC would have affected SSTs (and therefore also climate) in other parts of the world. Although the magnitude of related climatic shifts would have varied from place to place, it is likely that many such shifts were rapid rather than gradual, and this has important implications for the way we view environmental change during the Quaternary – particularly with respect to how the environment has influenced landscapes, flora and fauna, and our prehistoric ancestors. It underlines the fact that, in addition to usually being colder than now, the Quaternary has been a time of climate instability – with the relative stability seen over the last several thousand years of the Holocene being the exception rather than the norm.

As alluded to earlier, the asynchrony during the Last Glacial between Arctic and Antarctic temperatures can be explained by changes in the transfer of warm surface water from the South Atlantic to the North Atlantic. The AMOC has been referred to as a 'bipolar seesaw': when the THC is in its strong mode, NADW production is vigorous and warm surface water travels into the North Atlantic at the expense of the South Atlantic; whereas reduced THC causes less NADW production and more AABW production thereby making the seas warmer in the southern South Atlantic. There is some time lag involved in this process; but it is difficult to pin down the exact time it takes for changes in the THC in the North Atlantic to be felt in the south. An area of current research also involves looking at the extent to which AMOC changes are driven from the south as opposed to the north. Other global climate **teleconnections** involving variation in the AMOC, especially associated changes in the intensity of low-latitude monsoons and the location of tropical rain belts, are discussed in **Chapter 9**.

The vegetational conditions of full glacials in Europe

During the various full glacial stages of the Pleistocene, the vegetation of much of unglaciated, periglacial Europe was characterized by its open nature. Trees were relatively rare, and in many respects the plant assemblages displayed many characteristics one would expect in a cool 'steppe' environment (**Fig. 3.22**).

In Western Europe the pollen record of the last (Würm, Weichselian) glacial shows relatively little arboreal (tree) pollen, and traces of *Artemisia* and *Thalictrum* (characteristic of open habitats) are common. (**Table 3.3** lists common plant names, with botanical equivalents, used in the following sections.) In more maritime areas, such as Cornwall and Ireland, there was also some dwarf birch and willow during glacials, but even as far south as Biarritz in south-western France the proportion of tree pollen in full glacial sediments is low, though some oak and hazel may have existed in the Gascogne Lowlands. Thus, as with limits

of permafrost, the northernmost boundaries of the major zonal vegetational types were pushed far to the south of their present-day ones (**Fig. 3.23**).

Areas right at the ice fronts themselves were characterized by very sparse vegetation, being more arid and steppe-like further east in northern Europe. In the belt of loess which occurs further to the south, a more herbaceous flora seems to have been prevalent. In more favoured parts of Romania and Hungary there was even some woodland in full glacial times, possibly because precipitation was somewhat higher than in much of southern Europe. In fact, a wide range of tree species, principally coniferous but also some deciduous, were able to survive full glacial conditions in parts of central and eastern Europe south of 50° N latitude (Willis and van Andel 2004); although tree cover was open and probably confined to favourable microclimates. Most of Russia, on the other hand, from Poland across to the southern Urals, was covered by a salt-tolerant dry *Artemisia* steppe, and south of this there

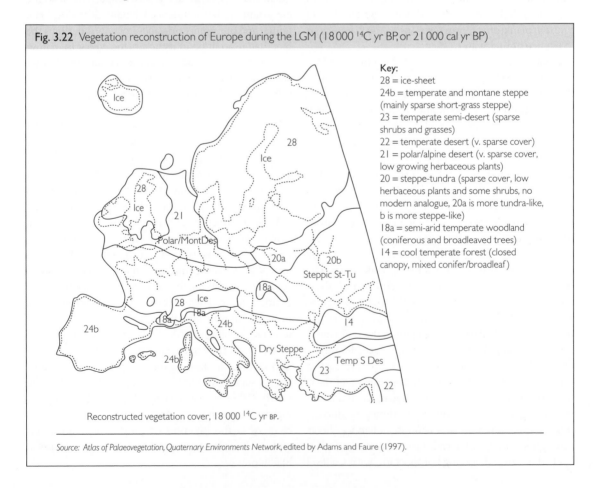

Fig. 3.22 Vegetation reconstruction of Europe during the LGM (18 000 ^{14}C yr BP, or 21 000 cal yr BP)

Key:
28 = ice-sheet
24b = temperate and montane steppe (mainly sparse short-grass steppe)
23 = temperate semi-desert (sparse shrubs and grasses)
22 = temperate desert (v. sparse cover)
21 = polar/alpine desert (v. sparse cover, low growing herbaceous plants)
20 = steppe-tundra (sparse cover, low herbaceous plants and some shrubs, no modern analogue, 20a is more tundra-like, b is more steppe-like)
18a = semi-arid temperate woodland (coniferous and broadleaved trees)
14 = cool temperate forest (closed canopy, mixed conifer/broadleaf)

Reconstructed vegetation cover, 18 000 ^{14}C yr BP.

Source: Atlas of Palaeovegetation, Quaternary Environments Network, edited by Adams and Faure (1997).

Table 3.3 Botanical and common names of glacial and interglacial plants referred to in the text

Botanical name	Common name	Botanical name	Common name
Abies	Fir	Najas	Naiads
Acer	Maple	Nothofagus	Southern beech
Alnus	Alder	Nyssa	Tupelo
Artemisia	Mugworts	Osmunda claytoniana	Interrupted fern
Azolla	Water fern	Ostrya	Hop-hornbeam
Betula	Birch	Picea	Spruce
Buxus	Box	Pinus	Pine
Carpinus	Hornbeam	Plantago	Plantain
Caryophyllaceae	Pink family	Poaceae	Grass family
Cedrus	Cedar	Pterocarya	Wingnut
Chenopodiaceae	Goosefoot family	Pyracantha	Firethorn
Corylus	Hazel	Rhododendron	Rhododendron
Cyperaceae	Sedge family	Quercus	Oak
Daboecia	St Dabeoc's heath	Salix	Willow
Erica	Heath	Salvinia	Floating fern
Fagus	Beech	Sequoia	Sequoia
Fraxinus	Ash	Taxodium	Bald cypress
Glyptostrobus	Cypress	Taxus	Yew
Hippophaë	Sea-buckthorn	Thalictrum	Meadow-rue
Ilex	Holly	Tilia	Lime
Juglans	Walnut	Trapa natans	Water chestnut
Juniperus	Juniper	Tsuga	Hemlock
Larix	Larch	Ulmus	Elm
Lemna minor	Duckweed	Vitis	Vine creeper
Liquidambar	Sweet gum	Xanthium	Cocklebur
Liriodendron	Tulip tree		

was woodland in the region of the Crimea, and along the shores of the expanded Caspian Sea. The frequent occurrence of salt-demanding and salt-tolerant plants also suggests that precipitation levels were low.

Writing about the vegetation of north-west Europe during the Last Glacial, Birks (1986) suggests that it was of a type unknown today. It had structural and floristic affinities with steppe and tundra, and contained a mixture of ecological and geographic elements. He maintains that this 'no-analogue' vegetation type suggests a 'no-analogue' environment, with relatively warm summers, extremely cold winters, highly unstable soils (caused by extensive frost churning), low precipitation, strong winds and strong evaporation.

A general impression of fluctuations in tree pollen through the last glacial cycle is illustrated in the Grande Pile pollen record (**Fig. 3.24**). Further south,

in the Massif Central region of France, long lake sediment records extending back some 450 000 years (to MIS 12) show pollen spectra similar to the Würm during previous glacial phases: in each glacial the only significant arboreal type is *Pinus* with proportionately more grass and steppe pollen types during the coldest times (Reille et al. 2000) (**Fig. 3.25**). In southern Europe and the Levant, around the Mediterranean Sea's northern shores, the vegetation was also characteristically steppe-like and arid (Bonatti 1966), with some areas of pine. Long pollen records from Greece show little to no arboreal pollen in each full glacial extending back to MIS 12 (e.g. Tzedakis 2005). This belt seems to have extended across into the Zagros Mountains of western Iran, with *Artemisia* again characteristic or dominant at lower altitudes, and a dry alpine flora at higher altitudes.

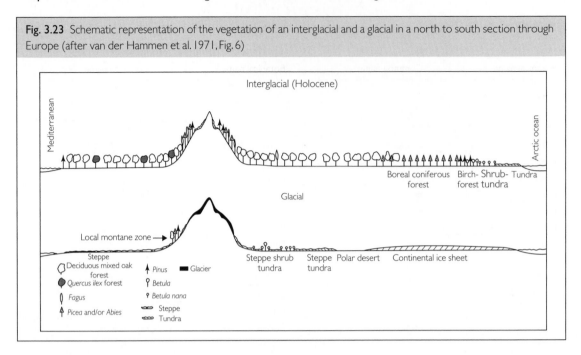

Fig. 3.23 Schematic representation of the vegetation of an interglacial and a glacial in a north to south section through Europe (after van der Hammen et al. 1971, Fig. 6)

The glacial vegetation of North America

While much of the country north of the European Alps during the Last Glacial Maximum supported tundra or, close to the ice, a cold rock desert, in America the available records indicate that much of the area south of the ice sheet (particularly in eastern parts of the continent) was covered with boreal forest rather than with tundra (**Fig. 3.26**). The reason for this difference is that the ice limit in Wisconsin times in America was much further south than it was in Europe, about 39° N in Illinois compared with 52° N in Germany. Further, the Alps, with their own large ice cap, reinforced the semi-permanent area of high atmospheric pressure associated with the Scandinavian ice sheet. This probably tended to divert warm, westerly air flows to the south of the Alps. There is no such mountain mass trending east to west in North America.

The widespread boreal forest of full glacial times in North America, dominated generally by *Pinus* and *Picea*, was not found everywhere, for there were patches of treeless tundra, but these were not as extensive as in Europe. The southern limit of the boreal forest is not known with certainty, but it was probably somewhere in the south-central United States, perhaps extending westward from Georgia. Thus boreal woodland may have formed a latitudinal belt as broad as it is today – 1000 km from Hudson Bay to the Great Lakes – though shifted far to the south. Compared with eastern North America, conifer woodland was more open in north-western areas where precipitation was less than it is at present (Clark et al. 2012). In the south-west, where precipitation was higher and pluvial lakes appear to have been synchronous with the main glacials (see **Section 4.5**), pollen evidence indicates high percentages of *Pinus* and other montane conifers during the Wisconsin glacial: now the same areas are characterized by semi-desert shrubs. In the Western Cordillera the treeline was lowered 800–1000 m, and the extent of alpine vegetation in the mountains was very greatly expanded (Orme 2002).

The interstadials of the Last Glacial

One of the problems of glacial correlation is that phases of lesser glaciation and relatively greater warmth occur during the course of a major glacial phase. Such interruptions are called **interstadials**, but because of frequent climatic variation during glacial cycles (over various lengths of time and to different degrees), the identification and definition of interstadials is somewhat arbitrary. Some workers suggest that the more minor phases of warming which left little signal in palaeobotanical records should be referred to as 'intervals' rather than as interstadials (e.g. Caspers and Freund 2001). Nevertheless, there are indications in many parts of Europe, and elsewhere,

Fig. 3.24 La Grande Pile record (France) showing changes in the percentage of tree and herb pollen through the last glacial cycle. The Eemian Interglacial, the LGM, and the onset of the Holocene are marked with arrows (modified from pollen data presented in Woillard and Mook 1982)

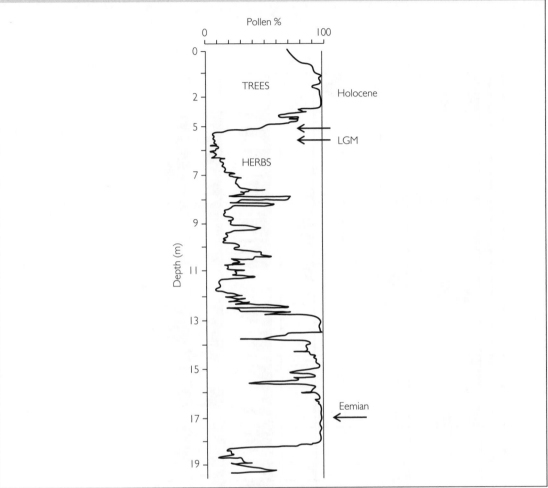

that the Würm–Weichsel–Wisconsin glaciation was interrupted by several phases of less intense glacial activity.

This is most strikingly evident in the high resolution ice core records from Greenland that show brief but high magnitude (sometimes on the order of 15 °C) shifts to warmer conditions – particularly after Heinrich events. As shown in **Fig. 3.27**, Greenland interstadials spanning the last glacial cycle are numbered from 1 to 24. These warm phases enabled soils and other distinctive sediments to develop, and quite a large number of these deposits have been dated by radiocarbon means. Variations in the quality and precision of dating of different interstadial deposits makes it difficult to correlate them with each other and to correlate the traditional land-based interstadial schemes with the Greenland record; but in some cases this has been possible. There is some clustering over a period from about 50 000 to 23 000 **radiocarbon years** BP. Greenland records indicate that this period was not a continuous phase of warmth, and there seems in many areas to have been a tendency for a particularly marked interstadial at the end of this time, notably around 30 000 years ago; for example, the Denekamp (Europe), Plum Point (North America), Olympia (British Columbia), Kargy (western Siberia) etc., that can be related to a late phase of warming in MIS 3. There were also some important interstadials near the

Fig. 3.25 Summary pollen diagram for Massif Central, France, extending back to MIS 12. Note the similar succession of tree taxa in each interglacial and the fluctuations in pine pollen during glacials (from Reille et al. 2000, Fig. 7). Reprinted with permission from Wiley

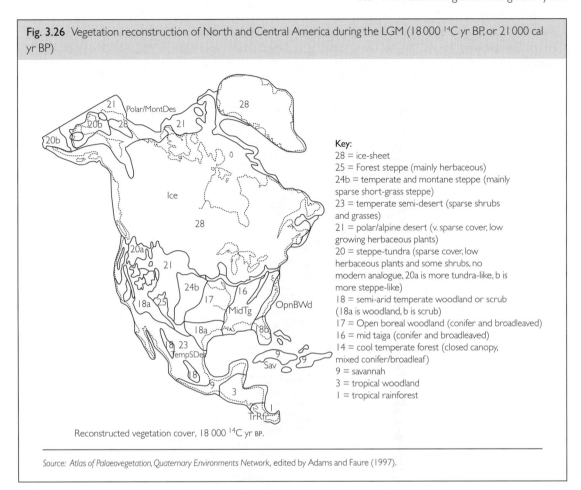

Fig. 3.26 Vegetation reconstruction of North and Central America during the LGM (18 000 ^{14}C yr BP, or 21 000 cal yr BP)

Key:
28 = ice-sheet
25 = Forest steppe (mainly herbaceous)
24b = temperate and montane steppe (mainly sparse short-grass steppe)
23 = temperate semi-desert (sparse shrubs and grasses)
21 = polar/alpine desert (v. sparse cover, low growing herbaceous plants)
20 = steppe-tundra (sparse cover, low herbaceous plants and some shrubs, no modern analogue, 20a is more tundra-like, b is more steppe-like)
18 = semi-arid temperate woodland or scrub (18a is woodland, b is scrub)
17 = Open boreal woodland (conifer and broadleaved)
16 = mid taiga (conifer and broadleaved)
14 = cool temperate forest (closed canopy, mixed conifer/broadleaf)
9 = savannah
3 = tropical woodland
1 = tropical rainforest

Reconstructed vegetation cover, 18 000 ^{14}C yr BP.

Source: *Atlas of Palaeovegetation, Quaternary Environments Network*, edited by Adams and Faure (1997).

beginning of the Würm–Weichsel–Wisconsin, and these may have been sufficient to lead to significant deglaciation; for example the Brørup followed by the Odderade in northern Europe, the St Germain I and II of France, the Chelford of England, and the St. Pierre followed by the Port Talbot of North America. These earlier interstadials identified in terrestrial deposits mostly relate to warming phases during MIS 5c and MIS 5a. **Table 3.4** summarizes widely recognized interstadials of Europe.

The period from after 26 000 years ago until the end of the Pleistocene (during MIS 2) saw a great expansion of glaciers, at least in the Northern Hemisphere, and this has been given a series of local names, including Hauptwürm in Europe, the Dimlington in England, the Woodfordian in the northern United States, and Pinedale in the Rockies. Following the LGM, the millennia spanning the Late Glacial were also marked by stadials and interstadials, and these are described in **Section 5.2**.

Pollen and faunal evidence has been utilized for assessing environmental conditions during the interstadials. In England, for example, the Chelford interstadial radiocarbon dated to around 60 000 years ago was characterized by boreal forest and a beetle fauna comparable to that found in south-east Finland at the present day. Similarly, across northern Europe the Brørup interstadial saw the expansion of birch and pine woodland across Germany and into Denmark. *Picea* pollen also achieved a similar distribution during the Brørup, but was much less abundant than *Pinus* pollen (Caspers and Freund 2001). The St Germain I interstadial as represented in long pollen diagrams from France (such as La Grande Pile) has been correlated to Greenland Interstadials 24–22, and the St Germain II interstadial to Greenland Interstadial 21. Both of these interstadials saw coniferous forests dominating across France with traces of more thermophilous deciduous species such as *Quercus*. Further south, a *Quercus*

Fig. 3.27 GRIP ice core δ¹⁸O record with numbered Greenland interstadials (IS 1 to 24) compared with named European interstadials (from Dansgaard et al. 1993, Fig. 1). Reprinted with permission of Macmillan Publishers Ltd: *Nature*

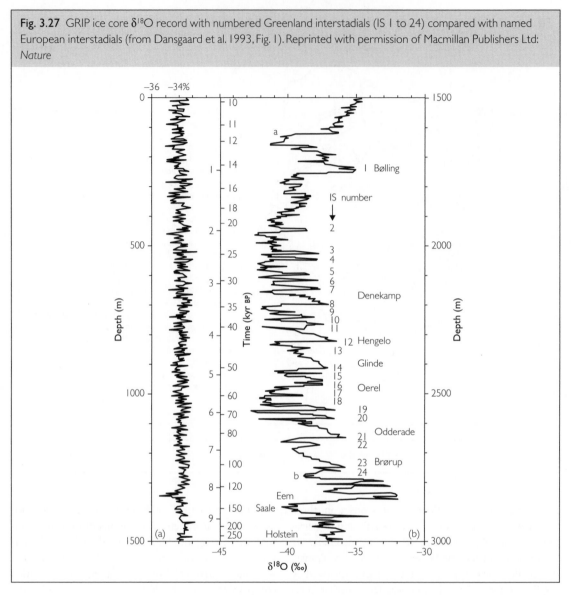

ilex vegetation was present in southern Spain, but a *Carpinus-Ulmus-Tilia* forest was present in Macedonia. Boreal forest re-expanded across northern Europe during the Odderade (Brimpton of England).

The fauna of the later Upton Warren interstadial in England suggests July temperatures at least 5 °C higher than those which existed during the full glacial conditions which followed. Indeed, in a review of the evidence provided by beetles, Coope (1975a,b) has suggested that at the thermal maximum of the Upton Warren interstadial, between 43 000 and 40 000 radiocarbon years BP, average July temperatures in central England were about 18 °C, which are a little warmer than today's.

He does, however, find that winter temperatures were somewhat lower than now, indicating a more continental climatic regime. The warm phase was relatively short-lived, possibly lasting only a thousand or so years. The thermal maximum of the Upton Warren, and the Hengelo recognized in the Netherlands, probably correlates with Greenland Interstadial 12.

In general, during the longer interstadials woodlands returned to quite wide areas of Europe, with boreal woodlands of spruce, pine, and larch bordering the North Sea and Baltic. Oak and hornbeam forests probably occurred in northern Italy, the former Yugoslavia, and Albania. Climatic reconstructions from

Table 3.4 Summary of European interstadials identified by biostratigraphy during the Würm–Weichselian glacial

MIS	England	Netherlands	France	NW Germany and Denmark	Norway
3		Paudorf			
		Denekamp		Denekamp	
	Upton Warren	Hengelo		Hengelo	Alesund
		Moershoofd		Moershoofd	
		Glinde		Glinde	Bø
		Oerel		Oerel	
5a	Brimpton	Odderade	St Germain II	Odderade	Torvastad
5c	Chelford	Brørup	St Germain I	Brørup	Fana
	Wretton	Amersfoort			

various central European **biostratigraphic** records indicate large temperature differences between inter-stadials and stadials of the early Würm: mean temperatures of the coldest month varied on the order of 10 to 15 °C (with actual values ranging between 0 and −20 °C at Les Echets, eastern France) and of the warmest month by some 5 to 10 °C (Klotz et al. 2004).

The nature of interglacials

In general terms the interglacials, to judge from the results of pollen analysis and other techniques, appear to have been essentially similar in their climate, flora, fauna, and landforms to the Holocene in which we live today. The most important of the characteristics of the interglacials was that they witnessed the retreat and decay of the great ice sheets, and saw the replacement of tundra conditions by forest over the now temperate lands of the Northern Hemisphere. Deciduous trees achieved much more northerly distributions across North America and Eurasia during interglacials than during the briefer and less optimal interstadial phases, and treelines reached higher altitudes.

As shown in the long EPICA Dome C deuterium record (see **Fig. 3.5**) the maximum temperatures attained in the most recent four interglacials appear to have been as high or higher than those of the present. During the last interglacial (MIS 5e, referred to as the Eemian in Europe and the Sangamonian in North America) the ice core data from Vostok, Antarctica, indicates a maximum temperature about 3 °C higher than today. During this time, global **eustatic** sea level was around 6 m, or possibly up to 9 m, higher than

today (see **Section 6.7**). Analysis of early Eemian-age Coleoptera remains from southern England indicates mean July temperatures about 4 °C higher than today (Coope 2001), and the presence of pollen from the sweet gum (*Liquidambar*) near Toronto, Canada, suggests that temperatures may have been 2–3 °C higher than those now experienced in that area. Summer sea ice in the Arctic was perhaps half as extensive then as it is today, and the northern treeline was further north. Similarly, in the Holstein Interglacial (within MIS 11) of Poland and Russia, the flora, particularly the distribution of beech, hornbeam, holly, and the Pontic alpine rose, suggest temperatures slightly warmer than at present.

While at first glance the definition and timing of an interglacial might seem easy to establish, the opposite is the case: there are different ways of marking the onset and ending of interglacials which lead to different estimates of their time span (for a detailed discussion see Forsström 2001). From terrestrial deposits, interglacials have traditionally been defined as those times when forests and thermophilous plant types were present across mid to high latitudes of North America and Eurasia. The problem here, however, is that due to relatively slow migration rates of trees in response to climate (as discussed further later in this section), the onset of an interglacial at higher latitudes will appear to be delayed relative to places further south. Likewise, the ending of an interglacial will appear delayed at lower latitudes where trees can survive for longer during the onset of a glacial stage. This is seen, for example, in European pollen records from the last interglacial, which show that forests in southern Europe continued to exist for some 5000 years after replacement of forest by tundra was underway in northern Europe (Tzedakis 2005).

From deep-sea sediment cores, by convention the onset and ending of interglacials are taken from the mid-points of decreasing or increasing δ¹⁸O in benthic forams representing phases of decreasing or increasing global ice volume. By correlating the mid points of oxygen isotope transitions with uranium–thorium dates on marine sediments and corals, MIS 5e has been dated to between approximately 130 000 to 115 000 years ago. The timing and lengths of earlier interglacials defined in this way can also be estimated by 'orbital tuning' of stacked δ¹⁸O records as described in **Section 2.5**. While this method makes it possible to identify and estimate the duration of times of lower global ice volume, it is important to be aware that the resultant MIS boundaries are not synchronous with the associated interglacial climate and vegetation changes. Again taking the example of the last interglacial, the shift to warm, interglacial climatic conditions as inferred from Antarctic ice cores occurred at least a couple of thousand years before the formal MIS 6/5e boundary. Moreover, the thermal maximum of the last interglacial was reached earlier than the time of minimum global ice volume dated to about 125 000 years ago, and climatic deterioration was underway before the formal MIS 5e/5d boundary. Vegetation change, on the other hand, lagged behind the initial warming with the full

interglacial forests of Europe not fully established until about 127 000 years ago – some 3000 years after the start of MIS 5e (Kukla et al. 2002).

The general relationship between the timing of changes in temperature, vegetation, and ice volume has been known since the 1970s, and this is illustrated in **Fig. 3.28**. Plainly, there would be very different response curves: ice sheets would tend to respond relatively sluggishly to a climatic amelioration because of their great mass, and because of their partial control of regional climates. Retreat rates of up to 3 km per 100 years for the Greenland ice sheet are much lower than the rates of floral advance. Using the beginning of the Holocene as an analogue for past interglacials, studies from Sweden suggest that interglacial Scots pine and pubescent birch spread at a rate of 205–260 m per year, alder at 175–230 m, elm and warty birch at around 190 m, and hazel at 130–190 m per year. As a whole it would seem that a rate of advance of some 200 m per year for trees with light seeds, and, rather less, say 160 m, for trees with heavier seeds, like hazel and oak, was characteristic. Thus advance of trees at the onset of interglacials could have taken place at about one kilometre in five years, or 1000 km in 5000 years. The response of fauna would tend to be even quicker. What has become more apparent recently is that the precise timing of such changes, and the respective time lags between them, has varied during different interglacials. This is fundamentally because

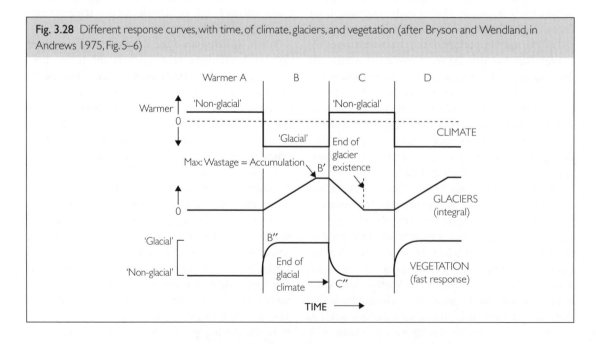

Fig. 3.28 Different response curves, with time, of climate, glaciers, and vegetation (after Bryson and Wendland, in Andrews 1975, Fig. 5–6)

the combination of orbital factors and other climatic variables that triggers interglacial warming (as well as the extent of ice sheets and location of glacial refugia from which plants spread) is never exactly the same at each glacial termination.

Notwithstanding the problems of measuring the length of interglacials, it can be generally said that over the past 900 000 years or so interglacials have lasted on average for about 11 000 years, or half a precessional cycle (see **Section 9.4** for a detailed discussion of the orbital cycles). However, there have been shorter interglacials such as MIS 7e which lasted less than 10 000 years and longer interglacials such as MIS 11. As referred to in **Section 3.2**, the Antarctic EPICA Dome C ice core shows that air temperatures associated with MIS 11 (and the Holstein interglacial of Europe) were at levels comparable to the Holocene for about 30 000 years (EPICA community members 2004).

While glacial stages were punctuated by abrupt, high magnitude climatic changes (the D–O cycles as described previously), by comparison interglacial climates appear to have been relatively stable. However, as discussed further in **Chapter 5**, our present interglacial has not been as stable as previously thought, and it is likely that earlier interglacials also contained significant short-term climate oscillations – if not as dramatic as those spanning the last glacial cycle. Another important characteristic of interglacial climates is that they begin and end abruptly. This is well illustrated in the EPICA Dome C record (**Fig. 3.5**) where temperature rises sharply with the onset of each interglacial, remains high for several thousand years (but usually showing a declining trend after an early peak), and then drops sharply at the beginning of each glacial stage. While orbital variation is involved in pacing their timing, the onsets and endings of interglacial warmth occur when thresholds are crossed and the climate system rapidly reorganizes. As described previously in relation to D–O cycles, shifts in the mode of the AMOC (and its ocean–atmosphere teleconnections) are undoubtedly involved; and there is increasing evidence that cooling at the end of the last interglacial occurred in an abrupt manner (e.g. Kukla et al. 1997). For instance, analyses of sediments from the Bermuda Rise area of the subtropical North Atlantic (see **Fig. 3.20**) indicate that SSTs declined in response to weakening of the THC system by about 2.5 °C within 100–200 years near the end of MIS 5e (Lehman et al. 2002). This change correlates with the replacement of broad-leaved trees by pine as shown

in the pollen record from La Grande Pile (Woillard 1978, 1979). In her analysis of the Grande Pile pollen core Woillard claimed that temperate forest of the last interglacial was replaced by a pine-spruce-birch taiga within approximately 150 ± 75 years, although other workers have argued for a more gradual forest transition in Europe (e.g. Frenzel and Bludau 1987).

The general sequence of vegetational development in temperate zones during the interglacials was rationalized by Turner and West (1968), who proposed the following type of pattern as being characteristic:

a) The first phase, one of climatic amelioration from full glacial conditions, can be called the Pre-temperate zone. It is characterized by the development and closing in of forest vegetation, with boreal types being dominant. *Betula* and *Pinus* are a feature of the woodlands, but light-demanding herbs and shrubs are also a significant element of the vegetation. Relics of the preceding late glacial periods, such as *Juniperus* and *Salix* may also be present.

b) The next phase, termed the Early-temperate zone, sees the establishment and expansion of a mixed oak forest with many shade-giving forest genera, typically *Quercus*, *Ulmus*, *Fraxinus*, and *Corylus*. Soil conditions are generally good, with a 'mull' condition, and this promotes dense, luxuriant cover.

c) In the next phase, termed the Late-temperate zone, there is a tendency for the expansion of late-immigrating temperate trees, especially *Carpinus*, *Abies*, and sometimes *Picea*, accompanied by a progressive decline of the mixed oak forest dominants. Some of these changes may be related to a decline in soil conditions, associated with the development of a 'mor' rather than a mull situation.

d) The fourth phase is called the Post-temperate phase, and is indicative of climatic deterioration. There is a reduction of thermophilous genera, and an expansion of heathland. The forest becomes thinner, with temperate forest trees becoming virtually extinct, and a return to dominance of boreal trees, such as *Pinus*, *Betula*, and *Picea*.

This general pattern can be seen in the summary pollen diagram from lake sediments from the Massif Central region of France that span the last 450 000 years (Reille et al. 2000) (**Fig. 3.25**). The last four interglacials are all marked by increases in *Quercus* and *Corylus* pollen, followed by increases in *Carpinus* and *Abies*. *Picea* and *Pinus* rise towards the end of each interglacial phase.

The scheme of Turner and West, whilst broadly applicable to the main interglacial phases, does vary between different interglacials and different places. There were probably climatic differences between the various phases, different barriers to migration, differing distances to glacial refuges from which genera expanded, changes in ecological tolerance, and variability within genera, and other changes consequent upon evolution or extinction (West 1972, pp. 315–24).

A broadly similar scheme for an 'interglacial cycle' was proposed by Iversen (1958) and has been used by Birks (1986) (Fig. 3.29). It has largely been applied to changing conditions in north-west Europe. The *cryocratic* phase represents cold glacial conditions, with sparse assemblages of pioneer plants growing on base-rich, skeletal mineral soils under dry, continental conditions. The *protocratic* phase witnesses the onset of the interglacial, with rising temperatures. Base-loving, shade-intolerant herbs, shrubs and trees immigrate and expand quickly to form widespread species-rich grasslands, scrub and open woodlands, which grow on unleached, fertile soils of low humus content. The *mesocratic* phase sees the development of temperate deciduous forest and fertile, brown-earth soils under warm conditions. Shade-intolerant species are rare or absent because of competition and habitat loss. The last, retrogressive phase of the 'cycle', produced by a combination of soil deterioration and climatic decline, is called the *oligocratic* phase, and is characterized by the development of open conifer-dominated woods, ericaceous heaths, and bogs growing on less fertile, humus-rich podzols and peats. This cyclic model can also be applied to areas like Central Florida and the Eastern Mediterranean.

Variations in the British and European interglacials

The interglacial temperate forests recorded in the early British interglacials, the Ludhamian and Antian (see Fig. 3.9 for the location of these phases in the local sequence), are mixed coniferous and deciduous, and the presence of hemlock and wingnut makes the vegetation assemblage different from that of any subsequent period in Britain. The relatively severe Baventian Glacial period led to the extinction from the British flora of hemlock (*Tsuga*), though it has remained in the vegetation associations of parts of North America until the present day. In northern Eurasia certain of the late Pliocene vegetation types do not appear to have reoccupied the area after the

first severe cold of the Pleistocene, though they were present in the Tiglian (Ludhamian). These plants included *Sequoia*, *Taxodium*, *Glyptostrobus*, *Nyssa*, *Liquidambar*, *Fagus*, *Liriodendron*, and several others. Thus a marked degree of impoverishment took place in the British flora as a result of the oncoming of the first cold phases of the Pleistocene (West 1972).

With regard to the later British interglacial floras, however, the Cromer Forest Bed of the Cromerian Interglacial on the Norfolk coast resembles the present-day British flora much more closely, and *Taxus*, *Quercus*, *Fagus*, *Carpinus*, *Ulmus*, *Betula*, and *Corylus* can be identified in the bed. The Hoxnian Interglacial (MIS 11) displays in its pollen profiles a high frequency of *Hippophaë* at the beginning of the sequence, then a late rise of *Ulmus* and *Corylus*, and the presence of *Abies* and *Azolla filiculoides*. Irish materials of the same age have a high percentage of evergreens, indicative of a high degree of oceanicity of climate (e.g. *Picea*, *Abies*, *Taxus*, *Rhododendron*, *Ilex*, and *Buxus*). The materials also contain certain Iberian species such as *Erica scoparia*, St. Dabeoc's heath (*Daboecia cantabrica*), and Mackay's heath (*Erica mackaiana*). The last two currently have a rather limited distribution in the Cantabrian mountains of Iberia.

The Ipswichian (Eemian, MIS 5e), on the other hand, seems to indicate rather more continental conditions. Its characteristic properties included an abundance of *Corylus* pollen in its early part, and then much *Acer* pollen; also present was *Carpinus*, but there was a scarcity of *Tilia* in the second part of the interglacial. Most of the Ipswichian sites contain a number of species not now native, including *Acer monspessulanum*, *Lemna minor*, *Najas minor*, *Pyracantha coccinea*, *Trapa natans*, *Xanthium*, and *Salvinia natans*. This suggests that conditions were somewhat warmer than during the Holocene climatic optimum (West 1972, p. 310), agreeing with ice core evidence. A warmer Ipswichian climate is also supported by the vertebrate faunal remains. In England remains of *Hippopotamus amphibius* and the pond tortoise *Emys orbicularis* have been found widely in Ipswichian-aged deposits, and both of these species require milder winters and warmer summers than have occurred in England during the Holocene. Other animals that occurred in England during the warmest part of the Ipswichian included the straight-tusked elephant, narrow-nosed rhinoceros, and the spotted hyaena (Jones and Keen 1993).

In the more continental parts of Europe the characteristic interglacial vegetation assemblages were a little

Fig. 3.29 The 'interglacial cycle' applied to various areas of north-western Europe, the eastern Mediterranean, and central Florida (USA) (from Birks 1986, Fig. 1.2). Reprinted with permission of Wiley modified, though the general sequence is not dissimilar. In the Likhvinian (Holstein, MIS 11) of Russia, for example, as in Turner and West's model, the first stage is represented by a high incidence of *Betula* and *Pinus* pollen, with *Picea* and *Salix* pollen only making up 1 per cent or less of the total (Ananova 1967). This birch and pine-birch forest stage was replaced by a dominantly pine and spruce-pine sequence, with *Pinus* and *Picea* predominant, with *Betula* making up about 5–10 per cent of the total pollen, and *Alnus* pollen constantly occurring with a frequency of about 15 per cent. This coniferous forest was replaced by a combination of coniferous, broad-leaved, and alder-thicket forests, with *Picea excelsa* dominant in the east, and *Pinus sylvestris* in the west. Both these types had about 40–60 per cent of the total pollen, with *Alnus* making up 25–40 per cent, and '*Quercetum mixtum*' (mixed oak) pollen about 10 per cent. *Abies* pollen was rare. In the next stage, *Abies* sometimes reaches 30 per cent, and *Carpinus* reached 20–30 per cent in some sections, though coniferous pollen was still dominant. Subsequently, there was a change back towards a more boreal flora, and this in turn was succeeded at the end of the interglacial by a reduction in tree pollen, the constant presence of various open vegetation plants (Poaceae, Cyperaceae, and *Artemisia*), and the arrival of plants of the periglacial type. Thus, although the sequence is comparable to that in western Europe, there are certain elements missing in northern and eastern Europe, including *Vitis, Pterocarya, Juglans regia, Abies alba, Carpinus orientalis, Buxus, Taxus, Tilia tormentosa*, and *Osmunda claytoniana*.

In southern Europe the interglacials may have been associated with moister conditions, in contrast to the glacials which were essentially drier. **Palynological** studies in southern Spain, for example (Florschutz et al. 1974) show that instead of a steppe-like vegetation such as characterized the glacial periods, the interglacials were characterized by a more humid assemblage of vegetation including *Fagus, Juglans, Quercus pubescens, Tsuga*, and *Cedrus*.

Faunal and floral fluctuations

The environmental changes of the Pleistocene, as already mentioned with regard to the changing nature of European vegetation in the various interglacials, led to a great impoverishment of flora, particularly of glaciated islands. It has been remarked, for example, by Pennington (1969, p.1), that the comparative poverty of the British flora, compared with that of continental Europe in similar latitudes, is the result of successive wiping-out of frost-sensitive species by repeated glacial episodes. After each glacial period, with its wholesale extinction of plants from Britain, migrating plants and animals followed northwards in the footsteps of the retreating ice, and combined with the descendants of the hardy species which had survived, to re-establish British flora and fauna.

A slightly more complex situation is illustrated by the Irish fauna, where both glacial and sea level fluctuations seem to have been important in determining the present types of animal encountered on the island. Ireland today lacks certain beasts which are encountered in England and Wales. These include the poisonous adder, the mole, the common shrew, the weasel, the dormouse, the brown hare, the yellow-necked field mouse, the English meadow mouse, and others. On the other hand, it does possess a certain proportion of the English fauna. The explanation for this seems to be that as the ice caps retreated animals from the continent (then connected by dry land to England because of the low eustatic sea level at the time), and from the unglaciated tundra area of southern England, crossed over to Ireland by a land bridge. In early post-glacial times however, when the climate had so ameliorated as to permit immigration of temperate species, the Irish Sea was in existence, and the dry passage to Ireland was disrupted. Thus many beasts were unable to cross.

A similar example of the role of the various post-glacial and Late Glacial events in creating the present pattern of fauna is provided by the distribution of bird species in the North American continent (Mengel 1970). During the late Wisconsin glaciation, which reached its peak around 20 000 years ago, the northern Rocky Mountains were covered by the Cordilleran ice, while to the east, the lower ground was covered by the massive Laurentide ice sheet (see **Section 3.3** for a further discussion of the extent of the American ice bodies). There is evidence that as these two ice sheets contracted in Late Glacial times a long arm of tundra and then taiga invaded the lower ground from southern Alberta to the Mackenzie River Delta. The north-west to south-east orientation of this corridor helps to explain what is a peculiar but recurrent feature in the distribution of North American birds and some other animals, namely the strong tendency for essentially eastern taxa, that had adapted to the taiga and its successional stages, to occur north-west to, or nearly to, Alaska, at the apparent expense of western montane kinds that had adapted to montane coniferous forest. The explanation seems to be that the western types were blocked by the persistent but dwindling Cordilleran ice sheet, enabling the eastern taxa to get there first, and to fill the niches: a situation which they have held since.

Broad patterns of biodiversity across Europe can also be related to episodes of geographical separation and subsequent pathways of northward migration with each glacial cycle (see Willis and Whittaker 2000 for a discussion). The three southern peninsulas of Europe (Iberian, Italian, and Balkan) were the main glacial refugia for many plants and animals; and with climatic amelioration and glacial retreat at the end of each cold stage, plants and animals re-expanded across the continent from these different locations. Isolation within the different refugia during glacials caused populations of the same species to become genetically different from each other, and these variations can be observed in some species today. For example, four clades of the European hedgehog have been recognized in different parts of Europe reflecting different centres of post-glacial dispersal of the animal. In France and Britain, for instance, hedgehogs belong to the 'Iberian clade' deriving from a population that was isolated in the Iberian peninsula. Similar effects have been observed in other European fauna such as brown bears, voles, house mice, newts, and grasshoppers, as well as in European flora such as oak, beech, and alder (Hewitt 1999).

The higher diversity of different trees in North American compared with northern European forests relates to the differing topography of the two continents. The presence of east–west trending mountains in Europe and the greater degree of fragmentation and isolation of European glacial refugia meant that migration routes were more constrained than in North America and fewer species were able to survive the cold stages and recolonize the continent after successive glacial cycles.

The question as to what degree the present flora and fauna were able to survive in areas that were glaciated is one of great interest. On the one hand some authorities maintain that the bulk of the fauna of, say, Iceland, is of post-glacial age, and has reached that island by post-glacial diffusion. On the other, there are authorities who consider that certain species were able to exist on small non-glacial peaks (nunataks), rising above the general level of the ice caps (Gjaeveroll 1963). Other people consider that in certain favoured coastal regions there were small 'refugia' where a hardy flora might be able to live through the glacial period. The last two concepts comprise the *Overvintring* concept of certain Scandinavian botanists. That such survival is possible is attested by the flora of present-day Greenland nunataks. Moreover, various Scandinavian and Irish geomorphologists have claimed to find evidence for refugia and nunataks. One line of evidence that has been used in Arctic Norway is the presence of block fields (*felsenmeer*) and other periglacial rather than glacial features on summits. Moreover, notably in Iceland, the present distribution of the flora often shows a bicentric or polycentric form, which ties in better with the idea of diffusion from internal refugia than with the idea that the whole flora was erased – the *tabula rasa* concept – and has been replaced by post-glacial migrations from overseas. If post-glacial migration were responsible for these plants arriving one might expect them to be more widely distributed.

The role of land bridges in the Pleistocene should not be exaggerated, though as will be discussed in **Chapter 6** the fall in relative sea level by around 130 m did expose large expanses of the continental shelf. Certain islands were therefore linked together, or to the mainland. Malta and Sicily, Capri and Italy, the Balearics, the Ionian Islands, and, possibly, Tunisia and Italy, are such examples. Other islands, on the other hand, remained isolated, and their fauna tends to this day to show a greater degree of endemism.

SELECTED READING

On the Pleistocene record:

The general nature of Pleistocene environments is treated in Williams et al. (1998) *Quaternary Environments,* and Wilson, Drury and Chapman (2000) provide a succinct account of Pleistocene climates and environments in *The Great Ice Age: climate change and life.*

The website http://www.quaternary.stratigraphy.org.uk/ provides detailed information on the current officially recognized subdivisions of the Quaternary and their history. The full version of **Fig. 3.3** is at http://www.quaternary.stratigraphy.org.uk/charts/

The Pleistocene of the British Isles is well documented in R.L. Jones and D.H. Keen (1993) *Pleistocene Environments in the British Isles*, and a recent compendium relating to the Pleistocene in America is *The Quaternary Period in the United States* (2004) edited by A.R. Gillespie, S.C. Porter, and B.F. Atwater.

Changes in the Southern Hemisphere are described in J.C. Vogel (1984) (ed.) *Late Cainozoic Palaeoclimates of the Southern Hemisphere*, and in C.J. Heusser (2003) *Ice Age Southern Andes – A Chronicle of Palaeoecological Events.*

On Pleistocene glaciations:

The past extent of Pleistocene glaciations in different parts of the world is treated in depth in parts I, II, and III of *Quaternary Glaciations – Extent and Chronology* (2004) edited by J. Ehlers and P.L. Gibbard, and more recent developments in this field are addressed in *Quaternary Glaciations – Extent and Chronology A Closer Look* (2011) edited by J. Ehlers, P.L. Gibbard, and P.D. Hughes.

Global reconstructions of glaciated areas at 1 kyr intervals since the LGM can be viewed at: http://www.ncdc.noaa.gov/cgi-bin/paleo/peltice.pl and related changes in global topography at http://www.ncdc.noaa.gov/cgi-bin/paleo/pelttopo.pl

Animations of the British Ice Sheet based on recent modelling studies can be accessed at http://www.aber.ac.uk/en/iges/research-groups/centre-glaciology/research-intro/britice-model/

On Quaternary climates:

Climatic aspects of the Pleistocene are treated in depth in H.H. Lamb (1977) *Climate: Present, Past and Future* Vol 2, and although becoming dated, this book still has much to offer the palaeoclimatologist.

P.A. Mayewski and F. White (2002) explain how studies of ice cores have advanced our understanding of Pleistocene climatic change in *The Ice Chronicles: The Quest to Understand Global Climate Change.*

The book *The Great Ocean Conveyor: Discovering the Trigger for Abrupt Climate Change* by W.C. Broecker (2010) provides a fascinating account of the development of ideas about how changes in ocean circulation are related to the abrupt climatic changes seen in the ice cores and other proxy data, and Chris Turney's (2008) *Ice, Mud and Blood: Lessons from Climates Past* offers a readable overview of many aspects of Quaternary climate.

The National Climatic Data Center website http://www.ncdc.noaa.gov/paleo/datavis.html contains graphics of various types of climate data, and maps for simulated climate variables during the LGM can be found at http://pmip2.lsce.ipsl.fr/ the PMIP2 home page.

On ecological characteristics:

Ecological aspects of the Quaternary are well treated in R.G. West (1977) *Pleistocene Geology and Biology*, and in H.J.B. Birks and H.H. Birks (2004), *Quaternary Palaeoecology.*

The Atlas of Palaeovegetation edited by Adams and Faure (1997) and global and regional maps reconstructing vegetation cover at different times since the last interglacial can be found at: http://www.esd.ornl.gov/projects/qen/adams1.html and http://geoecho.snu.ac.kr/nerc.html

? **REVIEW QUESTIONS**

1. Describe what the marine oxygen isotope record and Chinese loess sequences tell us about climate and environmental change through the Quaternary Period (see **Figs 3.2, 3.3,** and **3.4**).

2. In what ways do the ice core records augment our understanding of the long-term climate record of the Pleistocene?

3. Summarize the main changes that occurred with respect to glaciation, permafrost, and loess during the Middle and Late Pleistocene.

4. Using examples drawn from different forms of proxy data, contrast the climate of today with that of the Last Glacial Maximum.

5. Discuss the evidence for abrupt climatic changes over the last glacial cycle.

6. Contrast the ecological characteristics of the Northern Hemisphere mid latitudes associated with full glacial and full interglacial conditions, and explain why ecological changes (e.g. flora, fauna, soils) lagged the associated climate transitions.

7. In what ways is study of the Eemian interglacial relevant for predicting environmental changes associated with human-caused climate change (global warming)?

4

Pleistocene Environments of Lower Latitudes

Chapter overview

Lower latitudes underwent significant environmental changes in the Pleistocene. Although many deserts pre-date the Pleistocene, there appears to have been an accentuation of aridity in the Quaternary, so that the world became a very dusty place and one where, during dry phases (**interpluvials**), sand seas became very active. Large tracts of the tropics display ancient dune fields. However, within the Pleistocene there were both increases and decreases in water availability, and many lake basins and rivers (such as the Nile) show evidence for formerly wetter conditions (pluvials). These changes had important implications for fauna and flora, not least in areas like Amazonia.

4.1 The antiquity of deserts

While it is true that the landscapes and environments of high and mid latitudes were repeatedly and severely transformed by the **glacial** and **interglacial** events that we have just described, low latitudes also underwent major environmental changes during the course of the Quaternary. They were modified by changes in global temperatures, the movements of the great pressure and wind belts, and, above all, by changes in precipitation. Let us start off by a consideration of the age of deserts, and of the ways in which the extent and character of deserts has fluctuated.

Formerly many deserts were regarded as a result of Holocene (post-glacial) progressive desiccation. However, it is now clear that many of our present deserts are old (Goudie 2002). In the case of the Namib (southern Africa) and Atacama (South America) coastal deserts, their climatic development was closely related to plate tectonics and sea-floor spreading in that the degree of aridity must have been largely controlled by the opening up of the seaways of the Southern Ocean, the location of Antarctica with respect to the South Pole, and the development of the offshore, cold Benguela and Peruvian currents. Arid conditions appear to have existed in the Namib for some tens of millions of years, as is made evident by the Tertiary Tsondab Sandstone – a lithified sand sea (Ward 1988) that underlies the current Namib **erg**. Using cosmogenic dating Vermeesch et al. (2010) demonstrated that the residence time of sand within the sand sea is at least one million years. They concluded that, although the individual dunes may be only a few thousand years old, the area between the Sperrgebiet deflation area and the Kuiseb River Canyon has remained covered by sand through multiple cycles of Quaternary climate change in excess of one million years.

Likewise the Atacama appears to have been predominantly arid since at least the late Eocene and possibly since the Triassic (Clarke 2006), with hyperaridity since the middle to late Miocene. The uplift of the Andes during the Oligocene and early Miocene produced a rain shadow effect (Rech et al. 2010) while the development of cold Antarctic bottom waters and the Peruvian current at 15–13 million years ago created another crucial ingredient for aridification (Alonso et al. 1999; Garreaud et al. 2010).

In India and Australia, latitudinal shifts caused by sea-floor spreading and continental drift led to moist conditions during much of the Tertiary, but they entered latitudes where conditions were more arid in the late Tertiary. Isotopic studies in the Siwalik foothills of Pakistan illustrate increasing aridity in the late Miocene, where C_3/C_4 analyses show a change from a C_3 (mainly forested) setting to a C_4 (mainly grassland) setting at about 7 million years ago (Quade et al. 1989).

In China, Miocene uplift and a resulting transformation of the monsoonal circulation caused aridification

(Zhang and Sun 2011). The **aeolian** red clays and **loess** of China may have started to form around 8.5–7.2 million years ago (Qiang et al. 2001) (see **Section 3.5**). The Tarim basin also has aeolian dune sediments dating to c. 8 million years ago (Zheng et al. 2010). It is now recognized that some aeolian deposits in the far north west of China may even date back to c. 25–24 million years ago (i.e. the late Oligocene) (Qiang et al. 2011). Dupont-Nivet et al. (2007) have suggested that some aridification of the Tibetan Plateau may have been associated with widespread global cooling that took place around 34 million years ago at the Eocene–Oligocene transition. Another contributing factor may have been the Late Eocene retreat of the Paratethys Sea from the Tarim Basin (Bosboom et al. 2011). Indeed it is possible that the uplift of mountains and plateaux in Tibet and North America may have caused a more general change in precipitation in the Late Miocene, as is made evident by the great expansion of C_4 grasses in many parts of the world (Pagani et al. 1999).

Sediment cores from the Atlantic off the Sahara contain dust-derived silt that indicates that a well-developed arid area, producing **dust storms**, existed in North Africa in the early Miocene, around 20 million years ago (Diester-Haass and Schrader 1979). It is possible that uplift of the Tibetan Plateau played a role in this by creating a strong counter-clockwise spiral of winds that drove hot, dry air out of the interior of Asia across Arabia and northern Africa (Ruddiman 2001, p. 388). Schuster et al. (2006) have argued that the onset of recurrent desert conditions in the Sahara started 7 million years ago, with dune deposits accumulating in the Chad basin from that time.

The timing of desert initiation in North America is uncertain (see Wilson and Pitts 2010, for a review) though it is possible that some of its deserts may have existed as far back as 15 Myr ago. Uplift of the Sierra Nevada was an important factor in producing a rain shadow effect (Smith 2009).

4.2 Pleistocene intensification of aridity

Although most deserts were, therefore in existence before the Quaternary, in many regions aridity intensified in the late Pliocene and Pleistocene. The analysis of deep-sea cores in the Atlantic off the Sahara provides a picture of long-term changes in dust supply and wind activity. Some date back to the early Cretaceous (Lever and McCave 1983), and dust is present in **Neogene** sediments (Sarnthein et al. 1982). However, aeolian activity appears become more pronounced in the late Tertiary. As Stein (1985, pp. 312–313) reported:

Distinct maxima of aeolian mass accumulation rates and a coarsening of grain size are observed in the latest Miocene, between 6 and 5 Ma and in the Late Pliocene and Quaternary, in the last 2.5 million years.

They attribute this to both a decrease in precipitation in the Sahara and to an intensified atmospheric circulation. The latter was probably caused by an increased temperature gradient between the North Pole and the Equator due to an expansion in the area of Northern Hemisphere glaciation. From about 2.8–2.5 Myr ago, the great tropical inland lakes of the Sahara began to dry out, and this is more or less contemporaneous with the time of onset of mid-latitude glaciation. High dust loadings were a feature of the Pleistocene. Mean late Pleistocene dust inputs were two to five times higher than the values prior to 2.8 Myr ago (deMenocal 1995).

Likewise, loess deposition accelerated in China after around 2.5 Myr ago (Ding et al. 1992). The study of sediments from the central North Pacific suggests that it was around 2.5 Myr ago that there occurred the most dramatic increase in dust sedimentation (Leinen and Heath 1981).

4.3 Dust in the Pleistocene

Deflation by wind from dryland surfaces is, arguably, one manifestation of arid conditions. At certain times during the Quaternary, such as the **Last Glacial Maximum (LGM)** *c.* 21 kyr ago (Mahowald et al. 1999), the world was very dusty. This is indicated by its extensive deposits of loess, the presence of large amounts of aeolian dust in ocean, ice, lake and peat bog core sediments, and even accumulations of desert dust in **speleothems**. These natural archives have been intensively studied for their palaeoenvironmental significance (e.g. Muhs and Bettis 2000; Shichang et al. 2001; Pichevin et al. 2005).

The enhanced dustiness, especially during glacials, may relate to a larger sediment source (e.g. areas of glacial outwash), changes in wind characteristics both in proximity to ice caps and in the trade wind zone (Ruddiman 1997), and the expansion of low-latitude

deserts. It would be simplistic, however, to attribute all cases of higher dust activity to greater aridity in source regions, for as Nilson and Lehmkuhl (2001) point out this is but one factor, albeit important. Also important are changes in the trajectories of the major dust transporting wind systems, changes in the strength of winds in source regions, the balance between wet and dry deposition (which may determine the distance of dust transport), the degree of exposure of continental shelves in response to sea level changes, and the presence of suitable vegetation to trap dust on land. Moreover, it is possible that increased dust loadings during the LGM were not only a product of climatic change but also a contributory factor to that change, and this is something that is now being built into climatic models (e.g. Mahowald et al. 1999; Overpeck et al. 1996) (see **Section 9.3**). Bozzano et al. (2002) on the basis of their analysis of an ocean core off Morocco, found a correlation between dust supply and precessional minima in Earth's orbit. They argued that enhanced solar radiation in the boreal summer, driven by **precession of the equinoxes**, would have increased seasonal temperature contrasts, which in turn amplified atmospheric turbulence and stimulated storminess. In other words, they believe that a crucial control of dust storm activity is not simply aridity, but the occurrence of meteorological events that can raise dust from desert surfaces. However, changes in the hydrological and vegetative state of source regions will also have been very important (Werner et al. 2002).

There is evidence that the strength of the trade winds may have intensified in the Pleistocene as a whole, and also during particular phases of the Pleistocene. In the late Pliocene (from *c.* 3.2 Myr ago), there may have been an increase in atmospheric circulation driven by a steeper pole–equator temperature gradient owing to the development of the bipolar **cryosphere** (Marlow et al. 2000). Within the Pleistocene, analysis of ocean core sediments has shown variability in the grain size characteristics of aeolian inputs to the oceans which may be explained by variations in wind velocities (Parkin 1974; Kolla and Biscaye 1977; Sarnthein and Koopmann 1984). In addition, studies of pollen, **diatom** and **phytolith** influx have shown differences that have also been explained in terms of wind velocity changes (Hooghiemstra 1989). Moreover, changes in **upwelling** intensity and oceanic productivity, established by analyses of benthic **Foraminifera**, have been linked to changes in wind intensity (Loubere 2000).

Evidence from ocean sediment cores

Various studies based on these lines of evidence have indicated that north-east trade wind velocities were higher during glacials, probably because of an intensified atmospheric circulation caused by an increased temperature gradient between the North Pole and the Equator due to the presence of an extended Northern Hemisphere ice cap (Kim et al. 2003; Stein 1985; Ruddiman 1997; Broecker 2002). However, work off Namibia suggests that the south-east trades were also intensified during glacials compared to interglacials (Stuut et al. 2002). Off north-west Africa the highest wind velocities may have occurred during the last deglaciation rather than at the times of maximum ice conditions (Moreno et al. 2001). Moreno and Canals (2004) attribute this to the lowering of North Atlantic sea surface temperatures during deglaciation because of glacial water releases. This in turn strengthened the North Atlantic high-pressure system, caused a high temperature difference between land and sea, and enhanced the trade wind system. It is possible to obtain a long-term measure of dust additions to the oceans by undertaking studies of the sedimentology of ocean cores (Rea 1994). Working in the Arabian Sea, Clemens and Prell (1990) found a positive correlation between global ice volume (as indicated by the marine $\delta^{18}O$ record) and the accumulation rate and sediment size of dust material. Kolla and Biscaye (1977) confirmed this picture for a larger area of the Indian Ocean, and indicated that large dust inputs came off Arabia and Australia during the last glacial. On the basis of cores from the Arabian Sea, Sirocko et al. (1991) suggested that dust additions were around 60 per cent higher during glacials than in post-glacial times, though there was a clear 'spike' of enhanced dust activity at around 4000 years ago associated with a severe arid phase (Cullen et al. 2000). Jung et al. (2004) also report on Holocene dust trends in the Arabian Sea, and suggest that dry, dusty conditions were established by 3800 years ago.

Pourmand et al. (2004) refined this further and showed that high dust fluxes in the Middle East occurred during cold phases such as the **Younger Dryas**, **Heinrich events** 1–7 and cold **Dansgaard–Oeschger stadials**. They attributed this to a weakened south-west **monsoon** and strengthened north-westerlies from Arabia and Mesopotamia. Similarly, a core from the Alboran Sea in the western Mediterranean indicated an increase in dust activity during Dansgaard–Oeschger stadials and Heinrich events (Moreno et al.

2002). In the Atlantic offshore from the Sahara, at around 18 000 years ago the amount of dust was augmented by a factor of 2.5 (Tetzlaff et al. 1989, p. 198). Australia contributed three times more dust to the south-west Pacific Ocean at that time (Hesse and McTainsh 1999) and may have stimulated increases in planktonic productivity on the continental margin (Gingele and De Deckker 2005)

There is particularly clear evidence for the increased dust inputs at the time of the Last Glacial Maximum (LGM), when they appear to have been 2–4 times higher than at present (Kolla et al. 1979; Sarnthein and Koopman 1980; Tetzlaff and Peters 1986; Chamley 1988; Grousset et al. 1998). By contrast they were very low during the 'African Humid Period' (AHP). From 14 800 to 5500 years ago, the mass flux of dust off Cap Blanc, Mauritania, was reduced by 47% (**Fig. 4.1**) (deMenocal et al. 2000a). This is confirmed by analyses of the mineral magnetics record from Lake Bosumtwi (Ghana), which suggest a high dust flux during the last glacial period and a great reduction during the AHP (Peck et al. 2004).

In the Mediterranean, which derives much of its dust load from the Sahara, Larrasoaña et al. (2003) analysed a core from south of Cyprus, using its haematite content as a proxy for dust. It covered a period of three million years. They found that throughout that time dust flux minima occurred when the African summer monsoon attained a northerly position during times of insolation maxima. This increased vegetation cover and soil moisture levels, thereby dampening down dust activity in the Saharan source regions.

Cores from the Japan Sea (Irino et al. 2001) show the importance of dust deposition at the maximum of the LGM, and the amount and modal size of the silt being deposited indicate an intensification of dust supply at that time. In the mid-latitude North Pacific, which is also supplied with dust from Central Asia, dust deposition maxima during the last 200 000 years occurred in MIS 4 to latest MIS 5 and in the middle of MIS 6 (Kawahata et al. 2000). These were seen as times of reduced precipitation during the summer monsoon and of strengthened wind speeds during the winter monsoon.

Dust deposition as recorded in ice cores

Another major record of long-term information on rates of dust accretion is recorded in **ice cores** retrieved either from the polar ice caps or from high-altitude ice domes at lower altitudes. Because they are

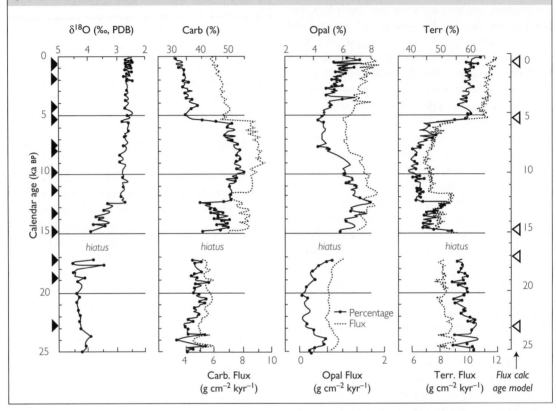

Fig. 4.1 Biogenic carbonate, opal and terrigenous sediment percentage, and flux records (dashed lines, in g cm⁻² kyr⁻¹) from Site 658C, off the coast of Cap Blanc, Mauritania. Terrigenous sediment percentage and flux measurements are used to show trends in dust input. Note the abruptness of the 14.8, 12.3, and 5.5 cal kyr BP sediment composition transitions, each of which were completed within several centuries (from deMenocal et al. 2000a, Fig. 3.) Reproduced with permission from Elsevier

generally far removed from source areas, the actual rates of accumulation of dust in ice cores are generally low, but studies of variations in micro-particle concentrations with depth provide insights into the relative dust loadings of the atmosphere in the last glacial and during the course of the Holocene. Thompson and Mosley-Thompson (1981) pointed to the great differences in micro-particle concentrations between the **Late Glacial** and the post-glacial. The ratio for Dome C Ice Core (East Antarctica) was 6:1, for the Byrd Station (West Antarctica) 3:1, and for Camp Century (Greenland) 12:1. Briat et al. (1982) maintained that at Dome C there was an increase in micro-particle concentrations by a factor of 10 to 20 during the last glacial stage, and they explain this by a large input of continental dust. The Dunde Ice Core from High Asia (Thompson et al. 1990) also shows very high dust loadings in the Late Glacial and a very sudden fall off at

the transition to the Holocene (**Fig. 4.2**). Within the last glaciation, dust activity both in Europe and in Greenland appears to have varied in response to millennial scale climatic events (Dansgaard–Oeschger events and **Bond cycles**) (Rousseau et al. 2002).

These early results are confirmed by the more recent study cores from Antarctica (Delmonte et al. 2004a) (see **Fig. 3.5**). In the EPICA core, the dust flux rose by a factor of ~ 25, ~ 20, ~ 12 in glacial stages 2, 4, and 6 compared to interglacial periods (the Holocene and MIS 5e) (**Fig. 4.3**) (Lambert et al. 2008). Delmonte et al. (2004b) found in the Dome B, Vostok, and Komsomolskaia cores that during the LGM dust concentrations were between 730 and 854 ppb, whereas during the **Antarctic Cold Reversal** (c. 14.5–12.2 kyr ago) they had fallen to 25–46 ppb, and from 12.1 to 10 kyr ago they were between 7 and 18 ppb. Isotopic studies suggest that the bulk of the dust was derived

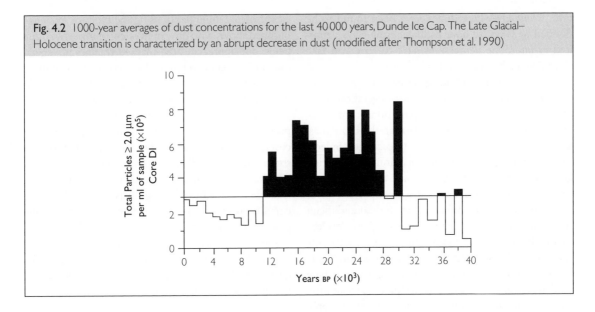

Fig. 4.2 1000-year averages of dust concentrations for the last 40 000 years, Dunde Ice Cap. The Late Glacial–Holocene transition is characterized by an abrupt decrease in dust (modified after Thompson et al. 1990)

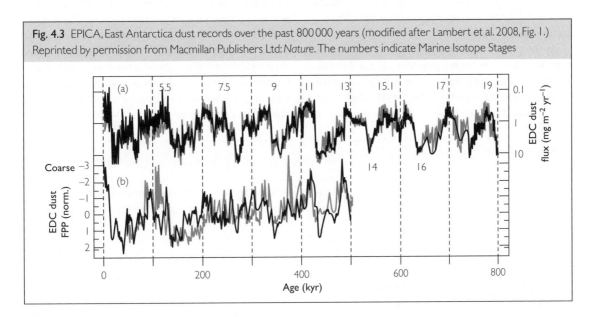

Fig. 4.3 EPICA, East Antarctica dust records over the past 800 000 years (modified after Lambert et al. 2008, Fig. 1.) Reprinted by permission from Macmillan Publishers Ltd: *Nature*. The numbers indicate Marine Isotope Stages

from Patagonia and the Argentinian Pampas (see also Iriondo 2000). For Greenland, a prime source of dust in cold phases was East Asia (Svensson et al. 2000).

4.4 Ancient sand seas

One of the best ways to assess the former extent of desert areas during the Pleistocene dry phases is by studying the former extent of major tropical and subtropical dune fields or sand seas (ergs). Indications that some dunes are indeed fossil rather than active are provided by features like deep-weathering and intense iron-oxide staining, clay and humus development, silica or carbonate accumulation, stabilization by vegetation, gullying by fluvial action, and degradation to angles considerably below that of the angle of repose of sand – normally *c.* 32–3° on lee slopes. Sometimes archaeological evidence can be used to show that sand deposition is no longer progressing at any appreciable rate, whilst elsewhere dunes have been found to be flooded by lakes, to have had lacustrine clays deposited in interdune depressions, and to have had lake shorelines etched on their flanks.

Sand movement will not generally take place over wide areas so long as there is a good vegetation cover, though small 'parabolic' (hairpin) dunes are probably more tolerant in this respect than the more massive 'siefs' (linear) and 'barchans' (crescentic). Indeed, dunes can develop where there is a limited vegetation cover, and vegetation may contribute to their development. Sediment availability is also an important factor in determining dune mobility. It is, therefore, difficult to provide very precise rainfall limits to dune development. Nevertheless, studies where dunes are currently moving and developing suggest that vegetation only becomes effective in restricting dune movement where annual precipitation totals exceed about 100 to 300 mm. These figures apply for warm non-coastal areas. Some opinions of workers from some major desert areas on the rainfall limits to major active dune formation are summarized in **Table 4.1**.

At the present time overgrazing and other human activities on the desert margins may induce dune reactivation at moderately high precipitation levels, and this is, for example, a particular problem in the densely populated Thar Desert of India. When one compares the extent of old dune fields, using the types of evidence outlined above, with the extent of currently active dune fields, one appreciates the marked changes in vegetation and rainfall conditions that may have taken place in many tropical areas. This is made all the more striking when one remembers that decreased Pleistocene glacial temperatures would have led to reduced evapotranspiration rates, and thus to increased vegetation cover. This would, if anything, have tended to promote some dune immobilization. Dune movement might, however, have been accentuated by apparently higher trade wind velocities during glacials and could in some circumstances have led to dune-building without any great reduction in rainfall (Wasson 1984).

Reviewing such evidence from the different continents, Sarnthein mapped the world distribution of ancient and modern ergs (**Fig. 4.4**) and summarized the situation thus (1978, p. 43):

Today about 10 per cent of the land area between 30° N and 30° S is covered by active sand deserts... Sand dunes and associated deserts were much more widespread 18 000 years ago than they are today. They characterized almost 50 per cent of the land area between 30° N and 30° S forming two vast belts. In between tropical rainforests and adjacent savannahs were reduced to a narrow corridor, in places only a few degrees of latitude wide.

Table 4.1 Rainfall limits of active and fossil dunes

Source	Location	Today's precipitation limit for formation of active dunes (mm)	Today's precipitation limit of fossil dunes (mm)	Dune shift (km)
Hack (1941)	Arizona	238–254	305–80	–
Price (1958)	Texas	–		350
Tricart (1974)	Llanos	–	1400	–
Tricart (1974)	NE Brazil	–	600	–
Grove (1958)	West Africa	150	750–1000	600
Flint and Bond (1968)	Zimbabwe	300	c.500	–
Grove and Warren (1968)	Sudan	–	–	200–450
Goudie et al. (1973)	S Kalahari	175	650	–
Lancaster (1979)	N Kalahari	150	500–700	1200
Mabbutt (1971)	Australia	100	–	900
Glassford and Killigrew (1976)	W Australia	200	1000	800
Goudie et al. (1973)	India	200–275	850	350
Sarnthein and Diester-Haass (1977)	NW/Africa	25–50	–	–
Sombroek et al. (1976)	NE Kenya	–	250–500	–

Fig. 4.4 The distribution of active sand dune areas (ergs): (a) today (b) at the time of the Last Glacial Maximum, c. 18 000 ^{14}C yr BP (21 000 cal yr BP) (modified after Sarnthein 1978)

The fossil dunes of northern India

Fossil dunes have been identified both in the Las Belas Valley area of Pakistan, in Gujarat, and in Rajasthan (Verstappen 1970). In Gujarat the dunes, showing calcification, deep gullying, and marked weathering horizons, are normally overlain by large numbers of small microlithic tools, suggesting that there has been relatively little sand movement since **Mesolithic** humans lived in the area. The fossil dunes include parabolics, transverse, longitudinal, and wind-drift types, and are now known to extend as far as Ahmedabad and Baroda in the south, and to Delhi in the east (Goudie et al. 1973) (**Fig. 4.5**). They occupy zones where the rainfall is now as high as 750–900 mm. In the Sambhar salt-lake area in eastern Rajasthan the dunes are overlain by freshwater lake deposits. The base of this lake bed had been dated at about 10 000 years ^{14}C yr BP, suggesting that dune movement may have ceased by that time (Singh 1971). In Saurashtra, the late Pleistocene dunes, locally called 'miliolite', are heavily cemented by calcium carbonate and are used as building stone

(**Plate 4.1**). In the Thar, Singhvi et al. (2010) found that dune accumulation has been highly variable through time, with no less than 12 cycles of dune accretion over the last 200 000 years.

The fossil dunes of Africa

A basically similar picture to India comes from southern Africa, where the Kalahari is dominantly a fossil desert, now covered by a dense mixture of woodland, grassland, and shrubs. Relict dunes are widespread in Botswana, Angola (Shaw and Goudie 2002), Zimbabwe, and Zambia (Thomas and Shaw 1991, Grove 1969) (**Plate 4.2**), and may well extend as far north as the Congo rainforest zone. Several phases of dune activity have occurred in the Late Pleistocene (Lancaster 1989).

North of the Equator, fossil dune fields extend south into the savannah and forest zone of West Africa. The so-called 'Ancient Erg of Hausaland' (Grove 1958), extends into a zone where present

Fig. 4.5 The former extension of the Great Indian Sand Desert in the Late Pleistocene. A: 250 mm mean annual isohyet; B: 500 mm mean annual isohyet; C: former extension of sand desert (after Goudie et al. 1973)

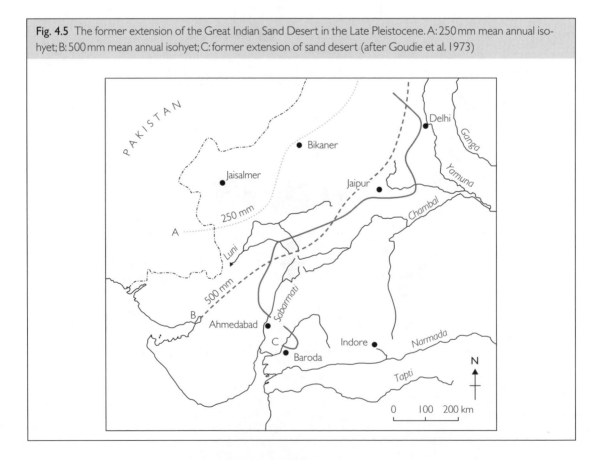

rainfall is as high as 1000 mm per annum (Nichol 1991). Many of the dunes in northern Nigeria are now cultivated, and in the vicinity of Lake Chad dunes have been flooded by rising lake waters. At one stage of the history of this area they blocked and altered the course of the River Niger. Indeed, in the middle Niger area there appear to be several ages of fossil dune, including old deeply-weathered linear dunes, and younger grey-brown and yellow dunes of lesser height. Details of age and pedogenesis are given in Völkel and Grunert (1990).

Further east, in the Sudan, west of the White Nile, a series of fixed dunes known locally as 'Qoz' covers most of the landscape. They extend as far south as 10° N, and merge northwards, locally, with mobile dunes at about 16° N. They succeeded in crossing the Nile, which thus probably dried up at the time of their formation. Again, as in West Africa and India, there appear to have been at least two phases of dune activity. These two phases were interrupted by a relatively wet phase when extensive weathering and degradation took place. The first phase suggests a shift in the wind and rainfall belts of about 450 km southwards, and the second phase of dune building in Holocene times represents a shift of about 200 km (Grove and Warren 1968).

The fossil dunes of North and South America

In the USA parts of the High Plains were formerly covered by large dune fields displaying the characteristic anti-clockwise wheelround features of the dune systems of Australia and southern Africa (Price 1958).

This American system includes the Rio Grande Delta erg. Another ancient American erg is called the Llano Estacado field, and this is outlined at least in part in the present landscape by the topographic grain of etched swales, remnant ridges, and deflated swale ponds and lakes, the latter being orientated along the swales. It has sometimes been suggested that the lineation of the dunes and lakes indicates a former wind pattern diverging as much as 90° from the present pattern, in addition to more arid conditions. These features probably indicate an expansion of the desert to the north and east by the order of 320 km. In Nebraska and South Dakota, the Sandhills, covering an area of 52 000 km², were also more active (Muhs et al. 1997). Some of the aeolian deposits are of considerable

antiquity, and the aeolian Blackwater Draw Formation may date back beyond 1.4 million years (Holliday 1989). Substantial dune development also took place in the Late Pleistocene (Wells 1983), and during the drier portions of the Holocene (Gaylord 1990).

In South America Tricart (1974) used remote-sensing techniques to identify the presence of ancient ergs. One area was in the Llanos of the Orinoco river, where fossil dunes dated to 36 000–11 600 years ago, partly fossilized by Holocene alluvium which was deposited 8700 to 4000 years ago, extend southwards as far as latitudes 6°30′ N and 5°20′ N (Vaz and Garcia-Miragaya 1992). Recent studies of late Pleistocene dunes in northern Amazonia and the Roraima-Guyana region include that of Latrubesse and Nelson (2001). Another erg was in the valley of the lower-middle São Francisco River in Bahia State, Brazil. At the time of its formation the river had interior drainage and did not flow throughout its length. In addition it is likely that aeolian activity was also much more extensive in the Pantanal and in the Pampas and other parts of Argentina (Tripaldi and Forman 2007). Details of relict dune forms in South America are given in Clapperton (1993) and their approximate distribution is shown in Fig. 4.6.

The fossil dunes of Australia

In Australia, fossil dunes associated with an enormous anti-clockwise continental wheelround are developed over wide areas (Nanson et al. 1995), and they have been mapped by Wasson et al. (1988: Fig. 9). They are particularly well displayed on the Fitzroy plains of north-western Western Australia, where they pass under (and thus pre-date) the Holocene alluvium of King Sound (Jennings 1975) (**Plate 4.3**). In the country to the south of the Barkly Tableland they are completely vegetated, have subdued and rounded forms, and appear broadly comparable to those of northern Nigeria. They probably represent a decrease of rainfall in that area of between 150 and 500 mm, representing an equatorward shift of the isohyets by about 8° of latitude or around 900 km (Mabbutt 1971).

Stratigraphic data from various areas indicates that a major phase of dune construction occurred in the late Pleistocene from c. 25 000 years ago (Wasson 1984). More recently, Fitzsimmons et al. (2007) have identified four periods of dune accretion at 111–106, 66–57, 22–11, and 4.3–2.8 kyr ago, while Lomax et al. (2011) found dunes that were at least 380 kyr old, and with phases of

Fig. 4.6 Dune fields and deflation basin areas in South America (modified after Clapperton 1993)

substantial dune deposition at 72–63 and 38–18 kyr ago (**Fig. 4.7**). In the Finke region of the Simpson Desert, regional linear dunes trending north were emplaced between 30 and 18 kyr ago. Between 18 and 10 kyr ago a change in palaeo-wind direction is observed in the dunes that indicate southwesterly or westerly sand-transporting winds. A shift to their present south-easterly wind direction occurred during the last 5000 years (Nanson et al. 1995). Lunette dunes on the lee sides of closed depressions have also had a lengthy history and provide many details about both dune and lake evolution (see, for example, Chen et al. 1995). It is possible that some Australian dunes date back to pre-Pleistocene times. Benbow (1990) suggested that

some from the Eucla Basin may have survived from the end of the Eocene, around 37 to 34 million years ago.

4.5 Pluvial lakes

Pluvial lakes accumulated in basins because of former greater moisture availability resulting from changes in temperature and/or precipitation (**Plate 4.4**).

Pluvial lakes of the Americas

In the USA, the Great Basin held some 80 pluvial lakes during the Pleistocene (**Fig. 4.8**), and they occupied

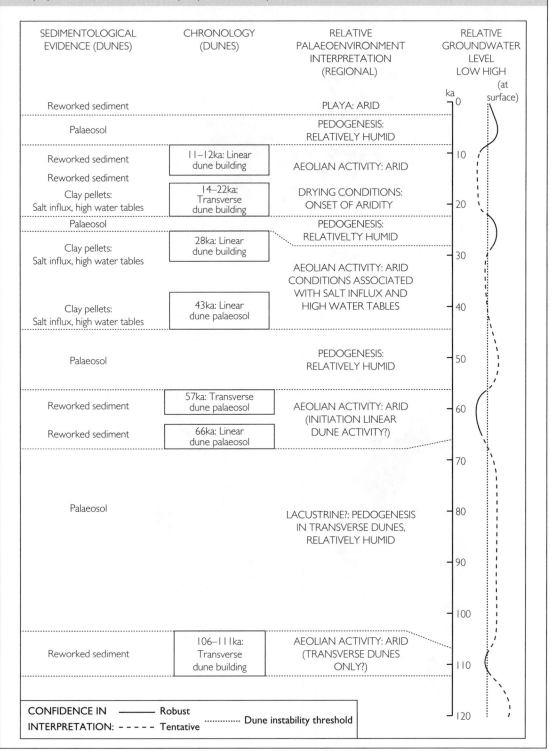

Fig. 4.7 Synthesis of dune activity and hydrology in the Lake Trome region, Strzelecki Desert, Australia, over the past 120 kyr (from Fitzsimmons et al. 2007). Reproduced with permission from Elsevier

Fig. 4.8 Lake basins of the Great Basin, (a) Distribution and extent of pluvial lakes (b) Lake Bonneville (c) Lake Lahontan (from Goudie 2002, Fig. 2.2). Reprinted by permission of Oxford University Press

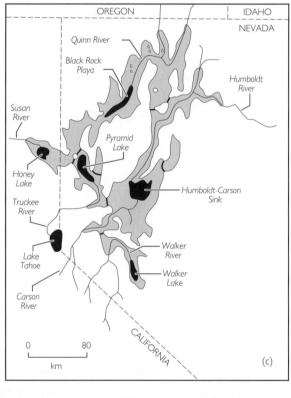

an area at least II times greater than the area they cover today. Lake Bonneville (**Plate 4.5**) was the largest of the late Pleistocene lakes in the Basin and Range Province and was roughly as big as present-day Lake Michigan, about 370 m deep and covering 51 640 km². It was the predecessor to the now modest Great Salt Lake of Utah. Studies from lake sediment cores and relict shorelines from Lake Bonneville (Patrickson et al. 2010; Benson et al. 2011) reveal dramatic changes in the evolution of this palaeolake system (**Fig. 4.9**). After 130 000 years ago there appears to have been a prolonged period of low lake level and **palaeosol** development until around 30 000 years ago when the lake began to rise. The rise may have stopped around c. 24 000 cal yr BP, forming one or more distinct shorelines resulting from the Stansbury highstand, with a brackish lake coving 24 000 km². After 21 000 cal yr BP the lake rose again, forming a freshwater lake which covered 51 640 km². Eventually, the lake level (Bonneville highstand) reached a spillway which entered the Snake River and eventually into the

ocean. It overflowed until c. 17 000 cal yr BP when a sudden failure in the weak deposits in the outlet channel caused the lake to release a superflood of 1 million m³ s⁻¹ into the Snake River, causing a rapid lake lowering of 100 m. By 15 000 cal yr BP, Lake Bonneville was a closed basin with no outlets and water levels were falling rapidly by evaporation alone. A small interval of lake level rise, corresponding in age to the Younger Dryas, formed the Gilbert shoreline, until c. 11 600 cal yr BP, when the lake level fell again. A further post-Gilbert highstand was recorded c. 8000 cal yr BP when lake levels rose c. 80 m (Patrickson et al. 2010).

Lake Lahontan was more complicated in form, covered 23 000 km², and reached a depth of about 280 m. It was nearly as extensive as present-day Lake Erie. River courses became integrated and lakes overflowed from one sub-basin to another. For example, the Mojave River drainage, the largest arid fluvial system in the Mojave Desert, fed at least four basins and their lakes (**Fig. 4.10, Plate 4.6**): Lake Mojave (including present day Soda and Silver lakes), the Cronese basin and the

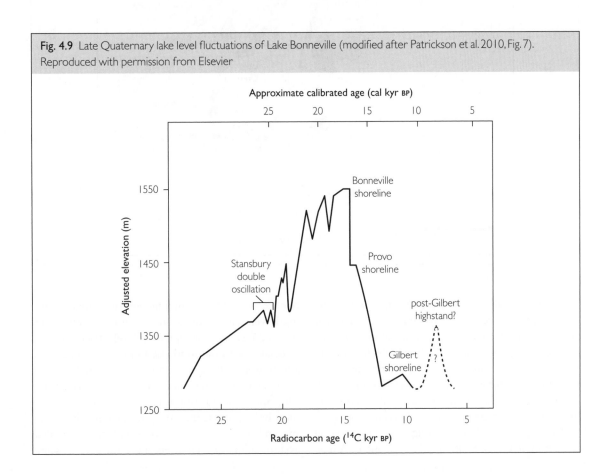

Fig. 4.9 Late Quaternary lake level fluctuations of Lake Bonneville (modified after Patrickson et al. 2010, Fig. 7). Reproduced with permission from Elsevier

Fig. 4.10 Present day and Quaternary lakes in the Mojave River and Owens Valley (California) watersheds and adjacent areas (from Tchakerian and Lancaster 2001, Fig. 2). Reprinted with permission from Elsevier

Manix basin (which includes the Afton, Troy, Coyote, and Harper sub-basins) (Tchakerian and Lancaster 2001; Enzel et al. 2003). The Eastern Californian Lake Cascade comprises a series of lakes which with sufficient discharge during the Quaternary may have linked to an eventual sump below sea level in Death Valley. Unlike Lake Bonneville and Lake Lahontan, however, this cascade never formed a continuous lake.

The high lake levels during the Last Glacial Maximum in the Great Basin region of the USA may

have resulted from a combination of factors, including lower temperatures and evaporation rates, and reduced increased precipitation levels. Pacific storms associated with the southerly branch of the polar jet stream were deflected southwards compared to today.

Other pluvial lakes occurred in the Atacama and Altiplano of South America (Lavenu et al. 1984). The morphological evidence for high lake stands is impressive and this is particularly true with regard to algal accumulations at high levels above the present saline crusts of depressions like Uyuni (Rouchy et al. 1996). There exists a great deal of confusion about climatic trends in this region, not least with respect to the situation at the Last Glacial Maximum and in the mid Holocene (Placzek et al. 2001). Nonetheless, various estimates have been made of the degree of precipitation change that the high lake stands imply. Pluvial Laguna Lejíca, which was 15–25 m higher than today at 13.5 to 11.3 kyr ago and covered an area of 9–11 km² compared to its present extent of 2 km², had an annual rainfall of 400–500 mm, whereas today it has only around 200 mm. Pluvial Lake Tauca, which had a high-stand between 25 and 12 kyr ago, covered c. 60 000 km² and reached depths in excess of 100 m. During this time its drainage basin was linked with Lake Titicaca via the Rio Desaguardero (**Fig. 4.11**). An annual rainfall of 600 mm compared with 200–400 mm today has been estimated (Cross et al. 2001). The water balance of the Altiplano is likely to correspond to changes in the adjacent Amazon Basin, the proximate source of water for the northern Altiplano basin (Sylvestre et al. 1999).

Pluvial lakes of Africa

In the Sahara there are huge numbers of pluvial lakes both in the Chotts of Tunisia and Algeria, in Libya (e.g. Lake Mega-Fezzan) (White and Mattingley 2006),

Fig. 4.11 Pluvial Lake Tauca and the Bolivian/Peruvian Altiplano (from Cross et al. 2001, Fig. 1). Reproduced with permission from Elsevier

in Mali (Petit-Maire et al. 1999) and in the south e.g. Mega-Chad (up to 800 000 km²) (Drake and Bristow 2006) (**Fig. 4.12**). In the Fezzan basin of Libya a lake with an area of 76 250 km² formed (Armitage et al. 2007). Lake Chad has fluctuated hugely in its extent during the late Quaternary (Leblanc et al. 2006). At present it stands at 282 m above sea level (asl), but at an earlier stage a Chad river formed a 40 000 km² delta in association with a lake at 380–400 m asl. The lake then shrank during an arid phase. The lake rose again to 320–330 m asl and formed a distinct shoreline over a distance of some 1200 km and the lake covered some

Fig. 4.12 The Chad Basin showing the extent of Mega-Chad (from Goudie 2002, Fig. 4.17). Reprinted by permission of Oxford University Press

350 000 km². Highstands have been identified between 40 000 and 22 000 years ago and during the early to mid Holocene period. During the early Holocene precipitation levels of 650 mm per year have been estimated, which is at least 300 mm per year higher than for the present day (Kutzbach 1980). Between 22 000 and 12 000 years ago lake levels were much lower, with a drier prevailing climate.

Bosumtwi crater lake in Ghana gives a good record of changes in climate during the Late Pleistocene and Holocene for that region. Low levels occurred at 110, 80, 60, and *c.* 16 kyr ago (Brooks et al. 2005) after which its level began to rise, but there was a **regression** synchronous with the Younger Dryas at 12.6–11.6 kyr ago (**Fig. 4.13**). The lake was at a high level and overflowed the crater in the early Holocene from change to 11.0 to 8.8 kyr ago. There were then various lesser oscillations in its level during the Holocene (Shanahan et al. 2006).

In the Western Desert of Egypt and the Sudan there are many closed depressions or playas (**Plate 4.7**), relict river systems, and abundant evidence of prehistoric human activity (Hoelzmann et al. 2001). Playa

sediments indicate that they once contained substantial bodies of water, which attracted **Neolithic** settlers. Many of these sediments have now been dated (Bubenzer et al. 2007) and indicate the ubiquity of an early to mid Holocene wet phase, which has often been termed the 'Neolithic pluvial' (see **Section 5.10**). A large lake, 'The West Nubian Palaeolake', formed in the far north-west of Sudan (Hoelzmann et al. 2001). It was extensive between 9500 and 5000 cal yr BP, and may have covered as much as 7000 km². If it was indeed that big, then a large amount of precipitation would have been needed to maintain it – possibly as much as 900 mm compared to the less than 15 mm it receives today.

In East Africa, the rift valleys are occupied by numerous lakes around which occur late Pleistocene and early Holocene shorelines. In Ethiopia, the biggest of the pluvial lakes occurred in the rift valley to the south of Addis Ababa. Today the basin is occupied by four smaller remnant lakes, Ziway, Langano, Abiyata, and Shala. At its maximum extent all four lakes were merged and stood up to 112 m above the present surface of Shala. In Afar, Lake Abhé covered an area of

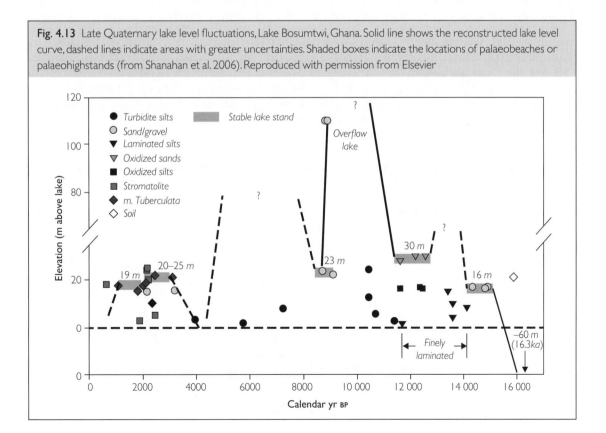

Fig. 4.13 Late Quaternary lake level fluctuations, Lake Bosumtwi, Ghana. Solid line shows the reconstructed lake level curve, dashed lines indicate areas with greater uncertainties. Shaded boxes indicate the locations of palaeobeaches or palaeohighstands (from Shanahan et al. 2006). Reproduced with permission from Elsevier

600 000 km² and had a depth of more than 150 m. Lake Chew Bahir, in the southern Ethiopian Rift, is now either dry or seasonally flooded (**Plate 4.8**). It did, however, attain a depth of 20 m. Palaeo-shorelines, spits and algal stromatolites indicate the former extent of this lake basin (Grove et al. 1975). Further to the south a large water body linked lakes Magadi and Natron (Roberts et al. 1993). Large lakes such as Tanganyika and Malawi have also fluctuated greatly, and notable was a period of severe drought between 135 and 75 kyr ago, when Lake Malawi's volume was reduced by at least 95%. This was followed by dramatically wetter conditions after 70 kyr ago (Scholz et al. 2007) (**Fig. 4.14**).

Between 30 000 and 22 000 years ago East African lake levels tended to be high, but during the LGM numerous lake levels fell (e.g. Lake Tanganyika was about 350 m lower than present) and some basins even became desiccated, including Lake Tana – the source of the Blue Nile (Lamb et al. 2007), Lake Victoria and Lake Albert. Lake Victoria dried out twice between 18 and 15 kyr ago (Stager et al. 2011), began to refill around 14.7 kyr ago and did not achieve open-basin status until about 11.2 kyr ago, when the Victoria Nile was formed (Johnson et al. 1996). Lake Tana also started to overflow into the Nile at 14.75 kyr ago suggesting that the

prior desiccation of both the Nile sources coincided with Heinrich event 1 (H-1) (*c.* 17 to 16 kyr), when the cessation of northward heat transport from the tropical Atlantic disrupted the Atlantic monsoon, causing drought in north tropical Africa (Lamb et al. 2007; Collins et al. 2011; Stager et al. 2011). Whilst Stager et al. (2011) argued that changes in African lake level status were synchronous across the tropics and low latitudes during the H-1 event, Thomas et al. (2012) have argued that lakes across Africa did not respond synchronously with complex spatial variability observed (**Fig. 4.15**). Lake levels were high during the early to mid Holocene, although there is evidence for some abrupt, short-lived lake lowering events during this period, for example at 8200 years ago (Gasse 2000). From 5000 years ago lake levels have generally fallen.

In the Kalahari of southern Africa, Lake Palaeo-Makgadikgadi encompassed a substantial part of the Okavango Delta, parts of the Chobe–Zambezi confluence, the Caprivi Strip, and the Ngami, Mababe, and Makgadikgadi basins (**Fig. 4.16**). It was over 50 m deep and covered 120 000 km², greater than the present area of Lake Victoria (68 800 km²). This makes Palaeo-Makgadikgadi second in size in Africa to Lake Chad at its Quaternary maximum. Its dating is, however,

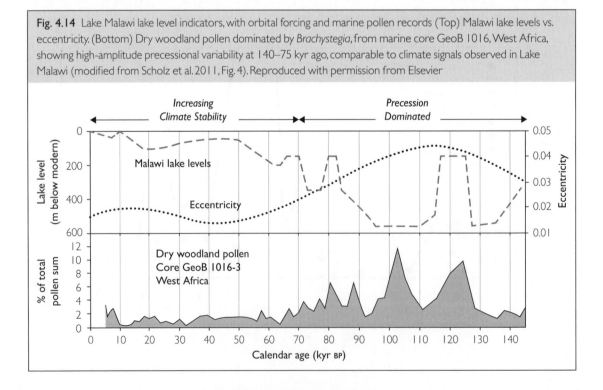

Fig. 4.14 Lake Malawi lake level indicators, with orbital forcing and marine pollen records (Top) Malawi lake levels vs. eccentricity. (Bottom) Dry woodland pollen dominated by *Brachystegia*, from marine core GeoB 1016, West Africa, showing high-amplitude precessional variability at 140–75 kyr ago, comparable to climate signals observed in Lake Malawi (modified from Scholz et al. 2011, Fig. 4). Reproduced with permission from Elsevier

Fig. 4.15 Lake level and marine records for Africa during the Heinrich 1 event. Note spatial variability for the records rather than synchronous drier conditions (modified from Thomas et al. 2012, Fig. 1). Reproduced with permission from Wiley

- ● Dry conditions
- ● Wet conditions
- ○ Intermediate status

problematic (Thomas and Shaw 1991) as is its source of water. Some of the water may have been derived when the now dry valleys of the Kalahari were active and much could have been derived from the Angolan Highlands via the Okavango. However, tectonic changes may have led to inputs from the Zambezi. OSL dates for highstands of Lake Ngami, Lake Makgadikgadi and the Mababe Depression, going back 300 000 years, are provided by Burrough et al. (2009). The Etosha basin was also filled by large bodies of water in the Pleistocene and at various stages in the Holocene (Brook et al. 2007).

Pluvial lakes of the Near East and Arabia

In the Levant and Near East maximum lake levels for the last glacial were achieved at *c.* 25 000 cal yr BP (Robinson et al. 2006). In Lake Lisan (Dead Sea), a major lowering was recorded at 24 000 years ago. During the LGM, lake levels were high but were falling during peak LGM conditions with an increase in regional aridity. Lake levels rose around 15 000 years ago with a highstand recorded in Lake Lisan at 13 000 years ago. During the Younger Dryas, lake levels fell with substantial halite deposition occurring between

Fig. 4.16 The Okavango delta, Makgadikgadi basin, and ancient shorelines (from Goudie 2002, Fig. 7.15). Reprinted by permission of Oxford University Press

13 200 and 11 400 cal yr BP. During the Holocene there was a highstand at 8500 cal yr BP followed by lower lake levels (**Fig. 4.17**).

Within the Rub'al-Khali desert of Arabia extensive lake deposits of Late Pleistocene MIS 5 age (130–74 kyr ago) are present (Rosenberg et al. 2011). These lakes attained depths of up to 10 m and it is thought that rainfall up to five times greater than today occurred in the central desert regions (Wood and Imes 1995). At Lake Mundafan, situated on the eastern slope of the Tuwaiq Escarpment, Saudi Arabia, fossilized floral and faunal remains from the lake deposits yielded an array of large vertebrates including: *Oryx, Gazella,*

Bos primigenius, Equus hemionus, Alcelaphus buselaphus, Bubalus, Hemitragus, Capra, Hippopotamus, Camelus, and *Struthio* (ostrich). Most common species found at Mundafan are Bovidae, which could only have existed if there were extensive grasslands over most of the region.

Evidence for earlier pluvial phases in Arabia from lake sediments is rare given the re-working of the dune fields under arid conditions. Speleothems from Oman record Pleistocene pluvial events at 325–300, 200–180, 135–120, 82–78, and 10.5–6.0 kyr ago. Each pluvial period was coincident with an **interstadial** or interglacial stage of the marine oxygen isotope record (Burns et al. 2001). Speleothems have also been

Fig. 4.17 An integrated, schematic average lake level curve (thick bold line) for the Lake Lisan/Dead Sea over the past 25 000 years (from Robinson et al. 2006, Fig. 6). Reproduced with permission from Elsevier

used to date wet phases in Yemen during the early Holocene and during MIS 5e (the Eemian), 5c, 5a, 7.1, 7.5 and 9. The wettest of these was that of the Eemian (130–123 kyr ago) (Fleitmann et al. 2011). Early to mid Holocene lake deposits are widespread across Arabia. However, they were smaller in size than the MIS 5 lakes. Notable examples have been described from the Highlands of Yemen (Wilkinson 2005), the Rub' al-Khali (Parker et al. 2006) and the Nafud of Saudi Arabia (Schulz and Whitney 1986; Engel et al. 2012).

Pluvial lakes of Central Asia and elsewhere

In Central Asia the Aral–Caspian system expanded (**Fig. 4.18**). At several times during the late Pleistocene (Late Valdai) the level of the Caspian Sea rose to around 0 m (present global sea level) compared to –27 m today and it inundated a huge area, particularly to its north. In the Early Valdai it was even more extensive, rising to about +50 m asl, linking up to the Aral Sea, extending some 1300 km up the Volga River from its present mouth and covering an area greater than 1.1 million km² (compared to 400 000 km² today). At its highest it may have overflowed into the Black Sea. In general, such **transgressions** have been associated with warming and large-scale influxes of meltwater (Mamedov 1997), but they are also a feature of glacial phases when there was a decrease in evaporation and a blocking of groundwater by **permafrost**. Regressions occurred during interglacials and so, for example, in

the Early Holocene the Caspian's level dropped to –50 to –60 m.

Large pluvial lakes also occurred in the drylands and highlands of China and Tibet, where levels were high from 40 000 to 25 000 years ago (Li and Zhu 2001). Similarly the interior basins of Australia, including Lake Eyre, have shown major expansions and contractions, with highstands tending to occur in interglacials (Harrison and Dodson 1993). Lake Eyre was large during the period from 130 to 80 kyr, attaining its peak at MIS 5e at *c.* 125 kyr. Other highstands were identified at 86, 64, 40, and 10 kyr ago (Webb 2009; Kafri and Yechieli 2012), but was largely dry during the LGM (**Fig. 4.19**). Another major palaeolake – Lake Mega-Frome – also developed in MIS 5 as a result of the coalescence of Lakes Frome, Blanche, Callabonna, and Gregory (Cohen et al. 2011). Generally wetter conditions prevailed between 30 000 and 24 000 years ago and were followed by general aridity at the glacial maximum until 12 000 years ago. During the early to mid Holocene lake levels rose again with peak moisture occurring at *c.* 7000–6000 years ago (Quigley et al. 2010).

Summary

As can be seen from these regional examples, pluvial lakes were widespread (even in hyper-arid areas), reached enormous dimensions, and had different histories in different areas. Pluvials were not in phase in all regions and in both hemispheres (Spaulding 1991;

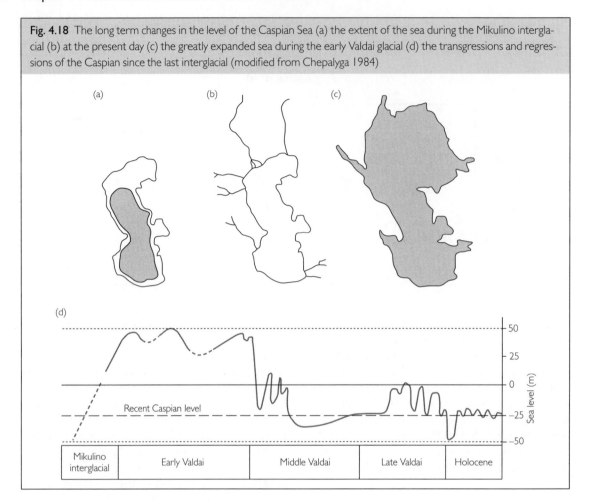

Fig. 4.18 The long term changes in the level of the Caspian Sea (a) the extent of the sea during the Mikulino intergla-cial (b) at the present day (c) the greatly expanded sea during the early Valdai glacial (d) the transgressions and regres-sions of the Caspian since the last interglacial (modified from Chepalyga 1984)

Thomas et al. 2012). In general, however, dry condi-tions during and just after the Last Glacial Maximum and humid conditions during part of the early to mid Holocene appear to have been characteristic of tropi-cal areas, though not of the south-west USA (Street and Grove 1983). Some lakes were also high during the Eemian interglacial (MIS 5e).

4.6 The Nile

The Nile represents the longest route of fluvial sedi-ment transport on Earth (Gazanti and Megid 2003) and provides a case study of how rivers in low latitudes have responded to environmental change. The Nile has a long history that over the Late Cainozoic has involved the amalgamation of various unconnected

systems (Butzer and Hansen 1968; Goudie 2005). Superimposed on that have been changes brought about by climate (and associated vegetational changes) of the Pleistocene.

The first clear evidence that the Egyptian Nile was linked to flow from the Ethiopian Highlands occurs in the so-called Pre-Nile sediments described by Said (1981). This phase may have lasted from c. 700 000 to 200 000 years ago. As Said reports (p. 51), 'The Pre-Nile represented a vigorous and competent river with a copious supply of water and a wide flood plain. Its sediments are coarse, massive and thick'. They are termed the Qena sands. The presence of pyrox-enes in appreciable quantities points to the Ethiopian Highlands as having contributed sediments to the Pre-Nile. The Nile cone appears to have been built to a large extent by the Pre-Nile sediments (Said 1981,

Fig. 4.19 Lake Eyre lake level curve for the past 150 000 years (from Kafri and Yechizli 2012, Fig. 8). Reprinted with permission of Elsevier

p. 55). Their thickness may exceed 1000 m. The Pre-Nile occupied a course to the west of the modern Nile. Its delta 'had an area three times as large as that of the modern delta ... The volume of sediment that was deposited by this complex Pre-Nile system exceeded 100,000 km³' (Said 1981, p. 58). The annual suspended matter load must have been around 0.2 km³, a figure which is almost five times that carried by the modern Nile prior to its control by dams and barrages.

Around 200 000 years ago, Said believes, a pluvial phase occurred in Egypt which resulted in the accumulation of thick, locally-derived gravels, which overlie the Pre-Nile sediments. Many of these are derived from the Red Sea hills of the Eastern Desert. It was a time when the Ethiopian contribution was small, possibly because of episodic uplifting of the Nubian Swell.

The formation and initial linking of Lake Victoria to the Nile seems to be the result of regional tilting and it may be only 400 000 years old (Johnson et al. 2000). Similar tilting also accounts for the form of Lake Kyoga (Doornkamp and Temple 1966). The establishment of the link of Lake Victoria with the Nile is important because the waters of the White Nile provide most of the Nile's water during the

Ethiopian dry season. During Pleistocene dry phases some of the East African lakes became closed saline systems and did not overflow into the Nile. Lake Victoria only became directly reconnected to the Nile system at *c.* 13 kyr ago (Beuning et al. 2002). Lake Albert has also from time to time been a closed lake, with no overflow to the Nile, as for several thousand years during the Late Pleistocene (Beuning et al. 1997; Talbot et al. 2000). Conversely some lakes currently cut off from the Nile (e.g. Chew Bahir and Turkana) overflowed into it during wetter phases via the Sobat (Nyamweru 1989). Further south, Lake Rukwa has in the past overflowed into Lake Tanganyika as a result of tilting and Holocene climatic changes (Delvaux et al. 1998).

Sapropel (organic-rich mud) layers in the eastern Mediterranean give a good picture of Nile flow variability in the Late Pleistocene. They were created by considerably enhanced flow, which caused collapse of deepwater ventilation and/or the elevated supply of nutrients, which fuelled enhanced productivity (Emeis et al. 2000; Kallel et al. 2000; Scrivner et al. 2004). Sapropel development appears to have coincided systematically with Northern Hemisphere insolation maxima related to the orbital cycle of precession, which intensified the African monsoon (Tuenter et al. 2003). Ethiopian rivers, fed by strong monsoon rains in summer, caused a strong seasonal flood in the Nile's discharge (Rossignol-Strick 1985).

Twelve sapropel events have occurred during the last 465 000 years: while most occurred during interglacials, two formed during glacial periods. Sapropel 1 belongs to MIS 1 (the Holocene), Sapropels 3–5 belong to MIS 5 (last interglacial), Sapropel 6 occurs within MIS 6, Sapropels 7–9 within MIS 7, Sapropel 10 in MIS 9, Sapropel 11 in MIS 11, and Sapropel 12 in MIS 12. Sapropels 6 and 8 occurred during colder, drier phases. Although these were cold times, they were still times of higher Northern Hemisphere summer insolation. Sapropel 6 has been dated to MIS 6.5 (at *c.* 175 kyr ago) (Masson et al. 2000). During each event, sapropel deposition was not necessarily continuous, and a *c.* 800-year disturbance has been identified in Sapropel 5 (at *c.* 125 kyr ago) (Rohling et al. 2002).

There may, however, have been lower sediment fluxes during sapropel events in spite of the greater flows. Krom et al (2002) have argued that higher monsoonal rainfall associated with the northward movement of the **Intertropical Convergence Zone** would have resulted in increased vegetative cover in the Ethiopian Highlands and to reduced sediment yield. The reduced inputs of Blue Nile sediment during times of saproprel formation contributed to the increased primary productivity by reducing the amount of phosphate removed on particles and to the observed change in N limitation in the eastern Mediterranean, which are important characteristics of sapropel deposition. Decreases in sediment flux from Ethiopia are significant in that in the modern Nile River, around 95 per cent of the sediment transported is derived from that area. The White Nile only contributes 3 per cent of the total load (Woodward et al. 2001).

The most recent period of sapropel deposition (Sapropel 1) started at *c.* 9000 years ago and ended after *c.* 6000 years ago; and like earlier sapropels, was caused by a massive input of Nile freshwater (Freydier et al. 2001), though it was not as intense as Sapropel 5 (Scrivner et al. 2004; Krom et al. 1999).

During the last glacial dry phase (from *c.* 20 000 years ago), the Ethiopian Highlands were cold and the treeline stood 1000 m lower than today. Large volumes of coarse and fine sediments were liberated and caused net aggradation downstream. Aeolian activity was extensive and many of the drainage systems, including the White Nile and its tributaries, suffered reduced discharge and were sometimes partially blocked by dunes (Williams and Adamson 1974). The White Nile may periodically have ceased to flow during all or part of the year (Williams et al. 2000).

Towards the end of the last cold stage, more humid conditions developed as the increase in Northern Hemisphere summer insolation (see **Fig. 5.6**) increased the intensity of the African monsoon and led to the greening of the Sahara. Thus African lake levels and Nile flows were higher between 12.5 and *c.* 5 kyr ago, although this period was punctuated by several short-lived dry periods. The main channel of the Nile shifted from its cold stage, sediment-charged braided system to a predominantly single channel as the Ethiopian uplands became stabilized by vegetation and sediment-yield declined. Large lakes formed in Egypt and the Sudan (Goudie 2003).

High **strandlines** along the White Nile indicate flood flow about 3 m above the modern maximum flood stage during the early Holocene (Williams and Adamson 1980). Tributaries, such as the Wadi Howar, contributed discharge from a large area of the Sahara (Pachur and Kröpelin 1987). From *c.* 6600 to 2700 years ago flow of the Nile was flashy, though intensified aridification after *c.* 4500 BP coincided with diminished

Nile flow and increased sedimentation (Stanley et al. 2003).

4.7 Faunal and floral changes in the tropics

The massive environmental changes described so far in this chapter have led to changes in floral and faunal distribution in the tropics, and to curious or anomalous patterns. The classic example of this is the distribution of the crocodile in Africa. It was formerly ubiquitous in the rivers of that continent from Natal to the Nile. Today it is found in pools in the Tibesti Massif in the heart of the Sahara, 1300 km from either the Niger or the Nile, and clearly isolated. There is no likelihood of natural migration there across the arid Saharan wastes, given present hydrological conditions, so that pluvial conditions presumably played a role.

Another example from Africa illustrates the way in which the flora of East African mountains has become isolated. The distinctive tree heath (*Erica arborea*) occurs in disjunct areas including the mountains of Ruwenzori, the Ethiopian mountains, the Cameroon Mountains in West Africa, and the peaks of the Canary Islands. In addition to this highly fragmented distribution in Africa, the plant has an extensive range in Europe and Iberia to the Black Sea. Once again it seems likely that post-glacial changes in temperature and rainfall have led to this position. In general, because of its height characteristics, the African continent would have been particularly severely affected by temperature depression in the glacials. The effect of a temperature depression of 5 °C would have been to bring down the main montane biomes from around 1500 m to 700 or 500 m (Moreau 1963, Olago 2001). Instead of occupying a large number of islands as it does now, and as it must have done in interglacials, the montane type of biome would have occupied a continuous block from Ethiopia to the Cape, with an extension to the Cameroons. The strictly lowland biomes, comprised of species today that do not enter areas above 1500 m, would, outside West Africa, have been confined to a coastal rim and to two isolated areas inland (the Sudan and the middle of the Congo basin).

The substantial degree of change in the altitudinal zonation of vegetation on tropical mountains during cold phases can also be demonstrated from the highlands of Latin America, tropical Africa, and southwest Asia and the western Pacific (**Fig. 4.20**). Detailed pollen analysis from numerous lakes and swamps (Flenley 1988) shows that over the last 30 000 years the major vegetation zones have moved through as much as 1700 m. The boundary between the forest and the alpine zone above it was low before 30 000 years ago, shows a slight peak (of uncertain height and imprecise date) between 30 000 and 25 000 years ago, reaches an especially low point at 18 000 to 15 000 years ago (more or less equivalent to the Last Glacial Maximum in higher latitudes), shows a steep climb as climate ameliorated between 14 000 and 9000 years ago, and reaches modern altitudes or slightly above them by about 7000 years ago.

Further massive changes in African biomes would have been occasioned by changes in humidity as well as of temperature. In West Africa, where the great sand ergs of the Sahara encroached as much as 500 km on the more moist coastal regions, the southward movement of the vegetation belts cannot have failed to have powerful effects on the flora and fauna, since at present the West African rainforests nowhere reach inland by so much as 500 km. If the entire system of vegetation belts had shifted south as much as did the Saharan dunes, then the whole of the West African forests would have been eliminated against the coastline.

An indication of the timing and degree of rainforest disruption in West Africa is given by a consideration of pollen analyses undertaken in Lake Botsumtwi in southern Ghana (Maley 1989). These show that at the time of the Last Glacial Maximum, and especially between 20 000 and 15 000 years ago, the lake had a very low level. Moreover, the principal pollen results show that before about 9000 years ago forest was largely absent in the vicinity. Indeed, between the present and *c.* 8500 years ago arboreal pollen percentages oscillated from 75 to 85 per cent, while before 9000 years ago they were generally below or close to 25 per cent. During the period from 19 000 to 15 000 years ago arboreal pollen percentages reached minimum values of about 4 and 5 per cent. Trees had in effect been replaced at that time by grassland and herbaceous taxa.

In the 1960s and 1970s it was believed that severe reduction in the extent of the Amazonian rainforest might have had major implications for fauna and flora. In particular a 'refugia hypothesis' was put forward in which it was suggested that the breaking up of the rainforest into small patches in a sea of savannah might have

Fig. 4.20 Summary diagram of the Late Quaternary low-latitude vegetational changes (after Flenley 1979) (a) New Guinea Highlands (b) Colombian highlands led to the development of endemism, thereby helping to explain the great biodiversity of the region (Haffer 1969; Prance 1973; Brown et al. 1974; Van der Hammen 1974).

However, it needs to be pointed out that the postulated explanation of species diversity in Amazonia by means of the forest refugia hypothesis is not without controversy. In particular, Colinvaux (1987) argued that there is little completely convincing evidence for very marked precipitation diminution in Amazonia in the Pleistocene, and he believes that a better explanation is that the species diversity results from ongoing geomorphological disturbance and fire in an area which is far from homogeneous. He concluded (p. 112):

That the Amazon basin supports the highest species diversity of any ecosystem on Earth is in part due to its great size which allows subtly varied climatic regimes in its different parts. Also this rainforest, more than the rainforests of Asia and Africa, may experience an added diversity-inducing system of disturbance because of the extraordinary erosion pressures of its great rivers driven by Andean precipitation.

Rossetti et al. (2005) suggested that the situation is more complex than this and that rather than being a monotonous place during the Quaternary, Amazonia was a place with frequent changes in landscape because of climatic and tectonic effects. They suggest that changes in the physical landscape stressed the biota, resulting in speciation and biodiversity.

The question as to the extent to which Amazonia dried up at the Last Glacial Maximum has been a hotly debated topic. On the one hand there are those who on the basis of geomorphological evidence (such as aeolian dunes and changes in river sedimentology) argue for sharply reduced moisture conditions (e.g. Filho et al. 2002; Latrubesse and Nelson 2001). They have been supported by some palaeoecologists (e.g. van der Hammen and Hooghiemstra 2000). On the other hand there are others, using evidence from the organic composition of Amazonian deep-sea fan sediments (Kastner and Goñi 2003) or other forms of palaeoecological analysis, who argue that extensive forested areas persisted during the LGM (e.g. de Freitas et al. 2001; Colinvaux and De Oliveira. 2000), and this view has been confirmed by some modelling simulations (Cowling et al. 2004).

Recent lake sediment and speleothem records show that the Atlantic rainforest area of Brazil experienced climatic changes during the Quaternary. **Stable isotope** analyses from Santana Cave and arboreal pollen from peat deposits at Colônia (Ledru et al. 2005) show that diminution in moisture and changes in the length of the dry season had a strong impact on rainforest distribution during the Quaternary (**Fig. 4.21**). They attributed this to changes in the position and incursion of polar air from the Atlantic across the region, affecting precipitation and cloud cover. These changes broadly correspond with records from marine and ice core sources. Lake sediment geochemical analyses in Northern Amazonia indicate that this region experienced dry conditions from 35 000 to 25 000 years ago, whilst during the LGM it was wet (Bush et al. 2002). However, precipitation did not necessarily change synchronously across Amazonia, because the vast watershed was undoubtedly influenced by local moisture sources.

In the subtropics, the climatic gradients across Arabia during the Quaternary, and the interface between mountain, desert and coast have had a profound effect upon the vegetation of this region. The flora of southeastern Arabia (northern Oman and the UAE) has a number of endemic species with strong similarities to the floras of Iran and south-west Pakistan. Floral elements that are found in the Makran region of Iran and in northern Arabia include *Euphorbia larica*, *Amygdalus*, and *Ficus*. Ghazanfar (1999) suggested that the floras of these regions were previously connected during the Quaternary when the Arabian Gulf region was dry and there was a land bridge. For example, during the Last Glacial Maximum global sea level was *c.* 120 m lower than present and the Gulf was free from marine influences. By 14 000 years ago the Straits of Hormuz had opened up as a narrow waterway and by 12 500 years ago marine incursion into the central basin of the Arabian Gulf had started. The present shorelines were reached by 6 000 years ago, and exceeded as global sea level rose 1–2 m above its present level. Thus during the Quaternary elements of the flora became separated during periods of high sea levels, which led to periods of geographical isolation, the most recent of which has occurred during the Holocene with the flooding of the Arabian Gulf via the Straits of Hormuz.

In addition to the above Asian influence upon the vegetation types, there are taxa which are associated with African elements of the flora (Ghazanfar and Fisher 1998). The composition of plant communities and the morphology and distribution of the endemic species indicates close floristic relations between these two areas. The flora contains several species that are of east African origins including *Acacia* spp. and *Zizyphus*.

Fig. 4.21 Comparison between arboreal pollen (%) from Colônia and the Santana speleothem record (both from Brazil) and the GRIP and Vostok ice core records and the marine ODP site 633 isotope record (from Ledru et al. 2005, Fig. 3). Reprinted with permission of Elsevier

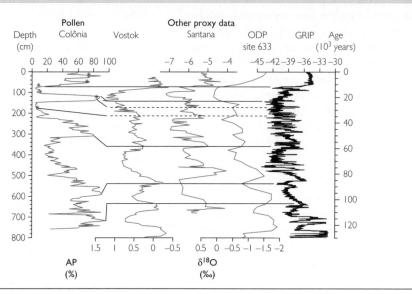

It is likely that past changes in climate across the Arabian Peninsula during the Quaternary have modified the composition, distributions and boundaries of these floras. As for former climates, the present day variations in temperature, humidity, water availability and salinity control the vegetation associations.

 SELECTED READING

Climate changes in low latitude arid areas are surveyed in: A.S. Goudie (2002) *Great Warm Deserts of the World: Landscape and Evolution.*

Fluctuations in tropical river discharges in the Quaternary are summarized in: K.J. Gregory and G. Benito (eds.) (2003) *Palaeohydrology. Understanding Global Change.*

The nature and consequences of climate changes in the tropics are well summarized by: M.F. Thomas (1994) *Geomorphology in the Tropics.*

A comprehensive review of desert climatology and hydrology past and present is contained in: S.E. Nicholson (2011) *Dryland Climatology.*

 REVIEW QUESTIONS

1. Give an account of the reasons for Pleistocene intensification of aridity at the global scale.

2. Describe and explain variations in atmospheric dust loading during the Pleistocene.

3. What evidence is there that parts of the tropics were drier at certain times during the Quaternary?

4. What evidence is there that some low-latitude lakes were more extensive in the Quaternary than they are now?

5. Were there biological refugia in Amazonia during dry phases, and what controversies surround explanations of biodiversity in this area?

6. What are the dates of the main pluvial phases in (a) the Late Pleistocene and (b) the Holocene?

7. Consider the reasons why the timing of pluvials in western USA was different from that of North Africa.

8. What similarities exist between low-latitude changes and the Pleistocene and Holocene record for the mid to high latitudes?

5

Environmental Change in Post-glacial Times

Chapter overview

It has become increasingly clear from research over recent years that climatic and environmental conditions of post-glacial times have been far from stable. Certainly the period following the Last Glacial Maximum (the Late Glacial) saw rapid and dramatic climatic oscillations; but we now know that there have also been a series of events that have been of considerable intensity during our current **interglacial** (the Holocene), including a major, short-lived fluctuation around 8200 years ago, various phases of renewed glacial advance termed neoglacials, and some phases when conditions were even warmer than now. In the tropics, there have been alternations of **pluvials** and droughts, with, for example, the Sahara being a savannah in the early to mid Holocene. Natural climatic changes had an impact on vegetation, but it is also a time when humans had an increasing impact on the environment.

5.1 A stable Holocene?

The ending of the **Last Glacial** period was not the end of substantial environmental change, though from time to time the existence of any major changes of climate has been doubted. For instance, Raikes (1967) put forward the hypothesis that 'From at latest 7000 BC, and possibly earlier, the worldwide climate has been essentially the same as that of today'. He believed that with the single exception of localized changes induced by large-scale, **eustatic–isostatic** marine **transgression**, all changes since about 7000 BC have been 'local, random, and of short duration'.

Raikes was especially sceptical about pollen, zoological, and historical evidence for Holocene climatic change, and pointed out correctly that human activity had influenced vegetation, that animals are often poor ecological indicators, and that some archaeological findings, which had been interpreted as resulting from climatic/environmental changes, could instead be explained by non-climatic factors. Such views were strengthened by Greenland **ice core** studies of the 1970s and 1980s that showed a relatively invariant Holocene $\delta^{18}O$ record compared with the preceding Pleistocene. Only in the past two decades or so have analyses of Greenland ice cores begun to reveal more of the complexities of Holocene climatic change (e.g. O'Brien et al. 1995) that have long been inferred from other proxy records.

There are a whole series of faunal and floral remains which indicate, for instance, the relatively higher temperature conditions of the **Hypsithermal** interval, there are a large number of radiocarbon dates which illustrate the fluctuations of glaciers, there are a large number of dates of lacustrine sediments indicating Holocene **lacustral** phases in the tropics and subtropics, and there are the more recent meteorological and hydrological records which indicate considerable changes and fluctuations in the last two centuries (**Chapter 7**). Such evidence taken as a whole shows clearly that the concept of a stable Holocene environment is quite untenable. To be sure, climatic changes were less severe than those of the Pleistocene, perhaps varying as a global average through a range of around 2 °C, but they were, nevertheless, significant. Study of Holocene environments has particular relevance for understanding present and future climatic change because through much of the epoch boundary conditions used in climate modelling (e.g. distribution of land and sea, area of polar ice, sea surface temperatures, etc.) have been similar to those of the present. Thus, the range of natural Holocene climatic variability must be considered in both detecting and predicting human-caused climatic change. Oldfield (2005, p. 2) makes the point eloquently:

> The Holocene period thus emerges not as a bland, pastoral coda to the contrasted movements of a stirring Pleistocene symphony; rather we now see it as a period of continuous change, the documenting and understanding of which becomes increasingly urgent as our concerns for future climate change grow.

This chapter is concerned with both the evidence for Holocene changes, and with the nature and influence of the changes themselves. It starts off by considering the nature and effects of the transition from a glacial to a non-glacial environment, and then it considers some of the major events of the Holocene itself. In comparison with the Pleistocene, chronologies of Holocene environmental change can often be established with more precision. However, it is important to be aware of the difference between the actual, calendar year ages of events and the ages for the same events derived by radiocarbon dating (aspects of radiocarbon dating and calibration of radiocarbon dates are discussed in **Section 2.5**). A potential source of confusion is that much of the older literature discusses Holocene environments in terms of uncalibrated radiocarbon-derived ages. As explained in **Chapter 2** and referred to later, the differences between ^{14}C dates and actual ages are particularly large for events in the **Late Glacial** and early Holocene.

In this chapter true ages are discussed as either 'years ago', as dates in BC or AD, or indicated by 'cal yr BP' (calibrated to calendar years before present). 'Years ago' implies an approximate estimate compared with usage of BC/AD or 'BP', and it is important to remember that BP is always taken as before AD 1950. Occasionally the notation 'b2k' is used for 'years before AD 2000'. When referring to radiocarbon-based ages that have not been calibrated, the form '^{14}C yr BP' is used. **Figs. 2.6** and **2.7** can be used to convert between ^{14}C and calendar year ages.

5.2 The transition from Late Glacial times

The **Last Glacial Maximum** (LGM), as we have already seen, occurred during **MIS** 2 between about 26 000 and

19 000 years ago. Studies of past sea level indicate that the global ice volume decreased by about 10 per cent within a few hundred years of 19 000 calendar years ago, thereby bringing the LGM to an end (Yokoyama et al. 2000). As discussed in **Chapter 3**, the sharp decreases in $\delta^{18}O$ seen in deep-sea sediment cores marking the end of glacial stages are referred to as **terminations**. Termination I marks the transition between MIS 2 and MIS 1, and is the marine expression of the most recent **glacial/interglacial** transition with the oxygen isotope signal in marine sediments responding to rapid deglaciation following the LGM. Studies of **eustatic** sea level have indicated two major meltwater pulses (MWP 1A/B) resulting from rapid ice-sheet wastage – the first dated to between 14 650 and 14 310 years ago (MWP 1A) (Deschamps et al. 2012), and the second centred *c.* 11 300 years ago (MWP 1B) (Stanford et al. 2011). Because of the relatively slow accumulation rates of deep-sea sediments, terminations (including Termination I) identified in long marine $\delta^{18}O$ records appear to indicate rapid and discrete shifts from higher to lower global ice volume. However, higher resolution **proxy data** from a variety of sources, including ice cores, lacustrine sediments and glacial deposits, have shown that Termination I was, in fact, much more complex than this. Over the course of the last glacial/interglacial transition, there were dramatic oscillations in climate, and glaciers did not retreat at constant rates. The years between the LGM and the beginning of the Holocene are usually termed the Late Glacial, and they were marked by various **stadials** and **interstadials**. In Europe, for example, a number of distinct cool phases interrupted the retreat of the Scandinavian ice margins, and there were times of glacial readvance. There were also a number of short warm phases during which glacial retreat was speeded up.

Study of the Late Glacial is particularly important because, spanning the most recent glacial/interglacial transition (Termination I), there is much more available evidence than is the case for previous transitions and the insights gained are relevant for understanding the nature of glacial terminations in general. Climatic and environmental changes over the course of the Late Glacial have been the focus of intensive study recently as more higher resolution records have become available and dating methods have been improved.

The traditional Late Glacial subdivisions

The classic threefold division (see **Table 1.5**) into two cold zones (I and III) separated by a milder interstadial

(II) emanates from a type section at Allerød, north of Copenhagen, where an organic lake mud was exposed between an upper and lower clay, both of which contained pollen of *Dryas octopetala* (mountain avens), a plant tolerant of severely cold climates. The lake muds contained a cool temperate flora including some tree birches, and the milder stage which they represented was called the Allerød interstadial. The interstadial itself, and the following **Younger Dryas** temperature reversal, are sometimes called the Allerød oscillation. The term Younger Dryas is used to distinguish this later cold phase from the 'Oldest Dryas' (a stadial postdating the LGM that underlies the sequence) and from the 'Older Dryas' that is sometimes recognized as a brief cold phase separating the Allerød from an earlier period of Late Glacial warmth known as the Bølling. As described in **Section 1.2**, the Jessen–Godwin **pollen zonation scheme** for Scandinavia and Britain was linked to this Late Glacial climatic scheme.

It is important to be aware that this scheme of subdivision was originally intended for changes recognized in Scandinavia, but because similar changes were identified in Late Glacial sequences in other parts of the world, the terminology became applied much more widely – and sometimes erroneously – in attempts to correlate various warm/cold events with the north-west European sequence. This led Mangerud et al. 1974 to propose strict chronological definitions in **radiocarbon years** BP for each of the terms as follows:

- Oldest Dryas and Bølling (pollen zones Ia and Ib) 13 000 to 12 000 ^{14}C yr BP

- Older Dryas (pollen zone Ic) 12 000 to 11 800 ^{14}C yr BP

- Allerød (pollen zone II) 11 800 to 11 000 ^{14}C yr BP

- Younger Dryas (pollen zone III) 11 000 to 10 000 ^{14}C yr BP

By using the terms to refer to chronozones, rather than to climatic or vegetational divisions which are often **time-transgressive** from place to place, it was hoped that there would be improvements in correlating fragmented and far flung Late Glacial deposits. Many workers continue to use the terms in this chronological sense, although problems with radiocarbon dating events during the Late Glacial mean that the ^{14}C year divisions devised by Mangerud et al. are far less precise than was once thought to be the case.

Dating events of the Late Glacial

Much progress has been made over recent years in calibrating radiocarbon dates into calendar years using tree rings, corals and **varved** sediments (see **Section 2.5**), and it has been found that radiocarbon dates on events during the Late Glacial are particularly imprecise. This is because there were large-scale fluctuations in the concentration of atmospheric $^{14}CO_2$ that have resulted in large mismatches between radiocarbon and calendar ages. Radiocarbon dates on Late Glacial deposits are thousands of years too young compared with the calibrated calendar ages, and there are several periods with a **radiocarbon plateau** – that is a lengthy time span over which radiocarbon dates give the same age. Radiocarbon plateaux are particularly marked during the early Bølling, during the Allerød and at the Younger Dryas/Holocene transition, and this makes it impossible to precisely date and correlate Late Glacial events using the radiocarbon method. It is ironic that the environmental changes of interest for dating are themselves responsible for the large changes in atmospheric carbon that make radiocarbon dating them so problematic.

As reviewed by Stuiver et al. (1991), there were largescale changes in ocean circulation that occurred in association with Late Glacial climatic changes that led to short-term increases or reductions in the ratio of atmospheric $^{14}CO_2$ to $^{12}CO_2$. For instance at the start of the Younger Dryas, the $^{14}CO_2$ to $^{12}CO_2$ ratio increased by about 50‰ within 200 years. This is thought to have been caused by rapid reduction in **thermohaline circulation** (THC), which reduced North Atlantic deepwater formation and hence also the process of ventilation that exchanges younger and older carbon between the atmosphere and deep ocean (Singarayer et al. 2008). The Earth's magnetic field was also relatively weak throughout the Late Glacial compared with the present, causing more ^{14}C to be produced in the upper atmosphere; thereby causing all radiocarbon dates on Late Glacial materials to be younger than the calendar year ages.

Because of these problems, it has been proposed that the radiocarbon-based chronozones should be replaced with a system of Late Glacial events and dates based on the Greenland ice core record: the climatic pattern is seen to be representative of the North Atlantic region in general, and the record is both continuous and annually resolved through the Late Glacial. The INTIMATE project (Integration of Ice-core, Marine and Terrestrial records of the North Atlantic) initially chose the GRIP ice core as the 'stratotype' for the Late Glacial, providing precise dates for the important events against which other proxy records can be compared (Björck et al. 1998; Lowe et al. 2001). As shown in **Fig. 5.1**, the Late Glacial portion of the GRIP record contains two stadials and an interstadial subdivided into 5 units. The first stadial (GS-2) represents cold conditions prior to Late Glacial climatic warming, with GS-2a representing the traditional 'Oldest Dryas'. GS-2a is also related to **Heinrich event** 1 which occurred c. 17 000 years ago. Late Glacial warming began with a rapid temperature increase about 14 700 years ago – the beginning of the interstadial represented by GI-1e. This phase of Bølling warmth seen in the GRIP record was interrupted briefly about 14 100 years ago by a cold phase that may represent the Older Dryas, and this was followed by a return to warmer conditions (though not as warm) during GI-1c (the Allerød). The second stadial (GS-1) represents the Younger Dryas phase which lasted around 1200 years spanning c. 13 000 to 11 700 years ago.

More recently, the NGRIP ice core has replaced GRIP as the stratotype for the last glacial termination, and the dates on the key Late Glacial events shown in the ice core record have been improved by integrating the records of the DYE-3, GRIP, and NGRIP ice cores. This integrated Greenland ice core chronology was completed in 2005 and so is referred to as GICC05 (Rasmussen et al. 2006). This record gives ages in years 'b2k' (before AD 2000). Using the GICC05, the INTIMATE group proposed a revised event stratigraphy for the last glacial termination spanning the period from 30 000 to 8000 years b2k (Lowe et al. 2008). The dates for events spanning the Last Glacial are summarized below and can be compared with **Table 1.5** and the traditional radiocarbon-based chronozones listed previously.

- Start of GS-2 (correlates with the LGM) at 22 900 years b2k

- Start of GI-1e (equivalent of the Bølling warm phase) at 14 692 years b2k

- Start of GI-1d (equivalent of the Older Dryas) at 14 075 years b2k

- Start of GI-c (start of the Allerød warm phase) at 13 954 years b2k

- Start of GI-b ('Intra-Allerød cold phase' or 'IACP') at 13 311 years b2k

Fig. 5.1 The Greenland GRIP ice core record of δ18O and methane spanning the Late Glacial. The δ18O data are from Dansgaard et al. (1993), and the methane data are from Blunier et al. (1993) and Chappellaz et al. (1993). Division of the Late Glacial into Greenland stadials (GS) and interstadials (GI) follows Björck et al. (1998). Revised, but largely similar, dating of this scheme against the NGRIP core and GICC05 chronology can be found in Lowe et al. (2008). Source: World Data Center for Paleoclimatology and NOAA Paleoclimatology Program

- Start of GI-a (latter part of the Allerød) at 13 099 years b2k
- Start of GS-1 (onset of the Younger Dryas) at 12 896 years b2k
- End of GS-1 (end of Younger Dryas) at 11 703 years b2k

One of the advantages of using this Greenland-based scheme for classifying the Late Glacial is that it also ties in with the system of **Dansgaard–Oeschger cycles (D–O cycles)** discussed in **Section 3.6**. For instance, GI-1 is recognized as the most recent of 24 Greenland interstadials that span the Würm–Weichsel glacial. For the full table of Last Termination events with GICC05 ages and depths in the NGRIP core, and for the position of various tephra layers in relation to the scheme, see Lowe et al. (2008).

Late Glacial climates

Since the macrofossil and pollen analyses of the early twentieth century, it has been known that the Late Glacial was a time of dramatic climate change; but on the basis of pollen data it was originally thought that temperatures gradually rose to a peak during the Allerød. However, this was not supported by the

Coleopteran (beetle) evidence in Britain (Coope 1975a). The Coleoptera indicate that the thermal maximum differed both in timing and intensity from that inferred from the floral evidence, and this can be explained by the much quicker response of Coleopteran assemblages to climate in comparison with vegetation. In Britain, the Coleoptera data show that temperatures rapidly reached their peak early in the **Late Glacial Interstadial** and that temperatures exhibit a declining trend until the Younger Dryas. The climate indicated by the Coleoptera during the thermal maximum was apparently warm enough to support a mixed deciduous forest (with mean July temperatures of 16 to 18 °C), but the pollen spectra at this time indicate a much more open country due to the longer lag time for the arrival of birch and other trees from glacial refugia. The pattern revealed in Coleoptera data, and other proxy records from the circum-North Atlantic region, corresponds well with the general pattern seen in the Greenland ice core records, notwithstanding differences in timing and problems with dating (Lowe and NASP Members 1995).

Following a lengthy period during the Late Glacial Interstadial when there were minor temperature fluctuations (such as the Older Dryas oscillation) superimposed on a declining trend, temperatures dropped precipitously at the start of the Younger Dryas (GS-1). Age estimates of the onset of the Younger Dryas from various European varve and tree ring records show good agreement with the estimates from the Greenland ice cores suggesting that Younger Dryas cooling was largely synchronous across Europe; although the magnitude of temperature decline varied spatially, being more subdued in northern Europe near the remnant ice sheet compared with areas further south (see the review in Anderson 1997). There is also evidence for cooling synchronous with the Greenland ice core record elsewhere in the Northern Hemisphere, for example, from isotope studies on a relict ice-wedge system near Barrow, northern Alaska (Meyer et al. 2010).

The Greenland δ^{18}O data indicate a mean annual temperature depression of about 10 °C relative to the maximum of the Late Glacial interstadial. Similarly in Britain and the Netherlands data from Coleopteran assemblages indicate a mean July temperature depression of between 7 and 8 °C (Atkinson et al. 1987; Lowe et al. 1994); and across Europe generally, the Younger Dryas cooling ranged between 5 and 10 °C (Clark et al. 2012). The mean annual temperature in Britain is estimated to have reached a minimum of between −5 and −2 °C during the Younger Dryas. A Coleopteran record from just north of Glasgow in Scotland indicates both that mean annual temperature was −5 °C and that mean winter temperatures plummeted to near −21 °C (Coope and Rose 2008). Chironomid evidence indicates a similar Younger Dryas temperature depression in Britain with, for example, mean summer air temperature reaching a minimum of 7.5 °C at Hawes Water, northern Lancashire (Bedford et al. 2004). On the other side of the Atlantic, in New Brunswick, chironomid-based reconstructions indicate a drop in mean summer lake surface temperatures of around 7 °C (Cwynar et al. 1994). Average air temperatures probably dropped by around 5 °C in eastern North America (Oswald et al. 2009).

The climate transitions of the Late Glacial were remarkable for their rapidity, being similar to the D–O cycles spanning the last glacial. The onset of Bølling warming was extremely fast. At this transition, Greenland δ^{18}O data show a c. 10 °C increase in mean annual air temperature inside of three years (Steffensen et al. 2008). Although there was a longer-term cooling trend through GI-1, there was a distinct 'cooling step' into the Younger Dryas which occurred over a period of a century or so. As with the D–O cycles

discussed in **Chapter 3**, there has recently been a great deal of research focused on determining the extent to which changes in other parts of the world resemble the dramatic changes seen in the Greenland ice core record and **stratigraphic sequences** from Europe. In general, it can be said that most of the Northern Hemisphere (even parts of the northern tropics) saw temperature and/or precipitation changes that are in phase with the Greenland changes – if not necessarily with exactly the same timing or magnitude. For instance, with the onset of the Bølling, glaciers in the northern Andes of Venezuela were retreating and sea surface temperature of the Cariaco Basin (north coast of Venezuela) was rising (Lea et al. 2003; Stansell et al. 2010). The **Intertropical Convergence Zone (ITCZ)** migrated northwards, increasing rainfall in lower latitude areas of the Northern Hemisphere that had been drier during the LGM and Oldest Dryas, and the African and Asian summer **monsoons** intensified. High resolution **speleothem**-based reconstruction of the strength of the Asian monsoon from the Dongge and Hulu caves in China shows very close timing between rapid monsoon strengthening and the Greenland temperature increase (Dykoski et al. 2005).

Antarctica and parts of the Southern Hemisphere, however, show an anti-phase relationship with the Late Glacial changes of the north (Blunier et al. 1998). The **AMOC** and bipolar seesaw, introduced in **Chapter 3**, explain why Antarctica and the Southern Ocean were warming at the times the north was cooling and vice versa (Fig. 5.2). As yet, the full picture for the Southern Hemisphere is unclear, but there is evidence for cooling in Tasmania (Tibby 2012) and glacier advance in south-west Patagonia (Moreno et al. 2009) and New Zealand (Newnham et al. 2012) contemporaneous with the **Antarctic Cold Reversal (ACR)**. The ACR spans the Bølling/Allerød (GI-1) and lasted until just before the onset of the Younger Dryas.

The Younger Dryas: global expression and causes

The global expression of the Younger Dryas cold phase (GS-1) was similar to that of the earlier Greenland stadials in that many parts of the Northern Hemisphere beyond Greenland, Europe and the north-east seaboard of North America became cooler and/or drier. Such change coeval with the Younger Dryas has been detected in deposits from western North America, for example from the Pacific Northwest (Mathewes 1993)

Fig. 5.2 Comparison of Greenland and Antarctic temperature variation through Late Glacial times to illustrate the 'bipolar seesaw' phenomenon. The 'Heinrich Stadial' refers to Heinrich event 1 c. 17 000 years ago and its reinforcement of cold conditions during the 'Oldest Dryas' in the north while there was warming in the south. The curves are drawn from data in Blunier and Brook 2001. Original GRIP data are presented in Dansgaard et al. 1993. Source: World Data Center for Paleoclimatology and NOAA Paleoclimatology Program

and from western Idaho (Doerner and Carrara 2001), and from the Andes (Clapperton 1998). Evidence for sea surface temperature decline has been found in high resolution ocean sediment cores from as far away as the western Pacific (Kudrass et al. 1991). In the northern tropics of South America mean annual temperature declined by up to 3 °C and glaciers advanced in the northern Andes (Stansell et al. 2010). Southward migration of the ITCZ and weakened monsoons caused reduced precipitation throughout Northern Hemisphere lower latitudes. Moreover, the decline in methane recorded in the ice cores (**Fig. 5.1**) indicates that conditions became drier over the majority of the land surface during the Younger Dryas. Yet at the same time, it has been shown that some Southern Hemisphere low-latitude regions, such as north-eastern Brazil (Wang et al. 2004a), saw increased precipitation because of this southward shift of the ITCZ. However, unlike the earlier stadials in the Greenland record, the Younger Dryas occurred during a time of rising Northern Hemisphere insolation receipt when deglaciation was already well underway. This makes it in some ways more enigmatic than the events of the last glacial and more challenging to explain.

A great deal has been written on the Younger Dryas and theories about its causation continue to be debated. What is not in doubt is that it is associated with a major change in the AMOC whereby a collapse in thermohaline circulation and reduced NADW formation in the North Atlantic caused a large decline in sea surface temperature and southward migration of the polar front to a position similar to where it was during the LGM (**Fig. 5.3**). (This figure demonstrates that throughout much of the last cold stage, the oceanic polar front was situated south of France.) Broecker (2006) has explained how this would have enabled winter sea ice to extend southwards across the Atlantic, far south of the British Isles, in turn contributing to much colder winters across the northern high to mid latitudes. He also suggested that prolonged snow and ice cover in the north accounts for the southward migration of tropical rain belts and the weakened low-latitude monsoons.

The key question, therefore, concerns the actual trigger that caused the Younger Dryas to happen when it did. In **Section 3.6**, the sensitivity of the Atlantic THC to freshwater input was discussed, and it is thought that such a meltwater pulse delivered to the critical regions of NADW formation was the primary cause (see **Fig. 3.20**). However, it has been difficult to identify the source of this pulse, or its magnitude. Indeed, neither of the major phases of deglacial eustatic sea level rise (MWP 1A and MWP 1B) correlate with the Younger Dryas. It was originally proposed

Fig. 5.3 Position of the North Atlantic Polar Front and limits of winter sea ice, during and immediately prior to the last glacial/interglacial transition (based on Ruddiman and McIntyre 1981a). 1) LGM position; 2) pre-Late Glacial interstadial position; 3) Late Glacial Interstadial position; 4) Younger Dryas position. PIL shows the southern limit of pack ice in modern times

that the meltwater pulse came through the Gulf of St Lawrence when meltwater from the **Laurentide ice sheet** was diverted away from its Mississippi/Gulf of Mexico outlet (Broecker et al. 1989) – the idea being that it was not the quantity but the *location* of freshwater delivery that was crucial.

It later became apparent that there is no clear geomorphic evidence for large-scale meltwater diversion via the St Lawrence coeval with the start of the Younger Dryas, despite $\delta^{18}O$ evidence in Gulf of Mexico sediments that meltwater input there was reduced (see the review in Broecker et al. 2010). Andrews et al. (1995) suggested an alternative source of meltwater from an iceberg discharge event via Hudson Strait which was similar to but smaller than the Heinrich events and has been labelled the H-0 event (see also Moore 2005). Other studies have continued to support freshwater routing via the St Lawrence at the onset of the Younger Dryas (e.g. Carlson et al. 2007). More recently it has been argued that a meltwater pulse associated

with the drainage of glacial Lake Agassiz entered the Arctic Ocean via the Mackenzie drainage system and that this caused sufficient freshening of the Arctic and North Atlantic to disrupt the AMOC (Tarasov and Peltier 2005; Murton et al. 2010). From geomorphic evidence of changes in the size of Lake Agassiz and **luminescence dating** on sediments above and below the flood erosion surface, Murton et al. estimate that c. 9500 km³ of water was rapidly discharged into the Arctic Ocean at a time close to the onset of Younger Dryas cooling seen in the Greenland ice records. **Figure 5.4** shows the Laurentide ice sheet and its proglacial lakes near the start of the Younger Dryas and the different meltwater routes.

It has also been proposed that the Younger Dryas was the result of a comet impacting North America (Firestone et al. 2007). Firestone et al. argued that many sites in North America and elsewhere contain evidence for an extraterrestrial impact at c. 12 900 years ago (such as elevated iridium and nanodiamonds), and

Fig. 5.4 The Laurentide ice sheet and its large proglacial lakes near the start of the Younger Dryas. Former Lake Agassiz was far larger than any of the Great Lakes today. The arrows indicate three different routes which varied in the proportions of meltwater they carried at different times during deglaciation: the Mississippi route to the Gulf, the St Lawrence route to the North Atlantic, and the Mackenzie route to the Arctic (from Murton et al. 2010, Fig. 1). Reprinted by permission from Macmillan Publishers Ltd: *Nature*.

they contend that this caused Younger Dryas cooling both through dust and **aerosols** blocking sunlight and through destabilization of the Laurentide ice sheet causing release of meltwater and icebergs into the North Atlantic. This hypothesis has undergone much scrutiny in recent years, with much of the alleged evidence put into question; although the evidence for an extraterrestrial impact at some sites has continued to be debated.

Regardless of the precise trigger for the Younger Dryas, it has become apparent through study of previous Terminations that this type of event is not unique to Termination I. Cheng et al. (2009a) compared reconstructions of the strength of the Asian monsoon from Chinese cave records and found that 'weak monsoon intervals' similar to the one that correlates with the Younger Dryas occurred during the previous three Terminations. In other words, each termination is characterized by a significant weakening of the monsoon (and a northern cold episode) while Northern Hemisphere insolation input was increasing. Furthermore, Termination III shares the 'tripartite' pattern of Termination I in that there is an equivalent Oldest Dryas – Bølling/Allerød – Younger Dryas like sequence prior to the onset of full interglacial conditions, both in northern temperature and in Asian monsoon strength. The similarity between Terminations I and III can be explained by the relatively slower rise and more limited peaks in summer insolation driving these terminations compared with Terminations II and IV (Carlson 2008). The relationship between insolation changes and glacial Terminations is discussed further in **Chapter 9**.

All this points to the idea that the Younger Dryas and the other climate oscillations characterizing the Late Glacial are the result of a complex interplay of feedbacks in the climate system that prevent deglaciations from occurring in the smooth manner that would be expected if climate responded in a linear way to the insolation changes that pace the glacial/interglacial cycles. Importantly, the Younger Dryas itself was not a period of uniform conditions: sediments from Lake Kråkenes, western Norway, suggest that from about 12 200 years ago onwards, climate in the eastern North Atlantic became increasingly variable, 'flickering' between warmer and colder states, prior to the abrupt establishment of interglacial temperatures in the region 11 700 years ago (Bakke et al. 2009).

Much research is currently focused on unravelling the timing and connections between the Late Glacial changes that occurred in different parts of the world. The 'bipolar seesaw' of the AMOC system no doubt plays a central role; but the relative importance of the north and south (and high versus low latitudes) in driving these changes continues to be debated. The NGRIP core suggests that the Younger Dryas cooling slightly preceded the increased flux in dust and other proxies for a weakened Asian monsoon, whereas the much more rapid Bølling and Holocene warmings of Greenland occurred *slightly after* the Asian monsoons strengthened (Steffensen et al. 2008). Evidence from lake sediments in Venezuela also suggest that warming had begun and the ITCZ had migrated northwards over the northern tropics of South America before the abrupt ending of the Younger Dryas as seen in Greenland, in Lake Kråkenes sediments, and in other European proxy records (Stansell et al. 2010).

Late Glacial environments

On the basis of flora studies attempts have been made to reconstruct the nature of the landscapes as they probably appeared in the Allerød interstadial. It is worth comparing **Figs 5.4 and 5.5** (showing ice limits in North America and Europe respectively) with the Last Glacial Maximum (**Fig. 3.6**). The ice sheets have been greatly reduced in extent in comparison with their maxima, but sea levels are still low. In North America much of Beringia (off the coast of Alaska) is not yet submerged and a large ice-free corridor is present between the Cordilleran and Laurentide ice sheets. In Europe, Britain is connected with the continent, Denmark is relatively unfragmented into islands, and tundra vegetation is less widespread. However, pine woodlands are known to have dominated the southern half of France, southern Germany, and northern Poland, and birch to have occupied much of northern France and northern Germany. Most of Fennoscandia was still glaciated.

Although overall area covered by ice was much reduced during the Late Glacial, cooling phases caused temporary readvances of glaciers in many parts of the world. In north-west Scotland, where glaciers existed throughout the times of Late Glacial interstadial warmth, major readvances occurred both during the Older Dryas and the Younger Dryas (Bradwell et al. 2008). The latter expansion, called the 'Loch Lomond Readvance' in Britain also saw glaciers reform in some upland areas. Over Rannoch Moor (north-west Scotland) the ice thickness may locally have exceeded 400 m. Smaller ice caps developed in the Grampians, on Mull and Skye, in the Southern Uplands of Scotland, in the English Lake District, in the mountains of North Wales, and in the Brecon Beacons. The end moraines from this short sharp cold period are still very clear in the present landscape, as are a whole range of **periglacial** landforms that developed down-valley from them, including blockfields, solifluction lobes, screes, rock glaciers, and protalus ramparts (Sissons 1976, 1980a, b). In Ireland, the equivalent Nahanagan Stadial saw the development of **cirque** glaciers in the Wicklow Mountains. There was also significant readvance of ice in Scandinavia, and various studies have shown correlatives in parts of North America, for example as far as British Columbia (Friele and Clague 2002).

The associated Late Glacial vegetational changes in Britain are well illustrated in lacustrine pollen data from the English Lake District. Through her classic work on sediment cores from Lake Windermere, W. Pennington (1947, 1969) identified gradual increases in the abundance of shrub (e.g. juniper, dwarf willow) and tree pollen (mainly birch) during the warm phases of the Late Glacial interstadial (known as the 'Windermere Interstadial' in Britain) followed by a

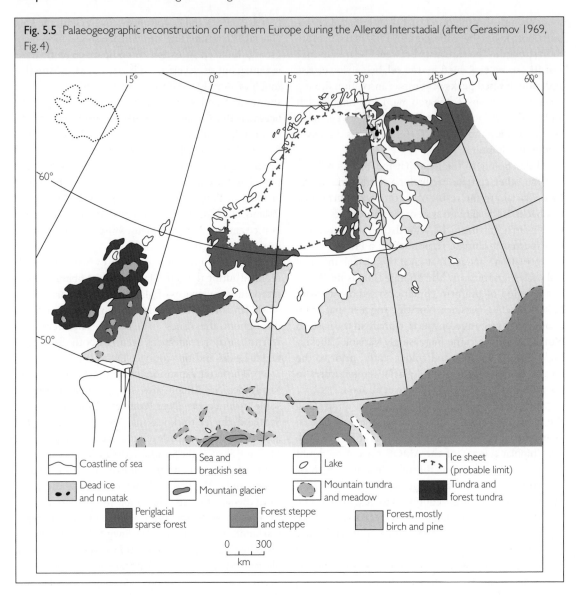

Fig. 5.5 Palaeogeographic reconstruction of northern Europe during the Allerød Interstadial (after Gerasimov 1969, Fig. 4)

reversion back to open tundra-type vegetation with the onset of cold conditions of the Loch Lomond Stadial (Younger Dryas). In addition to showing a decline in shrub and tree pollen, the Loch Lomond Stadial-aged lake sediments have also been characterized by a higher minerogenic content indicative of increased landscape instability and heightened erosion under periglacial and glacial conditions. Following the Loch Lomond Stadial, the sediments of Lake Windermere show increasing organic content and increases in shrub and tree pollen as vegetation re-established with the onset of Holocene warmth.

This kind of tripartite sequence is strongly reflected in Late Glacial sediment sequences in areas of northern Europe just outside the limits of Younger Dryas Stadial glaciation. A recent and high resolution record of Late Glacial environmental change that illustrates this well comes from Sluggan Bog, Northern Ireland (Walker et al. 2012). At Sluggan Bog, Late Glacial Interstadial ('Woodgrange Interstadial' in Ireland) sediments indicate two phases of birch–willow vegetation cover (equivalent to the Bølling and Allerød) interrupted by an episode of heathland cover (a rise in Ericaceae pollen) that correlates with the Older Dryas

(GI-1d) cold phase. At the onset of the Younger Dryas, the vegetation reverted from grassland with patches of woodland to arctic shrub tundra (with abundant *Artemisia* and Caryophyllaceae pollen). Further south in Europe, the Younger Dryas saw forest replaced by grassland communities rather than by tundra (Clark et al. 2012).

Late Glacial vegetation oscillation is less strongly reflected in other parts of the world where Younger Dryas cooling and drying was either less severe, or absent (as in the case of many Southern Hemisphere regions). In mid-latitude North America there was not such a clear oscillation between shrub and tree – to open tundra – to shrub and tree cover as in northern Europe. However, an oscillation in forest composition in eastern North America is well documented in pollen records (Yu 2007). For instance, a study from Connecticut shows a mixed woodland dominated by *Pinus* and containing *Ostrya-Carpinus*, *Fraxinus*, *Betula*, and *Quercus* giving way to one with more *Abies* and *Alnus* during the Younger Dryas (Oswald et al. 2009).

While vegetation changes in the British Isles, northwest Europe and eastern North America primarily reflect cooling during the Younger Dryas, pollen evidence strongly supports drying in southern Europe and much of the Near East, for example in the Zagros and Taurus Mountains (see Wright and Thorpe 2005 for a review). Rossignol-Strick (1995) suggested that across Turkey and in many areas of Greece, the Younger Dryas period was even more arid than the most extreme part of the last glacial with a dominance of Chenopodiaceae over *Artemisia*, and suggests that this indicates an annual precipitation of less than 150 mm across much of lowland Greece. Temperatures were also markedly lower, with species characteristic of the very continental Asian semi-deserts occurring in lowland northern Greece.

5.3 The onset of the Holocene

As the climate rapidly warmed, the recession of the ice sheets uncovered millions of square kilometres of land in higher latitudes, which thus became available for human occupation and colonization. The world population of migrant waterfowl possibly increased hugely with the addition of the great breeding and feeding grounds of the Northern Hemisphere. Further, the transgression of the sea, resulting from

the melting of the ice caps, though it may have led to a considerable inundation of the continental shelves (see **Section 6.8**), did in some ways improve the seashores for man. A more diversified and sinuous coastline would give a wider choice of environments, while the drowning of valleys to give rias (sinuous inlets of sea) would tend to lead to an increase in tidal ranges which would be most valuable for food-collecting peoples. The formation of quiet landlocked bodies of water might also provide a favourable setting for early trials in navigation. Many alluvial valleys grew in length and breadth and offered optimal sites for plant growth. As Sauer (1948, p. 258) has put it:

A new world took form, developing the physical geography that we know. The period was one of maximum opportunity for progressive and adventurous man. The higher latitudes were open to his colonization. In mild lands rich valleys invited his ingenuity. It was above all a rarely favourable time for man to test out the possibilities of waterside life, and especially of living along fresh water.

As we have already seen, radiocarbon dating put the end of the Younger Dryas, and the onset of Holocene warmth at *c.* 10 000 ^{14}C yr BP (start of the PreBoreal *sensu* Mangerud et al. 1974); and this, being a round and easily memorable number has long been taken as the base of the Holocene. However, as it became known that radiocarbon dates on Late Glacial and early Holocene material are too young compared with true calendar ages, it also became clear that another means of defining the base of the Holocene was necessary. As the climate transition is very clearly marked in the Greenland ice cores, various attempts were made in the 1990s to establish an age by counting annual ice layers. Initial estimates from the GRIP and GISP2 ice cores yielded ages of 11 550 ± 90 and 11 640 ± 250 cal yr BP respectively (Johnsen et al. 1992; Alley et al. 1993), and similar estimates were made through **dendrochronology** and study of annually laminated lake sediments in northern Europe (Gulliksen et al. 1998). With the completion of the high resolution NGRIP ice core and the establishment of the more accurate GICC05 timescale, it was decided in 2008 to establish a **GSSP** for the Holocene in the NGRIP core (Walker et al. 2009). Hence the Holocene Series/Epoch is now formally recognized as having started at 11 700 yr b2k (or 11 650 cal yr BP) with NGRIP providing the type section for the transition which in this ice core (borehole NGRIP2) is located at a depth of 1492.45 m.

Subdividing the Holocene

After the climatic flutter of the Late Glacial oscillation, the traditional division is made between the Late Glacial (Pleistocene) and the post-glacial (Holocene, Recent, or Flandrian). As introduced in **Section 1.2**, the classic terminology of the Holocene was established by two Scandinavians, Blytt and Sernander, who, in the late nineteenth and early twentieth centuries, introduced the terms PreBoreal, Boreal, Atlantic, SubBoreal, and SubAtlantic for the various environmental fluctuations that took place. These terms are still widely used for subdivisions of the Holocene (**Table 1.4**). The fluctuations which Blytt and Sernander and subsequent workers have established, though they have sometimes been contested by workers who believe the sequence of events has been less complex, with only a simple climatic optimum and then deterioration, have been remarkably durable. Nevertheless, it has to be remembered that it is essentially a scheme of vegetation change, and not a scheme of climate change. The bulk of the evidence used by Blytt and Sernander was provided by plant remains, and especially macrofossils. Thus in terms of climatic reconstruction certain inaccuracies may creep in because of non-climatic factors affecting vegetation associations. Such factors include the intervention of humans, the progressive evolution of soils through time, and the passage from pioneer to climax species during the course of succession. As already noted with regard to the Allerød oscillation, plants may not be able to respond with great alacrity to climatic change: migration and colonization take time. As explained by Moore et al. (1996) the spreading rates for trees in the Holocene were probably of the order of 300 to 500 m per year, a rate that is slow when compared to the rapidity of climate change. Given that a 1 °C rise in temperature is approximately equivalent to a shift of 100–150 km in spatial range, a 7 °C change in temperature at the start of the Holocene could demand a shift in the range of a tree species by 1000 km. Even a 'fast' tree species, with a spreading rate of 500 m per year, would take around 2000 years to arrive at its new limit, by which time the climate might well have shifted once more.

Consequently, although the terminology may still be used, there have been substantial changes in interpretation of the classic Blytt–Sernander model in recent decades. As with the Late Glacial terminology, Mangerud et al. (1974) suggested using the traditional terms as radiocarbon-based chronozones as follows:

- PreBoreal (pollen zones IV and V) 10 000 to 9000 ^{14}C yr BP
- Boreal (pollen zones VIa and VIb) 9000 to 8000 ^{14}C yr BP
- Atlantic (pollen zones VIc and VIIa) 8000 to 5000 ^{14}C yr BP
- SubBoreal (pollen zone VIIb) 5000 to 2500 ^{14}C yr BP
- SubAtlantic (pollen zone VIII) 2500 ^{14}C yr BP to the present

As radiocarbon dating is so commonly applied to Holocene deposits and materials, using the Blytt–Sernander scheme in this way provides a widely applicable framework for discussing and correlating Holocene sequences; although it must be remembered that the scheme suffers from the imprecision inherent in the radiocarbon dating method, and that the radiocarbon ages vary with respect to the calendar year age estimates (as explained in **Section 2.5**).

Early Holocene climate

The start of the Holocene is marked by a rapid increase in temperature at the end of GS-1 (the Younger Dryas termination) as shown in **Fig. 5.1**. The shift to interglacial conditions during the PreBoreal was remarkably quick: an early indication of this was based on δ^{18}O data from the Greenland Dye 3 ice core showing that mean annual temperature increased by 7 °C within about 50 years (Dansgaard et al. 1989). As mentioned previously, from the GISP2 record Alley et al. (1993) found that ice accumulation (snowfall) doubled in just three years. More recently, such a rapid warming at the start of the Holocene has been confirmed by analysis of the NGRIP core: a large shift in 'deuterium excess' (a parameter comparing δD with δ^{18}O), indicates that a major change in the moisture source region of precipitation over Greenland occurred within three years (Steffensen et al. 2008). (In fact, it is this very deuterium excess shift that is used as the GSSP for the Holocene.) Recent estimates of the magnitude of this change suggest that mean annual air temperature over Greenland rose by 10 ± 4 °C (Walker et al. 2009). All of this points to an abrupt and large-scale reorganization of Northern Hemisphere atmospheric circulation at the onset of the Holocene which was linked to the re-establishment of vigorous thermohaline circulation and the delivery of warmer waters to high northern latitudes (**Fig. 5.3**).

Furthermore, rapid climate warming is not only seen in the Greenland ice core record. A similar abrupt shift is also seen, for example, in temperature dependent $\delta^{18}O$ data from lake sediments from Lake Gosciaz, Poland (Goslar et al. 1993) and Lake Ammersee, southern Germany (von Grafenstein et al. 1999). Estimates of rapid warming on the order of 10 °C in Europe have also come from chironomids (e.g. Ilyashuk et al. 2009) and from Coleoptera-based reconstructions (e.g. Atkinson et al. 1987).

The Greenland ice cores also show that the concentration of atmospheric methane rose rapidly at the start of the Holocene in association with the increasing temperatures. In the GRIP core methane values increased from around 500 to nearly 750 ppbv in a matter of decades, suggesting a rapid expansion of low-latitude wetlands – a primary source for atmospheric methane (Blunier et al. 1995). Thus levels of precipitation in many lower latitude regions must have increased dramatically along with the warming of mid to high latitudes. Close analysis of the NGRIP record shows that dust concentration began to fall ten or so years *before* the deuterium excess and $\delta^{18}O$ data indicate warming over Greenland (Steffensen et al. 2008) – this suggests that low-latitude Asian deserts became wetter, reducing dust supply to Greenland, just before North Atlantic regional warming. This provides evidence that low-latitude changes may in some way have been the driver of, rather than simply responding to, the rapid warming of the northern mid to high latitudes.

In contrast with the rapid shifts in methane, the atmospheric concentration of carbon dioxide increased in a more gradual and steady manner from under 200 during the LGM to around 260 ppmv by the start of the Holocene.

Because of the anti-phase relationship of Greenland and Antarctic temperatures through the Late Glacial, warming took hold in Antarctica about 12500 years ago (Stauffer et al. 2004), nearly 1000 years before the PreBoreal warming recorded in Greenland and correlating with the onset of the Younger Dryas Event. After reaching a peak during the Younger Dryas, Antarctic temperatures dipped slightly as PreBoreal warming intensified in the Northern Hemisphere, but they remained at interglacial levels. Considering the Antarctic ice core deuterium data alone, it would seem that the Holocene started earlier than indicated by the Greenland records. However, the fact that the Greenland and Antarctic records agree in the timing of the sharp methane rise, and that this is synchronous

with the warming in Greenland, suggests that the timing of the start of the Holocene as inferred from Greenland ice cores is more globally representative.

Following the sharp temperature increase at the start of the Holocene, the Greenland ice core records show that temperature continued to increase (but at a slower rate) until reaching a thermal maximum between 9000 and 8000 years ago. Holocene temperatures probably reached their maximum in the early Holocene over most parts of the globe, with the exception of those regions that were in close proximity to the residual but decaying ice sheets (COHMAP 1988). For instance, temperature estimates derived from fossil Coleoptera indicate that temperatures in England were close to their maximum by around 9000 years ago (Atkinson et al. 1987); whereas in northern Fennoscandia, chironomid remains indicate that the maximum summer temperature was reached around 8000 years ago (Seppä et al. 2002). Averaged across Europe, pollen-based temperature reconstruction suggests that mean annual temperatures had reached present values shortly after 8000 years ago, with early Holocene summer temperatures having risen much more quickly than the winter temperatures (Davis et al. 2003).

This early Holocene warmth, following so quickly on from the rigours of the Younger Dryas, has been attributed to a particular combination of orbital conditions (Kutzbach and Webb 1993; Kutzbach et al. 1998). Summers were generally warmer than at present because the tilt of the Earth's axis was greater (thereby enhancing seasonal extremes) and because the Northern Hemisphere summer solstice occurred when the Earth was nearer the Sun on its orbit (at perihelion as opposed to at aphelion as occurs today). As illustrated in **Fig. 5.6**, this caused Northern Hemisphere solar radiation receipt in summer to be over 30 W m^{-2} or about 8% greater than it is today. This also had the effect of increasing the strength and latitudinal reach of summer monsoons: this is borne out in the ice core atmospheric methane data which show relatively high levels for the first two millennia of the Holocene, indicating a strengthened hydrological cycle and maximum extent of wetlands at lower latitudes (Blunier et al. 1995). Since peaking in the early Holocene, there has been a steady decline in Northern Hemisphere summer insolation towards the present. **Figure 5.6** also shows that the percentage of land ice and average sea surface temperatures (SSTs) had reached values similar to today (and that atmospheric

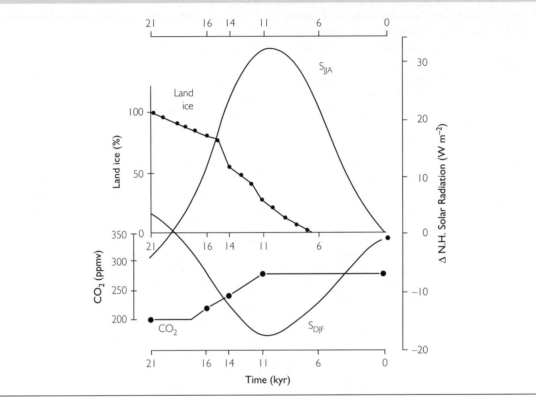

Fig. 5.6 Boundary conditions for the COHMAP simulation updated on a calibrated calendar year timescale extending from the LGM (21 000 years ago) to the present. External forcing is shown for the Northern Hemisphere solar radiation (S) in June–August (JJA) and December–February (DJF) as the difference in watts per square metre from present-day radiation receipts. Percentage land ice is expressed as a difference from the present, and CO_2 extending up to the present is shown as pre-industrial values (from Kutzbach et al. 1998, Figure 1). Reprinted with permission of Elsevier

CO_2 had approached pre-industrial values) between 9000 and 10 000 years ago.

Superimposed on the early Holocene warming trend were some brief and relatively minor temperature oscillations. The Holocene portions of Greenland ice cores show episodes of temperature decline at *c.* 11 400 (termed the 'PreBoreal Oscillation') and at *c.* 9300 years ago (Rasmussen et al. 2007). The Holocene portion of the NGRIP ice core is shown in **Fig. 5.7**. The PreBoreal cooling phase is thought to have lasted less than 200 years, but has been detected in various stratigraphic sequences from Europe and around the North Atlantic region. For instance, in a Swiss lake sediment sequence a July air temperature decline of *c.* 1.4 °C for the PreBoreal oscillation is inferred from chironomid evidence (Samartin et al. 2012). These oscillations, and others, are also detected in a high resolution chironomid

record from Hawes Water, north-west England: here a July air temperature drop of *c.* 1.6 °C is inferred for a 50–60 year span at *c.* 9300 years ago (Lang et al. 2010).

Thus it appears that the early Holocene contained climatic fluctuations of a similar nature, albeit of a much lower magnitude, to those of the Late Glacial. They involved variations in the strength of the thermohaline circulation which were related to changes in the quantity of meltwater entering the North Atlantic from the decaying Laurentide and Scandinavian ice sheets, and they are most evident in ice, marine and terrestrial sequences from the North Atlantic region.

Post-glacial spread of vegetation in Europe

The warming of climate in post-glacial times seems to have set off the successive return of species of

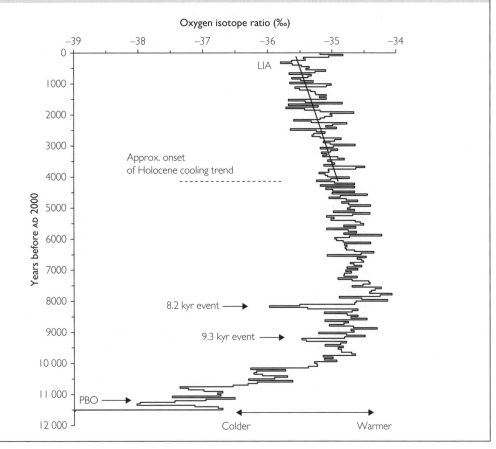

Fig. 5.7 The Holocene portion of the NGRIP ice core showing δ¹⁸O variability since 11 500 years b2k. The data are from NorthGRIP members (2004) and represent changes in air temperature in northern Greenland (coring site at 75.1° N, 42.3° W). Abrupt early Holocene cooling events are labelled (PBO = Preboreal Oscillation), the dashed line approximates the start of a long-term cooling trend (illustrated by the solid line) during the late Holocene, and the LIA (Little Ice Age) is also indicated. Source: World Data Center for Paleoclimatology and NOAA Paleoclimatology Program

tree with different tolerances of cold and different powers of colonization. As Roberts (1998, p. 100) has remarked:

…once climatic conditions permitted, there were few checks to a rapid spread of tree species across Europe, other than their own rates of dispersal. Between 11 500 and 9000 Cal. yr BP tree species moved in to occupy suitable vacant land with much the same self-interested zeal as the miners of the '49 Gold Rush. The result was no less chaotic, with short-lived, localized associations of convenience formed between species.

At first there were a relatively few arctic tree species in Britain; later larger numbers of species of more

tolerant type arrived and the woodlands became more complex in their composition (Rackham 1980). The various trees did not, in general, stream in from the south in successive waves of massive invasion. Rather, each crept up in small numbers and became widespread (though rare) long before it increased to its full abundance. Both the date of a tree's first known presence and the date at which it received full abundance are correlated with its 'arcticity' as measured by its present northern limit in Europe. In general the lag between the two dates was longer for the later arrivals such as ash and beech, for these had to displace existing trees and not merely to occupy vacant ground.

Birks (1990) synthesized the broad-scale changes in vegetation that occurred across Europe in response to Holocene warming. His work was based on consideration of radiocarbon dated pollen diagrams from many sites and the construction of **isopoll maps**. At 12000 ^{14}C yr BP (during the Late Glacial Interstadial) *Pinus* was mainly in southern and eastern Europe, but by 6000 ^{14}C yr BP was abundant in northern, central, and Mediterranean Europe but absent from much of the Western European lowlands. *Quercus* spread progressively northwards from southern Europe and reached its maximum range limits by 6000 ^{14}C yr BP. *Ulmus, Corylus,* and *Tilia* also reached their present-day range limits by the same time (see **Table 3.3** for botanical and common names). However, not all forest trees had reached these limits by 6000 ^{14}C yr BP. *Picea* has spread westwards through Finland, across Sweden, and into central and eastern Norway in the last 6000 years. *Fagus* had a rather small range in southern and central Europe by 6000 ^{14}C yr BP, which is in contrast to its present extensive range in western Europe today. Similarly, *Carpinus,* which is today widespread in lowland Europe was, at 6000 ^{14}C yr BP, very largely confined to Bulgaria, Italy, Yugoslavia, Romania, and Poland. **Figure 5.8** shows the spread pattern for *Pinus sylvestris* (Scots pine) and for *Quercus*. The post-glacial colonization of the British Isles also shows considerable complexity in terms of vegetation response (**Fig. 5.9**). The northward spread of trees varied as to direction, rate, and timing, and this may have been controlled by the place and time where colonization first occurred, which may in part have been a matter of chance occasioned by the vagaries of dispersal and establishment (Birks 1990, p. 142).

The transition to peak Holocene levels of tree cover in the eastern Mediterranean occurred more quickly than in north-western Europe, taking 1000 ^{14}C years following the end of the Younger Dryas, whilst in many parts of Europe the forest cover was still rather more open. In southern Spain closed oak forest consisting of both deciduous and evergreen oaks prevailed whilst deciduous oaks, and to some extent evergreen oak and pine, formed a savannah vegetation in many parts of Greece.

In Britain there are some good examples of how the warming of post-glacial times and the associated spread of forest has led to the fragmentation of the distribution of certain cold-loving flora types which covered a much wider area in the Pleistocene and early Holocene. Two of the most remarkable 'disjunct areas' are the Burren of County Clare and the Teesdale area of the northern Pennines (Seddon 1971). Although ecologically very different they both possess certain types of flora which are found scarcely anywhere else in the British Isles.

The effects of post-glacial warmth in creating relict clumps of certain plants in isolated areas is well displayed by the dwarf birch (*Betula nana*). It has been recorded at sites in many parts of Britain in Late Glacial and post-glacial deposits, but is now found

Fig. 5.8 The spread of *Pinus sylvestris* and *Quercus* in Europe in the early Holocene (modified after the work of Huntley and Birks 1983). Dates are in ^{14}C years BP, conversion to cal yr BP can be made with reference to **Fig. 2.7**

Fig. 5.9 Some examples of isopoll maps for the spread of various tree types through the British Isles (after Birks 1990): (a) Spreading of *Corylus avellana* (hazel). The contours in ^{14}C years before present (BP), show its likely range limit at 500-year intervals. The shaded area is where hazel was present prior to 9500 BP. (b) Spreading of *Quercus* (oak). The contours, in ^{14}C years BP, show its likely range limits at 500-year intervals. Note the slowing in its spread after 8000 BP. (c) Spreading of *Pinus sylvestris* (pine). The contours, in ^{14}C years BP, show its likely range limits at 500-year intervals in England and Wales and in Ireland, and at 1000-year intervals in Scotland. (d) Spreading of *Tilia* (lime). The contours, in ^{14}C years BP, show its likely range limits at 500-year intervals. Note the slowing in its spread after 7000 BP. (e) Spreading of *Fagus sylvatica* (beech). The contours, in ^{14}C years BP, show its likely range limits at 1000-year intervals. Note the absence of any slowing in its spread. Conversions to cal yr BP can be made with reference to **Fig. 2.7**. Reprinted with permission of IOS Press only in Upper Teesdale and on Scottish mountains. Equally, in north-western Europe there are similar relict areas in the French Jura, in the Harz Mountains and on peat from Luneberg Heath. It seems clear that this plant, which is from the Arctic-alpine group and is currently found at high altitudes in the Alps, was once widespread over the lowlands of Late Glacial, north-western Europe, but that it has been removed from all localities except on mountains and on some very special habitats with distinctive ecological and microclimatic conditions. This removal has been caused by the spread of forest trees into environments which were formerly unsuitable.

In addition to latitudinal changes in European vegetation types, there have also been altitudinal changes. A range of timberline studies from the European Alps shows that timberlines ascended rather rapidly at most sites during the early Holocene, reaching their maximum extent between *c.* 7000 and 3000 years ago (**Fig. 5.10**) (Burga 1988). In the southern French Alps treeline was about 300 m higher than at present *c.* 6800 cal yr BP (Talon 2010).

5.4 The 8200 cal yr BP event

Temperature records derived from central Greenland ice cores display a significant temperature anomaly centred on 8200 cal yr BP that lasted for about 200 years (Clark et al. 2001; Rasmussen et al. 2007). The negative excursion of $\delta^{18}O$ in the ice core data (**Fig. 5.7**) marks a decade-averaged drop in central Greenland annual temperatures estimated at $3.3 \pm 1.1\,°C$ in the GISP2 ice core (Kobashi et al. 2007). The generally accepted explanation for this event is due to a catastrophic meltwater release into the North Atlantic which curtailed the North Atlantic Deep Water formation (NADW), thereby disrupting the AMOC and its associated northward heat transport. This proposal has been supported from dates that record the final outburst drainage of proglacial Lake Agassiz/Ojibway during the terminal demise of the Laurentide ice sheet occurring within around 500 years of the event (von Grafenstein et al. 1998). A more recent estimate based on sea level reconstruction from the Netherlands places the drainage event *c.* 8470 ± 300 years ago (Hijma and Cohen 2010).

The Lake Agassiz/Ojibway system contained twice as much water as the present-day Caspian Sea (Teller et al. 2002), and it has been estimated that the influx of fresh water into the North Atlantic via Hudson Strait was as much as 4.3×10^{14} m³, causing a eustatic sea level rise of as much as 1.2 m (Törnqvist et al. 2004). Hijma and Cohen (2010) suggest that this may be an underestimate and that the global eustatic rise may have been closer to 3 m, perhaps also with an Antarctic

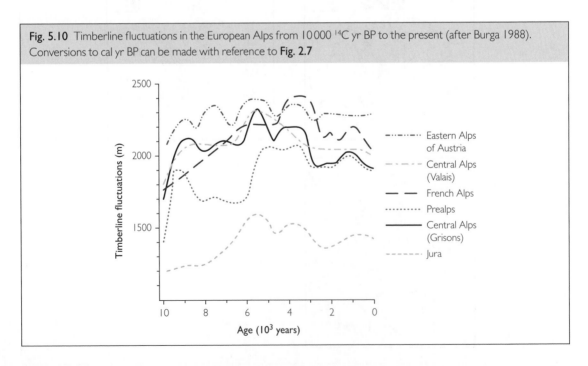

Fig. 5.10 Timberline fluctuations in the European Alps from 10 000 ¹⁴C yr BP to the present (after Burga 1988). Conversions to cal yr BP can be made with reference to **Fig. 2.7**

ice melt contribution. While it has also been proposed that the much more severe Younger Dryas cooling was triggered by a drainage event from Lake Agassiz, it is interesting that the 8.2 kyr event involved a much greater quantity of meltwater discharging suddenly into the North Atlantic – perhaps over 15 times as much.

Many shorelines around the world changed dramatically at this time, especially in gently sloping continental shelf areas. The shoreline of the Persian Gulf may have retreated back some 10 km within a year or so of the discharge event (Perkins 2002). Around this time the English Channel and the North Sea became connected to finally separate Britain from mainland Europe, and the shores of the Black Sea expanded as it became connected with the Mediterranean (Turney 2008).

As well as being recognized in the Greenland ice cores, anomalies have been observed around 8200 years ago in palaeoclimate archives across the globe. (Radiocarbon dates on this event are around 7400 ^{14}C yr BP.) For example, in deep-sea sediments from the Norwegian Sea the abundance of the cold-loving planktonic **foraminifer** *N. pachyderma* (sinistral form) increases sharply, suggesting a decline in sea surface temperature of at least 2 °C (Klitgaard-Kristensen et al. 1998). The oxygen isotope signals from the Irish Crag cave and from Lake Ammersee, southern Germany, also feature sharp anomalies *c.* 8200 cal yr BP (von Grafenstein et al. 1998). The 8200 cal yr BP event has also been detected in pollen data from eastern North America (Shuman et al. 2005).

Dating of a pronounced glacier advance in the western part of the Silvretta Mountains (western Austria) yielded ages ranging from 8010 ± 360 to 8690 ± 410 years with a mean age for moraine stabilization of 8410 ± 690 years. Kerschner et al. (2006) suggested that this may represent glacier response to the early phase of the 8200 cal yr BP event, as it is recorded in the Greenland ice cores. The glacier advanced to a position beyond its **Little Ice Age (LIA)** limit with the end moraines situated 750–1000 m further down valley. The corresponding drop of the equilibrium line altitude relative to the LIA datum was 75 m. The glacier advance was followed by a period of glacier recession and rock glacier development. In total, the climatic fluctuation as recorded in this region may have lasted for about 500 ± 200 years. Also in the Eastern Alps the oscillations recorded in the pollen curves for Swiss stone pine (*Pinus cembra*) and sedges (Cyperaceae) at

c. 8200 cal yr BP match, in time and magnitude, the δ^{18}O excursion in the Greenland GRIP ice core record as well as sea surface cooling in the North Atlantic (Kofler et al. 2005).

Studies of lake sediments in eastern France and the Jura mountains of Switzerland show higher lake levels during the 8200 cal yr BP event owing to cooler temperatures and more precipitation. Magny et al. (2003) also propose that the area of Europe that experienced wetter conditions *c.* 8200 cal yr BP extended between about 43 and 50° N latitude, whereas areas of Europe to the south and north of this belt were affected by drier conditions. This could have been the result of a strengthened Atlantic westerly jet stream that tracked more storm systems across mid-latitude Europe because of the steeper thermal gradient between high and low latitudes.

As well as in the temperate north, the 8200 cal yr BP event has also been identified from proxy records in the monsoon regions. For example, in the Cariaco basin marine core record, off the coast of Venezuela, the event is interpreted as an aridity peak due to a reduced northward migration of the Intertropical Convergence Zone (ITCZ) (Hughen et al. 1996). A broad Holocene minimum is also found in the Indian Monsoon domain, where the lake sediment record at Awafi, in the United Arab Emirates, shows an abrupt lowering in water level and the influx of **aeolian** sands in the site (Parker et al. 2006). Speleothem records from Oman record precipitation minima from the oxygen isotope signals highlighting a weakening of the monsoon system (Neff et al. 2001). A weak monsoon signal has also been recorded in China further highlighting the global extent of the 8200 cal yr BP event (Cheng et al. 2009b).

Since becoming recognized widely in the 1990s, the 8200 cal yr BP event has become the focus of intensive study. Understanding the event has particular relevance for climate modelling efforts that seek to model the sensitivity of the North Atlantic thermohaline circulation system to any future changes in fresh water input (e.g. from enhanced high-latitude snow and ice melt that might occur due to anthropogenic increases in atmospheric CO_2) and to predict the associated climatic effects (Schmidt and LeGrande 2005). Although similar in nature to the Dansgaard–Oeschger events and the Younger Dryas of the last glacial cycle, as an analogue for study the 8200 cal yr BP event has the advantage of having occurred during a time of 'warm world' climate boundary conditions, more similar to the present day (see the review in Shuman 2012). It represents the highest magnitude perturbation of the AMOC to have occurred during the Holocene.

5.5 Mid- to late Holocene vegetation in Europe

By about 6000 years ago most tree species had reached their maximum Holocene range limits in Europe, both in latitude and altitude. From then on the vegetation changes became more complex because of the increasing influence of people. Instead of the broad regional trends seen in tree spread during the early to mid Holocene, the later Holocene saw increased fragmentation of woodland and a great diversity of localized changes superimposed on a general trend of deciduous forest retreat from the earlier established range maxima. As described in **Chapter 8**, the transition from the **Mesolithic** to **Neolithic** culture had largely been completed across Europe by 6000 years ago. While Mesolithic hunter-gatherers did affect vegetation, particularly through burning, the pace of vegetation alteration and clearance accelerated with the spread of agriculture during the Neolithic. Hence from Neolithic times it becomes increasingly difficult to distinguish between climate-related and human caused vegetation changes inferred from pollen diagrams. The impact of people on Holocene vegetation is discussed further in **Section 5.8**. It can generally be said that by about 5000 years ago agriculture exerted significant effects on the landscape in most parts of Europe, although before about 3000 years ago large areas of forest in central and northern Europe were probably still untouched by farming.

Despite the difficulties in interpreting late Holocene pollen diagrams, Huntley and Prentice (1993) were able to draw on enough pollen sequences to identify changes in the distribution of different forest types in 3000-year time slices through the Holocene (**Fig. 5.11**). It is important to note that their reconstruction maps the *possible* distribution of different forest types under the prevailing climate of the time and in the absence of human clearance. In reality the distribution of, for example, deciduous forest was not contiguous at 3000 ^{14}C yr BP as it appears on **Fig. 5.11** because of farming as well as different physical/climatic factors operating at local scales. Instead, the figure is showing the broad areas within which particular forest types could grow. The figure shows that maximum range limits

of deciduous forest were reached by around 6000 ^{14}C yr BP. By 3000 ^{14}C yr BP mixed forest had expanded in the north, and boreal forests (coniferous) in the mountains, at the expense of deciduous forest. Compared with 3000 years ago, at the present time the range limit of deciduous forest is further south in Scandinavia and north-eastern Europe. However, mixed forest has expanded at the expense of boreal forest in the Alps since 3000 years ago, and over the same time period Mediterranean vegetation has expanded in southern Europe.

Another important vegetation trend through the late Holocene has been the expansion of peatland and heaths, particularly in the more oceanic, western parts of Europe. It has been argued by some, notably by Moore (1975) that peat initiation in many areas of Britain was due to human actions (see **Section 5.8**) that changed catchment hydrology (e.g. deforestation reduces evapotranspiration and clearance by burning reduces soil infiltration thereby increasing surface wetness). However, there is also evidence from various parts of Europe for peatland expansion caused by shifts to wetter and/or cooler climatic conditions – predominantly after 5000 years ago (Barber and Charman 2005).

5.6 The North American Holocene sequence

It is interesting to compare the North American sequence with that of Europe. Despite individualistic responses of different tree species to post-glacial warming and precipitation changes, there was rapid change in the distribution of major forest types: for instance the boreal forest zone migrated northwards to occupy areas of land vacated by the retreating ice sheets, with a maximum north latitudinal extent reached by about 7000 years ago. The northward limit of spruce forest was reached later in the north-east compared with the north-west due to the more delayed melting of the Laurentide ice sheet compared with the Cordilleran. As coniferous trees moved north, deciduous trees such as oak moved in to replace them from their glacial refugia in the south-east. These, and other patterns of post-glacial tree spread in North America, can be seen in the detailed set of isopoll maps at 1000 and 2000 year time slices extending from the present back to the LGM produced by Williams et al. (2004).

Fig. 5.11 Distribution of major vegetational units in Europe (at 9000, 6000, 3000 ^{14}C years BP, and the present) reconstructed from pollen data (after Huntley and Prentice 1993)

Reconstructed Vegetation
9000 ^{14}C yr BP

Reconstructed Vegetation
6000 ^{14}C yr BP

Reconstructed Vegetation
3000 ^{14}C yr BP

Reconstructed Vegetation
0 present day

Tundra   Birch woods   Birch-pine woods   Boreal forest

Mixed forest   Deciduous forest   Mediterranean vegetation   Steppe

Late Holocene climatic deterioration is inferred from southward shifts in the northern boundary of deciduous forests from 6.0 to 3.0 kyr BP and from 3.0 kyr BP to the present.

Overall rates of tree spread and ecosystem change were far higher in the early Holocene and slowed after *c.* 7000 years ago. Notwithstanding the effects native Americans undoubtedly had upon the landscape (see **Chapter 8** for a discussion of post-glacial human migration into North America), Williams et al. (2004) suggest that large-scale vegetation distribution underwent relatively less change, and was closely aligned with prevailing climatic conditions from *c.* 7000 years ago until European colonization from *c.* 500 years ago. Thus while there are similarities between North America and Europe in the rapid, individualistic spread of trees northwards in the early Holocene, during the mid to late Holocene the human impact on vegetation cover was more significant earlier on in Europe. As in northern Europe, $\delta^{18}O$ values of past precipitation indicate that conditions in eastern Canada during the mid Holocene were both warmer and moister than today (Edwards et al. 1996). Dean et al. (1996) proposed that a shift from drier Arctic air masses to moister Atlantic air masses in the mid Holocene was responsible for this change. Later in the Holocene, southward retreat of forest in central Canada after *c.* 2500 years ago appears to correlate with the onset of the cooler conditions of the European SubAtlantic (Scuderi 2002).

Bernabo and Webb (1977) produced a classic work on vegetational changes in north-eastern parts of North America by summing the observed changes in the pollen record from 62 sites, and this work has continued to be refined with the addition of further pollen records. One of the important features they identified is the fluctuating position of the boundary between prairie and forest (**Fig. 5.12**). The signs of prairie development in the western Midwest are visible in the pollen record over 11 000 years ago as the vast region formerly occupied by the Late Glacial boreal forest began to shrink. The largest eastward shift of the prairie took place rapidly between 10 000 and 9000 years ago in response to increasing aridity. Bartlein et al. (1984) proposed a major early Holocene precipitation decline in this area amounting to 10 to 25 per cent, which was combined with a mean July temperature increase of 0.5 °C to 2.0 °C. By *c.* 7000 years ago almost all of the lakes in the Midwest and south-east were at low levels (Harrison and Metcalfe 1985). After around 7000 to 6000 years ago there was a slower, general (and fluctuating) westward retreat of the prairie boundary, and an even more significant southward retreat in central Canada, towards the present position. The late Holocene retreat of the northern and eastern margins of the prairie can be explained

Fig. 5.12 Maps showing shifts in the prairie–forest boundary in central North America through time from Williams et al. (2009). Map A shows boundary changes in thousands of calendar years BP as inferred from reconstruction of woody cover. Map B shows the boundary reconstructed from the percentage of prairie forb pollen taxa from Webb et al. (1983) with numbers in bold being ^{14}C kyr BP and numbers in brackets being in cal kyr BP. Reprinted with permission of Elsevier

by both cooling and increasing moisture availability (Webb et al. 1983; Williams et al. 2009).

Other evidence of Holocene vegetational and climatic fluctuations has been derived from a study of aeolian stratigraphy in some of the drier areas of the USA. In the central Great Plains region, periods of rapid **loess** accumulation indicate episodes of extensive dune activity. Luminescence dating indicates that loess in this region was deposited in the early Holocene and ended shortly after 6500 years ago. This corresponds with other **proxy data** from the Great Plains indicating an early–middle Holocene dry period (Miao et al. 2005). In Wyoming, for example, Gaylord (1990) identified pronounced aridity and aeolian activity from 7545 to 7035 years ago and from 5940 to 4540 years ago. Indeed, one of the most interesting recent developments in the study of dune fields in the western USA (including the High Plains) is the appreciation that dry phases or very extended droughts have caused repeated dune reactivation episodes during the Holocene (see, for example, Arbogast 1986; Madole 1995).

5.7 Russia, Asia, and Australia in the Holocene

Following a sudden warming and moistening of climate at the onset of the Holocene tree cover quickly returned to northern Eurasia. By around 10 000 years ago open birch forest was widespread across western Siberia with open mixed temperate forest immediately west of the Urals. The central Asian steppe and forest-steppe zone seems to have extended further north than at present. In the far north, forest reached the Arctic coasts by about 10 000 years ago, due to a warmer-than-present climate. In the Kola Peninsula of Russia fossil pollen and stomatal evidence has demonstrated that *Pinus sylvestris* arrived as early as 9100 cal yr BP (8150 ^{14}C yr BP). Pine forest probably expanded northwards from sparse or small disjunct populations. *Betula* and other shrub and herb taxa declined at this time, possibly as a result of shading by the pine canopy. By the end of the middle Holocene warming the pine treeline was some 100 km north of its present position close to the Barents Sea coast. An approximate 2 °C increase in regional summer temperatures would be necessary to move the treeline to the present-day coastline. From about 7000 years ago the pine treeline gradually retreated southward to its present modern

position. Birch woodland replaced pine about 3000 years ago. The area today experiences tundra conditions (Gervais et al. 2002).

The Holocene climate optimum in central Siberia probably occurred within the traditionally defined Boreal period, when dense larch forests developed. The Atlantic period was characterized by warm conditions that favoured the establishment of larch-spruce forests, though a climatic deterioration is also recorded. During the SubBoreal, spruce fluctuated in importance, on the basis of which it is suggested that there were two cool periods with an intervening warm period. Since 3000 years ago, the climate has become considerably cooler and forests have degenerated. During the last 1000 years, unfavourable climate conditions have resulted in a forest tundra and widespread tundra communities developing (Hahne and Melles 1997).

In northern and central China forest cover returned shortly after the start of the Holocene and, as in Europe, the vegetation zones were shifted slightly northwards relative to the present due to a warmer climate. Vegetation also extended westwards into the present arid belt due to greater rainfall, with forest reaching its maximum inland position *c.* 7000 years ago (Ren 2007) (**Fig. 5.13**). After this time, forest cover declined in most areas of China except the north-east where this was delayed until *c.* 2000 years ago. A prominent trend through the mid to late Holocene has been the expansion of temperate steppe vegetation at the expense of forest in eastern China and southward retreat of subtropical forest in the south-east. Maps of vegetation reconstructions similar to those depicted in **Fig. 5.11** for China at various time slices can be found in Ren and Beug (2002). Vegetation records from the Xinjiang region to the north-west can be divided into three stages: a warming and dry early stage (from 11 000 to 10 000 years ago), a warm and wetter middle stage (from 8000 to sometime between 4500 and 3000 years ago) and a fluctuating cool and dry late stage thereafter.

The Holocene in the northern Tibetan Plateau can also be divided into three stages: a warming and wet stage in the early to mid Holocene, followed by a variable drying and probably warm stage (*c.* 5000–3000 years ago) and ending with a cool and dry stage (since 3000 years ago). In the Inner Mongolian Plateau, the early Holocene was warm and dry, and a warm and wet climate occurred from 7500 to 3500 years ago, reaching optimal conditions from *c.* 6000 years ago. The climate has been variably drying, and

Fig. 5.13 Outline map of China showing Holocene shifts in the boundary between forest to the south and east and more open terrain to the north and west. Lines mark 40% arboreal (tree) pollen isochrones with numbers referring to the time period in ×1000 ¹⁴C yr BP (from Ren 2007, Fig. 3). Reprinted with permission of Springer-Verlag

probably cooling, since *c.* 3500 years ago. In the north-western part of the Loess Plateau, several Holocene **palaeosols** have been identified. The best-developed palaeosol-equivalent in major valleys is a swamp-wetland facies deposited between 10 000 and 4200 cal yr BP (8885 and 3805 ¹⁴C yr BP) under an extremely wet regime. Feng et al. (2006) suggest that the Holocene Climatic Optimum occurred nearly contemporaneously at all sites in the Xinjiang region, in the Inner Mongolian Plateau and in the north-western part of the Loess Plateau. A warming and wet early Holocene in the northern Tibetan Plateau is most likely related to high effective soil moisture resulting from snow and ice melting.

In the early to mid Holocene there was probably also greater forest extent in south Asia due to a stronger summer monsoon in India and in Indo-China (Bishop and Godley 1994).

Pollen, dune, and lake level evidence across Australia generally suggest a moistening in climate from around 14 000 years ago. In Tasmania, pollen cores suggest that forest had returned during Late Glacial times. In central Tasmania, at Lake St Clair, a marked reduction in *Podocarpus* pollen and a strong increase of grasses suggest reduced precipitation levels during the period of widespread deglaciation (*c.* 18 500–11 000 years ago). The development of forest occurred from 11 200 cal yr

BP, characterized by expansion of *Nothofagus gunnii*, and by the establishment of *Nothofagus cunninghamii* with *Eucalyptus* spp. Cool temperate rainforest peaked at 7800 cal yr BP, under wetter climatic conditions than present. The maximum development of rainforest in the early–middle Holocene may indicate that the temperature was slightly warmer than present (Hopf et al. 2000). A stalagmite record from Lynds Cave, north-western Tasmania grew continuously between 9200 and 5000 cal yr BP (Xia et al. 2001). The record suggests that the warmest climate occurred between 8000 and 7400 cal yr BP, followed by a wettest period between 7400 and 7000 cal yr BP. These are broadly correlated with the so-called 'Mid Holocene Optimum' previously proposed using pollen and lake level records. Whilst in Tasmania, pollen cores suggest that forest had returned during Late Glacial times, forest development in both the north-east and the south-east of Australia lagged behind, taking until the early Holocene to develop fully.

In the wet tropic of the north-east Queensland region pollen evidence from Lake Euramoo shows the development of wet sclerophyll woodland with marginal rainforest in protected pockets between 16 800 and 8600 cal yr BP. From 8600 to 5000 cal yr BP warm temperate rainforest developed. Dry subtropical rainforest occurred from 5000 to 70 cal yr BP (Haberle 2005).

Isotopic analyses from a Holocene speleothem from Yudnamutana in the northern Flinders Ranges in arid south-central Australia showed wetter conditions from *c.* 11500 years ago with a peak in moisture at 7000 to 6000 years ago (Quigley e al. 2010). In south-east Australia, high lake levels recorded at crater Lake Keilambete were interpreted as evidence for wetter climate conditions due to the northward shift of the fronts and stronger westerlies during the early Holocene (Jones et al. 1998).

The onset of arid conditions in the Flinders Range, similar to those at present, started *c.* 5000 years ago. The onset of heightened aridity and apparent increase in large flood frequency at this time is interpreted to indicate the establishment of an **ENSO**-like climate in arid Australia in the late Holocene (Quigley et al. 2010). In contrast, in tropical northern Australia arid conditions are suggested to have occurred during the 'Climatic Optimum' based on thermoluminescence ages for dunes in tropical northern Australia between approximately 8200 and 5900 years ago (Nott et al. 1999). Shulmeister (1999) showed that northern Australia had a precipitation maximum after 5000 cal yr BP in contrast to south-east Australia.

5.8 Humans and the Holocene sequence

The previous sections have concentrated primarily on natural climatic and vegetational changes that characterized the Late Glacial and Holocene. As explained further in **Chapter 8**, these changes had profound implications for our prehistoric ancestors; but it has also become increasingly clear in the Holocene that humans were an increasingly potent agent of environmental change (Pennington 1969). Therefore, when seeking to explain Holocene changes, it is important to consider the role that prehistoric people may have played in producing or influencing various environmental and vegetational changes detected in Holocene deposits. This is particularly so from the mid Holocene onwards, but in many places humans also had significant effects on early Holocene and Late Glacial environments. A striking example concerns the extinction of many species of large Pleistocene mammals (and the associated ecological changes) that occurred as humans spread through the Americas during the early post-glacial. This, and other effects on animals, is discussed in **Chapter 8**: the rest of this

section looks at some examples of how humans have affected vegetation and soils in Europe.

For a considerable time it was believed that **Palaeolithic** and Mesolithic people were relatively ineffectual, either because of their small numbers during these stages, or because they did not possess the technological wherewithal. The characteristic Palaeolithic hand axe, for example, was apparently either a weapon or grubbing tool, and not until the development of the polished stone axe was humanity equipped with a tool to attack the forest cover of Europe and elsewhere (Smith 1970).

However, pre-Neolithic people did possess the so-called tranchet axe which could have been effective in forest clearance, but more importantly Mesolithic humans may have utilized fire deliberately for driving game and clearing woodland. In the British Isles a marked abundance of hazel appears in the Mesolithic and there is no doubt that the European hazel (*Corylus avellana*) is fire resistant: its high representation in pollen diagrams may owe in part to Mesolithic burning giving it an advantage over other trees.

On the other hand, there is also evidence that some tree taxa fared worse than others in response to human activity. The pollen of lime (*Tilia*) tends to disappear in many British sites at the same time as charcoal and other evidence of human activity appears (Turner 1962). There is an even more marked decline in *Tilia* pollen later in the Holocene as clearing for agriculture intensified. For example, a pollen diagram from Wytham Woods, Oxfordshire, shows a very clear relationship between declining *Tilia* pollen and increases in pollen associated with human clearance, such as *Plantago lanceolata* (Ribwort plantain) (Hone et al. 2001).

The decline of elm pollen in Denmark coincides with the arrival of the Younger Ertebolle culture (Troels-Smith 1956). In many parts of Europe the decline of elm, radiocarbon dated to about 5000 ^{14}C yr BP, has been attributed by some workers to the use of elm leaves for feeding stock – the leaves being collected for stalled animals. This phase was followed by more intensive clearances for shifting agriculture – the so called 'Landnam' clearances. Certainly, the elm decline was the most marked and rapid vegetational change in European post-glacial history with elm pollen falling drastically in a matter of just a few hundred years to levels that were typically reduced by half. There has been much debate about the causes of the elm decline, with disease (a pathogen carried by the elm bark beetle) currently being regarded as the main cause by many. Parker et al. (2002) scrutinized the elm decline as represented in pollen diagrams from 139 sites across Britain and Ireland. After calibrating the radiocarbon dates on the event, it was found that the event occurred within a time range of about 6300 to 5300 calendar years ago. They also proposed that reductions in elm pollen occurring over this interval were the result of a unique combination of interrelated factors, including climate change, disease, and human activity. The possible direct and indirect effects of various factors contributing to the elm decline are summarized in **Fig. 5.14**.

Changes in post-Atlantic vegetation (after 5000 ^{14}C yr BP or *c.* 6000 cal yr BP *sensu* Mangerud et al. 1974) owe relatively less to climate than the changes of the Boreal and the Atlantic, and humans and soil deterioration assume still greater importance. Taking the example of the British Isles, deforestation in most places post-dates the Neolithic transition and shows significant increase during the Bronze and Iron Ages as the land area under cultivation was increased (**Fig. 5.15**).

The role of soil deterioration in vegetation change is less easy to assess, though intense leaching of glacial sediments under the warm, wet conditions of the Atlantic may have led to the development of podzols and other soils relatively inimical to deciduous forest (Pearsall 1964). Hardpans of the podzolic type may also have led to some waterlogging of soils by impeding their drainage, and they would also have been relatively acid in type. This process would have aided peat growth as shown in **Fig. 5.15**, especially in the areas of Britain that receive higher precipitation and show earlier dates for the initiation of peat formation.

It is conceivable that new agricultural techniques could, from Neolithic times onward, have accelerated this podzolic condition (Mitchell 1972). Podzolic soils would in turn encourage blanket bog formation. Burning and ploughing would help to release minerals which would accumulate as hardpan. This hardpan, by impeding drainage, would give ideal conditions for blanket-peats to accumulate, and in Ireland and western Wales Mesolithic fields, occupation sites, and megalithic tombs are sometimes found buried by

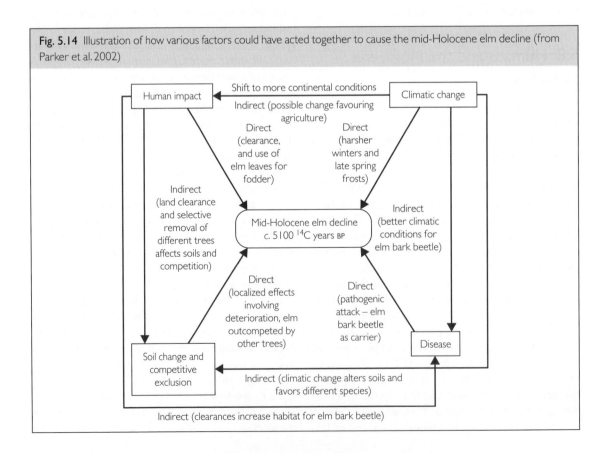

Fig. 5.14 Illustration of how various factors could have acted together to cause the mid-Holocene elm decline (from Parker et al. 2002)

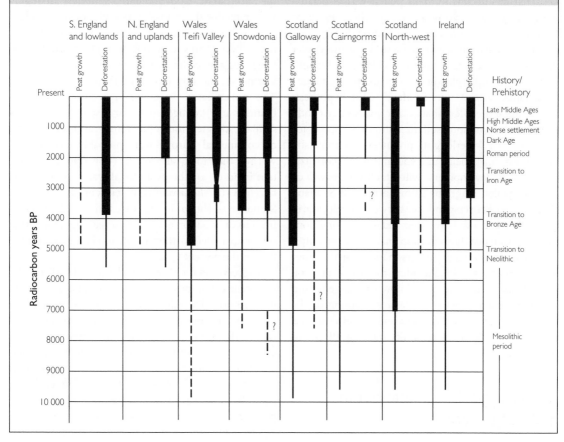

Fig. 5.15 A comparison of the timing of peat growth and deforestation during the Holocene in different regions of Britain and Ireland. The timescale is in ^{14}C yr BP, but can be converted to calendar years with reference to **Fig. 2.7**. Approximate positions of prehistoric/historic time periods against the radiocarbon timescale are also shown. Dashed lines indicate times of limited evidence, and the thickness of solid lines indicates relative magnitude. Note that deforestation reflects human activity, but the initiation and growth of peat may also be related to climate. Adapted from data in chapters by Mitchell et al., Greig, Chambers, and Birks in Berglund et al. (1996)

peat. However, peat bogs or blanket mires, the development of which coincides in some areas with the elm decline, do not always occur above well-developed podzol horizons (Moore 1975). Indeed, sub-peat profiles are often immature, suggesting that the soil is not always the predominant cause of mire development. Another possible factor in the expansion of peatland was the removal of the natural tree canopy as land became cleared for agriculture. This would reduce the transpiration demand of the vegetation and would also reduce the degree of interception of precipitation. Hence more water would be available to raise groundwater and soil-water levels, thereby favouring peat development. In the southern Pennines the

basal layers of all marginal peats (Tallis 1975) contain widespread evidence of vegetation clearance by burning, either in the form of microscopic carbon particles, small charred plant fragments, or larger lumps of charcoal.

Thus peat bogs, in that their development could be promoted by climatic change, soil maturation, and by human interference in the highland ecosystem, illustrate the complexity of factors that may be involved in Holocene environmental change.

It is indeed extraordinarily difficult to determine the importance of a great range of competing factors in determining particular vegetation changes. This becomes abundantly clear when one considers one of the most dramatic events in the British pollen record, which is the decline of Scots pine (*Pinus sylvestris*) pollen which occurred throughout the north-west Highlands of Scotland at *c.* 4000 ¹⁴C yr BP (Gear and Huntley 1991). Studies on the age and distribution of buried pine stumps in peatlands suggest that this decline in *Pinus* pollen was associated with a large-scale reduction in the northerly range of Scots pine forests. It has been postulated that this mid Holocene pine decline indicates a shift towards cooler and wetter conditions, possibly associated with a shift in the jet stream which brought a greater frequency of Atlantic storms into Scotland. This hypothesis hinges upon the effect that cooler and wetter conditions would have upon soil moisture. It is known, for example, that waterlogging produces anaerobic soil conditions which, being unfavourable for mycorrhizal and seedling development, will inhibit the growth of pine. In pollen diagrams from north-west Scotland there is a close correspondence between the decline in pine pollen and a shift to cooler and wetter conditions inferred from peat humification analyses shortly after 4000 calendar years ago (Anderson 1998). On the other hand, human activities, including the use of fire, soil deterioration, the role of unknown pathogens, and climatic consequences of a massive eruption of the Icelandic volcano Hekla, have also been discussed as possible triggers for the pine decline (e.g. Blackford et al. 1992).

5.9 Post-glacial climatic optima and neoglaciations

Whatever may have been the sequence of environmental changes which took place in different areas during the course of the Holocene, as identified by Manley (1966) one of the most contentious but interesting problems is that of the so-called climatic optimum. That climatic conditions were appreciably warmer earlier in the Holocene than in recent times, was discovered by Praeger (1892), who detected it in connection with his investigation of the fauna of the estuarine clays of the north of Ireland. From a comparison of the fauna of the Belfast estuarine clays with that of the present shores of Ireland, he gave the first proof of a definitely higher temperature during their deposition. Scandinavian workers subsequently showed that in Lapland the pine forest had at some stage in the post-glacial moved into zones which are now dominated by birch or alpine associations. This period of extended forest distribution has been called the post-glacial or Holocene Climatic Optimum. One of the most important faunal indicators of an optimum is the European pond tortoise (*Emys orbicularis*) which spread into Denmark, but disappeared in the SubAtlantic. Cool and damp summers are highly unsuitable for the animal, especially for the development of its eggs (Godwin 1956). In a classic study of ivy, holly, and mistletoe pollen in Denmark, Iversen (1944) suggested that summers there were 2 to 3 °C warmer in the mid Holocene than in recent times. The pollen-based vegetational reconstructions of Huntley and Prentice (1988, 1993) (see **Fig. 5.11**) yield similar estimates: mean July temperatures over most of continental Europe were probably around 2 °C warmer than today at 6000 ¹⁴C yr BP. Another important indicator of the post-glacial warm period is the hazelnut (*Corylus avellana*). Its present distribution in Scandinavia, shown in **Fig. 5.16**, is markedly different from its distribution 6000 to 5000 years ago, for at that time it extended further north, and higher in altitude than at the present time.

In Greenland too there is evidence that earlier in the Holocene conditions were more favourable to life than they are now. The edible mussel, *Mytilus edulis,* which now has a northern limit in Greenland waters at about 66° N, is found in raised beaches, which have been dated at about 7000–5000 ¹⁴C yr BP, which occur at a considerably higher latitude of 73° N. The warmer conditions caused the margin of Greenland's largest outlet glacier, Jakobshavn Isbrae, to recede well behind its present position during the mid Holocene (Young et al. 2011).

These and other sources of evidence therefore suggest a positive deviation over the recent mean of 1–3 °C for many areas in the temperate zone, but in high northern latitudes summer temperatures may have been 3–4 °C above modern values. In lower latitudes temperature increases appear to have generally been less and summer temperatures may often have been lower than today (Folland et al. 1990). The Paleoclimate Modeling Intercomparison Project (PMIP2) climate model simulations for the mid Holocene (*c.* 6000 years ago) suggest that when averaged globally, mean annual temperature and precipitation values differed little from today; however seasonality of climate was enhanced in the Northern Hemisphere due to the orbital configuration. The PMIP2 simulations support the inferences based on proxy data of summer air temperatures over much of Eurasia, North America,

Fig. 5.16 Indications of the post-glacial warm period illustrated by the present and post-glacial distribution of hazel (*Corylus avellana*) in Scandinavia. 1. Present general distribution. 2. Current records of individual occurrences. 3. Hazelnut fossils in sediments of the post-glacial warm period (from Frenzel 1973, Fig. 2)

and North Africa in the region of 2 °C warmer than recent (pre-industrial) values (Braconnot et al. 2007). Feedback effects related to the more northerly distribution of forest, reduced yearly period of snow cover, and less Arctic sea ice helped to amplify the early to mid Holocene warming at the high northern latitudes.

However, some dissatisfaction has been expressed from time to time with the use of the phrase 'Holocene Climatic Optimum', especially as in rather drier areas the increased temperature would be far from advantageous for plant growth. Various other terms have therefore been introduced including **altithermal** and hypsithermal. The latter was proposed in 1957 by Deevey and Flint as a term to cover four of the traditional pollen zones (V through VIII in the Blytt–Sernander system) embracing the Boreal through to the SubBoreal (c. 9000 to 2500 ^{14}C yr BP, see **Table 1.4**). However, the dates which other workers give for the 'optimum' or 'hypsithermal' do not always tally with this.

Another problem is that in different parts of the world the maximum temperature was recorded at different times. One factor was proximity of areas to residual ice masses during deglaciation. Different timings also apply for Holocene maxima or minima in precipitation. Although Northern Hemisphere summer insolation peaked around 10 000 years ago, showing a declining trend thereafter, levels remained significantly higher than today until four or five thousand years ago (see **Fig. 5.6**). Indeed, as early as 1970 Denton and Porter wrote that 'It is now known that rather complex low-order changes of climate characterized the hypsithermal interval, resulting in several early neoglacial episodes of glacier expansion. Therefore, in some regions at least, **neoglaciation** and the **hypsithermal interval**, as they are currently understood, partly overlap in time.' It may be better to think of the Holocene as containing multiple periods of 'optimal' conditions (during a lengthy span of higher Northern Hemisphere summer insolation) which had somewhat different timings in different places, rather than assuming that all regions fit within a generalized scheme. However, if generalizations are to be made, a good starting point is to consider the Holocene climate record as revealed in recent Greenland ice cores.

As displayed in **Fig. 5.7**, the record from NGRIP shows that air temperature over Greenland neared peak warmth about 9000 years ago and remained relatively high, though with some variation, until about 4300 years ago from which time temperatures have been in decline (Johnsen et al. 2001). The coolest conditions over Greenland since the 8200 cal yr BP event occurred between 2500 and 2100 years ago: dates which correspond fairly closely to a 'neoglacial' glacial advance, and the SubAtlantic deterioration as traditionally interpreted. Interestingly the inferred climatic cooling of the SubBoreal to SubAtlantic transition has proven to be one of the most robust inferences of the classic Blytt–Sernander scheme. Much recent palaeoclimatic research has added weight to the interpretation of widespread climatic change around this time; for example the evidence for cooling in north-west

Europe and elsewhere at c. 2700 cal yr BP presented by van Geel et al. (1996).

It should be remembered, however, that the Holocene portion of Greenland ice core δ^{18}O records is rather invariant compared with many other sources of palaeoclimate data – the quantity of which has grown enormously in recent years. Superimposed on the gradual late Holocene cooling as revealed in the ice cores were several brief phases of climate cooling that saw expansion of alpine glaciers ('neoglacial' phases) separated by periods of relatively warmer climate. Attempts to detect order in the complexity of Holocene neoglacial fluctuations have been the subject of much debate, and it is still difficult to say with certainty whether the Holocene glacial advances and retreats were contemporaneous in different parts of the world.

Neoglacials

In 1973, Denton and Karlén put forward a model in which they postulated that the Holocene had been punctuated by at least three phases of Alpine glacial expansion: the 'Little Ice Age' of the last few centuries, the period 3300–2400 ^{14}C yr BP, and 5800–4900 ^{14}C yr BP. They believed that in general the periods of glacier expansion lasted 600 to 900 years, and that they were separated by periods of contraction with duration of up to 1750 years.

More recent work has suggested that there were more episodes. On the basis of glacier and lake-level records, Holzhauser et al. (2005) reconstructed fluctuations of the Great Aletsch, the Gorner and the Lower Grindelwald glaciers in the Swiss Alps over the past 3500 years using tree-ring data, radiocarbon and archaeological data, in addition to historical sources. These glaciers experienced nearly synchronous advances at c. 1000–600 BC and AD 500–600, 800–900, 1100–1200, and 1300–1860. These fluctuations show a strong correspondence with lake-level variations reconstructed in eastern France (Jura mountains and Pre-Alps) and on the Swiss Plateau. They suggested that the impact of general winter cooling and an increase in summer moisture were responsible for feeding both glaciers and lakes in west-central Europe over the past 3500 years.

Intriguingly, the periodicity of neoglacial events originally suggested by Denton and Karlén (1973) resembles that of Holocene IRD layers found in North Atlantic deep-sea sediments. Bond et al. (1997)

identified Holocene peaks in IRD that they dated to 1400, 2800, 4200, 5900, 8100, 9400, 10 300, and 11 100 years ago. While minor compared with the IRD layers during the last glacial (the Heinrich layers as described in **Section 3.6**), these Holocene layers indicate times of sea surface cooling in the North Atlantic (amounting to SST depressions of 2 to 4 °C) when icebergs drifted further south. They display a periodicity of about 1500 years, similar to the pacing of the glacial D–O cycles; and Bond et al. suggested that the same millennial-scale climate cycle has continued to operate, albeit at lesser magnitude, through the Holocene. There is terrestrially-based evidence for climatic change correlating fairly closely with most of these IRD events, notably the 8200 cal yr BP event discussed in **Section 5.4**, although the precise way in which the climate of different regions responded to events in the North Atlantic Ocean during the Holocene remains an area of intensive study.

With respect to correlating phases of Holocene neoglaciation worldwide, J. M. Grove (1988) felt able to make certain generalizations, including some about differences in different hemispheres with respect to the timing of the extent of maximum Holocene glacial expansion (p. 359):

There seems to be a contrast between the Holocene behaviours of mid- to high-latitude glaciers in the northern and southern hemispheres. In both Patagonia and New Zealand, Little Ice Age advances were on a smaller scale than those of the Middle Holocene, whereas in Iceland, Greenland, and Spitzbergen, data at present available point to Little Ice Age advances, and notably those of the nineteenth century, as having been the greatest to have taken place since the end of the last glacial period.

A summary of glacier advances compiled by Wanner et al. (2008) (**Fig. 5.17**) broadly supports Grove's assertion: in contrast with South America, most records of glacier advance in the Northern Hemisphere show maxima reached late in the Holocene.

The Little Optimum, AD 800–1300

After the warmer phases of the mid Holocene, as we have just noted, conditions once again became cooler in many regions, but in early medieval times or pre-Renaissance times there was a return to more favourable conditions, particularly in Europe and parts of North America – the so-called **Little Optimum** or **Medieval Warm Period**.

The original proponents of this idea, including Lamb (1966), argued that from about AD 750 to 1200–1300 there was a period of marked glacial retreat which on the whole appears to have been slightly more marked than was seen in the twentieth century. The trees of this phase, which were eventually destroyed by the cold and glacial advances from about AD 1200 onwards, grew on sites where, in our own time, trees have not had time, or the necessary conditions, to grow again. In terms of a more precise date the medieval documents that are available place the most clement period of this optimum, with its mild winters and dry summers, at AD 1080–1180. At this time the coast of Iceland was relatively unaffected by ice, compared to conditions in later centuries, and Norse settlement was achieved in now inhospitable parts of Greenland. It is also believed that the relative heat and dryness of the summers, which led to the drying-up of some peat bogs, was responsible for the plagues of locusts which in this period spread at times over vast areas, occasionally reaching far to the north. For instance, during the autumn of AD 1195 they reached as far as Hungary and Austria. In northern Canada, west of Hudson Bay, fossil forest has been discovered up to 100 km north of the present forest limit, and four radiocarbon dates from different sites show that this forest was living about AD 870–1140.

One additional line of evidence that has been utilized to gain an appreciation of the nature of this phase is the presence of vineyards in various parts of Britain. The Domesday Book compiled in 1085 records 38 vineyards in England besides those of the king. The wine was considered almost equal to the French wine in quality and quantity as far north as Gloucestershire, Herefordshire, the London Basin, the Medway Valley, and the Isle of Ely. Some vineyards even occurred as far north as York. Lamb (1966a) regarded this as being indicative of summer temperatures 1–2 °C higher than in recent times. There was a general freedom from May frosts, and mostly good Septembers. There are correlatives of the European Medieval Warm Period in China (Hsieh 1976). At about the same time, lychees, sensitive trees which succumb at temperatures below −4 °C, were an economic crop in the Szechuan Basin in western China, but today they are limited to the south of Nanling.

However, some reconstructions of hemispheric and globally averaged climate for the past millennium, notably that by Mann et al. (1999) did not show evidence for a Medieval Warm Period. Doubts about

Fig. 5.17 Plots of glacier advance and retreat for different glacial regions over the last 6000 years. Shaded circles indicate times of proposed contemporaneous Neoglacial advances. The quality of the plots varies with the Alps and Scandinavia being the most studied and best constrained glacial areas (from Wanner et al. 2008, Fig. 4, drawn from various sources). Reprinted with permission from Elsevier

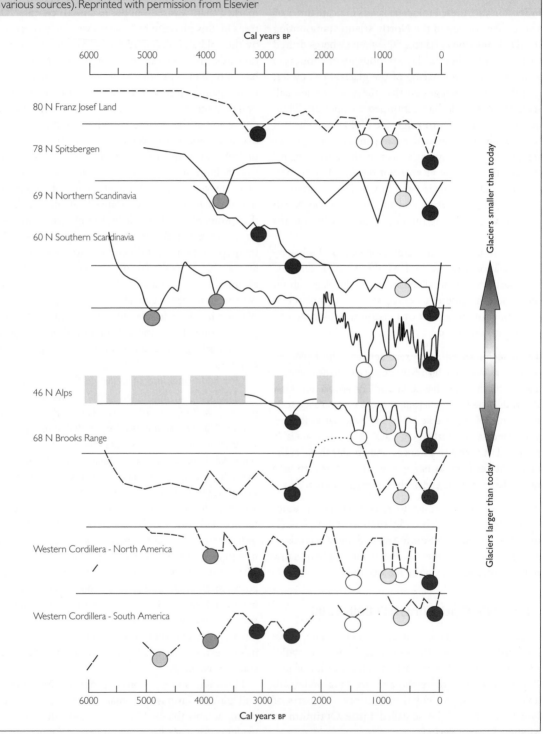

its significance were expressed during the 1990s and fundamental issues were addressed such as did the Medieval Warm Period actually exist? Was it a global phenomenon? Was it continuous or discontinuous? And did it span the entire period identified by Lamb and others? (Dean 1994.) In a review of the evidence existing at the time, Hughes and Diaz (1994, p. 109) came to a somewhat equivocal view:

... for some areas of the globe (for example, Scandinavia, Chile, the Sierra Nevada in California, the Canadian Rockies and Tasmania), temperatures, particularly in summer, appear to have been higher during some parts of this period than those that were to prevail until the most recent decades of the twentieth century.

These warmer regional episodes were not strongly synchronous. Evidence from other regions (for example, the Southeast United States, Southern Europe along the Mediterranean, and parts of South America) indicates that the climate during that time was little different to that of later times, or that warming, if it occurred was recorded at a later time than has been assumed. Taken together, the available evidence does not support a *global* Medieval Warm Period, although more support for such a phenomenon could be drawn from high-elevation records than from low-elevation records.

The global nature or otherwise of the Medieval Warm Period continues to be unclear and difficult to establish due to relatively few proxy records distributed world-wide that have sufficient climate sensitivity and chronological precision (Esper and Frank 2009). However, many still support the concept: Broecker (2001) put forward the case that it was a global event in the sense that changes synchronous with the warming in Europe have been identified in many other parts of the world, such as multi-decadal droughts in the western United States and an anti-phased cooling in Antarctica which supports the idea that the Medieval Warm Period was related to a phase of relatively stronger AMOC. Oceanographic studies have also shown that over this time span the **North Atlantic Oscillation (NAO)** was mainly in its positive mode (Trouet et al. 2009) and La Niña conditions of the ENSO were persistent in the Pacific (see **Section 7.4** for a discussion of the NAO). A more recent statistical analysis of proxy data spanning the past 1500 years by Mann et al. (2009) shows a warm anomaly of nearly 0.5 °C in the North Atlantic region (compared with the late twentieth century mean) between AD 950 and 1100; but when data are averaged

for the whole Northern Hemisphere, they find little difference. Many workers now use the term 'Medieval Climate Anomaly', to imply a global climate change at this time, but without a globally averaged warming.

The Little Ice Age (LIA)

One of the most significant Holocene environmental changes, not least because of its effects on the economies of highland and marginal areas in Europe, was the renewed phase of glacial advance since the 'Medieval Warm Period'. This phase has often been called the 'Little Ice Age' – a term first introduced by Matthes (1939) to describe an 'epoch of renewed but moderate glaciation which followed the warmest part of the Holocene', but it is now more widely used 'to describe the period of a few centuries between the Middle Ages and the warm period of the first half of the twentieth century, during which glaciers in many parts of the world expanded and fluctuated about more advanced positions than those they occupied in the centuries before or after this generally cooler interval' (Grove 1988, p. 3).

As with the Medieval Warm Period, the value and veracity of the Little Ice Age concept has been questioned (see Matthews and Briffa 2005, for a review). Bradley and Jones (1995, pp. 659–660) wrote:

... the last 500 years was a period of complex climatic anomalies, the understanding of which is not well-served by the continued use of the term 'Little Ice Age'. The period experienced both warm and cold episodes and these varied in importance geographically. There is no evidence for a world-wide synchronous and prolonged cold interval to which we can ascribe the term 'Little Ice Age'.

Nonetheless, Grove (1996) stoutly defended the concept and suggested that its effects have been noted all over the world. However, the dates at which the late-medieval Little Ice Age can be said to have begun and ended are variable from area to area, and also depend on whether the time span is being defined on the basis of glacier reconstructions or temperature reconstructions. Some of the earliest evidence for the onset of the LIA as an episode of renewed glaciation comes from Arctic Canada and Iceland where ice advance dates from as early as the thirteenth century, with intensification of the trend in the mid fifteenth century (Miller et al. 2012). However, some workers prefer to restrict the LIA term to the relatively cooler period

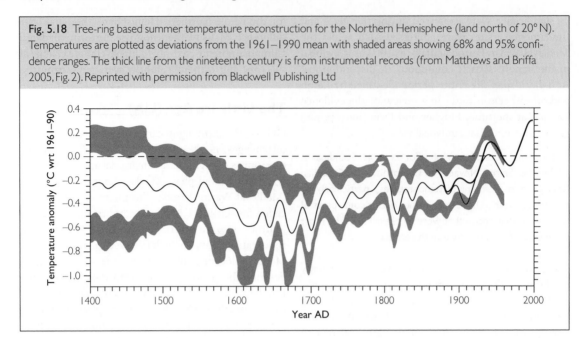

Fig. 5.18 Tree-ring based summer temperature reconstruction for the Northern Hemisphere (land north of 20° N). Temperatures are plotted as deviations from the 1961–1990 mean with shaded areas showing 68% and 95% confidence ranges. The thick line from the nineteenth century is from instrumental records (from Matthews and Briffa 2005, Fig. 2). Reprinted with permission from Blackwell Publishing Ltd

between AD 1550 and 1850, or even between AD 1550 and 1700 (Matthews and Briffa 2005) (**Fig. 5.18**).

Away from the North Atlantic region, this period also saw significant climate changes at various times: in China, for example, the LIA was well marked from 1650 to 1700 (Chu Ko-Chan 1973), but significant change occurred earlier than this. In south-west China, sediment geochemical analyses from Lake Erhai, indicate a cold-wet climate dated to AD 1550–1890 (Chen et al. 2005), and a shift to wetter conditions in the Tarim Basin, north-west China dates from AD 1560 (Liu et al. 2011). In general, evidence for a persistent cold anomaly relative to the late twentieth century mean is more strongly represented globally than the warm anomaly associated with the Medieval Warm Period, and Mann et al. (2009) suggest that the coldest LIA temperatures were reached between AD 1400 and 1700.

Yet, as previously argued by Bradley and Jones (1995), the climate was not continuously cold and year to year fluctuations were evident then as now. Thus the seventeenth century, a time of frequent severe winters in many parts of the world also experienced a number of hot summers such as that before the Great Fire of London in 1666. Moreover, the geographic expression of LIA climatic changes varied considerably. In most places average temperatures did not exceed 1 °C colder than the late twentieth century mean.

The patterns and timings of glacial advance during the LIA were also complex. For instance, glaciers in the European Alps responded differently to LIA climatic change than glaciers in Scandinavia, the former exhibiting some earlier maximal positions than the latter (Imhof et al. 2012). Grove (1988, p. 107) found no convincing evidence of expansion of Scandinavian glaciers before the seventeenth century. In Norway, where the glacial advances are relatively well chronicled by tax records and other sources, the advances appear to have begun between 1660 and 1700. The first half of the eighteenth century was marked by a general advance which amounted to several kilometres for some glaciers and culminated between 1740 and 1750. After this there was some recession, interrupted by readvances, notably in 1807–25, 1835–55, 1904–5, and 1921–5.

An examination of land rent assessments from the Jostedalsbre region of Norway, and of documents concerned with applications for their reduction, provided Grove (1972) with detailed information about the incidence of landslides, rockfalls, floods, and avalanches during the Little Ice Age in Norway (**Fig. 5.19**). The evidence makes it clear that there was a much increased incidence of major mass rock movements and floods in the late seventeenth century and on into the nineteenth century. Moreover, this environmental change began abruptly with a marked clustering

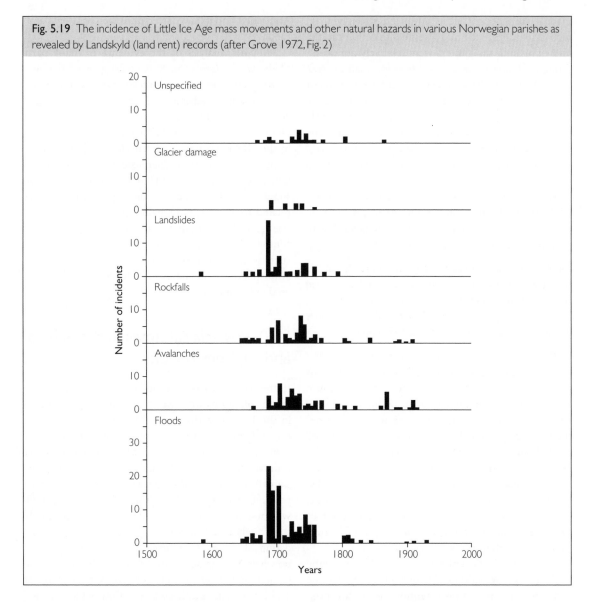

Fig. 5.19 The incidence of Little Ice Age mass movements and other natural hazards in various Norwegian parishes as revealed by Landskyld (land rent) records (after Grove 1972, Fig. 2)

of disastrous incidents between 1650 and 1760, and in certain years during that period, such as 1687, 1693, and 1702. Conditions for farming were thus very much less favourable than they had been in previous centuries.

The Icelandic glaciers and ice caps show a broadly similar history during this period. From the time of the first colonization of the island about AD 900 until at least the fourteenth century, the glaciers were considerably less extensive than they were after about 1700. There was a general advance in the early eighteenth century which reached its maximum around 1750. From 1750 to 1790 the ice tended to be relatively

stagnant or to be in a state of retreat, but it advanced again in the early nineteenth century and in some cases had, by 1840–60, reached a more forward position than the previous maxima of 1750. A general recession towards the present position took place after about 1890.

In the Alps the situation is extremely well documented compared to other areas. From about 1580 onwards tracts of cultivated land and forests were covered by advancing ice, and the local people were also subjected to greater flood risk. The local economies suffered and a series of supplications for tax relief were made. The Rhône Glacier advanced strongly from 1600 to 1680. From the mid seventeenth century to the mid eighteenth the glaciers were relatively quiescent, though still at a more forward position than in 1600. From the mid eighteenth century there was a major phase of advance, divided into three main stages: 1770–80, 1818–20, and 1835–55. Retreat was then fairly general between 1850 and 1880, only to be replaced by some advance from 1880–95. From 1895 to 1915 recession continued, but was temporarily reversed from 1915 to 1925.

The various stages are often visible as moraines, whilst the positions of hotels and other tourist facilities tell the story of retreat since the nineteenth-century maxima (**Plate 5.1**). As Grove (1966) said,

All over the Alps mountaineering huts stand high above the ice and must be approached by steep moraines, fixed ladders, or even ropes. These were not built to ensure an awkward scramble at the end of the day; their isolation is due to the wasting of ice during the last century.

The American Little Ice Age pattern shows that the glacial advances were broadly contemporaneous across the Northern Hemisphere. The maximum advance occurred in the middle 1600s and lasted until about 1700, from which there was some recession, then an advance, and then recession, which lasted until about 1785. Advances were characteristic of the 1880s, and in some areas the maxima of the 1700s were exceeded. In Alaska maxima were recorded between 1700 and 1835. In the North Cascade Range, Washington State, glacier termini are now hundreds of metres higher in altitude with areas reduced by over 50 per cent of their LIA maxima (**Plate 5.2**).

The glaciers of Greenland were also affected by the Little Ice Age, and advanced strongly between AD 1700 and 1850. The maximum extensions were, however, reached rather later than in many parts of the world. In south-west Greenland the maximum was around 1850. At the inland ice margin the maximum was reached about 1890–1900, whilst in the north-west the maximum was not reached until 1915–25.

The Antarctic seems to have escaped the cold epoch until rather late, a factor which doubtless helped the explorations of Captain Cook and others in the Southern Ocean. Between 1770 and 1830 the edge of the Antarctic ice appears to have been perhaps a degree of latitude south of its position in the years 1900–50 (Lamb 1967). On the other hand, the climatic deterioration seems to have persisted rather longer in Antarctica, and to have lasted until 1900 or later (Lamb 1969).

After her analysis of the evidence in many parts of the world, Grove (1988, p. 355) suggested that at a global scale 'there is a striking consistency in the timing of the main advances'. From analysis of 169 glacier length records from various parts of the world, Oerlemans (2005) showed that most glaciers, both in the Northern and Southern Hemisphere, were in retreat around AD 1850. Therefore, this date is usually taken as the end of the LIA. Oerlemans went on to suggest that this broad episode of glacier retreat was related to a *c.* 0.5 °C globally averaged temperature rise from the mid nineteenth to the early twentieth century. Various proposed causes of the Little Ice Age, such as volcanism, solar cycles, and internal ocean-atmosphere-ice oscillations and feedbacks are considered in **Chapter 9**.

5.10 Post-glacial times in the Sahara and adjacent regions

In the Sahara it has for long been suspected that climatic conditions were wetter at some stage or stages in the Holocene than now. This was deduced from the widespread distribution of rock paintings, and of stone and other tools in areas which are currently far removed from waterholes and today receive very low rainfall. Certain of the species represented in rock painting, notably elephant, rhino, hippo, and giraffe were regarded as being representative of a moderately to strongly luxuriant savannah flora (Holl 2004). The dating of the rock art itself has been problematic; however, Mori et al. (2006) dated peptides used in the paint binder from a prehistoric rock pictograph at Tadrart Acacus, southern west Libya, which yielded an age of 7157–6965 cal yr BP (6145 ± 70 ^{14}C yr BP) thus indicating an early Holocene age. This age was consistent with climatic and archaeological data from the central Saharan region.

Drake et al. (2011) have shown that a dense palaeo-river network formerly existed across much of the Sahara and that numerous closed basins supported some very large lakes e.g. Chad. When these lakes were full, they would have overflowed, thereby linking adjacent catchments. Analysis of the zoogeography of the Sahara shows that numerous species crossed the region during the Holocene. This dispersal of flora and fauna, including fish species, molluscs and amphibians, was possible because during the Holocene humid period the region contained a series of linked lakes, rivers, and inland deltas comprising a large interlinked waterway, channelling water and animals into and across the Sahara, thus facilitating these dispersals. As the environmental conditions ameliorated during the early Holocene the spread of wild fauna was rapid and there was a swift occupation of much of the Sahara by prehistoric populations (Kuper and Kröpelin 2006).

A wealth of ¹⁴C dates and an increasing number of OSL ages from aeolian and fluvial sediments provide detailed insight into the greening of the Sahara during the early Holocene as well as identifying periods of punctuated aridity and landscape instability (Gasse 2000; Bubenzer et al. 2007). By 15 000 cal yr BP the African monsoon front started to migrate rapidly northwards due to a shift in the position of the ITCZ driven largely by changes in solar insolation intensity across the Northern Hemisphere (deMenocal et al. 2000a). This was interrupted during the Younger Dryas when arid conditions briefly returned around 12 000 cal yr BP. During the early Holocene, **pluvial** conditions peaked around 9400 cal yr BP in the Saharan region (Hoelzmann et al. 2004) with evidence for the widespread development of carbonate lakes, siliceous mud deposits and fluvial sediment deposition. It has been suggested that the northern Red Sea region and Western Desert of Egypt were fed by rains originating from the Mediterranean region during the early Holocene climatic optimum (Arz et al. 2003). Radiocarbon dates of the base levels of palaeolakes and playas reveal the rapid onset of pluvial conditions between latitudes 16° N and 24° N at about 10 500 cal yr BP with a northwards shift in rainfall belts by as much as 800 km within just several decades (Kuper and Kröpelin 2006).

Palaeobotanical studies though the application of pollen, **phytolith,** and plant macrofossil analyses provide insights into the vegetation dynamics across the Saharan region during the Holocene. A good example of early to mid Holocene humidity in the dry heart of the eastern Sahara is provided by a study undertaken at Oyo by Ritchie and Haynes (1987). Their pollen spectra (**Fig. 5.20**) dating from c. 9500 cal yr BP (8500 ¹⁴C yr BP) until around 6800 cal yr BP (6000 ¹⁴C yr BP) show that there were strong Sudanian elements in the vegetation including Grewia tenax and Piliostigma, and they identified pollen of tropical taxa such as Hibiscus. During this phase Oyo must have been a stratified

lake surrounded by savannah vegetation similar to that now found 500 km to the south. After 6800 cal yr BP (6000 ¹⁴C yr BP) the lake became shallower and acacia-thorn and then scrub grassland replaced the sub-humid savannah vegetation. At around 5000 cal yr BP (4500 ¹⁴C yr BP) the lake appears to have dried out, aeolian activity returned and vegetation disappeared except in wadis and oases.

To the south, Wadi Howar flowed throughout the early Holocene crossing north-central Sudan, and draining into the Nile. During the onset of the Holocene climatic optimum, heavy rainfall delivered increased sediment loads into the channels of Wadi Howar until 7000–6000 cal yr BP (**Fig 5.21**) (Bubenzer et al. 2007). The discharge of the Nile during the early Holocene increased and there was a Nile-controlled lake at the Faiyum depression (Hassan 1986).

Roberts (1989) suggested that in effect the Sahara did not exist during most of the early Holocene. Bubenzer et al. (2007) presented data from numerous sites across the eastern Sahara, which combined palaeoenvironmental and archaeological data to reconstruct the environmental dynamics and landscape response to climate change during the early to mid Holocene. Whilst the onset of the Holocene pluvial was rapid, the onset of aridity and the weakening and southerly retreat of the African monsoon system was gradational. A difference of up to 2000 years has been suggested between the onset of aridity in the northern Sahara at c. 7000 cal yr BP compared with the southern Sahara c. 5000 cal yr BP (**Fig. 5.21**). This is a point of view that is supported by the work of Petit-Maire (1989) in the western Sahara (p. 652): 'Biogeographical factors implicate total disappearance of the hyperarid belt at least for one or two millenniums before 7000 BP. The Sahel northern limit shifted about 1000 km to the north between 18 000 and 8000 BP and about 600 km to the south between 6000 BP, and the present' (see **Fig. 5.20c**).

Whilst the early to middle Holocene across the Sahara was generally wet several studies have suggested that this pluvial period was punctuated by several arid events with drastic reduction in rainfall. Key periods during which lake levels drastically lowered, or in some cases desiccated completely, are around 8.5–7.8 kyr BP, 7.0–6.6 kyr BP, and 4.5–3.5 kyr BP (Gasse 2000).

Within the Arabian peninsula a similar picture for the early to mid Holocene has been observed with the presence of numerous relict palaeolakes within the

Fig. 5.20 (a) Pollen diagram from Oyo, in the eastern Sahara (after Ritchie et al. 1985) showing the main features of stratigraphy and pollen sequence in the Holocene. Reprinted by permission from Macmillan Publishers Ltd: *Nature*. (b) The location of Oyo. (c) The changing position of the Saharan-Sahel limit (after Petit-Maire 1989, Fig. 8)

Fig. 5.21 Synopsis of calibrated ¹⁴C and OSL chronologies from the Eastern Sahara. The OSL chronologies illustrate periods with stronger and weaker aeolian deposition. The ticks on the timescales represent the calibrated mean values of the individual dates. The curves are calculated using cumulative probability density histograms (Gaussian distributions) of luminescence dates and calibrated radiocarbon dates (cal yr BP). The onset of aridity during the Holocene is shown by the broken line showing a north-south progression through time (after Bubenzer et al. 2007, Fig. 2). Reproduced with permission from Elsevier

deserts of the Nafud, Rub' al-Khali and Wahiba Sands (Parker and Goudie 2008; Berger et al. 2012).

Speleothem records from Oman and Yemen record the incursion of Indian Ocean monsoonal rains into southern Arabia by 10 300 cal yr BP and northern Arabia by 10 100 cal yr BP (Fleitmann et al. 2007). At Mundafan lake, sediments show that monsoonal rains penetrated into the central interior of Arabia by 9600 cal yr BP (Rosenberg et al. 2011). At Awafi in south-eastern Arabia the main lake body formed at 8500 cal yr BP and persisted until 6000 cal yr BP. The region became vegetated with savannah-rich grassland with some *Acacia* and *Prosopis* tree elements. After 6000 cal yr BP conditions became drier, the lake shallowed and there was a general loss in vegetation cover though dry adapted grassland persisted. From around 4200 years

ago the lake desiccated and has only filled sporadically since (Parker et al. 2004).

In Yemen, extensive fluvial aggradation of sediments in the major wadi systems of the Hadramawt region occurred between 12 000 and 5400 cal yr BP, attributed to wetter conditions during the early to mid Holocene (Berger et al. 2012). Extensive incision and reworking of terraces and channel fills occurred from the mid Holocene, attributed to the onset of drier, more arid conditions.

In addition to these humid phases in the Holocene there may also have been some phases which were somewhat drier than today to the extent that dune reactivation took place in semi-arid areas. Thus, for example, studies of the **erg** in the Accra plains of Ghana (Talbot 1981) have demonstrated a phase of marked aridity combined with strong southwesterly winds at about 4500–3800 ^{14}C yr BP. In the eastern Sahara dune reactivation and accumulation occurred from 5000 cal yr BP in the Great Farafa Sand Sea (Bubenzer et al. 2007). In Arabia the onset of arid conditions began *c.* 6000 cal yr BP, with lakes in the central Rub' al-Khali desert and Yemen drying up. Further north, lakes persisted, though at much reduced levels, from moisture derived from winter westerly rainfall (Parker et al. 2004). Widespread dune reactivation occurred in the Liwa region of the Rub' al-Khali, with the deposition of 70 m high mega-barchan dunes from *c.* 3800 cal yr BP (Stokes and Bray 2005).

Under warmer Holocene conditions, monsoonal circulation was intensified, and in the Northern Hemisphere the Intertropical Convergence Zone would have shifted north, bringing rainfall into areas like West Africa, the Sudan, Ethiopia, Arabia and the Thar. As referred to in **Section 5.3**, the basis for this may have been increased summer solar insolation associated with the *c.* 22 000-year rhythm of orbital **precession**, for at around 10 000 years ago **Milankovitch** forcing led to Northern Hemisphere summers with almost 8 per cent more insolation than today (Kutzbach and Street-Perrott 1985) (see **Fig. 5.6**). Higher insolation caused greater heating of the land, stronger convection, more inflow of moist air and a higher summer monsoonal rainfall. In contrast, weaker insolation maxima around 60 000 and 35 000 years ago would have created weaker monsoons (Ruddiman 2001).

Changes in insolation receipts through the early post-glacial can help to explain the Northern Hemisphere low-latitude pluvial, but they have less direct relevance to the Southern Hemisphere (Tyson and Preston-Whyte 2000). Also important in determining the spatial and temporal pattern of precipitation change are sea surface temperature conditions associated with the build-up and disintegration of the great ice sheets (Shi et al. 2000; Rognon and Coudé-Gaussen 1996). In addition, changes in snow and ice cover over Asia, including Tibet and the Himalayas, could have had a major effect on the monsoon (Zonneveld et al. 1997).

5.11 Concluding points

The Holocene experienced abrupt and relatively brief climatic changes, which caused dune mobilization in the American High Plains (Arbogast 1996)

and alternations of pluvials and intense arid phases in tropical Africa. As with mid to high latitudes, the Holocene in lower latitudes was far from stable and it is possible that a climatic deterioration around 4000 years ago could have been involved in the near simultaneous collapse or eclipse of civilizations in Egypt, Mesopotamia, and North-west India (Dalfes et al. 1997) (see **Section 8.6**).

These brief events cannot be readily accounted for by orbital changes, so other mechanisms need to be considered, including changes in the thermohaline circulation in the oceans (e.g. Bond et al. 1997), or in land surface conditions (e.g. Gasse and van Campo 1994). Certainly the mechanisms causing changes in atmospheric circulation would have been both numerous and complex, and there will have been lagged responses (e.g. slow decay of ice-masses, gradual falls of groundwater, etc.). It is also apparent that there will have been differing hemispheric and regional responses to change (deMenocal and Rind 1993). For example, Arabia and North-east Africa may have been especially sensitive to changes in North Atlantic sea surface temperatures, while monsoonal Asia may have been especially affected by snow and ice conditions in the Himalayas and Tibetan Plateau. Causes of Holocene climatic variability, and the various **teleconnections** between higher and lower latitudes are considered in **Chapter 9**.

The change from the last glacial to the present Holocene interglacial was complex and involved large-scale and abrupt climatic changes: the Younger Dryas, for instance, demonstrates the instability of the Earth's climate and the importance that the oceans can play in this instability. The Earth warmed up very rapidly after the Younger Dryas though it took time for such phenomena as major vegetation belts, large ice sheets, and global sea levels to adjust. Overall, the early and mid Holocene were somewhat warmer than today, and, of no lesser significance, there were some major hydrological changes in lower latitudes, including a major pluvial event in some parts of the tropics. The human impact becomes more and more evident as the Holocene goes on, but even over the last 5000 years natural climatic changes have been of substantial importance. However, it has also been argued that the role of humans in influencing climate in pre-industrial times has been more important than previously thought. Ruddiman (2005) made the case that from *c.* 8000 years ago the spread of agriculture, and the deforestation that accompanied it, caused a significant enough upward trend in atmospheric carbon dioxide and methane to offset some of the cooling that would have occurred through the late Holocene owing to orbital trends (see **Section 9.6**). Notwithstanding a human-enhanced **greenhouse effect** to *reduce* the overall cooling that may otherwise have occurred over the last several thousand years, Greenland ice cores and many other proxy data sources show a late Holocene cooling trend; and importantly, we know that there have been complex fluctuations of climate at the decadal and century timescales superimposed on this trend as represented by such phenomena as drought intervals in the High Plains of America and the neoglacials that have affected valley glaciers.

SELECTED READING

On the Late Glacial and Holocene:

Undoubtedly the most useful general reference to the Holocene is N. Roberts (1998), *The Holocene: an environmental history* (2nd Edition). This book also contains an account of the Late Glacial.

The whole history of climate since the Last Glacial Maximum is discussed in H.E. Wright et al. (eds) (1993) *Global Climates since the Last Glacial Maximum*.

The Holocene in the USA is discussed in all its facets in H.E. Wright (1984) (ed.), *Late-Quaternary environments of the United States, ii, The Holocene*, and detailed pollen-based vegetational reconstructions for North America from the LGM through the Holocene can be viewed at: North American Pollen Viewer, http://www.ncdc.noaa.gov/paleo/pubs/williams2004/

A comprehensive summary of Holocene climatic changes in many parts of the world is contained in A.S. Issar (2003) *Climate changes during the Holocene and their impact on hydrological systems*, and the edited book by R. Battarbee et al. (2005) *Global Change in the Holocene* gives a strong overview and covers a wide range of the methods used to reconstruct Holocene changes.

The study of the Holocene climate record and its relevance for understanding contemporary anthropogenic climate change is the focus of the edited book by Battarbee and Binney (2008) *Natural Climate Variability and Global Warming*.

On late Holocene and recent changes:

The history of the world's climate since AD 1500 is covered in an edited work by R.S. Bradley and P.D. Jones (revised 1995) *Climate since AD 1500*.

J. M. Grove, in magisterial fashion, discusses the Little Ice Age in her (2004) *Little Ice Ages Ancient and Modern* (2nd edition).

A popular and accessible account of the Little Ice Age is found in B. Fagan (2000) *The Little Ice Age: How Climate Made History*.

There is also a journal devoted to the Holocene, called *The Holocene*.

REVIEW QUESTIONS

1. Consider the reasons why the start of our current interglacial is officially placed at c. 11 700 years ago rather than at c. 14 700 years ago when Greenland ice cores also show an abrupt and high magnitude warming.

2. Compare and contrast the Younger Dryas with the 8.2 kyr event. How are both of these events relevant to our understanding of climate change?

3. Why did vegetation associations with no 'modern analogues' arise in the mid latitudes during the early Holocene?

4. Describe when and why forests in the Northern Hemisphere reached their maximum Holocene latitudinal and altitudinal ranges.

5. Consider why research on the 'Medieval Warm Period' and the 'Little Ice Age' is of particular relevance for the attribution and prediction of anthropogenic climate change this century.

6. What drives the greening of the Sahara and Arabian desert regions during the early to mid Holocene?

6

Sea Level Changes of the Quaternary

Chapter contents

Chapter overview

During the Quaternary, sea levels have changed because of a range of factors. Some of these changes have been essentially worldwide and have been driven by changes in the volume of water held in the oceans and in the ice caps (**glacio-eustasy**). During **glacials** sea levels were low, whereas during **interglacials** they were at heights more similar to those of today. However, local factors, such as tectonic activity and the depression of the Earth's crust by ice (**glacio-isostasy**), complicate the global picture. In post-glacial times there has been a rapid rise in sea level, which has caused the drowning of many coastlines, though areas that were depressed by the mass of ice caps in the Pleistocene are continuing to rise now that their ice burden has largely gone.

6.1 The importance of sea level changes

The changes in climate and vegetation of the Quaternary were only equalled in importance by the relative sea level changes that took place, though these themselves were partially caused by climatic factors. Other contributing factors included tectonic and **orogenic** forces, local compaction of sediments, and loading of sediments into coastal basins of sedimentation. It is also possible to categorize the changes according to whether they were of a worldwide nature and involving sea level changes (**eustatic** changes) or of a local nature, and involving changes in land levels (tectonic changes).

The various types of sea level change were important, partly because they transformed the world's geography by such means as the opening and closing of straits and land bridges and the flooding and exposure of continental shelves, but also they influenced the migration and habitats of plants, animals and humans (Fernandes et al. 2006; Lambeck et al. 2011). Sea level changes may also have influenced regional and global climates, by, for example, modifying ocean circulation and the continentality of land areas.

6.2 Evidence of sea level change

The effects of such changes can be seen along most shorelines. Where there are stranded beach deposits (**Plate 6.1**), coral reefs, marine-shell beds, and platforms backed by steep cliff-like slopes and old sea caves one has evidence of emerged shorelines. One also often has evidence of submerged coastal features such as the drowned mouths of river valleys (rias) and glacial troughs (fjords), submerged dune-chains, notches and benches in submarine topography, and remnants of forests or peat layers at or below present sea level. Many coasts show evidence of both emergent and submergent phases in their history and numerous techniques are available to assist in their interpretation (Rose 1990).

One group of techniques for reconstructing past sea levels involves establishing the stratigraphy of submarine deposits. This can be done by seismic stratigraphy, which can be used to establish the existence of such phenomena as buried river channels, alluvial fills, and erosion surfaces (e.g. Giagante et al. 2011). Usually this is combined with the interpretation of sedimentary cores (e.g. Caruso et al. 2011) which reveal the subsurface existence of alternations of freshwater and saltwater facies. Precise palaeo-sea level information can be gained by micropalaeontological analyses, using such organisms as **Foraminifera, diatoms, ostracods**, and pollen (e.g. Gómez et al. 2005; Culver et al. 2011). These can often show alternations in, say, marshland or tidal flat environments (e.g. Gehrels and Newman 2004; Hayward et al. 2010; Josenhans et al. 1997; Engelhart et al. 2007). There may also be preserved macrofossils, such as peats (Mrani-Alaoui and Anthony 2011) and mangrove and sea-grass remains (Belperio et al. 2002). Fossil terrestrial soils have also been found (e.g. Hanebuth et al. 2011). The origin of organic material in cores can also be obtained through the use of carbon isotopes and C/N ratios (Lamb et al. 2006). (See **Section 1.6** for an introduction to stable isotope analysis.)

Various landforms, related in their development to sea level can also be used, including raised beaches, raised sabkhas (e.g. Gavish 1974), 'cheniers' and other barrier ridges (Chappell et al. 1982), planed-off submarine islands called '**guyots**' or seamounts (Paduan et al. 2009), submerged coral reefs (Webster et al. 2009), emerged coral reefs (Dumas et al. 2006), submerged aeolianites (Nichol and Brooke 2011) and other dunes (Lericolais et al. 2007), raised and submerged terraces (e.g. Cabioch et al. 2008), old river courses (e.g. Hanebuth et al. 2011), including incised rias (Garcia-Garcia et al. 2005) and huge braided systems under the English Channel (Gupta et al. 2007; Toucanne et al. 2010), tidal notches (e.g. Pirazzoli 2005), beachrock platforms (Desruelles et al. 2009), submerged **speleothems** (Surić et al. 2005), drowned karstic features (Mylroie et al. 1995), and drowned freshwater lakes (Murton and Murton 2012).

Another class of evidence is provided by what are called 'fixed biological indicators' (FBIs). These are sessile organisms which have a restricted vertical growth range, are confined to the intertidal zone, and so which on death become palaeo-sea level indicators. Ideal FBIs include oyster beds and barnacles (Lewis et al. 2008), and calcareous tubeworms, such as *Galeolaria caespitosa* (Flood and Frankel 1989).

Also much used have been microatolls. These are discoid corals, such as *Porites*, with flat upper surfaces that develop when upward coral reef growth is constrained by exposure at low tide. They have been employed because of their sensitivity to the seawater–air interface (Smithers and Woodroffe 2000). Other types of bioconstruction, including vermetid reefs, have also been productive (Chemello and Silenzi 2011). Living colonies of the fast growing coral *Acropora palmata* (Elkhorn coral) are restricted to within five metres of sea level and therefore fossil specimens (that can be dated by radiocarbon and uranium series methods) have been important for reconstructing sea level change over thousands of years in areas of its range in the western Atlantic and Caribbean (e.g. Toscano and Macintyre 2003). Evidence for post-glacial sea level rise based on studies of this coral is discussed in **Section 6.8**. Some marine organisms, such as the bivalve mussel *Lithophaga*, create borings, and these have been employed in such places as Pozzuoli in Italy to reconstruct sea level changes in historical times (Morhange et al. 1999).

Archaeological evidence can also be productive (Bailey and Flemming 2008). For example, some harbours may have become flooded (Martiner and Morhange 2007), while others may have been stranded by coastal uplift (Dominey-Howes et al. 1999) as is the case with Falasarna in Crete. Roman coastal fish tanks (*piscinae*) have also been used (Lambeck et al. 2004). Other archaeological evidence includes stone tools (e.g. Fedje and Josenhans 2000) and flooded quarries and millstones (Scicchitano et al. 2011).

Finally, evidence for sea level change is also provided by precise geodetic surveying and satellite altimetry, while archival evidence can be used to reconstruct sea level changes in cases like the rapidly uplifting areas of Fennoscandia (Steffen and Wu 2011).

6.3 Causes of sea level change

Table 6.1 attempts to categorize and list the various causes of sea level change according to whether they are dominantly worldwide or local. The eustatic types of sea level change will be discussed first in that they have general significance, while the 'anomalies' on this general pattern caused by local factors, such as **isostasy**, orogeny, and **epeirogeny**, will be discussed second. Thirdly, we will discuss the temporal pattern of sea level change for some key areas and key phases of the Pleistocene and Holocene.

6.4 Eustatic factors

Although glacio-eustasy (the response of sea levels to changes in the volume of ice stored in glaciers and ice

Table 6.1 Factors in sea level change

Eustatic (worldwide)	Local
Glacio-eustasy	Glacio-isostasy
Infilling of basins	Hydro-isostasy
Orogenic-eustasy	Erosional and depositional isostasy
Decantation	Compaction of sediments (autocompaction)
Transfer from lakes to oceans	Orogeny
Expansion or contraction of water volume because of temperature change	Epeirogeny
Juvenile water	Ice-water gravitational attraction
Geoidal changes	

sheets) is the most important of the eustatic factors that have affected world sea levels during the course of the Quaternary (see next subsection), it is worthwhile to look at some of the other eustatic factors which play a role, especially over the long term. Two very minor factors are the addition of juvenile water from the Earth's interior and the variation of water level according to temperature. The latter, often called the **steric effect**, could raise the level of the sea by about 60 cm for each 1°C in temperature of the seawater, while the former could probably add about 1 m of water in a million years. The evaporation and desiccation of **pluvial** lakes, some of which had large dimensions, would be unimportant in affecting world sea levels, adding a maximum of about 10 cm to the level of the sea, were they all to be evaporated to dryness at the same time (Bloom 1971). Some human activities, such as the storage of river runoff in large reservoirs, can also cause minor modification of global sea levels (Gornitz et al. 1997).

Another cause of eustatic changes of sea level, especially in the Holocene, is the process called 'isostatic decantation'. Isostatic uplift in the neighbourhood of the Baltic basin and of Hudson Bay has led to a reduction in the volume of these seas, and the water from them has thus been decanted into the oceans to affect worldwide sea levels. A comparison of the area and volume of the late-glacial precursor of Hudson Bay with Hudson Bay itself suggests that the volume of water decanted could only be sufficient to cause a rise in world sea level of about 0.63 m. The contribution of the Baltic Sea would be even less. This factor can thus be largely ignored.

These minor factors are of a very limited degree of importance in terms of a Quaternary timescale, especially when they are compared with the changes brought about by glacio-eustasy.

Rather more important may have been the infilling of the ocean basins by sediment, which would tend to lead to a sea level rise. With current rates of **denudation**, Higgins (1965) estimated that this could lead to a rise of 4 mm / 100 years. This is equivalent to a rise of 40 m in a million years. Changes in the rates of sedimentation associated with changes in rates of deposition of planktonic Foraminifera may also contribute to this mechanism (Harrison 1990, p 146).

Glacio-eustasy

As we saw in **Chapter 1**, during the first decades of the twentieth century **strandlines** of the Quaternary were perceived to be glacio-eustatic in origin (see Guilcher 1969). It was believed, correctly, that sea level oscillated in response to the quantity of water stored in ice caps during glaciations and deglaciations. It was also proposed that there was a suite of characteristic levels in Morocco and around the Mediterranean which could be related to different glacial events.

It was believed that the **transgressions** of the **interglacials** were succeeded by **regressions** during **glacials** and that the height of the various stages declined during the course of the Pleistocene (**Fig. 6.1**). Total melting of the two main current ice caps – Antarctica (about 30 million km³ of ice) and Greenland (almost 3 million km³) – would raise sea level a further 77 m or so if they both melted. Deep-sea core evidence, however, does not suggest that in previous interglacials of the Pleistocene these two ice caps did disappear, and without a general melting of them, sea level would only have been a few metres higher than now in the interglacials. This fact does not tie in too happily with the simple glacio-eustatic theory of progressive sea level decline during the Pleistocene. Some factors other than glacio-eustasy must be responsible for the proposed high sea levels of early Pleistocene times if indeed they are a reality at a global scale. Furthermore, because other factors have played a role, some local, it is not possible to correlate shorelines over wide areas on the basis of a common interglacial age.

Nevertheless, low Quaternary sea levels brought about by the storing up of water in the ice caps were quantitatively extremely important (Rohling et al. 1998a). Donn et al. (1962), on the basis of theoretical considerations from known ice volumes, reckoned that in the Riss (penultimate glacial), possibly the most extensive of the glaciations, sea levels might have been lowered by 137 to

Fig. 6.1 The classic sequence of Pleistocene sea levels, showing the downward trend in the elevation of raised beaches through time. The dashed line represents the approximate elevation of sea level if the ice in Greenland and Antarctica melted (modified after Frenzel 1973, Fig. 92)

Plate 2.1 The Mer de Glace in the French Alps near Chamonix, showing lateral and medial moraines *(ASG)*

Plate 2.2 Example of a till deposit. The lower part of the till section is covered in barnacles. (Photo taken on the southern shore of the Juan de Fuca Strait, Washington State) *(DEA)*

Plate 2.3 A glacial valley in north-central Portugal, showing a classic U-shaped form *(ASG)*

Plate 2.4 Patterned ground developed near Thetford, eastern England, with alternating stripes of grass and heather *(ASG)*

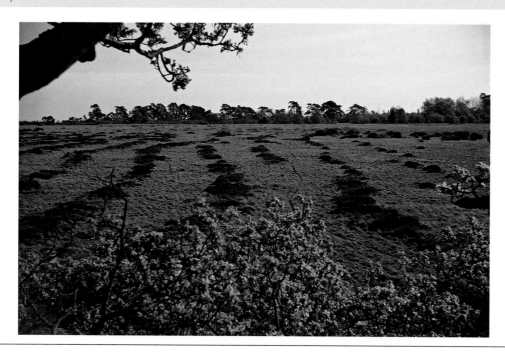

Plate 2.5 A loess and palaeosol section from Tajikistan, central Asia *(ASG)*

Plate 2.6 Example of a sediment core. This core shows a post-glacial sequence of basal clays grading into progressively darker lake sediments and eventually peat. (Photo taken at a bog in Glen Torridon, north-west Scotland) *(DEA)*

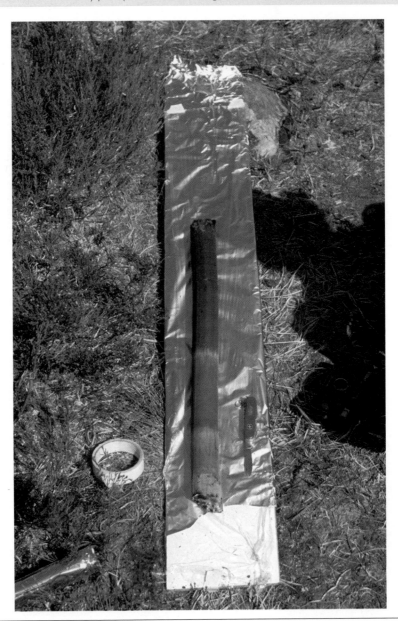

Plate 2.7 A buried pine stump being exposed by peat erosion in north-west Scotland. Layers of pinewood within peat have been interpreted as indicating periods of drier Holocene climate *(DEA)*

Plate 3.1 The glaciation of the Southern Alps of New Zealand created the great fjord (drowned valley) of Milford Sound *(ASG)*

Plate 4.3 Fossil dunes going under the alluvium of King Sound, West Kimberley district, Western Australia *(ASG)*

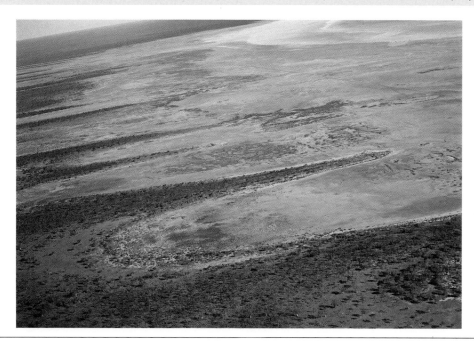

Plate 4.4 Lisan Shorelines on the side of the present Dead Sea *(ASG)*

Plate 4.5 Old shorelines of Lake Bonneville, Utah, USA *(ASG)*

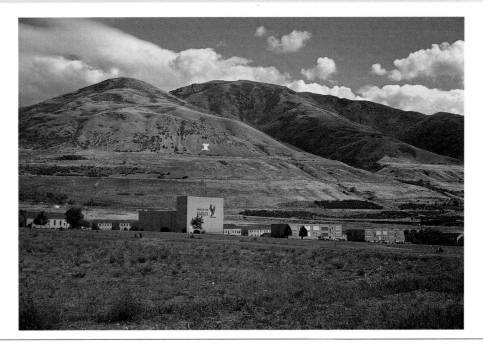

Plate 4.6 Tufa deposits (Trona Pinnacles) at Searles Lake, Mojave Desert. The tufa formed when Searles Lake, now dry, was one of many pluvial lakes east of the Sierra Nevada mountains *(ASG)*

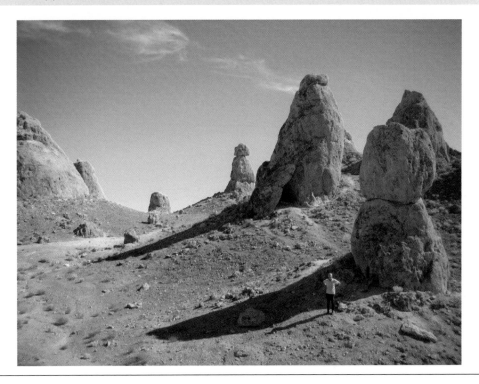

Plate 4.7 Holocene lake beds in the Dakhla oasis of the Western Desert of Egypt have suffered post-depositional deflation to generate large wind-eroded terrain (yardangs) *(ASG)*

Plate 4.8 The dry bed of a former pluvial lake in the Rift Valley of East Africa – Lake Chew Bahir, Ethiopia *(ASG)*

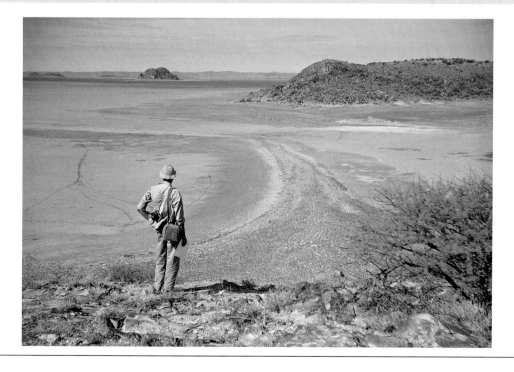

Plate 5.1 The Bossons Glacier near Chamonix, France. It has retreated and thinned during the twentieth century, as indicated by abandoned moraine *(ASG)*

Plate 5.2 The Nisqually Glacier, Mount Rainier, Washington State. Note the former height of the glacier as shown by lateral moraine compared with its present size. It has thinned and retreated over a mile up-valley from its maximum Little Ice Age position *(DEA)*

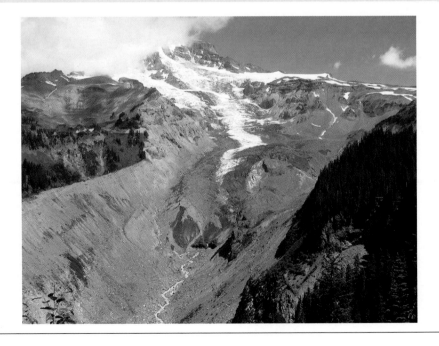

Plate 6.1 A raised beach at Portland Bill on the south coast of England *(ASG)*

Plate 6.2 Submerged Lycian tombs in south-west Turkey, and (left) abandoned shore platform, caused by local tectonic activity *(ASG)*

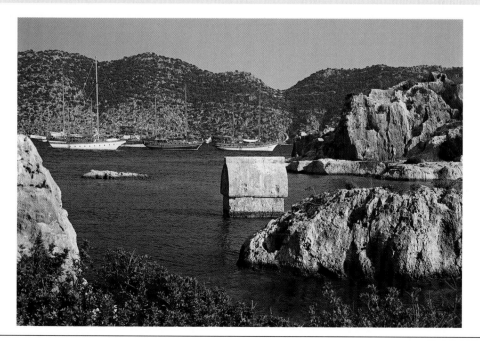

Plate 7.1 Nineteenth-century illustration of the Nygaard Glacier, Svalbard, by J.D. Forbes

Plate 7.2 Google Earth image of the Nygaard Glacier, Svalbard

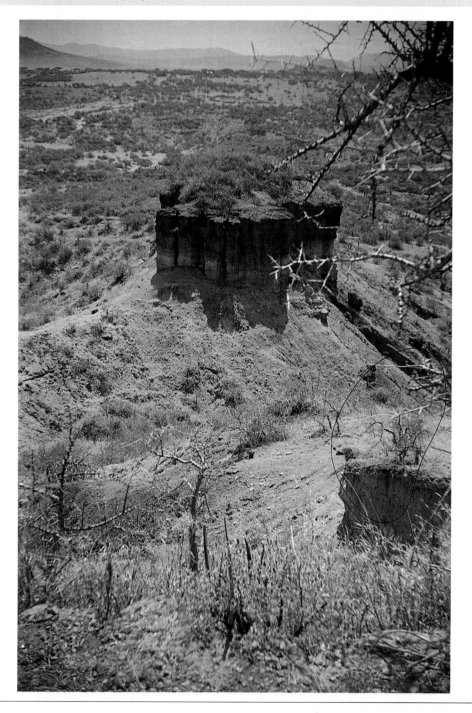

Plate 8.1 Olduvai Gorge, Tanzania, is an important site in the East African rift where evidence for several species of early of hominins has been found *(ASG)*

Plate 8.4 A Neolithic shell midden, Umm al-Qaiwain, United Arab Emirates. The Arabian Gulf Neolithic middens record the exploitation of marine resources as well as linking trade with Mesopotamia (Ubaid pottery) and the Oman mountains (chert flints) *(AGP)*

Plate 8.5 Sediment section through the lake deposits at Awafi, Ras al-Khaimah, United Arab Emirates. The sediments reveal desiccation at 4.2 kyr, during which time the lake dried out and filled with sand *(AGP)*

4.2 kyr

Plate 8.6 Terraced fields from the Dhamar region, Yemen (courtesy of Tony Wilkinson)

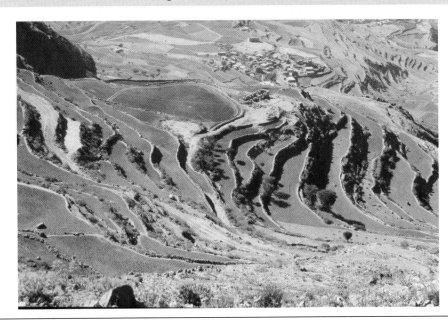

Plate 8.7 A donga in Swaziland. It is thought that this gullying was caused by the over-exploitation of wood for iron smelting from the eighteenth century *(AGP)*

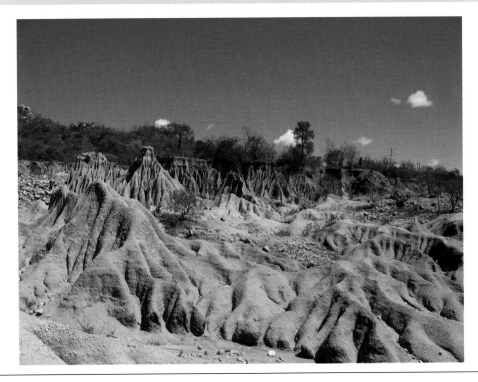

159 m below current sea level. In the Last Glacial they give a figure for lowering of rather less: −105 to 122 m. On the basis of dates for coral and associated material in the Great Barrier Reef (Australia), California, and south-east Caribbean sea areas, Veeh and Veevers (1970) favoured the conclusion that towards the end of the Last Glacial, sea level dropped universally to at least −175 m, some 45 m deeper than hitherto suspected, but this seems to be too high a value. Yokoyama et al. (2000) suggested from evidence from the Bonaparte Depression off Australia that the **Last Glacial Maximum** sea level depression was 130–135 m, while Clark and Mix (2002) indicated a figure between 120 and 135 m, whereas Hanebuth et al. (2009), working on the Sunda Shelf, suggested 123 m. From work in Barbados, Peltier and Fairbanks (2006) gave a value of *c*. 120 m.

The consequences of this low sea level included the linking of Britain to the continent of Europe (Gibbard and Lautridou 2003), of Ireland to Britain (Whittow 1970), of Australia to New Guinea, of Russia to Alaska across what is now the Bering Strait, and of Japan to China (Emery et al. 1971). Large parts of the floor of the Red Sea and Persian Gulf (Sarnthein 1972) were dry land. The Red Sea became reduced in size (Olausson and Olsson 1969) although it never seems to have been totally isolated from the Indian Ocean to become a hypersaline lake, though the Bab El Mandeb straits were considerably narrowed (Fernandes et al. 2006; Lambeck et al. 2011).

In the Mediterranean (**Fig. 6.2a**), large plains existed off the coast of Tunisia and fringed most of Italy, southern France, eastern Spain, and much of Greece. The Aegean and Ionian islands were linked up to themselves and to the mainland (Perissoratis and Conispoliatis 2003). Anatolian Turkey was connected to Europe by land bridges across the Bosphorus and the Dardanelles, while most of the Cyclades were

Fig. 6.2 Changes in coastal configuration at the time of the Last Glacial Maximum at *c*. 18 000 ¹⁴C yr BP (21 000 cal yr BP). The area in black shows the extent of the continental shelf that would have been dry land were sea level *c*. 120 m lower than at present. (a) The Mediterranean basin (b) the southern coast of South Africa (modified after van Andel 1989, Figs. 3 and 6)

merged into a single island. In southern Africa (**Fig. 6.2b**) there was a large area of land exposed between Cape Town and Port Elizabeth so that during glacial maxima the area of the coastal plain there expanded five-fold compared to today (Compton 2011). In the tropics, coral reefs were exposed to karstic processes and were subjected to sinkhole development (Woodroffe 2002; Dickinson 2004).

At certain times, as for example in the early stages of an interglacial, the rates of change occasioned by glacio-eustasy may be surprisingly great. This is because some ice sheets are potentially unstable in the face of a modest sea level rise. Through a process called 'decoupling', grounded ice starts to become buoyant and to float, whereupon the ice sheet starts to disintegrate (Anderson and Thomas 1991).

Orogenic eustasy

Although orogeny (mountain building) is normally regarded as being an essentially local factor of sea level change, and eustasy as being of worldwide nature, there is one class of process, **orogenic eustasy**, whereby a local change can have worldwide effects. It therefore acts as some sort of a link between these two main types of change.

Figure 6.3 represents the sort of picture that one can envisage and this is a situation that can easily be represented in the laboratory with simple materials. Two 'continents', represented by rectangular blocks of lead, float on a mantle of mercury. Water, representing the sea, is poured on to the mercury so that the 'continents' are just submerged. One of the continents can then be deformed and a mountain created by the simple process of turning one part of a continent upright, thereby effectively halving the width and doubling the thickness of the 'mobile belt'. The deformed 'continent' will displace the same amount of mercury as the undeformed continent, although the mountain will have a deeper root. Thus the level of the mercury will remain the same, but the water now has a larger area to spread over, so that it will spread out reducing the depth of the water. The stable 'continent' will thus emerge from the sea. In effect one is producing a worldwide regression of the sea by means of a local orogenic event (Grasty 1967). Equally, the variable volume of material that forms mid-ocean ridges could be significant. If the rate of sea-floor spreading increases, then the volume of the ridge crests increases, displacing water and causing additional inundation of continental areas. Likewise, a change in the pattern of subduction, as one plate descends beneath another,

Fig. 6.3 The production of worldwide (eustatic) changes of sea level as a result of orogenesis (modified from Grasty 1967)

can cause the ocean basins to become on average younger or older and so produce a similar effect to that caused by varying the amount of sea-floor produced per unit time (Harrison 1990).

It has been calculated that an increase of only 1 per cent in the area of the oceans would lower the sea level by about 40 m, assuming the average depth of the oceans to be 4 km. Over a long time-period this process could be significant, though it probably cannot explain the shorter amplitude sea level fluctuations of the Pleistocene. On the other hand, the gradual fall of interglacial sea levels during this epoch which has frequently been proposed could have been caused partly by this mechanism. Bloom (1971, p. 355), on the basis of information derived from studies of global tectonics, estimated that as ocean basins are spreading at rates of up to 16 cm per year:

The spreading of the ocean basins since the Last Interglacial could accommodate about 6 per cent of the returned melt-water, and the post-glacial shorelines would be almost 8 m lower than the interglacial shorelines of 100 000 years ago.

Geoidal eustasy

In recent decades the importance of a third type of eustatic change has been identified, notably by Mörner (1980). This is termed geoidal eustasy. The shape of the Earth is not regular, and at the present time the geoid (due to the Earth's irregular distribution of mass) has a difference between lows and highs of as much as 200 m. The ocean surface reflects this irregularity in the geoid surface. Ocean surface highs, for example, occur over the North Atlantic and lows over the equatorial Indian Ocean. The geoid surface varies according to various forces of attraction (gravity) and rotation (centrifugal), and will respond by deformation to a change in these controlling forces. The possible nature of such changes is still imperfectly comprehended, but they include fundamental geophysical changes within the Earth, changes in tilt in response to the asymmetry of the ice caps, changes in the rate of rotation of the Earth, and the redistribution of the Earth's mass caused by ice cap waxing and waning.

Two major consequences of the recognition of this cause of variability in sea level are that both sea level change and uplift/subsidence can only be measured relative to a fixed point – the Earth's centre – and that

apparent uplift could occur as the result of a lateral (tangential) movement of a geoid anomaly or of the land itself. Nunn (1986) has suggested that this may be a better explanation for some of the high-level fossil coral reefs of the tropical oceans than the normally invoked mechanism of local tectonic uplift.

Clark (1976) has postulated that the growth of massive Pleistocene ice sheets, such as that of Canada, would cause sufficient gravitational attraction for sea level to rise locally relative to the land. As the ice sheets melted and lost mass, the sea level would also fall in response to the reduced gravitational attraction. He suggests that this ice-water gravitational effect alone could cause raised beaches to occur 85 m above present sea level in Hudson Bay.

6.5 Isostasy

The Earth's crust responds when a load is either applied to it or removed from it. Thus, during the course of denudation erosion would remove a considerable mass of rock from on top of underlying strata and a certain amount of compensatory uplift would delay the point at which base level would be attained. However, it is unlikely that this particular isostatic effect would be of more than local significance during the relatively short time span of the Quaternary.

In areas of intense volcanic activity the loading of volcanic sediment on to the crust might cause some local depression which might in turn be compensated for by uplift at some critical distance away from the eruption (McNutt and Menard 1978) (**Fig. 6.4**). Isostatic depression is also a feature of areas of deltaic sediment loading. Some deltas show dated Holocene sediments at appreciable depths, and in such cases it is possible to calculate the rate at which loading-induced subsidence has occurred. For instance, Stanley and Chen (1993) have calculated Holocene subsidence rates for the Yangtze River delta of China and find at the seaward end that the rates are of the order of 1.6–4.4 mm per year. The Mississippi Delta has averaged a subsidence rate of c. 15 mm per year during the Holocene, while rates for the Nile are c. 4.7 mm per year and those for the Rhine up to 4.5 mm per year.

However, isostasy would be important on a broader scale in two main ways: by the application and removal of large masses of ice to certain parts of the Earth's crust (glacio-isostasy); and by the application and removal of large bodies of seawater, and, occasionally,

Fig. 6.4 A three-stage model of apparent sea level change through time, from initiation at (a) to the present at (c) brought about in the vicinity of coral reefs by volcanic loading onto the elastic lithosphere (after McNutt and Menard 1978)

lake water, from the continental shelves, and from lake basins.

The latter mechanism has recently been called 'hydro-isostasy' and is probably of less importance than glacio-isostasy, but its effects have often been forgotten. The hydro-isostasy theory can be summarized thus. Eustatic sea level changes brought about by the melting and freezing of the ice caps would alternately add and take away water to and from the ocean basins. This water would thereby add or remove a load from the ocean floor, and if one assumes that the density in the sub-crustal zone is nowhere less than 3.00 or 4.00 g cm^{-3} then total isostatic adjustments to the water loads resulting in depression of the seabed would be expected to range from one-third to one-quarter of

the effective depth of the water (Higgins 1969). In fact, however, the rate and amount of hydro-isostatic deformation would vary from place to place according to various factors. Coasts with nearby ocean water more than 100 m deep had the load of water from the post-glacial eustatic rise of sea level added early and close to the shore, whilst coasts which now border shallow seas had the load added late and, generally, far offshore. One would expect that the amount of submergence would be roughly proportional to the proximity of deep water. This has tentatively been confirmed by studies on the coast of America (Bloom 1967).

Often factors which might affect the degree of submergence consequent on the sea level rise would be the local sub-crustal density, its dynamic viscosity, and the degree of isostatic adjustment achieved before loading or unloading began. Estimation of the general effects of this mechanism, however, suggests that the melting of all the present Antarctic ice would raise sea level eustatically by around 60 m, but that compensatory hydro-isostatic sinking of the ocean floors would reduce the effective sea level rise to about 40 m, that is to around two-thirds.

Glacio-isostasy

The principle of glacio-isostasy is that during glacial phases, water loads were transferred from the oceanic 70 per cent of the Earth's surface to the glaciated 5 per cent. This led to depression of the crust, whilst the release of the weight of the ice resulting from melting led to uplift (**Fig. 6.5**). The degree of glacio-isostatic change, and the rate at which it occurred, is related to the differing volumes of the various ice caps (**Fig. 6.6**). This is illustrated when one looks at the amount of isostatic recovery that has taken place in the Holocene in North America, Fennoscandia, and the British Isles. It has been greatest in the area vacated by the **Laurentide ice cap**, the largest of the three ice caps under consideration, and least in the area vacated by the British ice cap, which was the smallest of the three. The depression near the centre of the Laurentide ice cap may have exceeded 900 m at the peak of glaciation.

Relatively little is known about the nature of the depression sequence, but as the emergence sequence is both evident in raised beaches and measurable at the present time, more is known about this (Andrews 1970). It is possible to identify three main stages of isostatic response: the period of restrained rebound as the ice sheet begins to lose mass; the period of post-glacial uplift during which ice-loss takes place at an accelerating rate giving a smooth acceleration of uplift; and the period of residual glacio-eustatic recovery when some coastal uplift still goes on in spite of total ice removal. In most areas this is the position that we are in today. Because of this sequence the gradients of tilt lessen on younger and younger shorelines in a way that is clearly related to the exponential form of post-glacial isostatic uplift. Some quantitative data are available from Scotland which suggests that in the eastern part of the country a degree of tilt of 18 mm $km^{-1} kyr^{-1}$ was established in the early Holocene, but that this rate decreases to a tilt of 10.9 mm $km^{-1} kyr^{-1}$ in the second half of the Holocene.

In areas peripheral to the ice sheets, such as parts of the eastern coast of the United States, the Baltic, and North Sea, there are indications that zones not subject to an ice-load bulged up during glacial phases (Newman et al. 1971), perhaps because of volumetric displacement of the upper mantle's low velocity layer, but that they have collapsed peripherally in post-glacial times, giving greater submergence than could

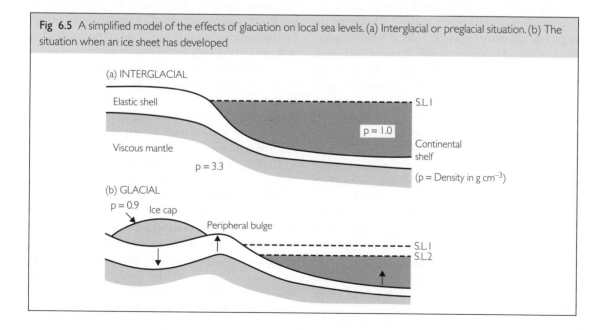

Fig 6.5 A simplified model of the effects of glaciation on local sea levels. (a) Interglacial or preglacial situation. (b) The situation when an ice sheet has developed

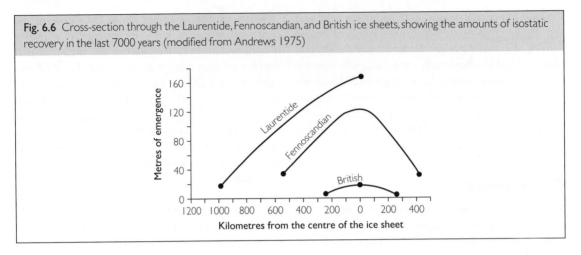

Fig. 6.6 Cross-section through the Laurentide, Fennoscandian, and British ice sheets, showing the amounts of isostatic recovery in the last 7000 years (modified from Andrews 1975)

be explained by the eustatic **Flandrian (Holocene) transgression** alone. This is thought to be the result of some compensatory transference of sub-crustal material. Thus in Europe, collapse of the **peripheral forebulge** produced a zone of subsidence that runs from the southern North Sea through northern Germany and central Poland to Lake Ilmen in north-west Russia (Steffen and Wu 2011). Similar effects of this can be seen in terms of the pattern of Holocene subsidence along the Atlantic coast of the United States with the amount of forebulge subsidence changing with distance from the centre of ice-loading in Hudson Bay. High rates of subsidence can be detected well into Virginia and the Carolinas (Engelhart et al. 2009).

In the areas that were covered by ice, however, the degree of maximum isostatic uplift has been considerable: around 300 m in North America and 307 m in Fennoscandia (**Fig. 6.7a**), but less in Great Britain (**Fig. 6.7b**). There has also been substantial uplift in Iceland (Le Breton et al. 2010). The ice caps of Greenland and Antarctica, moreover, are still exerting sufficient weight to create a considerable degree of isostatic depression (see **Fig. 6.7c**). Much of the bedrock surface of interior northern Greenland is currently at or below present sea level. Gravity readings and ice-thickness determinations obtained by trans-Greenland expeditions suggest that before the ice sheet formed, the presently low-lying bedrock areas of northern Greenland, some of which extend below sea level, were in the form of a plateau about 1000 m above sea level. If the ice were to be removed the bedrock surface would slowly rise up to this height once again.

In Greenland very rapid rates of adjustment to ice-load removal are evident in the early Holocene, and Pirazzoli (1991) reports on the available data. At Mesters Vig early Holocene emergence was at a rate of 90 m kyr^{-1}, decreased exponentially to about 6 m kyr^{-1} for the interval 9000 to 6000 ^{14}C yr BP, and has remained perhaps as low as 0.7 m kyr^{-1} since 6000 ^{14}C yr BP. On the west coast, at Søndre Strømfjord, the initial rate of uplift was at about 105 m kyr^{-1}.

The changes that have taken place in Arctic Canada have been the subject of a detailed monograph by Andrews (1970). **Figure 6.8a** shows the average rate of uplift for Northern and Eastern Canada for the whole Holocene, while **Fig. 6.8b** shows the present rate of uplift. At the south-east end of Hudson Bay the average rate has approached 30 m kyr^{-1}, and at present reaches 13 m kyr^{-1}. In the early Holocene rates were very much larger than either the average or present figures, for as in Norway and Greenland, the great bulk of the uplift took place in the first few thousands of years following deglaciation. Note also how closely the area of maximum rate of isostatic uplift corresponds to the area of maximum Laurentide ice thickness.

The effect of glacio-isostasy on the lakes of Finland is also of interest. It is known by tradition that the woodland and meadows on the low southern shores of the Great Finnish lakes had a tendency to become more swampy and marshy with time, whereas the northern shores tended to become drier. This is a consequence of the post-glacial tilting which has sometimes had spectacular results: fjords opening towards the north, to the Gulf of Bothnia, have been converted into lakes by the development of a threshold.

Fig. 6.7 The effects of glacio-isostasy on different land areas: (a) The degree of isostatic recovery (m) of Scandinavia for the last 10 000 years. (b) Generalized isobases (m) for the main post-glacial shoreline in Scotland (after Sissons 1976). (c) Cross-section of north Greenland showing the present shape of the ice sheet (a), the present level of the bed rock surface (c), and the estimated profile of the landmass with no ice load (b) (after Hamilton 1958)

Equally, many of the large lakes used to have their outlets on their northern sides, but they too have them diverted to the south shores by the same process. The rivers draining the lakes from the southern sides have not had time to fully adjust their profiles to this phenomenon and as a result they show rapids and falls which have proved to be popular sites for electricity development. Elsewhere some of the lakes have dried up through decantation. In the north of the country the uplift of the land has also caused difficulties for port authorities who have to deal with a progressive shallowing of their harbours. On the bonus side the uplift has provided further useable land for the nation. It was in fact the need to allot newly-emerged land to owners around the Gulf of Bothnia which led the director of the cadastral survey of Finland, Efraim Otto Runeberg (1722–70), to postulate in 1765 that small movements of the Earth's crust were responsible for many of the gains in land (Wegmann 1969). The degree of isostatic adjustment taking place in Finland at the present time has been estimated by analysis of tide-gauge data. Trends can be identified in rates, with minimum rates occurring away from the former centres of ice cap growth. During the twentieth century, rates in the south at stations like Helsinki and Hamina were about four times lower than those at the head of the Gulf of Bothnia (Oulu, Kemi, etc.) (**Table 6.2**). A full account of glacio-isostatic adjustments in Fennoscandia is provided by Steffen and Wu (2011).

Glacio-isostatic response has also been an important consequence of the late Holocene **Little Ice Age** and its waning, and helps to account for the very rapid current uplift rates of 32 mm per year in south-east Alaska (Larsen et al. 2005) and of 39 mm in Patagonia (Dietrich et al. 2010).

Fig. 6.8 (a) Average rate of Holocene uplift in metres per 100 years for northern and eastern Canada. The inset shows the thickness of the Laurentide ice sheet at the LGM (contours in m) (b) Present rate of uplift in metres per 100 years for northern and eastern Canada (modified from Andrews 1970, Figs. 2.7 and 8.10)

Table 6.2 Current rates of isostatic uplift as determined for Finland from tide-gauge data

Station	Mean rate (cm/10 yr)
Kemi	7.37
Oulu	6.53
Raahe	7.63
Pietarsaari	8.50
Vaasa	7.60
Kaskinen	7.03
Mantyluoto	6.20
Rauma	4.93
Turku	3.53
Degerby	4.20
Hanko	2.73
Helsinki	1.83
Hamina	1.80

(Determined by author from data in Lisitzin 1964.)

6.6 Miscellaneous causes of local changes in sea level

One of the prime causes of the observed changes of the land relative to the sea is orogenic activity, the process by which mountains are built. Signs of Quaternary volcanic activity and earth movements are visible in many parts of the world (Pirazzoli 1994) (**Plate 6.2**), and the term **neotectonics** is applied to current activity (Long et al. 2002). Current uplift can now be monitored through geodetic surveying, satellite altimetry and the use of Global Positioning Systems (GPS) (see, for example, Xu et al. 2000), while longer term changes can be determined through such techniques as fission track dating. Recently, a global analysis of dated sea levels for the last interglacial has suggested that many of the world's coastlines are currently uplifting, perhaps because long-term continental accretion has led to compression of continental plates and uplift of their margins (Pedoja et al. 2011).

The zones where orogenic activity has been most intense in the Quaternary have been recognized over recent decades. Seismic, mountain building and volcanic activity occur, for example, in a well-defined series of narrow belts (**Fig. 6.9**), with that surrounding

the Pacific Ocean being especially notable. Plate tectonics is crucial to understanding this pattern.

The most important type of plate boundary for mountain building (orogenesis) is the convergent plate boundary where plates move towards each other at a rate of several centimetres per year and subduction takes place. Mountains are thought to be created along convergent plate boundaries in at least four major ways: where a continent collides with an island arc; where a continent collides with another continent; where the oceanic lithosphere underthrusts a continental margin; and where the oceanic lithosphere underthrusts island areas.

Plate tectonic theory also helps to explain some zones of subsidence in the oceans. Flat-topped submarine mountains called 'guyots' appear at some point in the past to have been volcanic islands above sea level (Paduan et al. 2009). They formed at zones of sea-floor spreading, but as the spreading process continued they ceased to be active, became eroded by wave action, and gradually suffered subsidence, which brought the formerly wave-eroded surface to as much as 900–1200 m below sea level. If the subsidence rate has been sufficiently gradual the guyots may be capped by coral atolls.

On the other hand, some areas, the continental platforms, located away from plate margins, have suffered from relatively little mountain-building during the Pleistocene. They stand in contrast to the areas of new fold mountains, some of which may have been uplifted as much as 2000 m in the last few million years.

Indeed, it is possible to relate rates of recent vertical crustal movement (RVCM) to different geotectonic zones (Fairbridge 1981) in the following way:

Shields and platforms	1 mm yr⁻¹
Cenozoic orogenic belts	up to 20 mm yr⁻¹
Older Phanerozoic orogenic belts	up to 5 mm yr⁻¹
Intra-orogenic Basin-Range belts (with regional block faulting)	up to 10 mm yr⁻¹

Thus, the fastest rates of uplift occur where plate collision takes place, causing crustal shortening and producing orogenic belts such as the Alps, Andes, Himalayas etc. **Table 6.3** lists some long-term mean uplift rates in such areas.

However, some uplift is spasmodic in that rather than being a long-continued and relatively gentle process it takes place in association with one major seismic event. **Table 6.4** shows the amount of vertical

Fig. 6.9 The world's plate tectonic framework. (a) The major tectonic plates with their direction of movement. (b) The distribution of earthquakes (1957–1967)

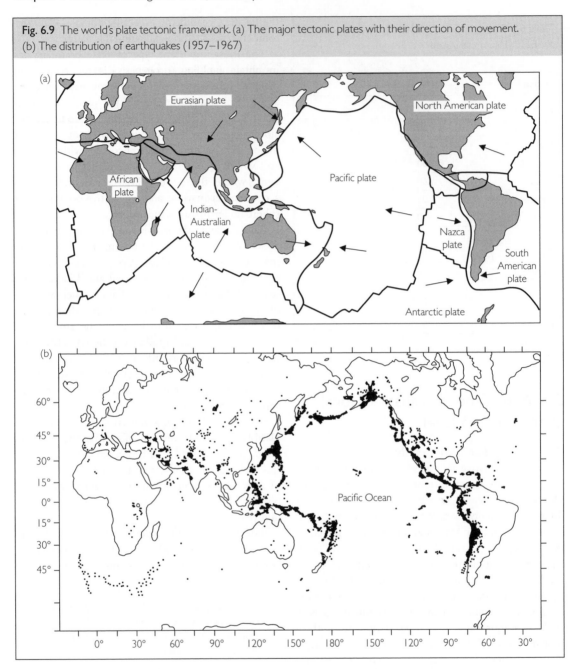

displacement that has been measured for some specific earthquake events over the last century or so. It is quite possible for the equivalent of some thousands of years of 'normal' uplift to take place in just one sudden spasm.

One of the largest recorded events is the Yakutat Bay earthquake, which caused over 14 m of vertical displacement in Alaska in 1899. Broadly similar amounts

of displacement also took place in the 1964 Alaskan earthquake, and doubtless at other times during the Holocene (Hamilton et al. 2005).

It is possible for some individual tectonic events to affect a large area. One of the finest examples of this is provided by an event which occurred in the eastern Mediterranean about 1550 years ago. It affected a considerable part of Crete, Karpathos, Rhodes, and the

Table 6.3 Long-term mean uplift rates in orogenic zones determined by various methods

Location	Method	Rate (mm kyr⁻¹)	Period
Central Alps	Apatite fission track age	300–600*	6–10 Myr BP
Central Alps	Rb and K-Ar apparent ages of biotite	400–1000*	10–35 Myr BP
Kulu-Mandi Belt, Himalayas	Apparent Rb-Sr ages of coexisting biotites and muscovites	700*	25 Myr BP to present
Southern Alps, New Zealand	Estimated ages of elevated marine terraces	5000–8000**	120 kyr BP to present
Huon Peninsula, Papua New Guinea	U-series and ¹⁴C dating of elevated marine terraces	1000–3000**	120 kyr BP to present
Bolivian Andes	Fission track dating	100–200*	20–40 Myr BP
Bolivian Andes	Fission track dating	700*	3 Myr BP to present

* Rock uplift
** Surface uplift
Source: from various sources in Summerfield (1991) table 15.1, and in Benjamin et al. (1987).

Table 6.4 Amounts of vertical displacement in specific earthquake events

Specific event's location	Date	Displacement (m)
Yakutat Bay (Alaska)	1899	14.3
New Zealand (Murchison quake)	1929	5
Adelaide (Australia)	1954	5–8
California	1872	7.01
Sonora (Mexico)	1887	6.10
Japan	1891	6.10
India	1897	10.67
California	1906	0.91
Taiwan	1906	1.83
Mexico	1912	0.61
Nevada	1915	4.75
Taiwan (Chichi)	1999	4–8
Izmit (Turkey)	1999	2
Corinth (Greece)	1981	0.45–1.3
Chi-Chi (Taiwan)	1999	8–10
Sumatra (Indonesia)	2004	15–20
Chile	2010	2.5
Tohoku (Japan)	2011	5–7

Epeirogeny is the uplift of what are usually large areas of the Earth's surface without significant folding or fracture. It is essentially associated with plate interiors rather than colliding plate margins. There is a range of processes that can account for plate centre movements including broad scale doming and rifting associated with hotspots, and subsidence associated with lithospheric cooling. Summerfield (1991, chapter 4) provides a review of the mechanisms involved.

Although data are still relatively sparse on long-term rates of uplift for zones of epeirogeny, one of the most fascinating aspects of this issue is the fact that rates of uplift can be appreciable in such situations. Summerfield (1991, p. 378) suggests that their rates of uplift, averaged over millions of years, appear to lie in the range of 10 to 200 m Myr⁻¹. These rates may be less than those for orogenic uplift, but they are still substantial.

On passive continental margins there are contrasting patterns of uplift and subsidence (Cronin et al. 1981). Long-term lithospheric flexural upwarping takes place on the continental margin, and subsidence occurs offshore, partly because of sediment loading. Along the eastern seaboard of the USA, Cronin found that the rates of upwarping averaged about 10–30 mm 1000 yr⁻¹, and that the rates of subsidence offshore averaged 20–40 mm 1000 yr⁻¹. These rates were based on the height relationships of dated marine terraces. Kukal (1990) provides a more general survey of the rates at which subsidence may occur in certain particular tectonic settings, such as continental margins.

south coast of Turkey near Alanya (Kelletat 1991). It was at its most intense in western Crete, where some 9 m of uplift appears to have taken place.

These are geological units where large thicknesses of sediment and relatively rapid subsidence are typical. In particular, passive continental margins are areas of long-term subsidence, and empirical curves of the mean subsidence rate over extended periods can be obtained using boreholes and geo-physical methods. Emery and Uchupi (1972), for example, calculated rates for the continental margin of the United States back to the Cretaceous. The shelf appears to be subsiding fastest in the Gulf of Mexico (at *c.* 400 mm kyr^{-1}) and slowest in Yucatan (at *c.*1 mm kyr^{-1}), but rates for the eastern seaboard appear to average 20–30 mm kyr^{-1}.

There are also areas of continental subsidence away from continental margins. Subsidence may take place in basins and rifts at an appreciable rate. Kukal (1990, p. 34) cites rates for various areas, and they typically range from 300 to 5000 mm kyr^{-1}.

Crucial to Darwin's model of atoll evolution was the idea that subsidence has occurred, and the presence of guyots and seamounts in the Pacific Ocean attests to the fact that such subsidence has been a reality over wide areas. In addition there are tide gauge records from the Hawaiian Ridge that demonstrate ongoing subsidence rates at the present day (1500 mm kyr^{-1} for Oahu and 3500 mm kyr^{-1} for Hawaii). Moreover, the occurrence of Holocene and Pleistocene submerged terraces and drowned reefs on the flanks of these two islands demonstrate the reality of this process. The thick coral accumulations that are superimposed on basaltic platforms that were once at sea level indicate that subsidence has continued on timescales of tens of millions of years. The coral cap at Eniwetok, which dates back to the early Eocene is 1400 m thick, and that at Bikini (of Miocene age) some 1300 m thick. Differences in the depths of wave-worn platforms along a volcanic chain, if the chronology of the seamount formation can be established by potassium-argon and other dating techniques, provide estimates of subsidence rates. In the case of the Tasman Sea chains off Australia rates of subsidence appear to have been of the order of 28 m Myr^{-1} (28 mm kyr^{-1}).

The causes of this subsidence are a matter of some debate (see Lambeck 1988, pp. 506–509). Some of it may be caused by the loading of volcanic material on to the crust, but some may be due to a gradual contraction of the sea-floor as the ocean lithosphere moves away from either the ridge or the hotspot that led to the initial formation of the island volcanoes.

In some localities, and on a much more limited scale, the flattening of sediments by the weight of overlying material (autocompaction) can led to subsidence of some consequence (Bartholdy et al. 2010). It is often apparent in peat and other such materials which have a very high porosity and a weak skeletal framework of vegetable fibres. Salt-marsh peats, for example, which make up a large part of many transgressive sedimentary deposits, have 80 per cent porosity, and frequently in section one sees logs that have become flattened from their original round shape to a more oval form. The subsidence caused by compaction of Holocene beds in Holland is estimated at 0.25 mm per year (Veenstra 1970). Values for estuarine areas on the east coast of England were 0.4 mm per year for the Fens and 0.6 mm for the Humber (Horton and Shennan 2009). Even higher values of 5 mm per year have been proposed for the peats of the Mississippi delta (Törnqvist et al. 2008). Subsidence caused by human activities (especially groundwater and hydrocarbon extraction from coastal aquifers) can lead to substantial and rapid falls in land level, leading in some cases (e.g. Venice) to accelerated coastal flooding problems. Finally, it is worth realising that local tectonic and seismic activity may be caused by crustal deformation associated with the application and removal of a glacial load (Stewart et al. 2000).

6.7 The nature of pre-Holocene sea levels

Although, as noted in **Section 6.4**, the classic sequence of Pleistocene sea levels involves a gradual reduction in the height of interglacial sea levels during the course of the Pleistocene, there is relatively little accurate dating of the sequence of sea level change. Recently, however, new dating techniques have enabled some generalizations to be put forward, though far more dates are required for any degree of certainty to be reached. Data are particularly sparse beyond about a quarter of a million years ago. Among the generalisations that can be made (Lambeck et al. 2011) are that over the last 800 000 years the dominant global sea level cycle has occurred every 100 000 years, with an amplitude of *c.* 130 m. Before this the dominant period was of a shorter duration and of a smaller amplitude. During successive interglacials the global sea levels have not exceeded their present positions to any great extent, and these interglacial and present shoreline positions are only representative of *c.* 10–15 per cent of the 800 000 year period.

The record of sea level change for the last 250 000 years has in recent decades been interpreted from two principal proxy sources. These are, firstly, the oxygen isotope record in deep-sea sediment cores and, secondly, the study of ancient shorelines and coral reefs that have been uplifted by tectonic activity (Matthews 1990).

Oxygen isotope ratios of Foraminifera vary with the temperature, salinity and $\delta^{18}O$ values of the ocean water from which their tests are deposited. Ocean water $\delta^{18}O$ values vary with the quantity of isotopically light ice stored on the continents so that the record derived from the forams is a blend of global ice volume and local water temperature and salinity components (see **Fig. 2.4**). For the late Pleistocene the temperature and salinity components of the signal are generally thought to be small compared to the ice volume signal.

The oxygen isotope record plainly implies that sea level has changed frequently, rapidly, and somewhat repetitively during the Pleistocene. Classic studies were carried out over four decades ago. For example, a sea level curve derived from Core V28-238 correlated extremely well with a curve derived from uranium series dates on Barbados emerged coral terraces (Mesolella et al. 1969), and both the sea level curves also appeared to correlate closely with the maxima on theoretical insolation curves (Veeh and Chappell 1970).

Oxygen isotope studies can also be used to ascertain whether or not glacio-eustasy caused sea levels to be higher than now in earlier interglacials. Shackleton (1987) doubted that sea levels were appreciably higher due to this mechanism (say by more than a few metres) during any interglacial of the past 2.5 million years.

Coral terraces from tectonically uplifted areas like the now classic Huon Peninsula of Papua New Guinea are like a continuous tape recorder; each coral reef developed when the rising sea level overtook the rising land, and the reef crests represent approximately the peaks of each sea level rise (Chappell et al. 1996; Yokoyama et al. 2001). Relative sea level can thus be estimated from the current heights of these suites of terraces. The assumption is made that sea level at around 125 000 years ago (the time of the last interglacial) was +6 m above present sea level. On the basis of this assumption, and that rates of uplift have been relatively constant, the sea levels can be calculated for other terraces.

Figure 6.10(a),(b), which are based on oxygen isotope and coral reef studies, show the classic model of the pattern of sea level change over the last 240 000 years. Comparable patterns have also been obtained for Barbados (Bard et al. 1990) (**Fig. 6.10c**). They show clearly that global sea levels have not been higher than those of the present (Holocene) interglacial except at around 125 000 years ago (the last interglacial). At that time global sea levels may have been as much as 9 m higher than at present (Hearty et al. 2007). Various oscillations in sea level interrupted the last interglacial highstand (Thompson et al. 2011).

The sequence of sea level change during the Last Glacial has been described by Cutler et al. (2003). They showed that towards the end of **Marine Isotope Stage** (MIS) 5e (*c.* 115 kyr ago), sea level was at −19 m. At MIS 5b (at *c.* 92 kyr ago) it was at −57, but rose to −24 m (during MIS 5a at *c.* 76 kyr ago). It then fell sharply to −81 m during MIS 4 (at *c.* 71 kyr ago). During MIS 3 sea level was between −85 and −74 m, before dropping further at the Last Glacial Maximum, to *c.* −120–130 m. Rapid changes within the Last Glacial occurred as a result of Heinrich events, which released large amounts of meltwater into the North Atlantic over short periods of time (Chappell 2002). For example, there may have been a rapid sea level rise of 10–15 m at around 19 000 years ago caused by a meltwater pulse from one or more Northern Hemisphere ice sheets (Clark et al. 2004). There is no evidence for a higher stand of sea level than the present during **interstadials** of the Last Glacial and earlier divergent opinions on this were produced because of unreliable radiocarbon dates (Bloom 1983, p. 218).

6.8 The post-glacial rise in sea level or Flandrian transgression

The low sea levels of glacial times were succeeded as the ice melted by a large rise in sea level: the **Flandrian transgression**. This transformed the continental shelves, deltas, and estuaries (Woodroffe 2000) and may have had such diverse effects as a stimulation of crustal instability and volcanic activity, and the triggering of submarine landslides (Smith et al. 2011).

At times the rate of rise was of a remarkable magnitude (Peltier 2005), perhaps because of the near catastrophic collapse of ice-shelves and ice caps, whether in the Arctic or Antarctic (Smith et al. 2011).

Fig. 6.10 (a) Oxygen isotope record for the past 240 000 years for the East Equatorial Pacific Core (V19–30); (b) sea level curve for the Huon Peninsula, recalculated to correlate with (a), modified after Chappell and Shackleton 1986; (c) sea level curve for Barbados, modified after Bard et al. 1990, Fig. 1

In Lancashire, north-west England, sea level may have risen by as much as 7 m in 200 years around about 8000 years ago, possibly because of the rapid final collapse of the massive Laurentide ice sheet of North America (Tooley 1978). In the Sunda Shelf region, around 14.6 kyr ago, sea level rose as much as 16 m in just 300 years (Hanebuth et al. 2000). This is termed Meltwater Pulse 1A (Clark et al. 2002a), though the relative roles of the Northern Hemisphere and Antarctic are the subject of debate (Peltier 2005).

Anderson and Thomas (1991) have argued that a continuous rise of sea level brought about by the melting of terrestrial ice sheets resulted in the spasmodic deterioration of marine ice sheets, which, in turn, caused high frequency (10^2 to 10^3 years), low amplitude

(less than 10 m) eustatic events. As sea level changes, marine ice sheets become decoupled from the sea-floor, resulting in their rapid collapse and in increased flow of the ice streams that feed them. These sub-millennial spasms of rapid sea level rise can be related to ice-rafting events (Heinrich events) during deglaciation (Locker et al. 1996).

One of the best records of the progress of the Flandrian transgression was provided by the study of radiocarbon dated samples drilled from submerged coral reefs on the Barbados shelf in the Atlantic (Fairbanks 1989; Toscano and Macintyre 2003) (**Fig. 6.11**). The coral that was involved, *Acropora palmata*, (as described in **Section 6.2**) grows within a few metres of sea level, and thus permits a relatively

Fig. 6.11 Sea level curve for Barbados and the Caribbean (from Toscano and Macintyre 2003, Fig. 6). Reprinted with permission of Springer-Verlag

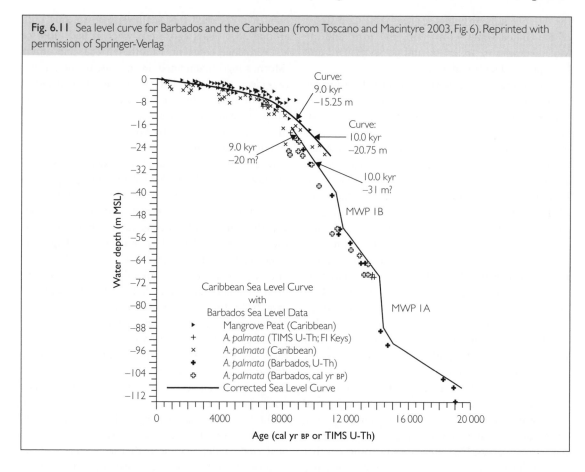

accurate determination of past water depths. The oldest and deepest of the samples indicates that at the time of the Last Glacial Maximum (though possibly as early as 26 000 years ago) the maximum depth of the shoreline was at −120 m (Peltier and Fairbanks 2006). A gradual rise started shortly afterwards and may, on the basis of the Tahiti record, have accelerated during the Bølling warming phase to a rate of up to 40 m in 1000 years between 14 650 and 14 300 cal yr BP (Meltwater Pulse 1A) (Deschamps et al. 2012). The rate of rise decreased during the **Younger Dryas** (Bard et al. 2010) and then accelerated again (Meltwater Pulse 1B) c. 11 300 cal yr BP until around 9000 years ago, when the shoreline was at a depth of around −25 m. The curve of sea level rise flattens out around 6000 years ago at −7 m. The steady deceleration of sea level rise evident in the Barbados curve after c. 8000 years ago reflects the fact that by that time glacier ice had long since disappeared from Britain, that only small amounts remained in Scandinavia, and

that the Laurentide ice sheet was restricted to residual remnants on Canadian Arctic islands. It needs to be noted, however, that there are some doubts about the Barbados eustatic record based on the possible isostatic and tectonic changes in that area (Gehrels 2010a). Study of a sediment core from Singapore, a relatively stable location in terms of tectonics, confirms a fast rate of sea level rise (1.8 m per 100 years) from 8900 to 8100 cal yr BP, a near cessation in the rate of sea level rise between 7800 and 7400 cal yr BP, followed by a renewed rise of 4–5 m that was complete by 6500 cal yr BP (Bird et al. 2010).

On morphological grounds, including the presence of steps on coastal shelves, it seems reasonable to postulate that the rapid transgression suffered some stillstands, and even slight regressions. This point of view has been put forward as a result of submarine notch and terrace studies for numerous areas in recent years including the Persian Gulf (61–64 m, 40–53 m, and 30 m below present sea level), the Bass Strait (60 m below sea level), the Gulf Coast of the United States (60 m, 32 m, and 20 m) and the Mediterranean (5, 10, 27, 55, and 96 m) (Ballard and Uchupi 1970; Flemming 1972).

The general Holocene trend of rapidly rising sea levels until around 8000 years ago meant that over low-angle shelves the sea must have advanced laterally at a fast rate. In the Persian Gulf region, for instance (Lambeck 1996), there was a shoreline displacement of around 500 km in only 4000–5000 years, a rate of no less than 100–120 m per year. This must have had profoundly disturbing results for inhabitants of the coastal plains. Comparable rates of transgression for the North Sea are 60 km kyr^{-1}, for the Bristol Channel 15 km kyr^{-1}, for the Sunda Shelf 20 km kyr^{-1} (Voris 2000), and for the Gulf of Mexico 18–32 km kyr^{-1} (Evans 1979). When rising sea levels reached the Bosphorus sill in the eastern Mediterranean, Ryan et al. (1997) have argued that water surged through into the Black Sea (then a lake) causing catastrophic flooding of more than 100 000 square km c. 7200 ^{14}C yr BP, though this flood hypothesis has been disputed by Görür et al. (2001). More recently the date of initial marine inflow to the lake has been dated at c. 9000 cal yr BP (Soulet et al. 2011), and it has been suggested that the area flooded as a result of reconnection with the Mediterranean was less than previously thought (Giosan et al. 2009). The alternative hypothesis is that there was a gradual overflow from the Black Sea into the Mediterranean (Hiscott et al. 2007).

Figure 6.12 illustrates the way in which the shorelines of 11 000 and 15 000 years ago compare to those of North America today. The shoreline at 15 000 years ago, being at a lower level, was a considerable distance across the continental shelf, and the various groups of islands of the east coast of Florida were linked up to form much larger land areas.

The main problem that arises with the interpretation of this Holocene transgression lies in what happened after about 6000 years ago. Four or five decades ago there were three fairly distinct schools of thought on this, though it has to be stated that the arguments were about possible changes of the order of only a couple of metres. It was generally accepted that in the last six millennia the rate of sea level rise, if present, had been far less than it was in the Early Holocene. One point of view had it that there had been a continuously rising sea level to the present time, though the rate of rise had diminished with time (Shepard's hypothesis (1963)). Godwin et al. (1958) on the other hand hypothesized that sea level rose steadily until around 4000 years ago, since when it had more or less remained constant. Fairbridge (1958) and others (Mörner 1971) maintained, in contrast to these other two ideas, that Late Holocene sea level oscillated to positions both above and below the present level. They suggested that the sea was at levels 1–4 m above the present about five times between 6000 years ago and the Middle Ages.

A considerable amount of evidence has been raised against Fairbridge's concept. Some people have reworked or reinvestigated some of the sites claimed by Fairbridge to illustrate high Holocene stillstands and they give a Pleistocene rather than Holocene age for the raised beaches and terraces. Jelgersma (1966) said that if the high sea levels had taken place one would expect that coastal plains would have been inundated on a very large scale. She said that data from the Gulf of Mexico, Florida, and the Netherlands failed to reveal such a degree of transgression. Very detailed archaeological researches in relatively stable parts of the Mediterranean, using diving techniques, have convinced Flemming (1969) that to within an accuracy of ±0.5 m there has been no eustatic change of sea level in the Mediterranean in the last 2000 years. Dating of freshwater peats in Australia, one of Fairbridge's field areas, by Thom et al. (1969) failed to indicate that sea level rose above its present position between 9000 and 2985 ^{14}C yr BP. Likewise, dating of chenier ridges in Queensland leads to a broadly comparable conclusion (Cook and Polach 1973). Similarly, after an expedition around some Pacific atolls, Newell and Bloom (1970) found no unequivocal evidence for recent higher sea levels, and abundant evidence that the characteristic morphology of the Indo-Pacific coral reefs is most probably in adjustment to the slow rise in sea level that has characterized the last 6000 years.

One of the main lines of evidence that has been used to substantiate the Mid-Holocene Highstand concept (Woodroffe and Horton 2005; Woodroffe 2005) is the presence of small raised terraces ('Daly levels') in many parts of the tropics. Radiocarbon dates for slightly elevated reefs of Holocene age, which cluster around about 4000 ^{14}C years BP do indicate the possibility of a slight Holocene transgression in some parts of the world. The concept of a high mid-Holocene sea level stand is certainly not dead, and evidence for it from the US Gulf Coast has recently been put forward by Törnqvist et al. (2004), from Pacific atolls by

Fig. 6.12 A comparison of the Atlantic shoreline of the USA at 15 000 years ago, 11 000 years ago, and the present. Confirmation that the continental shelf was once laid bare is found in the discovery of elephant teeth (triangles), fresh-water peats (dots), and the shallow water formations called oolites (circles) (modified from Emery 1969)

Dickinson (2004) and from the western Indian Ocean by Camoin et al. (2004).

Sea level change in the later Holocene

Behind the arguments as to whether global sea levels over the last 6000 years have been steadily rising, steady, or oscillatory, was a desire to try and achieve a worldwide eustatic curve (Jelgersma and Tooley 1995). Studies of the importance of geoidal eustasy, and the realisation that the geoid has not remained stable over time, have resulted in the recognition that there must have been regional differences in response to post-glacial sea level rise. Moreover, as part of this re-evaluation there has been a growing appreciation that crustal isostatic response to the removal of the weight of ice sheets has been accompanied by a consequential hydro-isostatic response, especially in the areas of the shelf seas. It has also been recognized that there are very few, if any, areas that have been sufficiently stable to use as 'eustatic dip-sticks'. Furthermore, the interpretation of past sea levels may have been complicated by changes in tidal levels through time as a response to changes in coastline position and topography.

In the light of considerations relating to the importance of changes involving hydro-isostasy and glacio-isostasy in determining the response of sea lev-els to glacio-eustasy, Clark and Lingle (1979) identified six different zones with certain distinctive sea level change characteristics (**Fig. 6.13**). Zone 1 consists of those regions beneath the ice sheets at the time of the glacial maximum. These would have curves that were characterized by immediate emergence during the shrinkage or disappearance of the ice, due to the elastic

uplift of the Earth and a reduction in the gravitational attraction exerted by the ice-mass on the surrounding ocean. Emergence would continue as viscous flow within the mantle caused the land formerly depressed beneath the ice to be displaced upwards. A transitional zone between Zone I and Zone II would be charac-terized by emergence after ice sheet thinning, but at a later time submergence followed by more gradual

Fig. 6.13 Typical Holocene relative sea level (RSL) curves predicted for each of the six global zones. Predictions are based on Northern Hemisphere deglaciation and a resulting eustatic sea level rise of 75.6 m between 17 000 and 5 000 years ago. Holocene deglaciation in Antarctica would modify the boundaries of the six zones (from Clark and Lingle 1979, Fig. 1). Reprinted with permission of Elsevier submergence would begin as a collapsing pro-glacial forebulge migrated towards the reduced ice sheets. Zone II would be characterized by submergence following ice sheet thinning. This would be caused by a flow of the mantle material into the uplifting areas from peripheral regions resulting in collapse of the pro-glacial forebulge. Zone III would be characterized by rapid submergence followed by more gradual submergence until a few thousand years ago, when slight emergence of less than 0.75 m began. Zone IV would be characterized by continuous submergence of 1–2 m during the last 5000 years, despite an assumed constant ocean volume during that period. Zone V is a region where initial submergence would be followed by slight emergence beginning when water was no longer added to the oceans. A Holocene emergence of up to 2 m is predicted. Finally, there is Zone VI, which consists of all continental margins except those adjacent to Zone II. Their emergence is due to the addition of meltwater to the oceans causing depression of the ocean floor. Material within the mantle then flowed from beneath the ocean to beneath the continents, causing upward displacement of the continents and tilting of the crust along coasts.

Following on from this Lambeck (1993) has established that different patterns of relative sea level response occur in different situations. In the near field (defined as sites within the limits of the former ice sheets), the dominant contribution to sea level change comes from ice-loading effects, with rapid uplift (as over Scandinavia or Hudson Bay) resulting in a general exponential fall of relative sea level to its present position. In ice margin sites (near the former ice margins and affected by the forebulge phenomenon, such as Newfoundland), the relative sea level change was characterized by an initially rapid fall during the **Late Glacial**, followed by stability and a more recent rise to its present position. Intermediate field sites correspond to the forebulge area, as in the Netherlands, with late glacial and post-glacial subsidence at decelerating rates. Far field sites, which cover much of the world, appear to have experienced a sea level curve that is essentially similar to the pattern of ice melting, rising to near present positions at about 6000 years ago. However, they have generally experienced a slight fall since then as a result of hydro-isostatic adjustment and the draining of water from the equatorial ocean into the collapsing forebulge basins – a process termed 'equatorial ocean siphoning'.

There is certain evidence from present tide-gauge records to suggest that following on from the current post-neoglacial (Little Ice Age) amelioration in climate and global warming there has been a corresponding rise of sea level (Gehrels et al. 2005). It is of course difficult to separate current tectonic submergences and other factors from the eustatic effect of current glacial melting and global warming, and Pirazzoli (1989) believes that some previous estimates may have been excessive as a result. Church et al. (2001, table 11.2) after reviewing the evidence, suggested that the best estimate is that there was a rise of 0.25 to 0.75 mm per year for 1910–1990, and between 0.60 and 1.09 mm per year for 1960–1990. Bindoff et al. (2007) reported that satellite altimetry showed a rate of sea level rise between 1993 and 2003 of 3.1 mm per year. **Figure 6.14** shows six long sea level records from major world regions. The observed trends range from 0.8 to 3.1 mm per year.

Figure 6.15, from Gehrels (2010b) building on work by Shennan (1992), shows an attempt to develop a spatial picture of the trends of crustal change in the British Isles. It is apparent from this that the area may be split into two provinces, one in the south-east of England, where there is a tendency for sea level rise, and the other in the north and north-west where sea level is falling (Shennan et al. 2011). These two provinces correspond to the Late Quaternary pattern of glacio-isostatic loading, with large ice masses developed over the north and west and a forebulge over the extreme south and east of England.

In some localities present-day subsidence, combined with eustatic rise, is of a sufficient magnitude to present a threat to low-lying settlement concentrations. In the case of London, for instance, historical records show that high-tide levels and surge levels relative to Newlyn Ordnance Datum are becoming progressively higher (**Fig. 6.16**), with an increase of the order of 1.3 m between 1791 and 1953. This increases flood risk, but it is not clear how much this is due to subsidence and eustatic rise alone, and to what extent embanking by man, changes in water temperatures affecting tides through changing viscosity of the water, and changes in climatic conditions (including rainfall and wind directions), have played a significant role (Bowen 1972; Horner 1972).

Elsewhere in the world, the relative stability of sea level in the last few thousand years may be one contributing factor to the widespread loss of sand from many beaches, with consequent erosion and threats to human activities. As long as the Flandrian transgression was taking place at an appreciable rate, new areas of coastal plain were being inundated and fresh

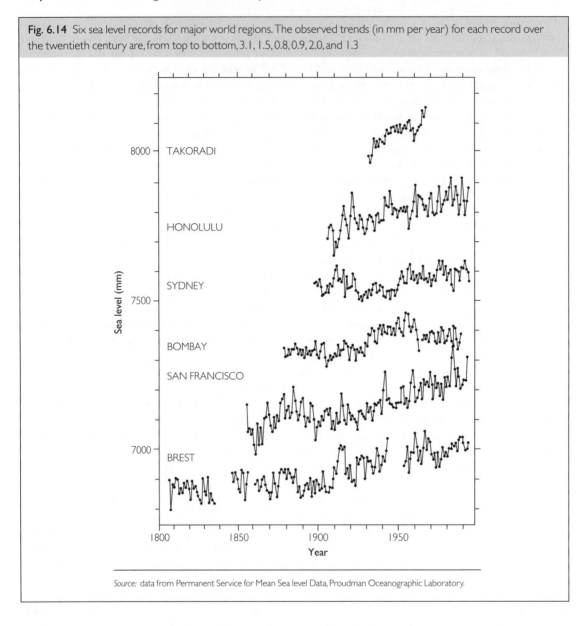

Fig. 6.14 Six sea level records for major world regions. The observed trends (in mm per year) for each record over the twentieth century are, from top to bottom, 3.1, 1.5, 0.8, 0.9, 2.0, and 1.3

Source: data from Permanent Service for Mean Sea level Data, Proudman Oceanographic Laboratory.

supplies of sediment were encountered, the coarser components of which were driven forward to produce beaches. An increase in beach volume took place so long as the rise of sea level continued. Surplus sand was blown downwind to form extensive coastal dune tracts. The beach-dune system probably reached its greatest volume as the stillstand was approached. However, once that level was attained, new sediment supplies were no longer encountered, and marine processes brought about a net loss to the system in many parts of the world. There is a pressing need for work to be undertaken on the relative importance of this factor compared to changes in wind conditions or storminess, and to humanly induced changes in sediment budgets.

6.9 Post-glacial sea level changes in northern Europe: the combined effects of eustasy and isostasy

In the case of the countries of north-western Europe, both glacio-eustasy on a worldwide basis and hydro- and glacio-isostasy on a more local basis have acted together

Fig. 6.15 Recent rates of crustal movements and relative sea level changes in Britain (mm yr⁻¹). Values in brackets indicate corrections accounting for changes in tidal range. Black circles near the upwards arrow indicate areas of emerging land, open circles near the downward arrow show areas of subsiding land (from Gehrels 2010b, Fig. 1). Reprinted with permission of Elsevier

in post-glacial times to give the observed sequences of sea level changes (Shennan et al. 2000; Shennan et al. 2006; Shennan et al. 2011; Bradley et al. 2011).

In essence the post-glacial eustatic changes have tended to lead to marine transgressions, whilst isostatic rebound from the weight of the ice caps has tended to lead to regression. In north-western Europe one can see how, through time, the rates of these two contrary processes have led to distinctive patterns.

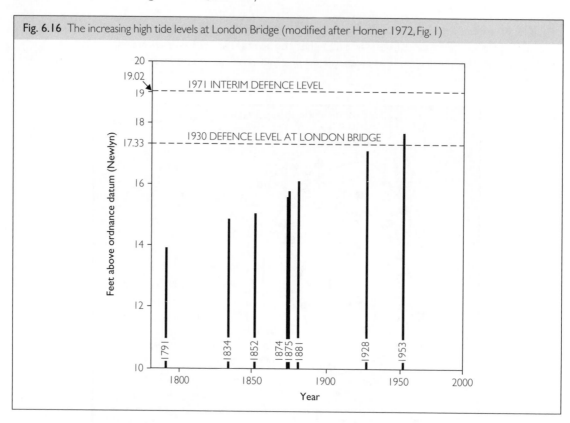

Fig. 6.16 The increasing high tide levels at London Bridge (modified after Horner 1972, Fig. 1)

In Scotland the rate and degree of isostatic rise was relatively small compared with parts of Scandinavia and Canada, so that at times when the eustatic sea level rise was going ahead with great speed (e.g. around 15 000 to 8000 years ago) it overtook the regression of the sea, interrupting the emergence of the coast. Thus in certain parts of Scotland, notably the Firth-Clyde lowlands, there are alternating deposits of marine and freshwater sediments that comprise a valuable record of land and sea level changes, and detailed survey, radiocarbon dating, and pollen analysis allow a comprehensive picture to be presented (Shennan et al. 2006; Smith et al. 2010).

Because of the weight of the ice the Late Weichselian situation was one of relatively high sea levels and submergence. It was during this phase that many of the high-level platforms were cut by the sea. The highest altitude of the Late Weichselian submergence is shown by raised beaches at 30–35 m above Ordnance Datum in Central Scotland. The Flandrian transgression, however, reached its maximum around 7800 cal yr BP (Smith et al. 2010) after a low position of sea level between 10 000 and 8000 cal yr BP. During that

regressive phase of low sea level some distinctive peats were formed and these now often underlie marine deposits of the Flandrian transgression, such as the Carse Clay, which produced another suite of raised beaches at a level generally lower than those formed during the maximum of the last glaciation. This main post-glacial shoreline reaches an altitude of 14.6–16.5 m in central Scotland, but because of the varied nature of the isostatic effect it declines in all directions from this peak. Current rates of uplift in Scotland are relatively modest and now appear to be being outpaced by global sea level rise (Rennie and Hansom 2011).

In Norway the pattern of sea level change in postglacial times is essentially similar to that in Scotland with high Late Glacial shorelines resulting from the ice-load effect being followed by a lowstand of sea level between about 12 000 and 10 000 years ago, and a transgression, often called the 'Tapes' transgression in Norway and the 'Nucella' transgression in Iceland.

Another situation where the effects of the post-glacial isostatic and eustatic sea level changes can be well seen is in the development of the Baltic Sea. At the maximum of the Weichselian Glaciation, around

20 000 years ago, the Baltic area was covered by a great ice body which deposited the Brandenburg and later moraines over the North European Plain. As this great ice sheet spasmodically retreated, a series of small ice-dammed lakes developed in the southern part of what is now the Baltic. The coalescence and expansion of these lakes produced the first major stage of the post-glacial evolution of the Baltic – the Baltic Ice Lake, dated at about 11 700 cal yr BP and to the termination of the Younger Dryas (Jakobsson et al. 2007). This lake was not totally landlocked and managed to overflow to the low Early Holocene ocean level by a series of channels, including one into the White Sea. The location of the overflow channels varied according to deglaciation, glacial re-advances, and isostatic updoming.

In due course the deglaciation of southern Sweden enabled the sea to open the Baltic basin (Björck 1995), and thus the 'Yoldia Sea' was established, lasting for less than 1000 years. However, the link between the Yoldia Sea and the ocean via southern Sweden was severed as isostatic updoming in that area raised the connection above sea level. This thus led to the formation once again in the early Holocene of a lake – 'Ancylus Lake' – which lasted for more than 2000 years, and its outlet was eventually through the Oresund, the channel that still separates Sweden from Denmark. Finally, the general worldwide rise in sea level exceeded the rate of isostatic uplift and the Oresund was submerged, thereby changing the Ancylus Lake into the so-called 'Littorina Sea' from about 9000 cal yr BP onwards (Meyer and Harff 2005). This sea was characterized by a relatively warm-water fauna, of a salt-water type, and was probably contemporaneous with the Tapes Sea recognized by Scandinavian workers outside the Baltic Basin. After a few thousand years, the 'Littorina Sea' merged into the current Baltic Sea, though faunal evidence suggests that as a result of the reduction in its outlet by updoming, the waters of the Baltic became progressively more brackish, with the associated constriction of the oceanic connection. Given present rates of uplift, however, and assuming no general rise of sea level, the Baltic, currently linked to the ocean by channels only 7–11 m deep, could become a lake again in 8000 to 10 000 years.

A combination of eustatic rise and subsidence led to the development of the southern North Sea as we know it now (Lambeck 1995; Behre 2007). **Figure 6.17** illustrates how 14 000 years ago the North Sea was largely dry land. However, by c. 8800 cal yr BP the situation had changed very radically, and the North Sea

was then linked through to the English Channel. The Dogger Bank became totally submerged by c. 8000 years ago (Shennan et al. 2000; Fitch et al. 2005). Compared to the present shorelines there remained extensive areas of low-lying ground at the mouths of the major estuaries, but these too were flooded as the Holocene progressed. The gradual sequence of inundation can be seen from an examination of the history of the British coastal lowlands, which now follows.

The rapid rise of sea level in the Holocene caused the reworking of sediments deposited on the present continental shelf by the rivers of glacial times. These sediments were combed-up as the transgression progressed to create some major sedimentary features such as Chesil Beach, Dungeness, Orford Ness, and Dawlish Warren in southern England. In addition, the sea flooded low-lying areas, including the lower reaches of river valleys, to create rias (especially in south-west England and Pembrokeshire) and embayments (e.g. the present sites of Romney Marsh, the Pevensey Levels, the Broads, the Fens, and the Somerset Levels). These have subsequently become filled with alluvial sediments or cut off from the sea by the growth of spits (e.g. Poole Harbour). The rising sea flooded the floor of the North Sea (Behre 2007) and the English Channel (c. 9200 cal yr BP), achieving the final separation from the Continent c. 8800 cal yr BP and the almost total inundation of the Thames Estuary shortly after 8000 years ago.

In very general terms the temporal pattern of Holocene (Flandrian) transgression is comparable in different areas of England and Wales. **Figure 6.18a** illustrates how in the last 10 000 ^{14}C years there has been a phase of very rapid sea level rise to around 6500 ^{14}C yr BP, followed by a rather gentle rise, or even stability, thereafter. However, in detail the pattern is more complex (as revealed by the stratigraphy of areas like the Fens) (Shennan 1986a; Shennan 1986b) and there are also differences between areas, caused by subsidence (in the south-east) and by isostatic uplift (in the north). **Figure 6.18b** shows sea level change for the southern North Sea on a calibrated calendar year timescale.

6.10 Future sea level rise

It is very likely that if temperatures climb in the coming decades (see **Section 7.9**), so will global sea levels. Globally, sea levels are currently rising (**Fig. 6.19**), and the reasons for this include the thermal expansion of

Fig. 6.17 The North Sea shorelines from 12 000 to 6000 cal yr BC (from Behre 2007, Fig. 6). Reprinted with permission of Wiley

seawater (the steric effect), the melting of the **cryo-sphere** (glaciers, ice sheets and **permafrost**) and miscellaneous anthropogenic impacts on the hydrological cycle (which will modify how much water is stored on land) (Gornitz et al. 1997).

As the oceans warm up, their density decreases and their volume increases. This thermal expansion causes sea levels to rise. Uncertainties arise as to the

rate at which different parts of the oceans will warm up in response to increased atmospheric temperatures (Gregory et al. 2001). Church et al. (2001, table 11.2) present data on global average sea level rise due to thermal expansion during the twentieth century. Between 1910 and 1990 rates ranged between 0.25 and 0.75 millimetres per year while those between 1960 and 1990 ranged between 0.60 and 1.09 millimetres per year.

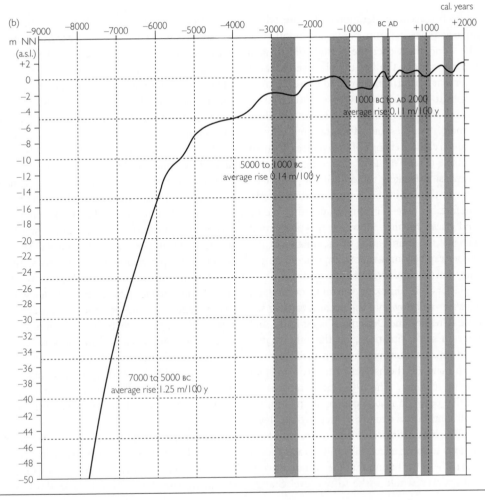

Fig. 6.18 (a) Curves of sea level change since 10 000 ^{14}C yr BP for different parts of England and Wales. 1 = Bristol Channel; 2 = English Channel; 3 = Cardigan Bay; 4 = Somerset Levels; 5 = North Wales. The datum level is MHWST (mean high water spring tide) (after Shennan 1983, Fig. 8). Reprinted with permission of Elsevier. (b) Simplified sea level curve for the southern North Sea (from Behre 2007, Fig. 7). Reprinted with permission of Wiley

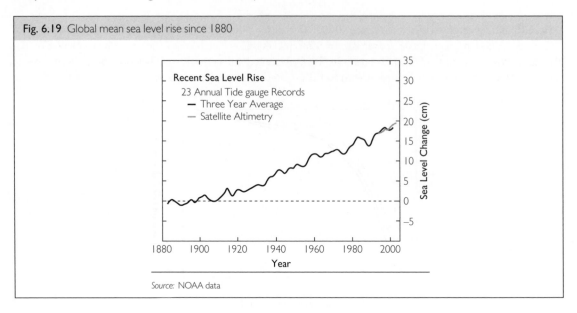

Fig. 6.19 Global mean sea level rise since 1880

Recent Sea Level Rise
23 Annual Tide gauge Records
— Three Year Average
— Satellite Altimetry

Sea Level Change (cm)

Year

Source: NOAA data

There has been a recent acceleration in the rate, which from 1993 to 2003 was *c.* 3.1 mm per year (Bindoff et al. 2007). The steric effect over the period 1910 to 1990 accounts for about one-third of the observed eustatic change over that period (Cazenave and Llovel 2010).

With regard to the future, sea level rise caused by thermal expansion between 1990 and 2100 is thought likely to be around 0.28 metres, out of a total predicted sea level rise due to climate change of 0.49 metres (Church et al. 2001, table 11.14). The IPCC (2007) suggested that thermal expansion is likely to account for 70–75% of sea level rise by the end of the century. In other words, it will comfortably exceed in importance the role of ice cap and glacier melting. However, this proportion might change if the contribution of the melting of the Greenland ice cap becomes a more important component of the total (Vermeer and Rahmstorf 2009).

Many glaciers will shrink as warming occurs and their meltwater will flow into the oceans, causing sea level to rise (Gregory and Oerlemans 1998; Radić and Hock 2011). In some situations, on the other hand, increases in precipitation may nourish glaciers causing them to maintain themselves or to grow in spite of a warming tendency. The IPCC (2001, table 11.14) suggested that between 1990 and 2100, the melting of glaciers will contribute *c.* 0.01 to 0.23 metres of sea level change, with a best estimate of 0.16 metres; second, therefore, only to steric effects. The IPCC (2001; 2007) believed that the contribution that the Antarctica and Greenland ice caps would make to sea level change

would be modest and less than those of glaciers or the steric effect. The Greenland contribution between 1990 and 2100 was predicted to be in the range −0.02 to 0.09 metres, and for the Antarctic −0.17 to 0.02 metres. However, current signs that the polar ice sheets are melting at an accelerating rate, suggest that their contribution to twenty-first century sea level rise may be greater than the IPCC had proposed (Rignot et al. 2011).

Over the years there has been a diversity of views about how much sea level rise is likely to occur by the end of this century. In general, however, estimates have tended to be revised downward through time (French et al. 1995; Pirazzoli 1996) (**Fig. 6.20**). They have now settled at best estimates of just under 50 centimetres by 2100. This implies rates of sea level rise of around 5 millimetres per year, which compares with a rate of about 1.5 to 2.0 millimetres during the twentieth century (Miller and Douglas 2004). However, with increasing evidence that the polar ice sheets are degrading rapidly, the rates should be revised upwards and by the end of the century sea level might be 50–140 cm above the 1990 level (Rahmstorf 2007), or even 75–190 cm (Vermeer and Rahmstorf 2009). Nicholls et al. (2011) believe that, with a 4°C or more rise in temperature, a credible upper boundary for twenty-first century sea level rise is 200 cm.

The consequences of increasing global sea levels will be magnified in those areas which are undergoing natural subsidence and those where, because of groundwater and hydrocarbon removal, the land is sinking.

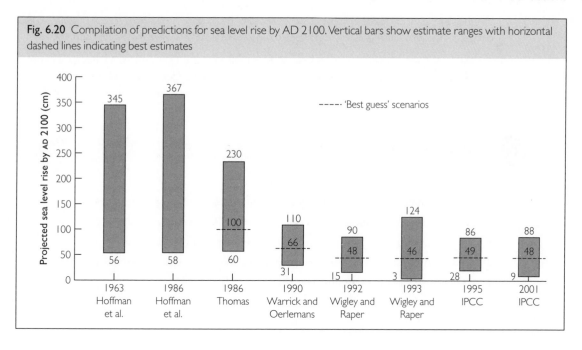

Fig. 6.20 Compilation of predictions for sea level rise by AD 2100. Vertical bars show estimate ranges with horizontal dashed lines indicating best estimates

6.11 Conclusions

It is clear that sea level changes during the Quaternary have been an important component of environmental change. It is equally clear that the causes of changes in the relative levels of land and sea are many and varied, and that in recent years the contrast or dichotomy between global and local factors has become somewhat blurred (Plag et al. 1996). Our information on sea level tendencies beyond about 250 000 years ago remains poor, but trends since the last interglacial and during the present Holocene (Flandrian) interglacial are relatively well established. Global sea levels are currently rising, but if global warming continues, rates of sea level rise may accelerate in coming decades, causing a whole suite of further environmental changes (Eisma 1995).

SELECTED READING

An enormous compilation which provides much useful data from many parts of the world, and explores the exciting theme of the role of geoidal changes is: N. A. Mörner (ed.) (1980), *Earth rheology, isostasy and eustasy.*

Techniques for sea level studies are expertly analysed by J. Rose in A. S. Goudie (1990) (ed.), *Geomorphological Techniques,* and O. Van der Plassche (1986) (ed.), *Sea level research: A manual for the collection and evaluation of data.*

A good analysis of isostatic effects is contained in D. E. Smith and A. G. Dawson (1983) (eds.), *Shorelines and Isostasy.*

A high-level survey of all aspects of isostasy is provided by: A. B. Watts (2001), *Isostasy and Flexure of the Lithosphere.*

Contemporary tectonic movements (neotectonics) are analysed by: C. Vita-Finzi (1986), *Recent earth movements: An introduction to neotectonics.*

The association between sea level changes and human activities is discussed in P. M. Masters and N. C. Flemming (1983) (eds.), *Quaternary coastlines and marine archaeology.*

A wide-ranging series of essays on many aspects of sea level change studies in many parts of the world is: R. J. N. Devoy (1987) (ed.), *Sea-surface studies.*

The complex history of Holocene changes is well reviewed by P.A. Pirazzoli (1991), *World Atlas of Holocene sea level changes*, and the same author gives a slightly longer time perspective (1996) in *Sea level changes: The last 20,000 years*.

Current and future changes of sea level are summarised in: Bindoff, N.L. and 12 others (2007), Observations: oceanic climate change and sea level. In S. Solomon and 7 others (eds.), *Climate Change 2007: the physical science basis*, pp. 385–432.

? REVIEW QUESTIONS

1.   What are the main causes of global sea level changes?

2.   Describe and account for the fluctuations in sea level that have occurred during the Pleistocene.

3.   What have been the main features of sea level change since the Last Glacial Maximum?

4.   What have been the main features of glacio-isostatic adjustments during the Quaternary?

5.   Drawing on regional examples, discuss how glacio-isostatic adjustment complicates local sea level histories.

6.   How and why are sea levels likely to change over the twenty-first century?

7

Environmental Changes During the Period of Meteorological Records

Chapter overview

For the last few hundred years, systematic meteorological observations can be used to give a picture of the changes in climate that have occurred in many parts of the world at a variety of different timescales. They indicate the nature of conditions at the end of the **Little Ice Age** and demonstrate the warming that has taken place in the twentieth century (except for a cooling episode at mid century). We can also establish that changes in rainfall regimes have occurred, that there have been changes in **dust storm** incidence, and that fluctuations in lake levels and in river flow have occurred. Finally, glaciers and sea ice have tended to retreat in recent decades.

7.1 Introduction

The great weight of data now available on the question of trends and fluctuations in the twentieth century has led to a great change in attitudes to climatology. As Lamb remarked (1966), 'Not so very long ago ... climate was widely considered as something static, except on the geological timescale, and authoritative works on the climates of various regions were written without allusion to the possibility of change, sometimes without mention of the period to which the quoted observation referred.' As Lamb and many others have now shown this static attitude of the 'old climatology' must now be replaced by the dynamic attitude of the 'new climatology'. The only unvarying thing about climate is its variability. Over the last century there have been changes in temperature, precipitation, the ocean-atmosphere systems, in runoff, in lake levels, in dust storm activity and in the state of glaciers and sea ice. These have had a whole suite of significant impacts both on the environment and on society.

Various pioneering scholars, especially Brooks, Ladurie, Manley, and Lamb, adroitly interpreted past climatic conditions from documentary records (see Bradley 1985, ch. 11 for a review of techniques, and Nash and Endfield (2002) and Endfield et al. (2004) on the use of archival materials to reconstruct climatic history in Southern Africa and Mexico respectively), thereby greatly adding to our knowledge of climatic conditions in earlier centuries. However, it was not until the

nineteenth century that there was an organized growth of instrumental observations from stations all over the world. It is on the basis of these relatively reliable instrumental records that most of our knowledge of the latest chapter of environmental evolution is founded.

Such records are not without their limitations: instruments need to be replaced and recalibrated from time to time, and sites and locations are liable to be affected by such factors as urbanization or vegetational change. However, by careful selection of the more reliable stations available, such as the Radcliffe Meteorological Station in Oxford, and by taking averages for several stations within an area, a valuable picture of changes can be obtained (e.g. Burt and Shahgedanova 1998).

7.2 Temperature changes during the period of instrumental record

Warming in the early twentieth century

In many parts of the world, a warm trend occurred in the late nineteenth century and the first decades of the twentieth century which effectively brought to an end the **Little Ice Age** (see **Section 5.9**). This warming was especially significant in the Arctic (Wood and Overland 2010).

Table 7.1 indicates the dates of the warmest and coldest decades, together with their temperature characteristics for Europe.

Table 7.1 Changes in the temperature conditions (April to June) at European stations between 1860 and 1960

Station	Warmest decade	Mean temperature (°C)	Coldest decade	Mean temperature (°C)	Difference (°C)
Angmagssalik	1926–35	2.14	1899–1908	0.09	2.05
Vestmanno	1889–98	7.69	1948–58	5.63	2.06
Spitzbergen	1951–60	−2.99	1912–21	−5.99	3.00
Haparanda	1945–54	6.22	1873–82	4.08	2.14
Bodø	1945–54	6.53	1873–82	5.29	1.24
Helsinki	1945–54	9.22	1873–82	7.13	2.09
C. England	1943–52	11.82	1879–88	10.46	1.36
De Bilt	1940–9	14.12	1951–60	11.71	2.41
Zurich	1942–51	13.85	1879–88	12.34	1.51
Milan	1943–52	18.98	1879–88	16.80	2.18
Barnaul	1938–47	11.36	1882–91	8.61	2.75

(After Harris 1964.)

One consequence of the warming trend can be seen when one looks at the dates of the first and last snowfalls in London from 1811 to 1960. As **Table 7.2 (c)** shows, whereas in the early years of the nineteenth century the mean dates of the first and last falls were separated by over 150 days, by the period 1931–60 this figure had declined to only 113 days. Even when one compares 1931–60 to 1901–30, the period during which one might expect snow was reduced by around four weeks. This may partly result from the effects of urbanization.

Another consequence of the greater warmth was that the length of ice-cover on rivers and lakes in high latitudes declined appreciably until the 1930s or later. Some data for Norway and Sweden are shown in **Table 7.2 (a & b)**, and of the lakes considered, Mjosa in Norway is the one which shows the greatest decline in ice-cover, with an average of 71.6 days of ice per year for the decade after 1910 falling to only 22.8 days in the 1930s.

Similar data are available for Oxford where, on a temperature basis associated with the first occurrence of five days with temperatures high enough for plant growth, an operational definition of spring has been made. For the first fifty years of the period 1869–1970

Table 7.2 Snow, ice, and frost frequencies in the nineteenth and early twentieth centuries

(a) Frost days, ice days, and cold days in Sweden

No. of	1861–70	1871–80	1881–90	1891–1900	1901–10	1911–20	1921–30	1931–40
Frost days[1]	121.8	122.3	123.4	124.7	125.0	115.4	117.2	103.4
Ice days[2]	55.9	58.2	57.2	57.9	56.3	55.9	57.2	47.1
Cold days[3]	43	48	24	33	8	19	21	19

1 = days with a minimum temperature <0 °C
2 = days with a maximum temperature <0 °C
3 = days with a maximum temperature <10 °C
(After Liljequist 1943.)

(b) Days of ice cover in Norway and Sweden

Lake	1900	1910	1920	1930	1940
Femund (Norway)	176.3	158.8	168.2	156.4	161.8
Mjosa (Norway)	–	71.6	65.2	22.8	49.8
Rossvatn (Norway)	–	164.4	159.0	138.9	144.6
Bolmen (Sweden)	–	105.2	89.4	86.0	93.9
Siljan (Sweden)	131.1	120.2	110.0	108.5	102.0
Storsjon (Sweden)	168.4	155.8	149.9	145.5	144.6
Mean value	–	129.33	123.62	109.68	116.12

(From data in World Weather Records processed by authors.)

(c) Dates of the first and last snowfall in London from 1811 to 1960

	Autumn	Spring
1811–40	18 Nov.	22 Apr.
1841–70	21 Nov.	17 Apr.
1871–1900	23 Nov.	12 Apr.
1901–30	25 Nov.	15 Apr.
1931–60	8 Dec.	1 Apr.

(After Manley 1964.)

the ten-year moving mean date of the first day of spring (as defined) fell between 13 and 22 March, but as a result of the warming trend of the 1930s and 1940s the date around 1940 was as early as 3 March.

The warming led to a general diminution in the ice-cover in the Arctic Sea, which had major implications for navigation. Off Iceland, both in the 1860s and 1880s, there had been on average between 12 and 13 weeks of ice in a year. By the 1920s the incidence was down to only 1.5 weeks per year, though by the decade 1947–56, because of the temperature fall already described, this had increased slightly to 3.7 weeks in the year. Equally, the area of drift-ice in the Russian sector of the Arctic was reduced by no less than 1 million km^2 between 1924 and 1944 (Diamond 1958). The ice also tended to become less thick, so that whereas Nansen found that the average thickness of ice in the Polar Sea was 365 cm in the period 1893–6, the Sedov expedition of 1937–40 found that it was down to 218 cm (Ahlmann 1948). Iceberg frequencies off Newfoundland also declined. The annual average for 1900–30 was 432 whereas for 1931–61 it was 351, a decrease of 19 per cent (Schell 1962). The coast of Greenland also became less subject to ice as illustrated by the frequency of years in which the polar ice, which comes around Cape Farewell, reached as far north as Godthaab. From 1870 to 1879 it was over 70 per cent; since 1910 it has always been less than 25 per cent (Beverton and Lee 1965). As a result of the improved ice conditions, the shipping season for the coalfields of West Spitzbergen lengthened from three months at the beginning of the century to about seven months in the 1940s.

The colonization of the west Greenland continental shelf by the cod (*Gadus morhua*) from Iceland is the best example of the response of fish to the general warming trend. Before 1917, except probably for short periods during the nineteenth century, only small local fjord populations of cod occurred in Greenland. After 1917 large numbers of adult fish appeared off the south-west coast as far north as Frederikshaab (62° N) and migrated north through 9° of latitude in 27 years (Ahlmann 1948). The haddock (*Melanogrammus aeglefinus*) and the halibut (*Hippoglossus hippoglossus*) showed a similar northward movement towards both Greenland and Novaya Zemlya. Between 1924 and 1949 swordfish, pollack, twaite shads, and dragonets were recorded for the first time off Iceland. On the other hand, there was a striking response of typically cold-water forms such as the white whale (*Delphinapterus leucas*) and the capelin

(*Mallotus villosus*). Their southerly limits contracted (Beverton and Lee 1965). The results of the amelioration may also have included the dramatic decline of the Plymouth herring fisheries of the English Channel, those of the Firth of Forth, and the haddock fisheries of the North Sea. The Channel fisheries were partially replaced, after 1935, by warmer-water forms, especially the pilchard (*Sardina pilchardus*) and the cuttlefish (*Sepia officinalis*).

The Baltic Sea also benefited from the climatic amelioration. It became more saline as a result of the increased frequency of southeasterly winds which tended to enlarge the outflow of brackish surface water from the Baltic and brought about a corresponding increase in the compensating inflow of saline North Sea water along the sea bottom. High salinity levels improve spawning conditions for cod, and this led to an enormous, perhaps twentyfold, increase in the abundance of the Baltic cod, which then supported a major fishery (Beverton and Lee 1965). Land flora and fauna of Northern Europe also showed changes in their distribution. In Finland the polecat (*Mustela putorius*) began spreading into the country about 1810 and by the late 1930s had occupied the whole south Finnish interior to about 63° N (Kalela 1952). The colder the winter and the greater its snowfall the more difficult it is for this mammal to find its natural food of small rodents, frogs, and the like. In north-east Greenland the musk ox had plenty of food from 1910 onwards, and its numbers increased markedly (Vibe 1967). Likewise, the roe deer, common in south and central Scandinavia in the years before the Little Ice Age, was almost extinct there by the early nineteenth century, only to reappear and spread strongly northwards after 1870. Also in Finland some permanently resident birds such as the partridge, to which severe snow is inimical, the tawny owl, and many species of tit, extended northwards (Crisp 1959).

In Iceland and Greenland the distributions of birds give an equally clear indication of the effects of the amelioration (Harris 1964). For example, the fieldfare (*Turdus pilaris*) was unknown in Greenland and Jan Mayen before 1937, but breeding became established there. Starlings (*Sturnus vulgaris*) arrived in Iceland in 1935 and became permanent in 1941. Swallows (*Hirundo rustica*) appeared in the Faeroes and Iceland in the 1930s. The white-fronted goose and the long-billed marsh wren began to breed in Greenland and some species like the mallard and long-tailed duck, which were summer visitors, remained throughout the year. However, the reduction in numbers of the little auk provides a clear example of how an improving climate may have a directly adverse effect on species. It feeds on small crustacea which are particularly plentiful in the surface water near the sea ice front. As the sea ice front retreated northwards from Iceland the birds in the Icelandic colonies had to fly over larger distances for food, and the colonies were thereby gradually deserted (Crisp 1959).

In various parts of the world, many subalpine meadows, which lie at high altitudes between closed forest and treeless alpine tundra, experienced a massive invasion by trees and an increasing density of trees at the treeline. Although it might be argued that changes in humanly induced fires, or a reduction in grazing by domestic animals, might be contributing factors, it is probable that the bulk of the invasion by trees can be attributed to climate. Tree-ring studies indicate that the most intense phase of invasion coincided with the temperature peak from the early 1920s to the late 1940s. The increased temperatures and diminished depth and duration of winter snow increased the growing season to the benefit of the trees (Heikkinen 1984). The global history of treeline advance since 1900 is given in Harsch et al. (2009).

The atmospheric conditions which gave the period of warming over Britain and Scandinavia in the first half of this century were highly zonal in type, with westerly winds being dominant. Thus the mean yearly number of days with general westerly winds over the British Isles (**Fig. 7.1**) was 109 in the 1920s and 99 in the 1930s (Lamb and Morth 1978).

A cooling episode at mid century

When one compares temperature conditions of 1900–19 with those of 1920–39 one finds that about 85 per cent of the Earth's surface experienced increases in mean annual temperature, whereas when one looks at temperature data for the period between 1940 and 1960 about 80 per cent of the total Earth surface was probably involved in a net annual cooling (Mitchell 1963). Only a few areas continued to show a net warming. Parker and Folland (1988) suggest the cooling affected the Northern Hemisphere from about 1945 to 1970 over land and from about 1955 to 1975 over the oceans.

The cooling in the middle years of this century had a series of consequences including the development of snow banks and small glaciers in the Canadian Arctic (Bradley and Miller 1972); an increase in snowfall frequencies and quantities in New England; a decrease in the length of the growing season in Oxford (Davis 1972); and an increase in sea ice-cover in the Baltic.

The warming episode of the late twentieth century

Some parts of the world seem never to have experienced the brief cooling episode of the mid twentieth century. Warming continued without interruption in New Zealand (**Fig. 7.2**) (Salinger and Gunn 1975) and over much of Australia (Tucker 1975).

In the last three decades, however, there is mounting evidence that warming is once again a rather general feature of the world's climate. This is shown in **Fig. 7.3** and discussed by the IPCC (2007). There is no

Fig. 7.1 Number of days with general westerly winds over the British Isles from 1861–1976. The bold line shows the 10-year averages at 5-year intervals (after Lamb and Morth 1978, Fig. 11). Reprinted with permission from Wiley: *Geographical Journal*

Fig. 7.2 Observed mean annual New Zealand temperatures over a 130-year period to the mid 1990s, smoothed using a 3-year moving average. The horizontal line gives the average temperature over the 1961–1990 period (after Salinger and McGlone 1990, and Salinger personal communication)

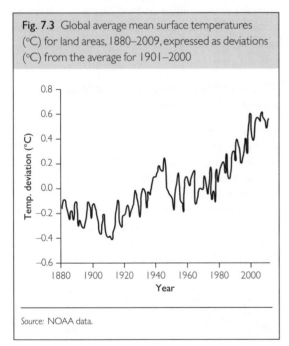

Fig. 7.3 Global average mean surface temperatures (°C) for land areas, 1880–2009, expressed as deviations (°C) from the average for 1901–2000

Source: NOAA data.

doubt that the 1980s and 1990s were particularly warm decades in many areas. Indeed, the average temperature for the 1980s was about 0.2 °C above the mean for the period 1950–79. The greatest warming occurred over the continents between 40° and 70° N. The overall increase in temperature during the twentieth century was about 0.6 °C and is likely to have been the largest of any century during the last thousand years. The greatest increases in temperature took place in the Arctic, where over the last 100 years the increase was almost twice the global average (Trenberth et al. 2007).

Associated with the warming has been a tendency for reduced diurnal temperature ranges over land. Minimum temperature increases have been about twice those in maximum temperatures. Nights have warmed more than days. The reason for this is that cloud cover, which reduces heat loss from the ground at night and obstructs daytime sunshine, has increased in many of the areas with a reduced diurnal temperature range (Nicholls et al. 1996, pp. 144–146).

The warming trend has not been constant and regular. The Mount Pinatubo eruption of 1991, for example, caused a decrease in global mean surface air temperatures by about 0.5 °C for several years, but warmer surface temperatures reappeared in 1994 following the removal by natural processes of Mt. Pinatubo **aerosols** from the atmosphere. ENSO events may also explain some of the variability.

Warming has had a number of major environmental impacts. If we consider cold regions, snow cover declined in the Northern Hemisphere between 1922 and 2010 (Brown and Robinson 2011), with the rate of decrease accelerating over the past 40 years. In the western United States, there has been a substantial decrease in snow depth, particularly at lower altitude sites, over the period 1949–2005 (Grundstein and Mote 2010). Changes in the amount and timing of snowfall have caused greater runoff in spring and less runoff in the midsummer for the Arctic rivers of Eurasia (Tan et al. 2011). Similarly, the effects of recent warming on the date of occurrence of the last hard frost of spring (defined by a minimum 24 hour air temperature less than or equal to −2.2 °C) is made evident for New England (USA) in **Fig. 7.4**. This suggests that over

Fig. 7.4 Area weighted date of last hard spring-freeze date for the New England region, 1961–1990 (modified from Cooper and Leduc 1995, Fig. 2)

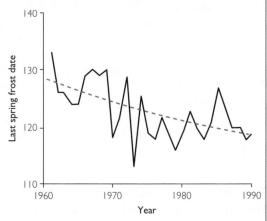

that region as a whole the frost-free season began 11 days earlier in 1990 than it had 30 years earlier (Cooper and Leduc 1995). Warming may have contributed to **permafrost** melting in the high-latitude regions of North America. Kwong and Tau (1994), working on the Mackenzie Highway, found that between 1962

and 1988 the southern fringe of the discontinuous permafrost core had moved north by about 120 km in response to an increase over the same period of 1 °C mean annual temperature.

Another consequence of recent warming may be the phenomenon called **coral bleaching**. Although pollution and other anthropogenic stresses may play a role in this form of coral reef degradation, anomalously warm water conditions have been implicated in it in areas like the Caribbean (Goreau and Hayes 1994).

Global warming has also contributed to an increasing drought occurrence since the 1950s, most notably in Africa, east and south Asia, the Mediterranean region, eastern Australia and mid-latitude Canada (Dai 2011a, b).

There has been considerable debate about the causes of the warming tendencies that have been observed during the period of instrumental observation. The degree of change and temporal pattern cannot be accounted for solely in terms of the enhanced **greenhouse effect** caused by emissions of greenhouse gases, but when allowance is made for the cooling effect of sulphate aerosols in climate simulations a close agreement between the simulations and surface temperature trends becomes evident (**Fig. 7.5**) (Houghton et al. 1996, p. 33).

Fig. 7.5 Simulated global annual mean warming from 1860–1990 allowing for increases in greenhouse gases only (dashed curve) and greenhouse gases and sulphate aerosols (solid curve), compared with observed changes over the same period (after Houghton et al. 1996, Fig. 15)

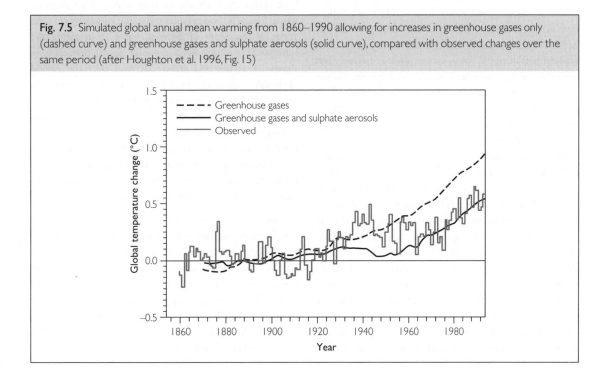

7.3 Precipitation changes during the period of instrumental record

Overall, global land precipitation has increased by about two per cent since the beginning of the twentieth century (Houghton et al. 2001, p. 142), with increases being dominant over land north of 30° N but with downward trends dominating the tropics since the 1970s (Trenberth et al. 2007).

Bradley et al. (1987) have attempted to provide a general picture of precipitation changes over Northern Hemisphere land areas since the middle of the last century, using available instrumental records. They describe the general patterns of change for a series of major areas. In Europe they found that annual precipitation totals had increased steadily since the middle of the nineteenth century, with well above average precipitation since a dry spell in the 1940s. Most of the upward trend was evident in the winter precipitation, with lesser amounts in autumn and spring. Summer rainfall, on the other hand, showed a slight decline. In the former USSR they found that rainfall had increased dramatically since the 1880s, with most of the change occurring before 1900 and after 1940. The increase in annual totals was mainly accounted for by increases in autumn, winter and spring. Summer totals displayed very little trend. In the USA precipitation totals declined from around 1880, reaching a low in the 1930s, and generally increasing thereafter. In North Africa and the Middle East very little trend was evident until the 1950s, when, after a relatively wet episode, precipitation declined drastically, especially in summer. In south-east Asia a relatively wet episode in the 1920s and early 1930s separated two dry periods: the former centred on 1900 and the latter from the mid 1960s through to the present.

More recently the global geography of twentieth century precipitation change has been reviewed by Nickl et al. (2010). They suggest that taking the world as a whole a consistent increase in annual land surface average precipitation of approximately 0.2–0.5 mm per year occurred during the first half of the twentieth century. This increase was followed by over four decades with decreases of 0.3–0.6 mm per year. In the early 1990s there was once again an increasing trend of about 0.75–2.1 mm per year between 1992 and 2002. However, there was considerable spatial variability. Over the last phase, there was, for example, marked drying over much of North America, south Asia

and North Africa as well as parts of South America. Conversely, sizable increases in precipitation occurred over some other parts of South America, southern Africa, northern Australia, and south-east Asia. Similarly, Zhang et al. (2007), analysing data for the twentieth century, found that anthropogenic forcing had contributed significantly to observed increases in precipitation in Northern Hemisphere mid latitudes, drying in the Northern Hemisphere subtropics and tropics, and moistening in the Southern Hemisphere subtropics and tropics.

We will now consider the more detailed picture from the British Isles, low latitudes, and desert margins.

Records of precipitation change for Britain and other mid-latitude areas

The changes in rainfall that have taken place in Britain during the period of instrumental records are as difficult to generalize about as are temperatures (Gregory 1956; Barrett 1966). Wigley and Jones (1987), after analysis of precipitation data back to the 1760s (Fig. 7.6), concluded (p. 245):

In spite of popular perceptions that precipitation values are changing, none of the regional, annual or seasonal time series show trends on time scales of 30 years or more. There is, however, considerable variability on time scales of the order of decades, more so in spring and summer than in autumn and winter. Spring precipitation showed a strong and steady increase over the period 1956–1969, especially in the NW and NE regions. In recent years, spring precipitation has been highly variable, with remarkably wet seasons occurring in 1979, 1981 and 1983. In contrast, spring 1984 was unusually dry in the NW region. Summer precipitation showed a decline from the late 1960s to the mid 1970s, with no trend since then. There has, however, been an unusual number of dry summers in the past 10 years (1976–1985).

One of the most important controls of agricultural activity is the incidence of drought conditions, and the particularly severe drought of 1976 caused a reassessment of the significance and frequency of these phenomena in Britain. The incidence of major soil moisture deficits during the growing season of British crops (May to August inclusive) has been plotted from data for Kew (London) that go back to the end of the seventeenth century (Wigley and Atkinson 1977). From these data

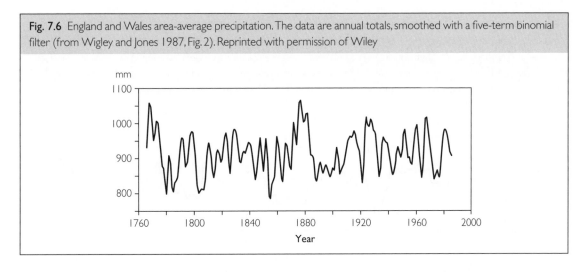

Fig. 7.6 England and Wales area-average precipitation. The data are annual totals, smoothed with a five-term binomial filter (from Wigley and Jones 1987, Fig. 2). Reprinted with permission of Wiley

(**Fig. 7.7a**) it can be seen that there have been fluctuations of some importance, and this is brought out further in **Fig. 7.7b**. In particular it is noteworthy that of the 14 worst soil moisture deficit years, 8 have occurred since 1900. Moreover, more serious droughts occurred in the period 1960–76 than in any preceding 20-year period. It is likely that the temperature increase that has prevailed from 1880 has caused some of the drier soil conditions of the last century. One consequence of drought has been extreme low flows of some English and Welsh rivers, notably in 1870, 1887, 1921, 1933/4, 1976, 1984, 1989, 1990, 1994, and 1995 (Jones et al. 2006).

Conversely, another important aspect of rainfall variability in Britain is the incidence of rainfall-related flooding. There is some evidence that serious floods have increased in frequency in recent decades. Rodda's work (1970) showed that an increase in intense rainstorms occurred in central England (as represented by climatic data from the Oxford). For the period 1881–1905 the return period for a storm of just over 50 mm was about 30 years, but for the period 1941–65 the return period of the same size of daily fall had dropped to little more than 5 years.

This climatic explanation for increasing flood levels has been confirmed for South Wales by Walsh et al. (1982). However, Lawler (1987) sounded a note of caution about overgeneralizing about this particular climatic trend and argues not only that no simple synchroneity across the country exists in any changes identified but that since 1968 there has in some areas been a reversal in the trend of increasing storm rainfalls and associated floods.

There is some evidence of increased high rainfall events in various countries over recent warming decades. Examples are known from the United States (Karl and Knight 1998), Canada (Francis and Hengeveld 1998), Australia (Suppiah and Hennessy 1998), Japan (Iwashima and Yamamoto 1993), South Africa (Mason et al. 1999), and Europe (Forland et al. 1998). In the UK there has been an upward trend in the heaviest winter rainfall events (Osborn et al. 2000). Heavy precipitation events appear to have been intensified over very large parts of Northern Hemisphere land areas (Min et al. 2011). A possible consequence of this is changes in flood characteristics. In their analysis of flood records for 29 river basins from high and low latitudes with areas greater than 200,000 square kilometres, Milly et al. (2002) found that the frequency of great floods had increased substantially during the twentieth century, particularly during its warmer later decades.

Precipitation change in lower latitudes

We now turn to a consideration of rainfall tendencies in lower latitudes, where in recent decades great concern has been expressed that climatic deterioration may be contributing to the phenomenon of **desertification**. The situation appears to be complex. When one examines rainfall data for the arid areas of the Sudan-Sahel in Africa, central Australia, northwest India and Arizona, USA (**Fig. 7.8**), for the period between the mid 1940s and the mid 1970s, it is apparent that some areas showed relatively little evidence of a downward trend whereas others did.

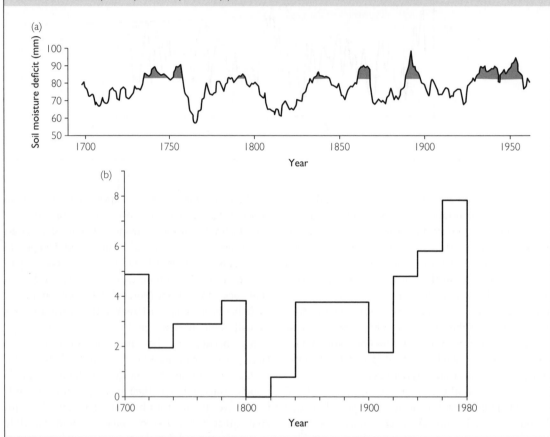

Fig. 7.7 Soil moisture deficits at Kew, London (after Wigley and Atkinson 1977). (a) 10-year running means of soil moisture deficit averaged over the growing season at Kew. An arbitrary datum level of 84 mm has been shown to accentuate the periods of higher deficits. (b) Number of times the soil moisture deficit averaged over the growing season exceeded 100 mm in 20 year intervals. Note the low incidence in the early nineteenth century and the very high incidence in the very latest period. Reprinted by permission from Macmillan Publishers Ltd: *Nature*

The Sahel (**Fig. 7.9**) and Sudan belts to the south of the Sahara suffered a major drought from the mid 1960s onward. One of the prime drivers of this drought was a change in sea surface temperatures in the North Atlantic Ocean (Giannini et al. 2003; Hagos and Cook 2008). It is also possible that the drought was reinforced by increasing levels of dust in the atmosphere causing a diminution of rainfall (Hui et al. 2008) and by land cover changes which altered surface **albedo**, temperatures, and convectional activity (Lauwaet et al. 2009). Climatic deterioration was certainly severe in the Sudan (**Fig. 7.10**) where, in White Nile Province, annual rainfall in 1965–84 was 40 per cent below 1920–39 levels, and wet season length contracted by 39–51 per cent (Walsh et al. 1988). The dry epoch which started in the mid 1960s has continued

and intensified in the 1980s. Further west, the dry epoch has had dramatic effects on Lake Chad. Its area declined from 23 500 km² in 1963 to about 2000 km² in 1985, which was probably the lowest level of the twentieth century (Rasmusson 1987: 156). Whether the desiccation of Lake Chad had an adverse effect on rainfall amounts is uncertain but probably unlikely (Lauwaet et al. 2012). In southern Africa, analyses of long-term precipitation changes were made by Tyson (1986). He could find no statistical evidence that southern Africa is becoming progressively drier (as has often been popularly maintained), but he did find evidence for a number of quasi-periodic rainfall oscillations. He reported (p. 197) that the most noteworthy of these 'is that with an average period of about 18 years. Since the turn of the century eight approximately 9-year spells of either

Fig. 7.8 Rainfall variations in selected arid zone stations between 1945 and 1974

Fig. 7.9 Sahel precipitation anomalies for 1900–2007 for June–October inclusive

Source: NOAA Global Historical Climatology Network Data

predominantly wet years, which average to show above-normal rainfall during the spell, or predominantly dry years, showing below-normal rainfall, have occurred.' A general review of African rainfall trends is provided by Hulme (1992).

In the Rajasthan Desert of India the trend of rainfall in the twentieth century appears to be very different from that in the Sahel. Analyses of **monsoonal** summer rainfall for the Rajasthan desert (Pant and Hingane 1988) indicate that there was a modest upward trend in precipitation levels between 1901 and 1982. The overall pattern of rainfall trends in India as a whole is extremely complex, though some broad contiguous areas showing statistically significant trends have been identified (Kumar et al. 1992). Nevertheless, if one looks at India as a whole some interesting temporal trends are evident (**Fig. 7.11**). A decreasing trend is evident between 1889 and 1902 and again during 1958–1969. The decrease in precipitation in those periods was relatively sudden, while in the intervening period the increase was slow (Subbarmayya and Naidu 1992).

In the drought-prone region of north-east Brazil, Hastenrath et al. (1984) have undertaken an analysis of rainfall records since 1912 (**Fig. 7.12a**). The incidence of runs of dry years is thereby highlighted, but there is no very conspicuous evidence of any long-term trend, either upwards or downwards.

Rainfall over Australia exhibits a high degree of spatial and temporal variability (Hobbes 1988) and during the twentieth century may have shown a general increasing trend (**Fig. 7.13**). However, there were regional differences. The geographical pattern of rainfall trends in Australia during the twentieth century was of decreased winter rainfall over much of south-western Australia and increased summer rainfall over much of eastern Australia (Nicholls and Lavery 1992), but Australian precipitation, which is heavily influenced by ENSO, has also been characterized by large inter-annual fluctuations. A more recent analysis of Australian rainfall from 1970 to 2006 (Taschetto and England 2007) drew a distinction between the west, where except along much of the coastline the rain has been increasing, and the east where precipitation has been decreasing, particularly along the north-east coast.

Zhang and Zhou (2011) undertook an analysis of trends on monsoonal rainfall in the twentieth century. They suggested that there was an increasing tendency from 1901 to 1955 and a decreasing trend since the 1950s.

In lower latitudes one potentially very important cause of change in the occurrence of extreme rainfall events is the incidence of tropical cyclones (hurricanes). However, this is one of the most interesting, and, it has to be said, least comprehended features of climatic change in low latitudes in the twentieth century. The Intergovernmental Panel on Climate Change (Houghton et al. 1990, p. 232–233) was wary of interpreting changes in hurricane frequency in terms of late twentieth century warming:

Current evidence does not support this idea, perhaps because the warming is not yet large enough to make its impact felt. In the North Indian Ocean the frequency of

Fig. 7.10 (a) The shifting position of the mean annual rainfall in the central Sudan between 1920 and 1939, and 1965 and 1984 (from Walsh et al. 1988, Fig. 4). (b) Changes in the duration of the wet season at three locations in central Sudan over the same period. The bars indicate the inter-quartile ranges of wet season onset and termination dates (from Walsh et al. 1988, Fig. 5). Reprinted with permission from Wiley: *Geographical Journal*

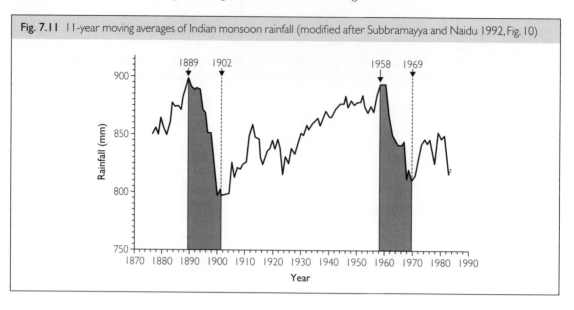

Fig. 7.11 11-year moving averages of Indian monsoon rainfall (modified after Subbramayya and Naidu 1992, Fig. 10)

tropical storms has noticeably decreased since 1970 … There is little trend in the Atlantic … There have been increases in the recorded frequency of tropical cyclones in the eastern North Pacific, the south-west Indian Ocean, and the Australian region since the late 1950s. However, these increases are thought to be predominantly artificial and to result from the introduction of better monitoring procedures.

It took a similar view in 2001 (see Houghton et al. 2001, p. 33).

Walsh (2004) has reiterated that currently there appear to be no detectable changes in observed tropical cyclone characteristics that can genuinely be ascribed to global warming and that there is little evidence of substantial changes in the poleward extent of active tropical cyclones, once they leave their tropical regions of formation. Moreover, as Goldenberg et al. (2001) have pointed out, although the sea surface temperatures in the North Atlantic have exhibited a warming trend over the last 100 years and cyclone activity was high in the last years of the twentieth century, Atlantic hurricane activity has not exhibited trend-like variability over the last century, but rather distinct multi-decadal cycles. In other words, factors other than or additional to sea surface temperature play a role. Trenberth (2005) has argued that these factors include an amplified high pressure ridge in the upper troposphere across the eastern and central North Atlantic and reduced vertical wind shear over

the central North Atlantic (which tends to inhibit vortex formation). On the other hand Emanuel (2005) has suggested that tropical cyclones have become increasingly destructive over the last 30 years and that this can be related to warmer sea surface temperature conditions. The relationship between sea surface temperature increase and increasing global hurricane activity has been confirmed by Hoyos et al. (2006) and Saunders and Lea (2008). Santer et al. (2006) have demonstrated that human factors have caused the increase in sea surface temperatures and cyclone development in both the Atlantic and Pacific regions, while Webster et al. (2005) and Elsner et al. (2008) have shown that the strongest tropical cyclones have become more intense in recent decades. There is also the possibility that tropical cyclones, at present relatively unimportant in South America could become more significant, as presaged by Hurricane Catarina, which hit southern Brazil in March 2004 (Pezza and Simmonds 2005).

Analysis of long-term precipitation data for the western Mediterranean area (Maheras 1988) enables an assessment of fluctuations between 1891 and 1985 (p. 187): 'On an annual basis, two principal moist periods occur (one from 1901 to 1921, another from 1930 to 1941) and also two secondary ones; between, we have the dry periods, primary and secondary, the second being the most marked (from 1942 to 1954). Another important feature is the decrease in precipitation beginning in 1980 and lasting up to 1985.' Maheras believes that annual precipitation in the western

Fig. 7.12 The trend of annual rainfall in selected dryland areas in the twentieth century. (a) North East Brazil from 1913 to 1985 showing the average annual rainfall expressed as the normalized departure from the mean. (b) Africa from 1901–1987 showing rainfall expressed as a percent departure from the long-term mean. (c) Western Australia showing yearly winter rainfall expressed as the normalized departure from the mean

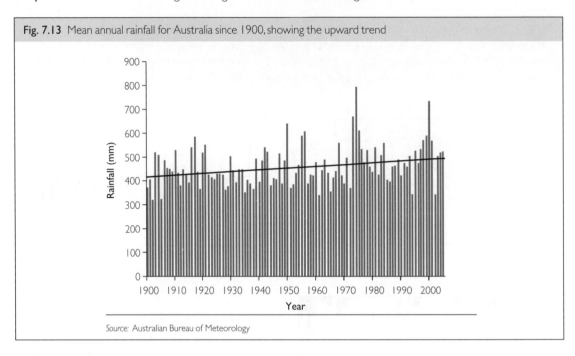

Fig. 7.13 Mean annual rainfall for Australia since 1900, showing the upward trend

Source: Australian Bureau of Meteorology

Mediterranean over the past century has shown an approximate periodicity of 20 years. Further east, in the Balkans the 1930s also emerge as a wet decade (Maheras and Kolyva-Machera 1990) and the early 1980s as a very dry phase.

7.4 The various scales of climatic fluctuation

Some of the variability in precipitation that is observed in the historical record is associated with major changes in the interactions between the ocean and the atmosphere (Viles and Goudie 2003) at less than century and decadal timescales:

Antarctic Circumpolar Wave

Antarctic Oscillation (also known as Southern Annular Mode)

Arctic Oscillation (also known as Northern Annular Mode)

Atlantic Multidecadal Oscillation

El Niño

Interdecadal Pacific Oscillation

La Niña

Madden–Julian Oscillation

North Atlantic Oscillation

Pacific Decadal Oscillation

Pacific-North American Teleconnection/Pattern

Quasi-Biennial Oscillation

Southern Oscillation

The El Niño–Southern Oscillation

The most interesting illustration of these is the *El Niño*. Although the term originally referred to a local warm current that runs southwards along the coast of Ecuador in the eastern Pacific around the Christmas season (hence, 'The Boy-Child'), it has now become employed to describe much larger-scale warmings of the eastern equatorial Pacific that last for one or two years and occur in intervals ranging from 2 to 10 years (**Fig. 7.14**).

The relationship between El Niño and global climate needs to be seen in the context of its atmospheric counterparts, the 'Walker Circulation' and the 'Southern Oscillation'. Under 'normal' conditions of the Walker Circulation, there is a great longitudinal cell readily observable across the Pacific. Near the coast of South America the winds blow offshore, causing **upwelling** and the subsequent cold, nutrient rich, offshore waters. However, under El Niño conditions, the circulation pattern is reversed so that

Fig. 7.14 The Southern Oscillation Index since 1876

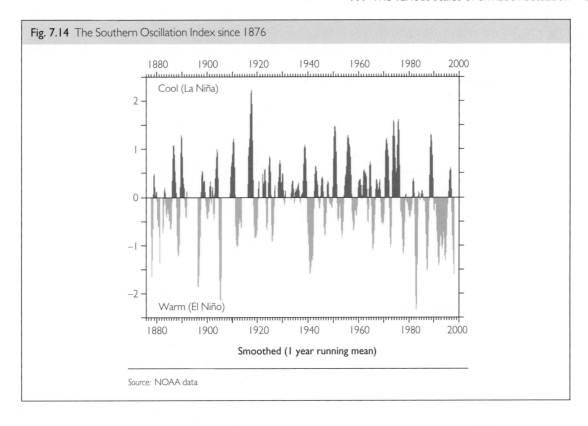

Smoothed (1 year running mean)

Source: NOAA data

water temperatures off the coast of South America in the equatorial belt rise substantially. This see-saw in atmospheric conditions is called the Southern Oscillation (**Fig. 7.15**).

This oscillation seems to have significance beyond the Pacific coast of equatorial South America, where the presence of warm water may be associated with anomalously high rainfalls over Peru and Ecuador, and with fish mortality in the sea. At times of El Niño numerous persistent climatic anomalies occur elsewhere on Earth, such as droughts in Australia, Indonesia and north-east Brazil, severe winters in the USA and Japan, and cyclones in the central Pacific.

A measure of the strength of the Walker Circulation is the Southern Oscillation Index (SOI) (**Fig. 7.14**) derived from the pressure difference between Darwin (in northern Australia) and Tahiti (in the South Pacific). The SOI is high when the subtropical high over the south-eastern Pacific is at its strongest and when the pressure over Indonesia is at its lowest. When the SOI is low (negative) one tends to have the El Niño phenomenon and when it is strongly positive one has La Niña conditions. Reviews of the ENSO phenomena are provided by Philander (1990).

Precipitation and temperature anomaly patterns appear to characterize all El Niño warm episodes. These can be summarized as follows:

- The eastward shift of thunderstorm activity from Indonesia to the central Pacific usually results in abnormally dry conditions over northern Australia, Indonesia and the Philippines.

- Drier-than-normal conditions are also usually observed over south-eastern Africa and northern Brazil.

- During the northern summer season, the Indian monsoon rainfall tends to be less than normal, especially in the north-west.

- Wetter-than-normal conditions are usually along the west coast of tropical South America, and at subtropical latitudes of North America (the Gulf Coast) and South America (southern Brazil to central Argentina).

- El Niño conditions are thought to suppress the development of tropical storms and hurricanes in the Atlantic but to increase the numbers of tropical storms over the eastern and central Pacific Ocean.

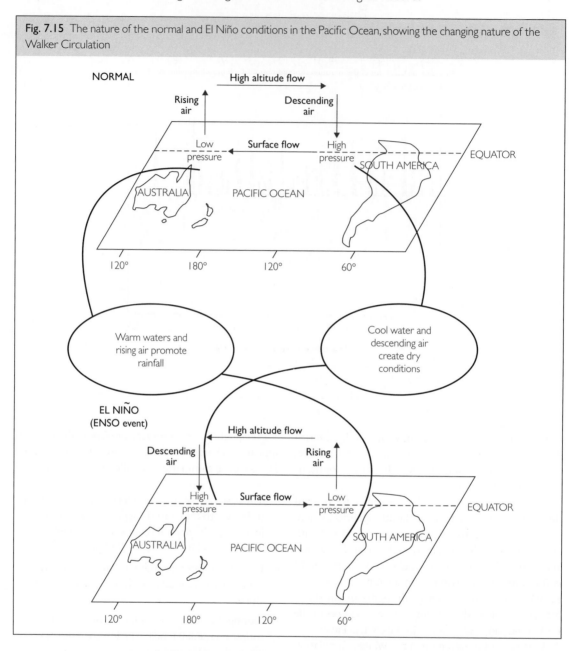

Fig. 7.15 The nature of the normal and El Niño conditions in the Pacific Ocean, showing the changing nature of the Walker Circulation

In all there were around 25 warm events of differing strengths in the twentieth century, with that of 1997–8 being seen as especially strong (Changnon 2000). ENSO was relatively quiescent from the 1920s to 1940s (Kleeman and Power 2000). Indeed, ENSO may display quasi-regular cyclic behaviour with a cycle of 50–90 years (Li et al. 2011). Severe El Niños, like that of 1997/8, can have a remarkable effect on rainfall amounts. This was shown with particular clarity in the context of Peru (Bendix et al. 2000), where normally dry locations suffered huge storms. At Paita (mean annual rainfall 15 mm) there were 1845 mm of rainfall while at Chulucanas (mean annual rainfall 310 mm) there were 3803 mm. Major floods resulted (Magilligan and Goldstein 2001).

Besides ENSO fluctuations, there are various other climatic oscillations with a scale of years and decades:

The Pacific Decadal Oscillation

The Pacific Decadal Oscillation (PDO) is a long-lived El Niño-like pattern of Pacific climate. It differs from ENSO in that PDO events in the twentieth century persisted for 20–30 years (**Fig. 7.16c**), while typical ENSO events persisted for 6 to 18 months. Cool PDO regimes prevailed from 1890–1924 and again from 1947–1976, while warm PDO regimes dominated from 1925–1946 and from 1977 onwards. There is evidence from tree-ring studies that decadal-scale reversals of Pacific climate have occurred throughout the last four centuries (Biondi et al. 2001) and using tree-ring chronologies MacDonald and Case (2005) have reconstructed the PDO for the last millennium. Variability in the PDO is important for understanding changes in precipitation in the western United States (McCabe and Dettinger 1999).

The Arctic and Antarctic Oscillations

The Arctic Oscillation (AO) is one in which there is a seesaw-like oscillation in atmospheric pressure at polar and mid latitudes in the Northern Hemisphere (Zhao et al. 2010). In its negative phase higher-than-normal pressure occurs over the polar region and lower-than-normal pressure at about $45°$ N. The positive phase brings opposite conditions that steer ocean storms further north, bringing wetter conditions to Alaska, Scotland and Scandinavia, but drier conditions to more southerly locations, such as California, Spain, and the Middle East.

North Atlantic Oscillation

The **North Atlantic Oscillation** (NAO) is also one of the dominant modes of Northern Hemisphere climate variability (Perry 2000). Its positive phase sees below-normal pressure in the region of the Icelandic Low and above-normal pressure in the Azores. This leads to strong southwesterly winds over northern Europe and north-western Asia, with above-average temperatures in these regions. In its negative phase, the Icelandic Low and Azores High pressure weaken and migrate southward, shutting off the southwesterly surface flow (Ottersen et al. 2001). This produces severe winters in northern and western Europe, as was the case, for instance, in the winter of 2009/10 (Osborn 2011).

Changes in the strength of the North Atlantic Oscillation have been traced back for over a thousand years by means of the analysis of a stalagmite from NW Scotland (Proctor et al. 2000). It may also be possible to reconstruct past NAO behaviour through tree-ring analysis (Briffa 2000). The NAO index exhibits considerable long-term variability (**Fig. 7.16b**). The 1960s displayed an extreme negative phase whereas a prolonged positive phase occurred in the late 1980s and early 1990s (Hurrell 1995). Among the recent responses to changes in the NAO are the distribution, intensity and prevalence of storms, wave climate (Wang and Swail 2001), sea ice volume and iceberg flux (Dickson et al. 2000).

The Atlantic Multidecadal Oscillation

The Atlantic Multidecadal Oscillation (AMO) has a 50–80 year cycle with a $0.4°C$ temperature range (Wyatt et al. 2011) (**Fig. 7.16a** and **Fig. 7.17**), and has a major role in Northern Hemisphere climate variability. During AMO warming, most of the USA experiences less-than-normal rainfall, and between AMO warm and cool phases Mississippi River outflow varies by 10 per cent (Endfield et al. 2000). It is a major control of droughts in the Great Plains, including the Dust Bowl of the 1930s (Nigam et al. 2011).

7.5 River discharge fluctuations

The changes in rainfall and temperature that have taken place over the past century or so would be expected to have consequences for river flow, though it is important to recognize that anthropogenic effects (e.g. land cover change, dams and inter-basin water transfers) will have confused this picture. In extreme cases such as the Colorado River of the south-west USA and the rivers flowing into the Aral Sea, river discharges may have been very substantially reduced by these mechanisms.

North American rivers (**Fig. 7.18**) showed in almost all cases a general decrease during the first three or four decades of this century, with the lowest discharges being attained during the so-called 'Dust Bowl' years of the 1930s. This was a time of higher-than-average temperatures and lower-than-average precipitation over much of North America. North American river discharges appear to have varied in response to El Niño/Southern Oscillations (Dracup and Kahya 1994) and to the Atlantic Multidecadal Oscillation (Nigam et al. 2011).

Fig. 7.16 Climatic oscillations. (a) The Atlantic Multidecadal Oscillation index since 1856. (b) The North Atlantic Oscillation index for December–March since 1865. (c) Monthly values of the Pacific Decadal Oscillation index since 1900

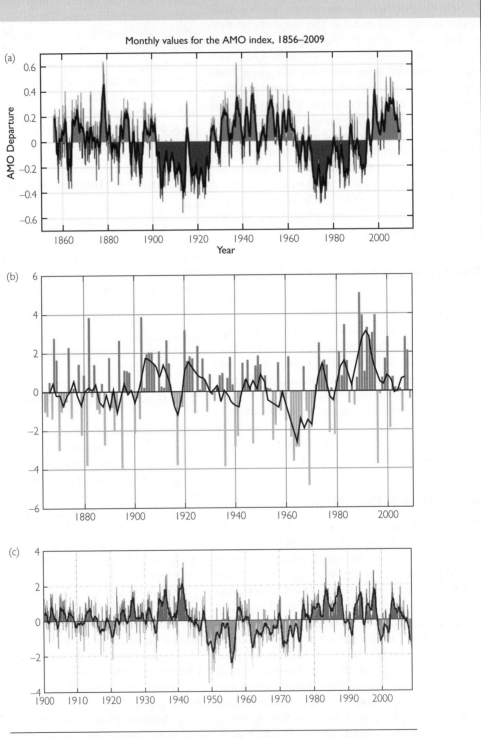

Source: NOAA data and Wikipedia Commons

Fig. 7.17 The Atlantic Multidecadal Oscillation index from 1567 to 1990 expressed as a 10-year moving average of annual sea surface temperature values reported as standard deviations (after Gray et al. 2004)

Fig. 7.18 Data on river discharge for selected stations expressed as 10-year moving means of the mean annual monthly discharge. Periods of low discharge are shaded (after Goudie 1972)

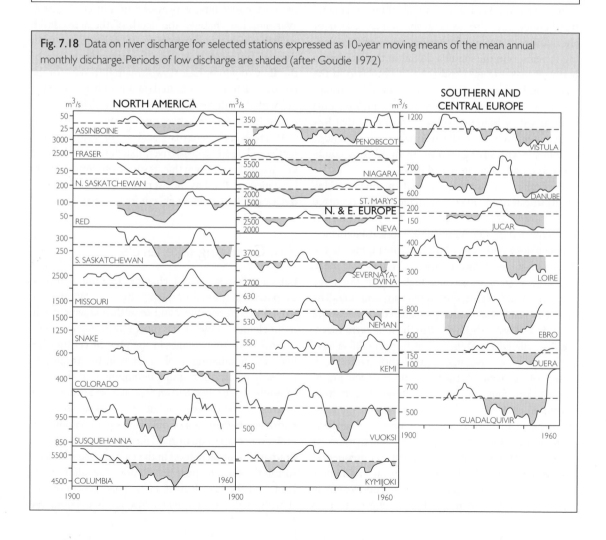

Information on long-term discharge fluctuations in Europe was provided by Probst (1989), and Probst and Tardy (1987) analysed discharge trends for a large selection of the world's rivers. Overall they found that global runoff had increased by about 3% over the 65-year period from 1910–1975. More importantly, they characterized river flow in different regions to show periods of low flow in dry periods and of high flow in humid periods. They found, for example, that North American rivers tended to have low flows around 1930–1940 and around 1960 and that European rivers showed low flows around 1940–1950.

Analyses of long-term records of river flow have also been undertaken for South America (Marengo 1995). A significant downward trend was evident in three catchments (Chicama, Peru; São Francisco, N.E. Brazil; and Paraiba do Sul, S.E. Brazil), while four catchments (Rio Negro, Brazil; Orinoco, Venezuela; Chica, Peru; and Parana, Argentina) show no general trend either way. The rivers all showed the effects of alternations of wet and dry periods, some of them lasting 15 years as in the Paraiba do Sul in the dry period between 1968 and 1982 and the São Francisco in the wet period between 1938 and 1951. The relationships between stream flow and ENSO events in Amazonia is discussed by Marengo (1992), who found that low river water levels in northern Amazonia have tended to coincide with strong or very strong El Niño conditions, as in 1912, 1926, 1983, and 1987.

Particular attention has focused on flow variability in African rivers. Major fluctuations have occurred in the flow of the Nile (**Fig. 7.19**). The record for the Blue Nile, fed by rainfall in the Ethiopian Highlands, between 1900 and 1989, showed a mean annual discharge of 52 km³. However, the highest 10-year mean discharge at 56.7 km³ was 9% higher than this long-term mean and occurred between 1903 and 1912. The lowest 10-year mean discharge occurred between 1978 and 1987 and was 19% lower than the long-term mean. For the main Nile at Dongola for the period 1890–1990 the mean annual discharge for the whole period (corrected for upstream abstractions and reservoir losses) was 89.9 km³, but 10-year mean flows ranged from 109 km³ from 1890–1898 (a 21% increase over the long-term mean) to only 74 km³ from 1978–1987 (an 18% decrease over the long-term mean). Fluctuations of such a magnitude plainly have immense consequences for water resource development (Conway and Hulme 1993). Nicholson (2011, p. 222) noted the complexities in assessing rainfall from stream flow in exogenous rivers

such as the Nile. For example, Nile flow downstream from Khartoum, Sudan, integrates flow from both the White and Blue Nile. This poses problems when trying to interpret the rainfall conditions which produced the flow. In addition, the level of Lake Victoria also controls the flow of the White Nile in its headwaters. Yin and Nicholson (2002) highlight that its flow is not indicative of rainfall conditions because lake levels are not directly correlated with rainfall at Lake Victoria.

No less dramatic changes occurred in the rivers of West Africa. Thus whereas the mean annual flow of the Senegal River at Bakel over the 84- year period to 1985 was 22.3 × 10⁹ m³, over the period 1968–1985, the time of the Sahel Drought, the equivalent value was a mere 13.7 × 10⁹ m³. A similar picture is shown for the Niger.

A particularly interesting analysis of recent temperature and stream flow trends in the permafrost region of Northeast China is provided by Liu et al. (2003). Warming has changed the depth of the active layer, enabling enhanced transport of subsurface water through unfrozen soils and a greater contribution of this flow to winter base flow. Stream flow data for the period 1958 to 1998, indicated significantly greater runoff, with flow during February and March in the 1990s increasing by 80–100% over prior values.

Peterson et al. (2002) have analysed discharge records for the six largest Eurasian rivers that flow into the Arctic Ocean and have shown that as surface air temperatures have increased so has the average annual discharge of freshwater. It grew by 7 per cent from 1936 to 1999.

7.6 Changing lake levels

Lake levels respond to changes in rainfall inputs, evapotranspirational outputs, which are largely controlled by temperature, and to anthropogenic catchment changes. One of the most interesting examples of environmental change in the present century has been the fluctuating level of lakes in the tropics. In the early 1960s, prior to the development of the Sahel Drought, Lake Chad had an area of 23 500 square kilometres, but by the 1980s it had split into two separate basins and had an area of only 1500 square kilometres. However, many equatorial lakes in Africa showed a dramatic increase in level in the early 1960s, which led to the flooding of port installations, deltaic farming land, and the like (Butzer 1971). This rise contrasted sharply with the frequently low levels encountered

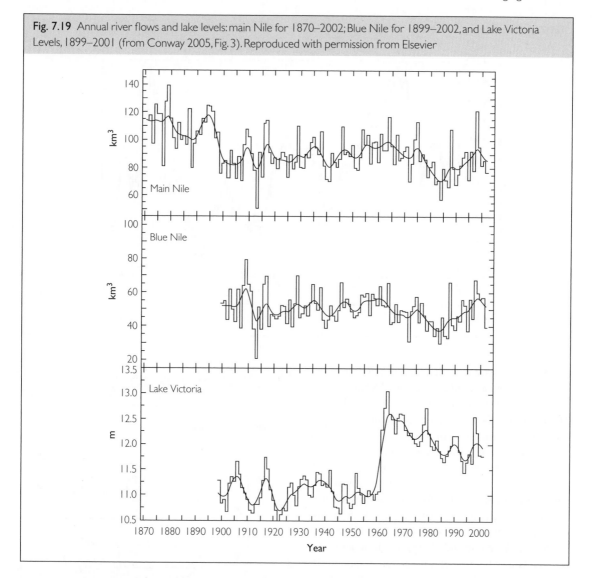

Fig. 7.19 Annual river flows and lake levels: main Nile for 1870–2002; Blue Nile for 1899–2002, and Lake Victoria Levels, 1899–2001 (from Conway 2005, Fig. 3). Reproduced with permission from Elsevier

in the previous decades. Lake Malawi (Nyasa) seems to have reached a minimum around 1927–9, Lake Tanganyika was very low in the 1920s and again between 1948 and 1956, Lake Victoria's lowest level was reached in 1922, while Lake Naivasha showed a very sharp progressive fall after 1938. The relatively dry phase which seems to have been the dominant reason for these low levels in the 1920s and following decades started to develop in the 1880s in the basins of Lakes Nyasa, Tanganyika, and Victoria but rather later (around 1898) in the Turkana and Chew Bahir basins. The lakes studied by the colonial scientists were thus very different from those described by the great explorers and may have contributed towards the

concept of progressive desiccation and **desertification** which caused concern in Africa between the wars.

In the 1960s the level of the lakes of East and Central Africa rose sharply, Lake Tanganyika stood about 3 m higher in 1964 than it had in 1960, Lake Victoria rose by over 1.5 m and rises of 2.3 m have been recorded for Lakes Baringo, Nakuru, and Manyara. Lake Turkana began to rise 4 m in late 1961 and submerged over 300 km² of the Omo Delta.

The Great Salt Lake in Utah, USA, has an especially impressive record of fluctuating levels dating back to the middle of the last century. The lake rose from an elevation of about 4200 feet in 1851 to a peak of around 4210 feet in 1873. Thereafter it declined very markedly,

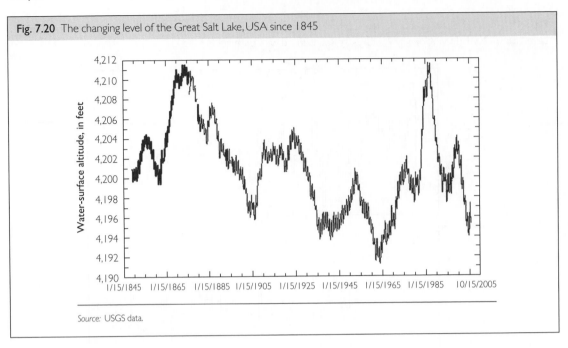

Fig. 7.20 The changing level of the Great Salt Lake, USA since 1845

Source: USGS data.

reaching the lowest recorded level in 1963. Since then it has risen from around 4194 feet in elevation back to the sort of elevation achieved in the 1870s (**Fig. 7.20**). There has thus been a total fluctuation of this great water body of the order of 5 m over that time period (Stockton 1990). The latest rise is a consequence of a run of very wet years since the mid 1970s.

One demonstration of the complex relationship between changes resulting from the activity of man, and changes resulting from natural causes can be seen in the recent history of the Caspian Sea (**Fig. 7.21**). In the 1930s this large inland sea fell by 2 m in level. This resulted largely because of a reduction in inflow from the River Volga, which, on average, contributes 80 per cent of the overall surface discharge to it. The causes of this phenomenon were both natural and man-induced. The climatic factor may well have been predominant, particularly in the earlier phases. In the years before 1929 the airflow over the European part of the former USSR was predominantly westerly but this changed to a chiefly meridional and easterly pattern during the 1930s and 1940s. As a result the number of depressions penetrating from the Atlantic dropped, whereas the frequency of dry anticyclones from the Arctic and Siberia went up fairly substantially during the winter season. However, reservoir formation, irrigation, and municipal and industrial withdrawals have also been of major importance. In the 1970s, 1980s, and 1990s the Caspian has seen a restoration in its levels, caused by a decrease in

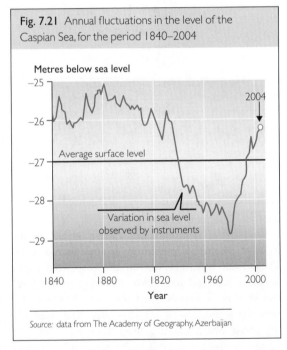

Fig. 7.21 Annual fluctuations in the level of the Caspian Sea, for the period 1840–2004

Source: data from The Academy of Geography, Azerbaijan

the difference between evaporation and precipitation, by a runoff decrease and, to a certain extent, the damming of the Kara Bogaz bay. In any event, an amelioration of climate since the late 1970s has caused some two metres of recovery in the level of the lake (Sapozhnikov et al. 2010). But for anthropogenic effects its level would have returned to pre-1930 levels (World Meteorological

Organization 1995, p. 124). Anthropogenic effects have been more severe in the case of the Aral Sea, which has suffered and continues to suffer severe desiccation. Between 1960 and 1990, largely because of diversions of river flow, the Aral Sea lost more than 40 per cent of its area and about 60 per cent of its volume, and its level fell by more than 14 metres. By 2002 its level had fallen another 6 metres. By 2008, the area of the Aral Sea was just 15.7% of that in 1961 (Kravstova and Tarasenko 2010; Micklin 2010) (**Fig. 7.22**).

From the early 1950s to the middle 1980s the total area of lakes in China with an individual area of over 1 square kilometre declined from 2800 to 2300 square kilometres and the whole area of China's lakes has been reduced from 80 600 square kilometres to 70 988 square kilometres (Lin and Fu 1996, pp. 133–152). An increasingly warm and dry climate was the principle cause of the reduced lake area on the Qingzang Plateau, Northwest China, the Inner Mongolian Plateau and the North China Plain.

Taking western China, Kazakhstan, and Kyrgyzstan, there has been an almost 50% loss in area of lakes between 1975 and 2007 (Bai et al. 2011), largely because of a decrease in precipitation.

7.7 Changing dust-storm frequencies

The changes in temperature and precipitation conditions in the twentieth century have had an influence on one further element of the environment: the development of dust storms. These are events in which visibility is reduced to less than one kilometre as a result of particulate matter, such as topsoil, being entrained by wind. This is a process that is most likely to happen when there are high winds and large soil moisture deficits. As with lake-level fluctuations it is not always easy to separate the importance of climatic change and human influence (especially vegetation removal and surface disturbance) in affecting the incidence of dust storms. Probably the greatest incidence of dust storms occurs when climatic conditions and human pressures combine to make surfaces susceptible to wind attack (Goudie and Middleton 1992). Some areas have shown increasing trends (e.g. the Sahel zone of Africa and the eastern Mediterranean), while others have shown declining trends in the late twentieth century (e.g. China and Australia) followed by a spike of activity in the early years of the present century (e.g. Mitchell et al. 2010).

Various twentieth-century examples serve to show how dust-storm incidence has been affected to a very marked degree. The first is from the High Plains of the USA in the 'Dust Bowl' years of the 1930s when 'black blizzards' caused great distress (**Fig. 7.23**). There was a great spike in the curve of dust-storm days at Dodge City, with over 100 a year occurring in the mid 1930s. Dust was transported over huge areas and distances, and was noted in New England. This era was precipitated by a spasm of sod-busting, caused by the

Fig. 7.22 The changing extent of the Aral Sea, 1960–2009 (from Micklin 2010, Fig. 5). Reprinted with permission of Wiley

Fig. 7.23 Frequency of dust storm days at Dodge City, Kansas (1922–1961) and for Big Spring, Texas (1953–1970) (modified after Gillette and Hanson 1989)

increasing use of mechanized agriculture in response to increasing wheat prices after the First World War, and to the increasing availability of the internal combustion engine to power trucks and tractors. In some areas, partly because of sounder land management practices, dust storm occurrence has declined in recent decades (**Fig. 7.24**).

The second example comes from the Sahel belt in West Africa since the late 1960s. Because of the great Sahel Drought (see **Section 5.10**) and the ever-increasing population levels in the area, dust storm incidence showed a massive increase in a great belt from Mauritania in the west to the Horn of Africa in the east. Many stations show a great increase in the 1970s and 1980s. The variation in frequency of annual dust-storm days and annual rainfall totals for Nouakchott (Mauritania) is shown in **Figure 7.25**. The increase in dust-storm days after 1968 is dramatic.

These trends (**Fig. 7.26a**) have been reflected in rising concentrations of Saharan dust monitored at Barbados since 1965 (**Fig. 7.26b**) (Chiapello et al. 2005). The Barbados dust concentrations are inversely related to the previous year's rainfall in Sahelian Africa (Prospero 1996), but winter transport to that island is related to the North Atlantic Oscillation (NAO) as well (Chiapello et al. 2005).

Observations in north-western Italy show an increasing trend of Saharan dust events since 1975 (Rogora et al. 2004). Data from the Mediterranean coast of Spain over the period 1949–1994 also showed a

marked increase in the number of dust-rain days since the 1970s (Sala et al. 1996). The long-term average there was approximately 2 dust-rain days per year, but from 1985 to 1994 the annual total averaged 6.5 dust-rain days, with 9.0 dust-rain days per year recorded for the period 1989–1994. In Mallorca, there has also been an increase in dust rains over the period from 1981 to 2003, except for a decrease between 1991 and 1996 (Fiol et al. 2005). Dessens and Van Dinh (1990) noted a marked increase in the frequency of Saharan dust outbreaks in southern France, over the period 1983–1989. Similarly, a significant increase in the quantities of Saharan dust falling over the French Alps since the early 1970s, with very high inputs occurring after 1980 (De Angelis and Gaudichet 1991), was detected from an **ice core** that yielded dust deposition data over a 30-year period (1955–85).

Additional evidence for recent increasing Saharan dust-raising activity comes from the eastern Mediterranean, where in the 1970s some 25 million tonnes of Saharan dust reached the East Mediterranean Basin each year, most settling into the Mediterranean Sea, rising more recently to 100 million tonnes per year (Ganor and Foner 1996). The increase reflects the steady rise in frequency of Saharan dust episodes over Tel Aviv (Israel) from 10 per year in 1958 to 19 per year in 1991 (Ganor 1994). Variability in Sahel dust emissions may be related not only to droughts, but also to an increase in the frequency by which wind exceeds the threshold wind velocity for dust entrainment. It also seems to be related to changes in the North Atlantic Oscillation (Engelstaedter et al. 2006), North Atlantic sea surface temperatures (Wong et al. 2008), and the Atlantic Multidecadal Oscillation (Foltz and McPhaden 2008; Jilbert et al. 2010). The steadily increasing trend of dust storms from Africa in the Eastern Mediterranean over the last five decades in association with changing synoptic conditions has been analysed by Ganor et al. (2010). The slope of the increase is 0.27 days per year.

A third example of changing dust storm frequencies comes from east Asia. In Mongolia, Natsagdorj et al. (2003) analysed data for the period 1960 to 1999 and identified an increasing trend from the 1960s to the 1980s, with an approximately threefold increase over that period, followed by a downward trend in the 1990s (**Fig. 7.27a**). They believe that human activities accounted for the first of these two phases, but that an increase of precipitation may have caused the reversal in trend during the latter phase.

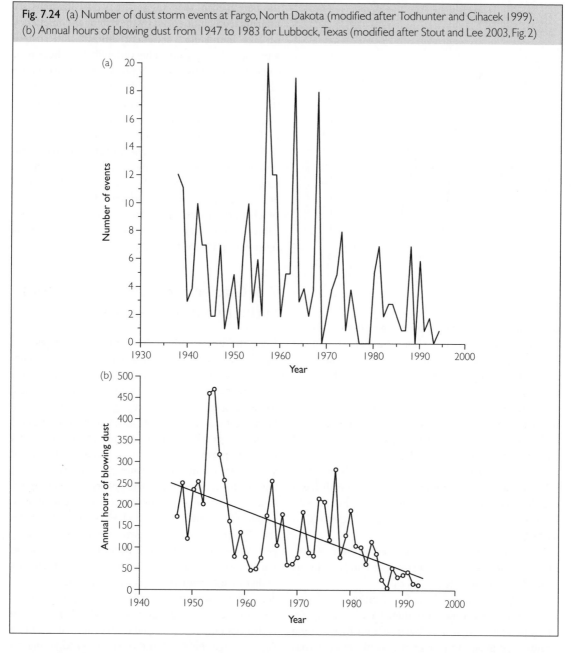

Fig. 7.24 (a) Number of dust storm events at Fargo, North Dakota (modified after Todhunter and Cihacek 1999). (b) Annual hours of blowing dust from 1947 to 1983 for Lubbock, Texas (modified after Stout and Lee 2003, Fig. 2)

Zhou and Zhang (2003) and Wang et al. (2004b) analysed the frequency of severe dust storms for the period since 1954 in China. They found that the highest frequency of such events occurred in the 1950s, but was lowest in the 1990s (**Fig. 7.27b**). Similarly Qian et al. (2002) found high levels of dust activity in the 1950s and a steady decline at Beijing and Baotou thereafter. They suggested that in the 1950s–1970s dust storms were twice as prevalent as they were from the mid 1980s.

They attribute this to a reduced meridional temperature gradient, resulting in reduced cyclone frequency in Northern China (see also Ding et al. 2005) On the other hand, Parungo et al. (1994) attributed the negative trend in dust storm frequency to the planting of a vast belt of forests – 'The Great Green Wall' – across the northern arid lands of China. They asserted that when in the 1980s and 1990s dust storms were rare in Beijing there were not statistically significant changes in wind speed

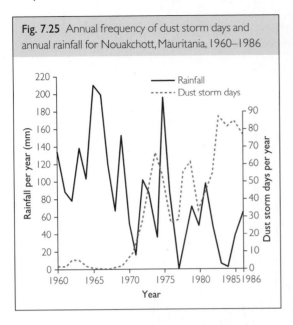

Fig. 7.25 Annual frequency of dust storm days and annual rainfall for Nouakchott, Mauritania, 1960–1986

the period 1980–2000 suggest that dust loadings are intermittently influenced by the extent and frequency of lake inundation, sediment inflows, and surface wind speed variability. The variability of these is in turn influenced by the El Niño–Southern Oscillation and Indian Ocean sea surface temperature anomalies (Bryant et al. 2007). Lake flooding and desiccation is also an important control on dust storm frequencies in the Seistan Basin of Iran (Miri et al. 2010).

Further information on changing global dust storm frequencies is presented by Goudie and Middleton (2006). Using a variety of data sources, Mahowald et al. (2010) have tried to estimate the global picture of changes in dust storm activity for the twentieth century. They suggest a doubling of desert dust took place over much of the globe.

or precipitation. An analysis by Wang et al. (2004b) confirmed that for China as a whole, the highest frequencies of dust storms occurred in the 1960s and 1970s, though they recognized that in some regions they were increasing, and they attributed this to localized desertification brought about by human pressures on the land. In the early twenty-first century reduced precipitation and a concomitant decrease in vegetation cover caused a resurgence of dust events (Zhou and Zhai 2004; Yang et al. 2008; Kurosaki et al. 2011). Links have been established between dust emissions from the Tarim basin and the Arctic Oscillation (AO) Index, with dust activity being high during the negative phase of the AO (Gao and Washington 2010; Mao et al. 2011).

There also appears to have been greater atmospheric instability, leading to stronger winds and thus more dust storms (Gao et al. 2003).

Recent research in China has demonstrated that over the last five decades or so there have been changes in wind activity that have been related to warming trends. The general message is that as warming has occurred wind velocities have fallen (e.g. Xu et al. 2006; Wang et al. 2007). It has also been suggested that as a consequence dunes in northern China have become more stable (Wang et al. 2007).

Dust trends in Australia are shown in **Fig. 7.28**. There was a high incidence in the early 1960s and a generally low incidence in the 1970s and 1980s.

In southern Africa fluctuations in the emission of dust from the Makgadikgadi Pans of the Kalahari over

7.8 Glacier and sea ice fluctuations in the twentieth century

Since the nineteenth century, many of the world's alpine glaciers have retreated up their valleys as a consequence of the climatic changes, especially warming, that have occurred in the last hundred or so years since the ending of the Little Ice Age (Oerlemans 1994). Studies of the changes of snout positions obtained from cartographic, photogrammetric and other data therefore permit estimates to be made of the rate at which retreat can occur (**Plates 7.1 and 7.2**). The rate has not been constant, or the process uninterrupted. Indeed, some glaciers have shown a tendency to advance for some of the period. However, if one takes those glaciers that have shown a tendency for a fairly general retreat (**Table 7.3**) it becomes evident that as with most geomorphological phenomena there is a wide range of values, the variability of which is probably related to such variables as topography, slope, size, altitude, accumulation rate, and ablation rate. It is also evident, however, that rates of retreat can often be very high, being of the order of 20 to 70 metres per year over extended periods of some decades in the case of the more active examples. It is therefore not unusual to find that over the last hundred or so years alpine glaciers in many areas have managed to retreat by some kilometres.

A global analysis of the state of the world's glaciers is provided by the World Glacier Monitoring Service (WGMS) (2008) (**Fig. 7.29**). Some glaciers at high-altitude locations in the tropics have displayed fast

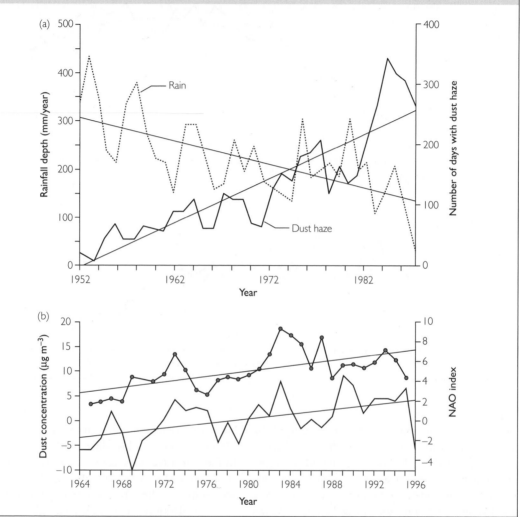

Fig. 7.26 (a) The relationship between annual dust haze days and annual rainfall for the Sahelian station of Gao between 1952 and 1987 (modified after N'Tchayi et al. 1994, Fig. 6). (b) Comparison of the North Atlantic Oscillation (NAO) index (bold continuous line) and dust concentrations at Barbados in the West Indies (modified after Moulin et al. 1997, Fig. 4)

rates of decay over the last hundred or so years, to the extent that some of them now only have around one-sixth to one-third of the area they had at the end of the nineteenth century (Kaser 1999). In the mountains of New Guinea nearly all the ice caps have disappeared since the late nineteenth century and the cessation of the Little Ice Age (LIA). In East Africa, the ice bodies of Mount Kilimanjaro have shrunk from *c.* 20 km² in 1880 to *c.* 2.5 km² in 2003. At present rates of retreat the glaciers and ice caps of Mount Kilimanjaro will have

disappeared by 2020. The reasons for the retreat of the East African glaciers include not only an increase in temperature, but also a relatively dry phase since the end of the nineteenth century (which led to less accumulation of snow) and a reduction in cloud cover (Mölg et al. 2003). New Zealand glaciers lost between one-quarter and almost half of their volume between the LIA maximum and the 1970s. A further net ice volume loss between 1977 and 2005 has also been reported. In the Pyrenees, since the first half of the nineteenth

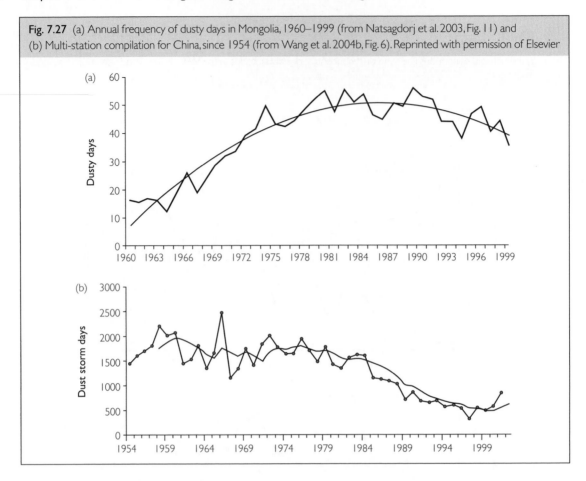

Fig. 7.27 (a) Annual frequency of dusty days in Mongolia, 1960–1999 (from Natsagdorj et al. 2003, Fig. 11) and (b) Multi-station compilation for China, since 1954 (from Wang et al. 2004b, Fig. 6). Reprinted with permission of Elsevier

century, about two-thirds of the ice-cover has been lost, while in South America, except for a few cases in Patagonia and Tierra del Fuego, there has been a general retreat, with enhanced rates in recent decades. In the tropical Andes, the Quelccaya ice cap in Peru also has an accelerating and drastic loss over the last 40 years (Thompson 2000). In neighbouring Bolivia, the Chacaltaya glacier lost no less than 40 per cent of its average thickness, two-thirds of its volume and more than 40 per cent of its surface area between 1992 and 1998. Complete extinction was expected within 10 to 15 years (Ramirez et al. 2001) and indeed occurred, earlier than expected, in 2009. In China, the overall glacier loss is estimated at 20% since the LIA maximum. Over the twentieth century glacier area is estimated to have decreased by 20–25% in the Tien Shan, by 30–35% in the Pamirs and by c. 50% in northern Afghanistan. In North America glaciers have shown a general retreat after the LIA maximum, particularly at lower elevations and southern latitudes. In the Rocky Mountains the area lost since the LIA is about 25%. In the Sierra Nevadas, ice volume loss from 1903 to 2004 averaged no less than 55% and showed a rapid acceleration in the early 2000s (Basagic and Fountain 2011). In Alaska (Arendt et al. 2002), glaciers appear to be thinning at an accelerating rate, which in the late 1990s amounted to 1.8 metres per year. In the European Alps the glacier cover is estimated to have diminished by about 35% from the 1850s to the 1970s and by another 22% by 2000 (see also Fitzharris 1996, p. 246), though there were interruptions observed in the mid 1960s, late 1970s, and early 1980s. Globally the WGMS suggests that the average annual mass loss of glaciers during the decade from 1996 to 2005 was twice that of the previous decade (1986–1995) and over four times that of the decade from 1976–1985. The change in global glacier mass balance since 1945 is shown in **Fig. 7.30**.

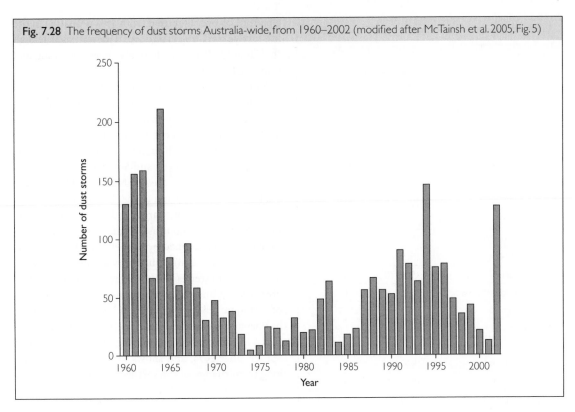

Fig. 7.28 The frequency of dust storms Australia-wide, from 1960–2002 (modified after McTainsh et al. 2005, Fig. 5)

Not all glaciers have retreated in recent decades. In Scandinavia, glaciers close to the sea have seen a very strong mass gain since the 1970s, but mass losses have occurred with the more continental glaciers. The positive mass balance (and advance) of some Scandinavian glaciers in recent decades, notwithstanding rising temperatures, has been attributed to increased storm activity and precipitation inputs coincident with a high index of the North Atlantic Oscillation (NAO) in winter months since 1980 (Nesje et al. 2000; Zeeberg and Forman 2001). In the case of Nigardsbreen (Norway), there is a strong correlation between mass balance and the NAO index (Reichert et al. 2001). A positive mass balance phase in the Austrian Alps between 1965 and 1981 has been correlated with a negative NAO index. Indeed, the mass balances of glaciers in the north and south of Europe are inversely correlated (Six et al. 2001).

However, since 2000, the Norwegian glaciers, which were still advancing in the 1990s, have started to retreat at a rapid rate. If this continues, then by 2100 about 98% of Norwegian glaciers are likely to disappear and the glacier area may be reduced by around a third (Nesje et al. 2008).

The recent response of Himalayan glaciers shows great regional variations depending on such factors as the amount of debris cover that occurs on glacier surfaces. In the west of the region, around half of the westerlies-influenced Karakoram glaciers have been advancing or have been stable (Hewitt 2005), while the great bulk of the monsoon-influenced glaciers in the central Himalaya have been retreating, especially those with a limited debris cover (Scherler et al. 2011).

Glaciers that calve into water can show especially fast rates of retreat. Indeed, retreat rates of greater than 1 kilometre per year are possible (Venteris 1999). In Patagonia, in the 1990s, retreat rates of up to 500 metres per year were observed. This rapid retreat is accomplished by iceberg calving with icebergs detaching from glacier termini when the ice connection is no longer able to resist the upward force of flotation and/or the downward force of gravity. The rapid retreat is favoured by the thinning of ice near the termini, its flotation and its weakening by bottom crevasses. The Columbia tidewater glacier in Alaska retreated around 13 kilometres between 1982 and 2000. Equally the Mendenhall Glacier, which calves into a proglacial lake, has displayed rapid rates of retreat, with

Table 7.3 Retreat of glaciers in metres per year in the twentieth century

Location	Period	Rate
Breidamerkurjökull, Iceland	1903–48	30–40
	1945–65	53–62
	1965–80	48–70
Lemon Creek, Alaska	1902–19	4.4
	1919–29	7.5
	1929–48	32.9
	1948–58	37.5
Humo Glacier, Argentina	1914–82	60.4
Franz Josef, New Zealand	1909–65	40.2
Nigardsbreen, Norway	1900–70	26.1
Austersdalbreen, Norway	-	21
Abrekkbreen, Norway	-	17.7
Brikdalbreen, Norway	-	11.4
Tunsbergdalsbreen, Norway	-	11.4
Argentière, Mont Blanc	1900–70	12.1
Bossons, Mont Blanc	1900–70	6.4
Oztal Group, Austria	1910–80	3.6–12.9
Grosser Aletsch, Switzerland	1900–80	52.5
Carstensz, New Guinea	1936–74	26.2

Source: tables, maps and text in Grove (1988).

Location	Period	Mean Rate
Rocky Mts.	1890–1974	15.2
Spitzbergen	1906–1990	51.7
Iceland	1850–1965	12.2
Norway	1850–1990	28.7
Alps	1850–1988	15.6
Central Asia	1874–1980	9.9
Irian Jaya	1936–1990	25.9
Kenya	1893–1987	4.8
New Zealand	1894–1990	25.9

Source: Oerlemans (1994).

3 kilometres of terminus retreat in the twentieth century (Motyka et al. 2002). Certainly, calving permits much larger volumes of ice to be lost to the glacier than would be possible through surface ablation (van der Ween 2002). The glaciers of the Antarctic Peninsula, all of which calve into the sea, also show

a pattern of overall retreat, with 87% retreating in the last fifty years (Cook et al. 2005). Tidewater glaciers on the Antarctic Peninsula have also been retreating and thinning fast (Pritchard and Vaughan 2007).

Since the 1970s, sea ice in the Arctic has declined in area by about 30% per decade (Vinnikov et al. 1999; Stocker 2001; Comiso 2002). Its thickness may also have reduced, with a near 40% decrease in its summer minimum thickness, though the data to ascertain this are imperfect (Holloway and Sou 2002). **Figure 7.31a** shows the decrease in area of Northern Hemisphere sea ice between 1979 and 2010. The rate of decline is −2.7% per decade for March, and −11.6% per decade for September. By contrast, changes in Antarctic sea ice have been described by the IPCC as 'insignificant' (Stocker 2001, p. 446). It has proved difficult to identify long-term trends because of the limited length of observations and the inherent inter-annual variability of Antarctic sea ice extent, but it is possible that there has been some decline since the 1950s (Curran et al. 2003). **Figure 7.31b** illustrates the extent of Antarctic sea ice since 1979, and shows little evidence for any long-term change of the sort seen in the Arctic.

7.9 The future

There are some key areas where humans may be capable of making significant changes to global climates. The most important categories of influence are in terms of the chemical composition of the atmosphere and the albedo of the Earth's surface, though, as the following list shows, there are some others that also need to be considered.

Possible mechanisms
Gas emissions
   CO_2 – industrial and agricultural
   methane
   chlorofluorocarbons (CFCs)
   nitrous oxide
   water vapour
   miscellaneous trace gases
Aerosol generation
   dust, smoke, etc.
Thermal pollution
Albedo change
   dust addition to ice caps
   deforestation
   overgrazing

Fig. 7.29 Historical glacier length changes. The number of advancing (dark tone) and retreating (light tone) glaciers are plotted as stacked columns in the corresponding survey year

Source: World Glacier Monitoring Service, 2008, Fig. 5.1.

Miscellaneous

extension of irrigation

alteration of ocean currents by constricting straits

diversion of fresh waters into oceans, affecting **thermohaline circulation system**

impoundment of large reservoirs

That warming of the Earth's climate has taken place over the last four to five decades is unequivocal, and the IPCC (2007, p. 10) asserts that 'Most of the observed increase in global average temperatures since the mid twentieth century is *very likely* due to the observed increase in atmospheric greenhouse concentrations'.

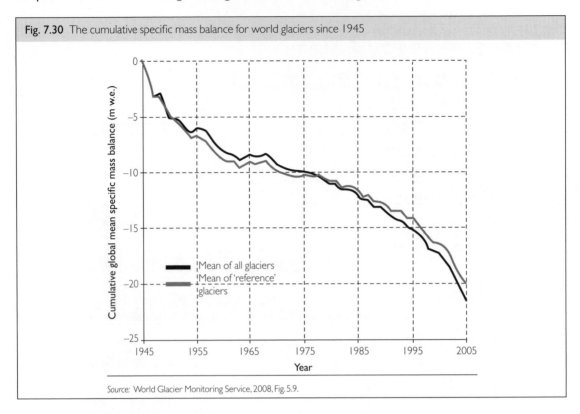

Fig. 7.30 The cumulative specific mass balance for world glaciers since 1945

Mean of all glaciers
Mean of 'reference' glaciers

Source: World Glacier Monitoring Service, 2008, Fig. 5.9.

Humans have been enhancing the greenhouse effect by increasing carbon dioxide and methane levels in the atmosphere. This may have started with deforestation in the Holocene (Ruddiman 2003) but has accelerated in recent centuries. Since the beginning of the industrial revolution humans have been taking stored carbon out of the earth in the form of coal, petroleum and natural gas, and burning it to make carbon dioxide (CO_2), heat, water vapour and smaller amounts of sulphur dioxide (SO_2) and other gases. The prime cause of increased carbon dioxide emissions is fossil fuel combustion and cement production (*c.* 0.5–5.5 GtC yr^{-1} in the 1980s), but with the release of carbon dioxide by changes in tropical land use (primarily deforestation) being a significant factor (*c.* 1.0–1.6 GtC yr^{-1}).

The pre-industrial level of carbon dioxide may have been as low as 260–70 parts per million by volume (ppmv). The present level is over around 380 ppmv, and at the present rate it would reach 500 ppmv by the end of the twenty-first century.

In addition to carbon dioxide, it is probable that a cocktail of other gases will contribute to the greenhouse. Individually their effects may be minor, but as a group they may be major. Indeed, molecule for molecule some of them may be much more effective as greenhouse gases than CO_2.

One of the more important of the trace gases is methane (CH_4). Ice core studies and recent direct observations suggest that until the beginning of the industrial revolution levels were relatively stable at around 600 parts per billion by volume (ppbv) although they may have been increased prior to that by rice farming and other agricultural activities (Ruddiman and Thomsen 2001). They rose steadily between AD 1700 and 1900, and then increased still more rapidly, attaining levels that averaged 1300 ppbv in the early 1950s and 1600 ppbv by the mid 1980s (Khalil and Rasmussen 1987) and over 1774 ppbv in 2005. This increase of 2.5 times over background levels results primarily from increased rice cultivation in waterlogged paddy fields, the enteric fermentation produced in the growing numbers of flatulent and belching domestic cattle, and the burning of oil and natural gas. In some countries, including the UK, emissions have recently declined.

Since the Second World War, chlorofluorocarbons (CFCs), used as refrigerants, foam makers, fire control agents and propellants in aerosol cans, despite their relatively trace amounts in the atmosphere, increased

Fig. 7.31 (a) Average monthly sea ice extent in the Arctic in February since 1979. (b) The changing extent of Antarctic sea ice for September (the month of maximum ice extent) and February (the month of minimum ice extent)

Source: NOAA data.

very markedly in terms of their emissions and their concentrations until production was curtailed through the Montreal Protocol. They have a very strong greenhouse effect even in relatively small amounts and also have a long residence time. Nitrous oxide (N_2O) can also contribute to the greenhouse effect, primarily by absorption of infrared at the 7.8 and 17 μm bands. Combustion of hydrocarbon fuels, the use of ammonia-based fertilizers, deforestation and biomass burning are among the processes that lead to an increase in atmospheric N₂O levels. Atmospheric N_2O concentrations increased from around 275 ppbv in pre-industrial times to 319 ppbv in 2005.

Human activities have also increased atmospheric aerosol loadings. One of these is industrial emission of smoke and dust particles. Industrialization is not, however, the sole source of particles in the atmosphere. Intensive agricultural exploitation of desert margins could create a dust pall in the atmosphere by exposing larger areas of surface materials to deflation in dust storms. This pall could so change atmospheric temperature that convection, and thus rainfall, would be reduced.

Dust loadings affect air temperatures through the absorption and scattering of solar radiation (Durant et al. 2009), modify cloud formation (Toon 2003) and convectional activity, influence sulphur dioxide levels in the atmosphere, either by physical absorption or by heterogeneous reactions (Adams et al. 2005), and influence marine primary productivity and thus atmospheric carbon dioxide levels (Ridgwell 2003).

It is possible that aircraft contrails and the development of thin cirrus clouds could lead to warming of the Earth's surface. Observations of cirrus cloud cover over the last 50 years near to major USA flight corridors show a clear upward trend that seems to correlate with US domestic jet fuel trends over the same period (Yang et al. 2010).

There are an increasing number of modelling experiments that suggest vegetation removal can have important regional and even global effects, though there are considerable divergences between different models (Nobre et al. 2004). Nonetheless, most show *decreases* in mean evapotranspiration between 25.5 and 985.0 mm per year, *increases* in mean surface temperatures of 0.1 to 3.8 °C, and *reductions* in regional precipitation (Sampaio et al. 2007). It is even possible that the effects of Amazonian deforestation on precipitation could extend some distance away from there, to the Dakotas and the Midwest Triangle in the USA, as modelling by Werth and Avissar (2002) has shown. Similarly, the clearing of native vegetation in eastern Australia seems to have led to an increase in the number of dry and hot days, thereby intensifying drought conditions (Deo et al. 2009).

Albedo effects may occur in higher latitudes as well. As Betts (2000) has pointed out, in a snowy environment, forests are generally darker than open land, because trees generally remain exposed when cultivated land can become entirely snow-covered.

Snow-free foliage is darker than snow. This means that forest has a smaller surface albedo and so may exert a warming influence. Thus land cover changes in the boreal forest zone can have substantial climatic implications. Betts (2001) suggested that mid-latitude agricultural regions would be approximately 1–2 K cooler in winter and spring in comparison with their previously forested state, due to deforestation increasing the surface albedo during periods of snow cover.

The Intergovernmental Panel on Climate Change (IPCC 2007, p. 30) concluded that 'The impacts of land use change on climate are expected to be locally significant in some regions, but are small at the global scale in comparison with greenhouse gas warming'.

As regards future trends in climatic conditions, the IPCC (2007) suggests that global average surface temperatures will rise by 1.8 to 7.1 °C compared to pre-industrial levels by 2100, with a best estimate of 2.5 to 4.7 °C. This represents a much larger rate of warming than that observed over the twentieth century and, based on palaeoclimatic data, is very likely to surpass anything experienced over the last 10 000 years. In its turn, increased temperatures will transform the behaviour of the atmospheric heat engine that drives the general circulation of the air and the oceans, leading to changes in the nature, pattern and amount of precipitation. In particular it is likely that northern high latitudes will warm more rapidly than the global average, probably by more than 40 per cent. There may also be marked changes in precipitation, with a tendency for increases to occur globally, but particularly over northern mid to high latitudes and Antarctica in winter. At lower latitudes some areas will see reductions in precipitation (which will exacerbate soil moisture losses caused by increased evapotranspiration), while others may become wetter. In general climate models predict that precipitation will generally increase in areas with already high amounts of precipitation and generally decrease in areas with low amounts of precipitation (New et al. 2011). Rainfall intensity is a major factor in controlling such phenomena as flooding, rates of soil erosion, and mass movements. Under increased greenhouse gas concentrations some GCMs exhibit enhanced mid-latitude and global precipitation intensity and shortened return periods of extreme events (Jones and Reid 2001; New et al. 2001). (See **Section 1.6** for a discussion of environmental modelling and GCMs.) The IPCC (2007, p. 91) suggests that 'Extremes of daily precipitation are *likely* to increase in many regions. The increase is considered to be *very likely* in northern Europe, south Asia, East Asia, Australia and New Zealand'. There are various reasons to expect increases in extreme precipitation if and when significant warming occurs. There will be more moisture in the atmosphere, and there is likely to be greater thermodynamic instability (Kunkel 2003).

Droughts may become more prevalent (Dai 2011a, b). The models of Sylla et al. (2010) suggest that this will be so in the West African Sahel, partly because of a weaker monsoon flow and reduced easterly wave activity. While it is likely that many areas that are currently dry may see enhanced aridity because of reductions in precipitation, even within an area such as the Middle East, some parts may become wetter and some may become drier. The Saudi Arabian desert may, for instance, become wetter because of a more northerly intrusion by the **Intertropical Convergence Zone (ITCZ)** (Evans 2010). There is a wide consensus that the south-western USA will become more arid (Seager and Vecchi 2010). Zeng and Yoon (2009) have suggested that as conditions become drier and vegetation cover is reduced, there may be vegetation-albedo feedbacks which will serve to enhance any aridity trend. By 2099 their model suggests that globally the warm desert area may expand by 8.5 million km² or 34%. Munson et al. (2011) also argue that with predicted increases in droughts in the western USA, perennial vegetation cover will be reduced and **aeolian** activity will be increased.

As we saw earlier in this chapter (**Section 7.3**) the relationship between sea surface temperature changes and changes in hurricane activity has been controversial. However, the IPCC report (2007, p. 15) concludes thus on tropical cyclones:

Based on a range of models, it is *likely* that future tropical cyclones (typhoons and hurricanes) will become more intense, with larger peak wind speeds and more heavy precipitation associated with ongoing increases in tropical sea surface temperatures.

 SELECTED READING

Climatic changes during the period of meteorological observations (and earlier) have been discussed by H.H. Lamb (1977) *Climate: Present, Past and Future* and (1996) *Climate, History and the Modern World* (2nd edition).

A valuable series of essays, which takes a worldwide perspective, is: S. Gregory (1988) (ed.) *Recent climatic change*.

Possibly the most authoritative treatment of the question of climatic changes of the last century or so is provided by: J.T. Houghton et al. (eds.) (2001) *Climatic Change 2001: The Scientific Basis*.

An update on the views of the IPCC is provided in: Trenberth, K.E. and 11 others (2007) Observations: surface and atmospheric climate change. In *Climate Change 2007: The Physical Science Basis*, S. Solomon and 7 others (eds.), pp. 235–336.

Recent glacial fluctuations in low latitudes are discussed by G. Kaser and H. Osmaston (2002) *Tropical Glaciers*, and a more general treatment of the response of glaciers to climate change is: A. Nesje and S.O. Dahl (2000) *Glaciers and Environmental Change*.

The World Glacier Monitoring Service (WGMS) website http://www.geo.uzh.ch/microsite/wgms/ hosts a great deal of information and data about recent glacier fluctuations.

A good review of future climate changes is provided by: J.T. Houghton (2009) *Global Warming: The Complete Briefing*.

A comprehensive overview of climatic changes in the low latitudes is given by: S. Nicholson (2011) *Dryland Climatology*. Cambridge University Press.

? REVIEW QUESTIONS

1. What were the main consequences of the warming phase in the early twentieth century?

2. How have global temperatures changed since the 1970s and what have been the main environmental consequences?

3. Assess the importance and nature of climatic fluctuations such as ENSO during the twentieth century.

4. What have been the causes and characteristics of the Sahel Drought in West Africa since the mid 1960s?

5. Consider the reasons why there are likely to be regional differences across the Earth in how the climate changes over the next hundred years.

8

Links between Environmental Change and Human Evolution and Society

Chapter contents

Chapter overview

The environmental changes described in the foregoing chapters have had a marked impact on human evolution and on societies. They have contributed to our evolution, to our dispersal across the face of the Earth, and to certain key developments in our history, including the development of agriculture and the rise of urban societies. Severe climate events, such as major droughts or extended cold phases, have presented severe challenges. However, the other side of the coin is that, increasingly, humans have been having an impact on environmental change.

8.1 Earliest hominins

The earliest fossil **hominins** date to the Miocene Epoch *c.* 7 to 6 million years ago (Myr) and show a varied mix of features, some similar to primates and others to humans. **Fig. 8.1** shows the primate phylogeny (family tree). Hominins, or human species, began to evolve from the **hominids**, or African Apes (Toth and Schick 2005). Comparison of blood and DNA shows that chimpanzees and humans last shared a common ancestor around 8 to 6 Myr. During the Miocene the Earth's temperature cooled (see **Section 1.4** on the

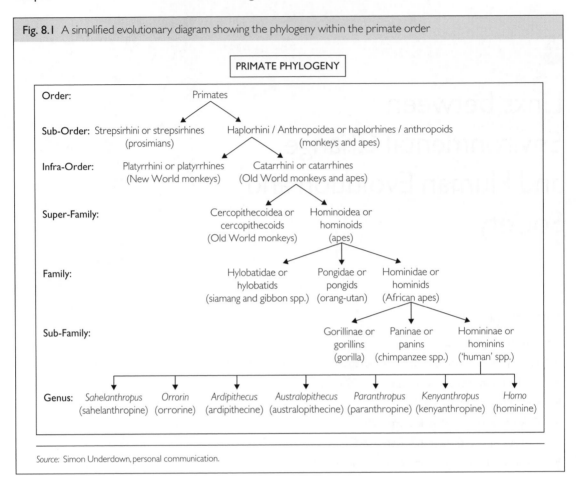

Fig. 8.1 A simplified evolutionary diagram showing the phylogeny within the primate order

Source: Simon Underdown, personal communication.

Cainozoic climatic decline), and in the African trop-ics rainfall decreased and woodlands and grasslands expanded. It has been suggested that the changes in climate and vegetation forced some primates down from their tree habitats to occupy the savannah grasslands and the sparser woodlands, where they moved around bipedally (on two feet instead of four). Contemporary hominins (humans) share a number of derived traits that make them unique from their hom-inoid and some hominin ancestors, such as bipedal locomotion, a more parabolic jaw arrangement with smaller canines and thick enamel, a very large brain with regard to body size, slow juvenile development, and a dependence on culture and language. The early hominins, however, looked and probably behaved more like apes. Their lineage is shown in **Fig. 8.2**.

Africa provides the setting for the study of human evolution with the majority of remains being found within sediments from the Rift Valley of East Africa, or

in cave deposits in South Africa. The earliest remains are from small-brained, bipedal hominins, which have been identified from Chad, Kenya, and Ethiopia. The oldest recognized is *Sahelanthropus tchadensis* found in the Djurab Desert of Chad. This is the westernmost early hominin site in Africa and the hominin lived in a savannah environment around 7–6 Myr (Brunet et al. 2002). At 6.0 Myr *Orrorin tugenensis* also dates to the time when apes and humans diverged. Remains of this fossil hominin were found in the Tugen Hills in the highlands of Kenya and indicate that it possessed small teeth with thick enamel like later hominids. The femora show evidence of bipedal locomotion. Associated faunal remains from the same strata indi-cate that *O. tugenensis* lived in a mixed woodland and savannah landscape (Senut et al. 2001). In the Middle Awash basin in Ethiopia two species of *Ardipithecus* have been recognized namely *A. kadabba* and *A. rami-dus*. *A. kadabba* is a distinct species from *A. ramidus* and

Fig. 8.2 Summary diagram of important palaeoclimatic and hominid evolution events during the Pliocene–Pleistocene over the past 4.5 Myr (A. = *Australopithecus*). Palaeoclimatic records indicate changes in the dominant orbital signals from precessional driven (23–19 kyr) before c. 2.8 Myr, to obliquity-driven (41 kyr) global glacial variability from c. 2.8 to c. 1.0 Myr, to eccentricity-driven (100 kyr) glacial-interglacial cycles at c. 1.0 Myr. During the Plio-Pleistocene the African climate became progressively more arid with punctuated humid phases. These events are associated with changes toward dry-adapted African faunal compositions, including important steps in hominid speciation, adaptation, and behaviour. Soil carbonate carbon isotopic data from East African hominid fossil localities document the Pliocene–Pleistocene progressive shifts from closed woodland forest C₃-pathway vegetation to arid-adapted C₄-pathway savannah grassland vegetation (from deMenocal 2004, Fig. 11). Reprinted with permission of Elsevier.

is only known in the fossil record from a few post-cranial bones and sets of teeth (Haile-Sallasie et al. 2004). *A. kadabba* has been dated to 5.8–5.2 Myr. Remains of *A. ramidus* were found sandwiched between two volcanic layers dated to 4.4 Myr (White et al. 1994). This species shows evidence for bipedalism and the teeth of *A. ramidus* are also more human-like, with a reduced canine, but their molar teeth are more ape-like. Carbon isotope analysis of the tooth enamel of *A. ramidus*, as well as from other large mammals associated with the *Ardipithecus* bearing sediments, suggest they lived in C_3 dominated woodland (White et al. 2009). The anatomy and behaviour of *A. ramidus* indicates that early hominids originated and persisted in more closed, wooded habitats. It should be noted however, that C_4 resources were available but that *A. ramidus* focused almost exclusively on C_3 foodstuffs. It was not until the emergence of *Australopithecus* that we see a change in ecological conditions with adaptive responses associated with more open, savannah grassland conditions with a strong C_4 floristic signal.

From 4.0 Myr there is better representation of hominins within the fossil record. Over the next 2 million years there are least four to seven species of hominids present at any given time. The most well known of these belong to the Australopithecines, which appear in the fossil record over the next three million years. The earliest discovery of *Australopithecus*, meaning 'southern ape', was made by Raymond Dart in 1924 after the discovery of an early hominin, *Australopithecus africanus*, in cave breccia at Taung in South Africa. For many years the notion that the 'Taung Child' was a human precursor was contested. The australopithecines were small bipeds, with shorter legs and longer arms than modern humans. Since the first discovery of *Australopithecus africanus* a number of other species of *Australopithecus* have been recognized within Africa. The most famous specimen, known as Lucy, belongs to the species *A. afarensis* (Johansen and Edey 1981). It occupied East Africa between 4.0 and 3.0 Myr with notable sites at Hadar in the Ethiopian Rift and Laetoli in Tanzania. Also from the latter site fossil footprints, dated to *c.* 3.6 Myr, were found preserved in volcanic ash and support the evidence for bipedalism within hominins at this time. To date a number of species have been identified: *A. anamensis, A. afarensis, A. africanus,* and *A. garhi*. However, it should be noted that some workers place *Homo habilis* and *H. rudolfensis* in the same genus as *Australopithecus*. These latter two forms are present from *c.* 2.4 Myr – around

the time stone tools begin to appear in the archaeological record. This is represented by a technology known as the Oldowan tool industry. Oldowan tools were simple, hand-held tools with a sharpened edge. *Australopithecus sediba* from Malapa, South Africa existed *c.* 2.0 Myr. A mix of australopithecine and *Homo*-like traits make *Australopithecus sediba*, or a species very similar to it, a possible direct ancestor of our own genus, *Homo* (Berger et al. 2010).

The genus *Paranthropus* includes three known species, namely: *P. robustus, P. aethiopicus,* and *P. boisei*. From the neck down *Paranthropus* is similar to the australopithecines. However, it differs by having jaws and large teeth adapted to heavy chewing of tough plant materials. As a result it has been referred to as 'Nutcracker Man'. The genus *Kenyanthropus* contains one species, *K. platyops,* and has been found on the west side of Lake Turkana, Kenya. This is a relatively new fossil find, dated between 3.5 and 3.2 Myr, and the features of this hominin are unique when compared with both australopithecines and paranthropines, including its comparably small molars and a broad face.

Isotopic evidence from tooth enamel of australopiths demonstrates a significant difference in diet from earlier hominins such as *A. ramidus,* which was not only a denizen of woodland but also focussed on an exclusive C_3 diet. Sometime between 3.0 and 1.8 million years ago, the diets of *Australopithicus africanus* from South Africa and *Paranthropous boisei* from East Africa demonstrate important changes in food consumption with a shift to foodstuffs originally fixed by C_4 photosynthesis, including tropical/savannah grasses, sedges and/or animals eating C_4 foods (Lee-Thorp et al. 2010). Lee-Thorp et al. (2010, p. 3389) argue 'evidence for engagement with C_4 food resources may mark a fundamental transition in the evolution of human lineages'. **Figure 8.3** shows the changes in woodland versus more open grassland mammal taxa from the Omo Basin, Ethiopia, between 4.0 and 2.0 Myr (Bobe et al. 2002).

8.2 Pleistocene environments and hominin speciation

The importance of changes in climate and landscape in hominin evolutionary processes since the Miocene have resulted in intense academic debate. The analysis of palaeoclimate, geomorphological, and tectonic

Fig. 8.3 Proportion of faunal abundances during the Plio-Pleistocene showing the change from closed woodland and forest dominated mammal taxa versus woodland and bushy grassland taxa from the Omo Basin, Ethiopia. The decline in woodland taxa begins c. 3.4 Myr and falls steadily to 2.3 Myr. Woodland and bushy grassland taxa increase between 2.8 Myr and 2.4 Myr (after Bobe et al. 2002, Fig. 5). Reprinted with permission of Elsevier

records track changes in the landscape evolution of East Africa during the past 5 million years. As cooling in the Cainozoic proceeded, records show a gradual retreat of equatorial rainforests and the emergence of drier savannah grassland in East Africa (**Fig. 8.2**) (deMenocal 2004). After c. 3 million years ago, the Northern Hemisphere witnessed the development of ice sheets. Marine sediment cores (see **Section 3.2**) record the onset of clearly defined alternating **glacials** and **interglacials**. Within Africa, the Northern Hemisphere cold glacials tended to be characterized by arid conditions, whilst wetter (**pluvial**) conditions defined the interglacial (see **Chapter 4**). The alternation between wet and dry phases would have led to the fragmentation of habitats and the potential for isolating hominin populations with a variety of possible responses ranging from minimal adaptations through to local and regional extinctions (**Fig. 8.2**). From 2.0 Myr it has been recognized that hominin speciation accelerated, leading to a period of great hominin diversity.

Key changes in the East African palaeoclimate records appear to coincide with changes in the hominin lineages. Three main hypotheses have been forwarded regarding climate-induced evolutionary processes in hominins, and they are as follows:

- The *turnover pulse hypothesis* suggests that abrupt changes in climate caused progressive changes in habitat forcing the adaptation of existing species and the evolution of new ones (Vrba 1985).

- The *variability selection hypothesis* suggests that the simultaneous emergence of species is driven by environmental conditions becoming increasingly unstable (Potts 1998).

- The *pulsed climate variability* hypothesis promotes the notion that the key driver of environmental selection is caused by spikes of more variable climate (but unrelated to high-latitude glacial cycles) (Trauth et al. 2005; deMenocal 2011).

Within this time of Plio-Pleistocene gradual drying, a series of distinct, punctuated phases of increased humidity and wetness occurred which led to phases of increased lake levels along the East African Rift. Several notable highstands have been identified within the records at c. 2.6, 1.8, and 1.0 Myr. The triggers for these wet events are not fully understood; however, it has been suggested that changes in ocean circulation after 3.5 Myr (caused by the closure of the Isthmus of Panama) (Bartoli et al. 2005), the intensification of Walker Circulation c. 1.8 Myr (Ravelo et al. 2004), and the switch from **obliquity**-driven (41 kyr) global glacial variability to **eccentricity**-driven (100 kyr) glacial-interglacial cycles at *c.* 1.0 Myr were key mechanisms for these changes (see **Fig. 8.2**) (Mudelsee et al. 1997). See **Section 9.4** for discussion of the orbital cycles.

Africa

The earliest documented member of the genus *Homo* is *Homo habilis*. From the Hadar Formation, Ethiopia, *H. habilis* has been dated by faunal associations and **radiometric dating** to 2.3 Myr (Kimbel et al. 1997). *H. habilis* remains from Olduvai and Koobi have been dated to 1.9–1.5 Myr; but recently the age of some important specimens has been revised from 1.88–1.90 to 1.65 Myr (Gathogo and Brown 2006).

A much more human-like hominin evolved in Africa approximately 1.8 million years ago. This has been named *Homo ergaster*. Fossil specimens representing *H. ergaster*, which had a larger and more complex brain than its predecessor *H. habilis*, have been found in a number of places, including Olduvai Gorge in Tanzania (**Plate 8.1**), Lake Turkana in Kenya, and Swartkrans in South Africa (McHenry and Coffing 2000). *H. ergaster* represents a significant departure from the australopithecines in a number of ways. Anatomically they had longer legs and shorter arms, and it has been suggested that they may have reached heights up to two metres tall. It is likely that they had evolved to eat meat, may have controlled fire, and practised complex foraging.

Homo ergaster continued to use Oldowan tools, though between 1.6 and 1.4 Myr new tools begin to appear in the fossil record. Known as 'bifaces', these tools were both more complex and more standardized in form than the Oldowan tools. The Acheulean hand axe suggests a higher degree of behavioural sophistication owing to its bilateral symmetrical form. Acheulean tools, and handaxes in particular, would have been used in a variety of ways including: animal butchery, missiles against prey and/or predators, stripping bark, digging roots and for tool manufacture. This technology lasted for more than 1 million years until *c.* 200 kyr ago without any distinctive changes. Early evidence for the controlled use of fire comes from Koobi Fora, Kenya, dated to *c.* 1.6 Myr and from Swartkrans Cave, South Africa, where multiple layers with evidence for burning occurred between 1.5 to 1.0 Myr. It has been suggested that the use of fire for warmth and protection may have helped small groups of *Homo ergaster* to move out of Africa and into Europe and Asia, where colder conditions prevailed and more effective shelter was required.

Asia

In Asia, *c.* 1.8 Myr, a different hominin appears to have evolved, *Homo erectus*. It is assumed that *H. ergaster* is the African ancestor of a population that moved into Asia, giving rise to *H. erectus*, although some analysts argue that there is no basis for such a separation. The key difference lies with the technologies used. From Dmanisi in the Caucasus Mountains, Republic of Georgia, two skulls of *Homo ergaster*, dating to around 1.75 Myr, were discovered, suggesting their presence in Eurasia at about the same time they emerged in Africa (Gabunia and Vekua 1995). Anatomically the shape of these skulls is very similar to that of the African finds dated to the same period. The Dmanisi finds are important as they represent a turning point in hominin evolution as evidence for migration out of warm African environments to cooler, drier areas. Gabunia et al. (2002) have proposed that a mandible from Dmanisi should be assigned to a new taxon, *Homo georgicus*, which they considered to be an early form of the genus *Homo* that dispersed from Africa to Eurasia between 2 and 1.8 Myr.

H. ergaster, as described above, used the Acheulean technology. This industry is not found associated with *H. erectus*. Important sites occur in Java (Mojokerto and Sangiran), which are dated between 1.8 and 1.65 Myr for *H. erectus* (Swisher et al. 1994). From China, lacustrine sediments at Majuangou uncovered four layers of indisputable hominin stone tools. Magnetostratigraphy dates the lowest layer to *c.* 1.66 Myr and the upper to 1.32 Myr. This provides the oldest record of stone tool processing in East Asia and implies that a long yet rapid migration from Africa enabled early human populations to inhabit northern latitudes of East Asia over a prolonged period (Zhu et al. 2004). Other key sites include Xiaochangliang 1.36 Myr (Zhu et al. 2001) and Donggutuo 1.0 Myr (Lanpo and Qi 2001). These sites contain stone artefacts and very poorly-preserved fossil vertebrate remains.

Until recently it was widely accepted that only one hominin genus, *Homo*, was present in Pleistocene Asia, represented by two species, *Homo erectus* and *Homo sapiens*. Both species are characterized by greater brain size, increased body height, and smaller teeth relative to Pliocene *Australopithecus* in Africa (Brown et al. 2004).

Excavations at Liang Bua, a large limestone cave on the island of Flores in eastern Indonesia, have yielded evidence for a population of tiny hominins (dubbed the Hobbit), which were sufficiently distinct anatomically to be assigned to a new species, *Homo floresiensis*. They were only about one metre tall, and their brains were also tiny (in hominin terms) with a cranial capacity of around 380 cm³ (**Plate 8.2**). Dating by ¹⁴C, luminescence, uranium-series, and electron spin resonance (ESR) methods indicates that *H. floresiensis* existed from before 38 kyr until at least 18 kyr. Associated deposits, dated between 90 to 13 kyr ago, contain stone artefacts, including simple flakes, points, perforators, blades and microblades, and animal remains, including Komodo dragon and an endemic, dwarfed species of *Stegodon* (Morwood et al. 2004, 2005).

Brown et al. (2004) suggested that the most likely explanation for its existence on Flores is long-term isolation, with subsequent endemic dwarfing, of an ancestral *H. erectus* population. More detailed analysis of the skeletal remains suggests traits more archaic than *H. erectus* and more similar to australopithecines, *H. habilis* or *H. ergaster* (Argue et al. 2006). Importantly, *H. floresiensis* shows that the genus *Homo* is morphologically more varied and flexible in its adaptive responses than previously thought. Flores has always been isolated from the mainland, and therefore only a very few kinds of large mammals populated it.

The suggestion that the Hobbit is a separate species is not universally accepted though. As modern humans (*H. sapiens*) have been known in this region for at least the last 40 thousand years, these individuals have also been suggested to be genetic human dwarfs due to conditions possibly including hypothyroid cretins (Oxnard et al. 2010) or microcephaly (Martin et al. 2006).

In the remote Denisova Cave in the Altai mountains, Siberia, a finger bone fragment of a juvenile female of an unknown *Homo* species that lived *c.* 41 kyr was found. **Mitochondrial DNA** (mtDNA) analysis of the finger bone showed it to be genetically distinct from the Neanderthals and modern humans (Krause et al. 2010). Further DNA studies showed that Denisovans share a common origin with Neanderthals as well as evidence for interbreeding with modern humans. Up to 6 per cent of the DNA of Melanesians and Australian Aborigines derives from Denisovans (Reich et al. 2011). This work has indicated that archaic Denisovans must

have lived over an extraordinarily broad geographic and ecological range, from Siberia to tropical Asia.

Europe

In the first half of the Middle Pleistocene, another group of hominins emerged in Africa and Europe. These hominins are called *Homo heidelbergensis* after the German town of Heidelberg where some of their fossils were first found. It is present at Atapuerca, Spain (0.78 Myr), Boxgrove, England (0.5 Myr), and Mauer, Germany (0.5 Myr). It is unclear, though, whether *all* fossils from Africa and western Eurasia should belong to the same species.

During the later Pleistocene, *H. erectus* and *H. heidelbergensis* may have coexisted in East Asia. Also at this time tool manufacture became more complex, based on a flaking technique called the Levallois. Production of tools in this fashion is difficult compared to Acheulean industries and requires a much higher degree of preparation and planning.

From *c.* 200 kyr, Neanderthals (*Homo neanderthalensis*) appeared in south-west Asia and Europe. The Neanderthals represent a highly specialized, highly derived, European/Near Eastern offshoot that coexisted with populations of anatomically modern *H. sapiens*. Neanderthal characteristics included: very large brains (larger than modern humans), very robust musculature, a protruding facial skeleton and a large nose (**Plate 8.2**). These traits are thought to have developed as adaptations to the cold conditions of glaciation among a geographically isolated population. Neanderthals utilized Mousterian tools, which exhibit a technological advance over Lower **Palaeolithic** technologies in the efficient use of raw materials. Some insights into Neanderthal behaviour reveal that they buried their dead and cared for their old and sick. They appear to have lived short lives with a high incidence of bone trauma. A *Homo neanderthalensis* footprint dating to sometime before 62 kyr was found in Vartop Cave, Romania (Onac et al. 2005).

Whilst Neanderthals evolved gradually over at least 200 000 years of successful adaptation to the glacial climates of north-western Eurasia, populations declined abruptly between 40 and 30 kyr, to be replaced by populations of modern humans. The final evidence for Neanderthals comes from southern Iberia *c.* 24 kyr (d'Errico and Goñi 2003).

The question of the coexistence and potential interaction between the last Neanderthals and the earliest intrusive populations of anatomically modern humans in Europe has recently emerged as a topic of lively debate. Interaction between Neanderthals and modern humans has been suggested on the basis of findings in the Middle East. Two of the oldest modern human sites (Skhul and Qafzeh) were discovered in close spatial (and overlapping temporal) proximity to Neanderthal sites (Tabun, Amud, and Kebara). However, it remains uncertain what type of real interaction occurred between the groups. Recent genetic evidence suggests that gene flow between populations of modern humans and Neanderthals did occur (Green et al. 2010). This may shed some light on the issue of the rather sudden disappearance of Neanderthals in Europe, an occurrence that has traditionally been attributed to the arrival and colonization of *H. sapiens*.

8.3 Modern human origins and dispersals

Two major models are suggested in addressing the emergence of anatomically modern humans (though there are other models that borrow from both extremes). The first model is the Multiregional Evolution Hypothesis (**Fig. 8.4**), which argues that there was more-or-less simultaneous evolution in modern humans across the Old World and that ancient regional populations maintained species continuity with modern humans through gene flow. This is the reason why, it is argued, that modern humans never diverged to the point of being different species (Wolpoff et al. 1994).

The second hypothesis, known as the Recent African Origin or Out of Africa 2 or Replacement model, suggests that natural selection and gene flow led regional hominin populations along distinct evolutionary pathways after the emergence of *Homo* from Africa during the Lower Pleistocene (coined Out of Africa 1). Thus, only in Africa should there be continuity between archaic and modern hominins, making it the only and relatively recent home to modern humans prior to their dispersal across the rest of the world (**Fig. 8.4**) (Stringer and Gamble 1993).

In addition, two other models have been proposed (**Fig. 8.4**). The 'Hybridization and Replacement' model is a variant of Recent African Origin but allows for a greater or lesser extent of hybridization between

Fig. 8.4 Evolutionary models of modern human origins: (a) Recent African Origin; (b) (African) Hybridization and Replacement Model; (c) Assimilation Model; (d) Multi-regional Evolution (from Stringer 2002, Fig. 2). Reprinted with permission of the Royal Society of London

the migrating population and the indigenous premodern population (Bräuer 1992). The 'Assimilation' model combines elements of Recent African Origin and Multiregional Evolution but also accepts an African origin for modern humans, emphasizing the importance of gene flow, admixture, changing selection pressures, and resulting directional morphological change (Smith 1992).

Key arguments and questions have been raised as to which model, if any, is correct. Opinion differs as to whether interbreeding ever occurred between resident archaic populations. It should be noted that the hypotheses were forwarded prior to the development of routine genetic studies, which today play a crucial role in characterizing the development of humans. Increasing genetic evidence lends support to the replacement model. For example, mitochondrial DNA data indicate that all humans living today descended from a common population that lived in Africa between 200 and 100 thousand years ago. In addition, it is important to note that contemporary Africans are the most genetically variable of all human groups. For example, two Africans are more likely to be genetically different from each other in terms of their mitochondrial DNA than any two other people on the Earth. This evidence places the original human population explosion in Africa.

Until recently a weakness highlighted in the multi-regional model related to the fact that if earlier populations did indeed interbreed with modern groups then novel genes would be expected in these regions. The recent genetic evidence from Denisovans has, however, shown that interbreeding between species, including Neanderthals and modern humans occurred. Neanderthal DNA has also shown some interbreeding with *H. sapiens* as well.

Africa

The first modern *Homo sapiens* evolved in Africa between 200 and 100 kyr. They were much more slender and lighter built than earlier groups such as Neanderthals (**Plate 8.2**) and other hominins of the time – *H. erectus* and *H. heidelbergensis*. The earliest reliably dated *H. sapiens* fossils include those from Omo-Kibish, Ethiopia (195–150 kyr), Singa, Sudan (150–140 kyr), and Mumba Cave, Tanzania (120–100 kyr). The archaeological record is supported by mtDNA evidence used as a tracer for the origins of modern humans. This suggests that a small ancestral population evolved in sub-Saharan Africa sometime between 200–100 kyr ago. Another useful marker in the timing of evolutionary events is the use of Y-chromosome markers, microsatellites and chromosome-21. These collectively suggest an East African origin between 180–125 kyr ago.

Outside of East Africa *H. sapiens* remains have been dated to *c.* 160 kyr in North Africa at Jebal Irhoud in Morocco (Smith et al. 2007). By as early as *c.* 140 kyr the earliest Aterian stone tools are found at Ifri n'Ammar (Richter et al. 2010) and these are thought to belong to modern humans. In order to reach North Africa and into the Maghreb by this time humans would have had to cross the Sahara, rather than via the Nile corridor, and a green Sahara at some point during **MIS** 6 would

have provided the opportunity to do so (Osborne et al. 2008). In southern Africa, Klasies River provides the largest single-site assemblage of early *H. sapiens* fossils (Rightmire et al. 2006). Klasies River preserves two distinct stratigraphic horizons. The lower member is attributed to MIS 5e, with a depositional age of 120 kyr, while the overlying member was deposited between 100 and 80 kyr.

The first appearance of explicitly symbolic objects in the archaeological record marks a fundamental stage in the emergence of modern social behaviour in humans. Until recently this has been associated with the Upper Palaeolithic and only after 60 kyr. Recent finds in Africa have shown evidence for enhanced cognitive ability and the onset of 'modernity' during the Middle Palaeolithic/Middle Stone Age (MSA). For example, at Grotte des Pigeons (Taforalt), Morocco, perforated *Nassarius gibbosulus* shell beads dated to *c.* 82 kyr represent some of the earliest evidence for ornamentation (**Plate 8.3**) (Bouzouggar et al. 2007). At the other end of the African continent perforated shell beads of *Nassarius kraussianus* dated between 78 and 75 kyr have been found at Blombos Cave located on the Western Cape region of South Africa (d'Errico et al. 2005). In addition, further evidence for symbolism was found in the form of engraved ochre pieces recovered from contexts dated between *c.* 100 and 75 kyr (Henshilwood et al. 2009).

Out of Africa: Arabia and the Near East

It is likely that the first pulse out of Africa of anatomically modern humans occurred sometime during the onset of the last interglacial (MIS 5e) *c.* 125 kyr (Ambrose 1998). (Environments of the preceding interglacial, the Eemian, are discussed in **Section 3.6**.) Prior to this, during the penultimate glaciation, North Africa would have been characterized by intense aridity, restricting the migration and movement of human populations out of Africa. The modern human remains from Qafzeh and Es Skhul in the Near East have been dated between 110 and 90 kyr and provide evidence for a pulse of dispersal during more humid conditions through the Levantine corridor (Shea 2008).

In addition to the Levantine corridor there are at least two other plausible routes out of Africa: the Straits of Gibraltar into southern Iberia and the Bab al-Mandeb into southern Arabia (Rose 2004). Evidence for modern human dispersals via the southern route, across the Red Sea at the Bab al-Mandeb has been found in the Arabian Peninsula. At Jebel Faya, UAE, stone tools with an East African affinity were dated to *c*. 125 kyr suggesting the presence of modern humans in Arabia during the last interglacial (MIS5e) (Armitage et al. 2011). Further evidence from Oman has suggested the presence of modern humans in southern Arabia at 106 kyr (MIS 5c), where stone tools of the Nubian Complex, which are found only in the Horn of Africa, were found (Rose et al. 2011). Palaeoclimatic reconstructions from Arabia indicate that the dispersal of modern humans into Arabia occurred during wet, pluvial conditions in which numerous lakes and extended drainage systems developed across the landscape, which supported rich savannah grasslands (**Fig. 8.5**). The existence of Lower and Middle Palaeolithic stone tools from Arabia indicates that the Arabian corridor was utilized for earlier phases of migration into this region during the Quaternary (Petraglia 2003).

Europe

Europe became colonized by modern humans from the Middle East initially into eastern and then into western Europe. The appearance of anatomically modern humans in Europe and the nature of the transition from the Middle to Upper Palaeolithic are matters of intense debate. Radiocarbon dates from modern human fossil remains including infant teeth from Grotta del Cavallo, Italy and a jawbone and teeth from Kents Cavern, Devon, England, place modern humans in Europe by 45 kyr ago (Higham et al. 2011; Benazzi et al. 2011).

Asia and Australia

The timing of early human dispersal to Asia is a central issue in the study of human evolution. The Toba volcanic super-eruption occurred 74 kyr. Whilst the eruption occurred in the island of Sumatra, Indonesia, the impacts were extensive with Toba ash being found in India and the Arabian Sea. It has been suggested that the Toba eruption had a profound impact on modern human populations and caused a bottleneck for dispersal in which *H. sapiens* could only leave Africa and colonize the globe after this event (Ambrose 1998). This view is, however, debated.

Archaeological evidence from India indicates that Middle Palaeolithic industries in stratified contexts are present prior to and after the Toba super-eruption (Petraglia et al. 2007). Middle Palaeolithic hominins

appear to have survived the negative effects of the volcanic eruption and climatic fluctuations in the Late Pleistocene. It has been suggested that these lithics were most probably manufactured by *Homo sapiens*.

The periodic lowering of sea levels during the Pleistocene created a large dry-land extension of the SE Asian landmass, encompassing the present islands of Sumatra, Java, Borneo, and Bali. The ancient landmass is termed Sundaland. In Sarawak, Borneo, modern human skeletal remains and lithic artefacts were found in sediments dated from *c*. 45 kyr ago (Barker et al. 2007).

At periods of lower sea level, Australia and New Guinea would have also been connected via the Sahul Shelf (see **Chapter 6**). This region is termed Greater Australia. The Philippines and eastern Indonesia remained separated from these two landmasses owing to the depth of the sea in this region. These islands are coined Wallacea.

The migration of modern humans to Australia during the Late Pleistocene, restricted by the sea, across to the islands of Wallacea occurred by 50 kyr. Upper Palaeolithic technologies spread to many regions, including Northern Australia, by 50 kyr, Western Australia by 40–35 kyr and Lake Mungo in semi-arid New South Wales at the same time. Whilst Papua New Guinea, Australia, and the Solomon Islands were colonized during the Pleistocene *c*. 40 kyr, the colonization of the Pacific islands beyond occurred during the late Holocene period with people moving by canoe. As *Homo sapiens* colonized Australia, and the islands of the Pacific much later (New Zealand was reached around AD 1000), they left numerous species extinctions in their wake. By *c*. 50 000 years ago humans had certainly become resourceful and highly skilled hunters, and many large mammals and birds that had no prior experience of humans were easy prey. Australia is estimated to have lost over 80 per cent of its large mammals, including a two ton giant wombat and a 2.5 m high kangaroo.

There are two schools of thought, based on the method of dating employed, relating to the timing of the colonization of Greater Australia. The first group argues for an age between 60 and 50 kyr based on **luminescence dating**. For example, engraved rocks at Jinmium, Northern Territory, Australia were dated by TL to approximately 58 kyr (Fullagar et al. 1996) and those from the Malakunanja II rock shelter to around 50 kyr (Roberts at al. 1990). It has been argued on the basis of ESR and U-series dates that the Lake Mungo

Fig. 8.5 Comparison of the Jebel Faya, United Arab Emirates, chronology with regional and global proxy records. Vertical grey bars represent pluvial periods. (a) OSL ages from Jebel Faya, showing the presence of early modern human artefacts dated to the last interglacial c. 125 kyr. (b) Phases of aridity and aeolian dune accumulation in the Wahiba Sands, Oman. (c) Timing of wet phases in Oman from Hoti and Qunf Cave speleothem records. (d) Phases of lacustrine development, Mudawwara, Jordan. (e) Indian Ocean Monsoon Index. (f) Oxygen isotope record from Soreq Cave. (g) Red Sea eustatic changes. (h) SPECMAP stacked oxygen isotope record (from Armitage et al. 2011, Fig. 3). Reprinted with permission of AAAS: *Science*

3 burial may extend back as far as 60 kyr. However, on geological grounds an age of 40 kyr has been suggested. The dates from this site are still a matter of dispute. **Fig. 8.6** shows some of the key sites studied in Australia.

The second group suggest a younger age of no more than *c.* 45 kyr based on radiocarbon dating (**Fig. 8.7**). The currently oldest dated site, at Devil's Lair, is 47 kyr (Turney et al. 2001). At Carpenter's Gap in the central southern Kimberley range, on the edge of the Western Desert, the site has been dated to *c.* 43 kyr

and contains palaeobotanical material and stone artefacts (O'Connor and Fankhauser 2001). Dates from this region show consistent, though at times sparse, human occupation between 30 and 11 kyr. The site at Swan River in Western Australia was occupied by 40–35 kyr (Pearce and Barbetti 1981). The Willandra Lakes region, part of the Murray–Darling system draining eastern Australia, includes the Lake Mungo site where dates *c.* 40 kyr have been measured on materials including charcoal and emu eggshell (Bowler and Price 1998).

Fig. 8.6 Location of major desert areas in Australia and key Pleistocene sites with human occupation (from O'Connor and Veth 2006, Fig. 2.1). Reprinted with permission from Blackwell Publishing Ltd

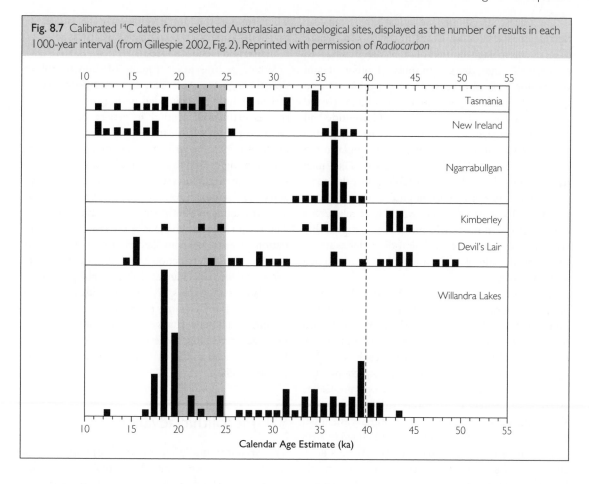

Fig. 8.7 Calibrated ^{14}C dates from selected Australasian archaeological sites, displayed as the number of results in each 1000-year interval (from Gillespie 2002, Fig. 2). Reprinted with permission of *Radiocarbon*

The oldest well-dated skeletal remains come from the Lake Mungo 1 cremation, which is approximately 25 kyr old. Cranial material from the terminal Pleistocene in Australia is highly variable, which has led some palaeoanthropologists to argue that the Australian continent was populated by two hominin groups each with a distinct evolutionary origin (Thorne 1980).

From around 3500 years ago, seafaring, pottery-making farmers, the Lapita culture, extended into remote Oceania from the Bismarck Islands, off New Guinea. Many of the Pacific islands were not colonized until after 2000 years ago.

Colonization of the Americas

Around 15 kyr Siberia and Alaska were linked by a land bridge (known as Beringia), which formed during the last glacial period. Global sea level reached a low point of *c.* 120 m lower than present during the **Last Glacial Maximum (LGM)** (see **Section 6.4**). Although sea levels fluctuated over the last glaciation, the two continents remained linked until climatic amelioration during the Holocene resulted in their separation and the formation of the Bering Strait. The colonization of the Americas is fiercely contested. Some researchers believe that the first arrivals by humans took place as far back as 40 kyr, others 20 kyr and some are happy with a date as late as 15 kyr. Archaeological, genetic and linguistic research shows that the first Americans originally came from Asia. It is assumed that the first migrants out of Asia hunted and foraged across the steppe/tundra landscape, finding their way across the Bering land bridge. It is likely that settlement was sporadic and may have taken several generations to achieve. These original people were highly mobile and their legacy of migration has left almost no archaeological signature. Their original migration routes are now mostly submerged under the sea of the Bering Strait or lie in the frozen expanse of Alaska.

The oldest dated archaeological site from the north-east is at Bluefish Caves in the Yukon, where a small assemblage of microlith blades is dated to c. 15.7 kyr (Yesner 2001). A number of other sites in this region have also been dated between 13.7 and 13.1 kyr. It has been postulated that humans passed along ice-free corridors between the **Laurentide** and Cordilleran ice sheets and/or they may have migrated southwards along the coastline from Beringia to the Pacific Northwest (Mandryk et al. 2001). They may have even used skin boats to paddle down the coast (although there is no archaeological evidence to support this). These early peoples would have relied on a diet of terrestrial mammals, birds, fish, and probably sea mammals. Evidence for their tools includes stone-tipped spears and slotted antler spear heads tipped with microliths.

Once beyond the ice margins, the nomadic hunter-gatherers migrated across much of North America. A number of sites have emerged which show evidence for pre-Clovis human occupation. At Buttermilk Creek, central Texas, over 15 000 stone tools and debitage, dated between >15.5 and 13.2 cal kyr BP by OSL have been interpreted as an undisturbed pre-Clovis assemblage (Waters et al. 2011a). At the Manis site, Washington State, evidence for hunting was shown by a tip of a projectile point found embedded in a rib of a mastodon dated to 13 800 cal yr BP (Waters et al. 2011b). The Meadowcroft rock shelter in Pennsylvania has provided traces of human occupation between 14 to 12.5 kyr (Adovasio et al. 1990). These early occupations pre-date the Clovis tradition by up to 1500 years.

The Clovis tradition, characterized by hunter-gatherers, is found across the USA and Canada, with the densest concentration in south-eastern USA. Clovis technology is characterized by distinctive stone spear points. It is generally thought that during the onset of climatic amelioration following the LGM, humans migrated rapidly through North America, into Central America and beyond to South America. However, doubt has been cast on the theory that the colonization of the Americas stemmed from a progressive southward migration of peoples. The excavation of a small settlement at Monte Verde, Chile, has been dated to between 15.5 and 15 kyr, long before many other areas further to the north were occupied (Dillehay 1997). The Monte Verde finds indicate that people lived in a forested landscape, unlike the great plains of North America, and their artefacts were mainly made of wood. Other sites have emerged which support the Monte Verde early age, suggesting that South America was colonized as early as 40–35 kyr (Guidion and Delibrias 1986).

Whatever the exact timing of human migration through the Americas, as with Australasia the arrival of humans coincided with a major phase of species extinction: around 30 genera of mammals went extinct in North America and over 40 genera in South America. According to the 'blitzkrieg theory', over-hunting caused the demise of many well-known Pleistocene megafauna such as mammoths, elephants, and camels in North America and giant ground sloths and armadillos in South America (Southwood 2003). The abrupt climatic changes at the last glacial **termination** have also been invoked as a cause of this major late Pleistocene extinction event. Although climatic shifts would have changed habitats and put animal populations under pressure, as described in **Chapter 3**, there were many previous terminations characterized by similar climatic changes. These earlier terminations did not result in the level of extinction seen when humans arrived in the Americas towards the end of the last glacial.

8.4 Mesolithic advances

The **Mesolithic** traditionally covers the period from the onset of the Holocene until the adoption of a predominantly **Neolithic** farming economy. In some regions there is little evidence for a transition between the two (e.g. in the Fertile Crescent) and the Mesolithic may not exist in such areas where agriculture developed very early in the post-glacial. In north-west Europe the end of the **Younger Dryas** saw rapid warming and this led to widespread changes in vegetation, with open **Late Glacial** environments being replaced by forests of birch and pine, followed by thermophilous taxa including oak, elm and lime (**Chapter 5**). Sea levels rose, gradually inundating low-lying plains, lakes, coasts, river valleys, and estuaries of the North Sea basin (**Chapter 6**).

The economic basis for the Mesolithic was hunter-gathering. In Europe, hunting would have focussed on large terrestrial mammals including elk, red and roe deer, wild pig, and aurochs. Smaller mammals would have included rabbits, badger and otters which would have provided furs in addition to meat. The burial of dogs in Mesolithic contexts suggests that not only had they been **domesticated** but that they were held in high esteem. It is likely that dogs would have been used for the tracking and hunting of prey. In addition to the hunting of terrestrial mammals, lakes, rivers, and coastal regions would have provided important elements of the diverse subsistence base of the Mesolithic. The hunting of birds, often in vast numbers and sometimes of a single species, has been recorded from a few Mesolithic sites e.g. Aggersund and Øgaarde in Denmark. Marine resources, including sea mammals, fish, and shellfish formed an important component of Mesolithic coastal communities, as shown by a large number of midden sites (Bonsall et al. 2002).

The gathering of plants would have formed an important component of the Mesolithic diet. In Europe the early to mid post-glacial forests would have been a rich source for edible berries, fruits, nuts, and fungi. In British Mesolithic sites the preservation of hazelnuts is by far the most common plant material. From Franchthi cave, Greece, the gathering of barley, oats, lentil, bitter vetch, pear, and wild almond has been identified from carbonized remains (Hansen and Renfrew 1978).

The use of fire, especially from upland sites in Britain, has been noted from a number of Mesolithic sites in, for example, the North York Moors, Pennines, Lake District, and Wales (Simmons 1993). It has been suggested that the deliberate use of fire would have prevented the development of closed woodland and maintained open conditions, perhaps to attract large herbivores which could then be hunted.

As with the earlier Palaeolithic periods, stone tools dominate the archaeological record of the Mesolithic (Mithen 1994). It appears that flint was the preferred choice. However, other stone types were used in its absence such as quartz and quartzite. Other important materials would have included antler, bone, and wood. In Britain, from 10.3 kyr ago long-blade lithic technology prevails and appears to represent the first colonization from the continent (Barton 1997). Microliths are often directly associated with the Mesolithic although macroliths still continued to be used in some regions (e.g. long-blades continued to be used during the late Mesolithic in parts of Ireland). As well as hunter-gathering, there is some evidence from across Europe for settlements or dwellings, though it should be noted that these are rare. These tend to be restricted to building floors and postholes.

A Mesolithic economy persisted along the north-western extremities of Europe long after a Neolithic economy had been established in the Near East and many parts of continental Europe. The transition to farming on the North European Plain along with the Channel coast of France occurred between 7.4 and 6.9 cal kyr BP but was not widely adopted until 6.1 to 5.8 cal kyr BP in Britain, Ireland, Denmark, and southern Sweden – a delay of around 1000 years (Bonsall et al. 2002; Rowley-Conwy 2011). Dietary tracing of human bones indicates an abrupt transition from Mesolithic to Neolithic in terms of material culture and economy. Carbon isotope analysis of human skeletons from coastal sites in Scotland and Denmark indicates a distinct decrease in $\delta^{13}C$ about 6000 years ago. This reflects a change in diet from the late Mesolithic to the Neolithic, with a switch from subsistence based on marine resources that could be gathered to a subsistence economy based on agriculture as the dominant mode of food production (**Fig. 8.8**). (Due to carbon fractionation during plant photosynthesis, the $^{13}C/^{12}C$ ratio is less in terrestrial food sources compared with marine.) It has also been suggested that the relatively rapid switch from a hunter-gatherer to an agrarian economy c. 6000 years ago in the maritime regions of north-west Europe could have been related to a climatic shift towards a drier, more continental type of climate that would have enhanced the potential for cereal-crop cultivation in coastal areas that were previously marginal for early cultivation (Bonsall et al. 2002).

In India, the amelioration in climate and increase in humidity during the early Holocene permitted the spread of hunter-gatherer communities. They occupied cave and rock shelters in Central and South India and the Eastern Ghats, together with the sand dunes of the Thar Desert. As in Europe, microliths dominate the stone tool technology. Rhino, elephant, deer, cattle, buffalo, deer, fox, and jackal, as well as a wealth of aquatic fauna were hunted and consumed. In the central Indian hills, Mesolithic rock paintings depict people hunting game, gathering plant resources, and trapping animals. The first farmers gradually succeeded Mesolithic communities ushering in the Neolithic, with pastoralism and agriculture supplementing hunter-gathering as the prevalent mode of subsistence.

8.5 Origins and spread of agriculture

Whilst modern humans (*Homo sapiens*) have existed for some 150 000 to 200 000 years, agriculture is a relatively recent development, dating from *c.* 11 kyr.

Fig. 8.8 Variations in carbon isotope values from human bones from Scotland showing the shift in diet across the Mesolithic and Neolithic periods. The change is thought to be related to a change in human diet from a marine-rich source to one dominated by terrestrial sources (from Bonsall et al. 2002, Fig. 2)

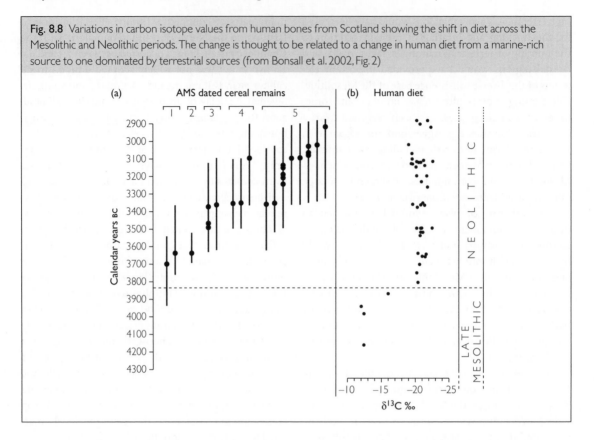

During the Pleistocene humans had depended upon hunting wild animals and gathering plants for food. The early hominid diet (from about four million years ago), evolving as it did from that of primate ancestors, consisted primarily of fruits, nuts and other vegetable matter, and some meat items that could be foraged for and eaten with little or no processing. The diet of pre-agricultural but anatomically modern humans (from c. 30 kyr) diversified somewhat, but still consisted of meat, fruits, nuts, legumes, edible roots, and tubers, with consumption of cereal seeds only increasing towards the end of the Pleistocene. However, from approximately 11 000 years ago there was a major shift to the domestication of animals and plants (**Fig. 8.9**), although hunting and gathering still continued long after the advent of agriculture.

The practice of agriculture first began around 11 kyr in the Fertile Crescent of the Levant (part of present day Iraq, Turkey, Syria and Jordan). At this time the region was characterized by open woodlands and grassland, with stands of wild grasses and legumes. They had to undergo few genetic changes to be of use

to farmers. These plants were all readily storable and were easy to grow. By 10 kyr free-threshing wheat and barley had been developed slightly before the pea, lentil, and chickpea (**Table 8.1**). Between 8.5 kyr and 6 kyr ago the farming that had developed in the Levant had spread across North Africa, into Europe and Iberia, and eastwards into India. The earliest domesticated animal was the dog with morphological changes recognized in both the Old and New Worlds from c. 13 kyr. Other early domesticated animals across the Old World include goat (*Capra hircus*), sheep (*Ovis aries*), cattle (*Bos taurus*), and pig (*Sus scrofa*).

In eastern Asia agricultural **domestication** is focussed on the middle and lower regions of the Yangtze and Yellow rivers (**Fig. 8.9**). The region lies within the monsoonal belt and produced a series of summer cereals of which rice has evolved as the most important one. The timing for the domestication of crops in this region is comparable with that of the Levant. However, recent research from China has suggested that rice domestication may have taken place as early as 13 kyr along parts of the Yangtze River

Fig. 8.9 The regions, sites, and dates at which some of the most important domestications took place (after Simmons 1989, Fig. 4.1). Reprinted with permission of Blackwell Publishing Ltd

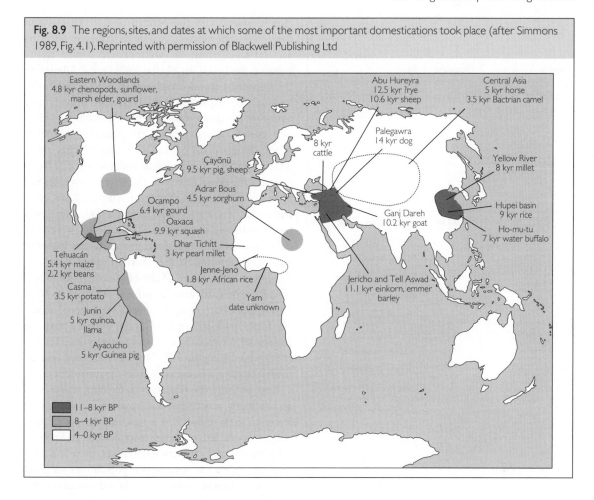

(Lu et al. 2002). Rice and millet were certainly domesticated by 9.5 kyr, followed by mung, soy, and aduki beans. In south-east Asia there is no evidence for food production before 5500 years ago.

In the Sahel region of Africa local rice and sorghum were domesticated by 7 kyr. Local crops were domesticated independently in West Africa and possibly in New Guinea and Ethiopia.

In southern Africa the rapid spread of farming, distinctive ceramics, and metallurgy occurred through the expansion of the Bantu speaking population over the past two millennia. Pearl millet (*Pennisetum* sp.), finger millet (*Eleusine coracana*), and sorghum (*Sorghum bicolor*) were the principal crops. In addition, pulses, cowpeas, and curcurbits were also grown (Mitchell 2002).

Three regions of the Americas independently domesticated corn, squashes, potato, and sunflowers. Apart from the high Andes of Peru, where llama,

alpaca, and guinea pigs were domesticated, few other regions of the Americas saw the domestication of herd animals. Turkeys were domesticated in Middle America and dogs were eaten by the Maya. The Americas also lacked highly productive cereals. The exception to this is maize (*Zea*), which became widespread across Mesoamerica from *c*. 4000 years ago. The earliest domesticates were mainly condiments, fruits and fibres including chilli pepper, gourd, avocado, and cotton.

So why did agriculture originate? This is a contested theme and there is no generally accepted explanation. A number of hypotheses have been put forward (**Table 8.2**): cultural progress, changing climate, population pressure, food abundance, and coevolution.

A number of proxy sources can be used to give insight into the timing and nature of agricultural practices (**Table 8.3**).

Table 8.1 Dates for the domestication of plants

Region	Dates kyr	Plant
Near East	11–9.3	Emmer wheat (*Triticum*)
		Two-rowed barley (*Hordeum*)
	9.8–9.6	Einkorn wheat (*Triticum*)
		Pea (*Pisum*)
		Lentil (*Lens*)
		Vetch (*Vicia*)
Far East	from 11.5	Rice (*Oryza*)
		Soybean (*Glycine*)
		Millet (*Pennisetum*)
		Sugarcane (*Saccharum*)
		Rape (*Brassica*)
		Hemp (*Cannabis*)
New World: Mexico and Peru	from 10	Corn (*Zea*)
		Squash (*Cucurbita*)
		Tomato (*Lycopersicon*)
		Lima beans (*Phaseolus*)
		White potato (*Solanum*)
		Sweet potato (*Ipomoea*)
		Chili peppers (*Capsicum*)
		Peanut (*Arachis*)
		Guava (*Psidium*)
		Avocado (*Persea*)

Within the archaeological record there are basically two types of agriculture:

- seed-crop (cereal grains) e.g. wheat, maize, rice, that occur in simple ecological communities and are very productive;
- vegeculture (root crops and tree crops) e.g. manioc, yams, taro, avocado, potato, that are from complex plant communities and are less productive than seed crops.

The former tend to have developed in relatively dry regions with good archaeological preservation, while the latter originated in the moist tropics with poorer preservation. Thus more is known about the origins of seed crops than root crops.

In the Near East it has been suggested that the origin and expansion of cereal-crop cultivation can be closely linked with the pattern of Late Glacial and early Holocene climatic change described in **Sections 5.2** and **5.3**. During the warmer and wetter conditions of the Late Glacial (the Bølling and Allerød of the European **chronostratigraphy**) wild cereal grains and nut trees expanded in the Near East, providing an abundant enough resource for people to adopt a sedentary lifestyle. Wright and Thorpe (2005) proposed that as the climate dried out in the Younger Dryas, these wild resources became scarcer and it was necessary to cultivate the cereal grains that had long been collected in order to maintain the sedentary culture. Thus, environmental stress associated with the Younger Dryas probably helped to stimulate the development of crop cultivation in the Levant and in the Zagros foothills. After its establishment, the return of more favourable conditions at the onset of the Holocene made it possible for crop cultivation to expand out from its Near East origins. For example, Willcox et al. (2009), drawing from work in northern Syria, contend that the change to warmer, wetter, and more stable climatic conditions during the early Holocene enabled agriculture to become a sustainable basis for subsistence.

Until recently, the transition to farming was seen as an inherently progressive one in which people, having learnt to grow crops, had a new and improved food source which led to larger populations, technological advances, and civilization. However, there is a wide body of evidence to suggest that populations adopting a sedentary lifestyle based around cereal agriculture suffered from deterioration in health (because of lack of a varied diet), returning to pre-agricultural levels of nutrition only in recent history. The spread of infection can in part be attributed to the congregation of people living in close proximity and also to a decline in dietary quality that accompanied intensive cereal farming (Cohen 1989).

Another aspect that has been the focus of debate is the extent to which agriculture was actively spread by migrant groups ('migrationism') as opposed to spreading by indigenous people adopting the methods from neighbouring agriculturalists ('adoptionism'). Rowley-Conwy (2011) provides a review of this debate in relation to the spread of agriculture westwards across Europe.

The rise of civilization

Within a few thousand years of the adoption of cereal agriculture, the old hunter-gatherer style of social organization began to decline. Agriculture permitted

Table 8.2 Hypotheses for the origins of agriculture

Hypothesis	Proposal	Limitations
Cultural progress	Is the oldest hypothesis and assumes that humans would develop an agriculture subsistence as part of progress from simpler to more complex lifestyles, from an arduous nomadic life to a comfortable sedentary one, from wild to a more and more 'civilized' or 'settled' state.	Agriculture not adopted everywhere. Observations from multiple hunter/gatherer societies don't support this hypothesis as some groups know about agriculture but choose not to adopt this strategy e.g. !Kung bushmen of southern Africa.
Changing climate	There appears to be a correlation of agriculture origins with the climate changes at the end of Pleistocene/early Holocene. For example, climate in the Near East resulting in a drying trend brought people together in regions where water remained, leading to the need for increased food production and a less mobile lifestyle.	Whilst climate change evidence suggests this may be a factor in origins of agriculture, it is not sufficient in itself to explain it.
Population pressure	Notion that population growth forces foragers to adopt agriculture, owing to limited availability of wild resources.	Why did it take so long before population pressure came to a head? Was population growth cause or effect?
Food abundance	Idea that agriculture did not arise from need so much as it did from relative abundance. This led to permanent settlements and time for sedentary lifestyle and experimentation with agriculture.	
Coevolution	People became dependent on artificially selected plants and the plants became dependent on people for propagation. For example, wheat mutation caused grain to remain on stem.	

the growth of larger and denser populations, which in some regions coalesced into larger complex concentrated settlements (cities) and more institutionalized social formations (states). With the rise of civilization and the state came highly organized societies exhibiting a high degree of urbanization, social stratification and socio-economic classes, job specialization, and centralization of power with governments and armies (Trigger 2003; Brooks 2006). An important feature of early states is the development of writing, which appears to have developed independently in a number of different regions. Trigger (2003) suggests that early states fell into two principal categories: those that developed around cities (city-states) and those that fell within blocks of land (territorial states) in which cities formed administrative, political centres.

Early states developed in areas which permitted high agricultural productivity and so tended to be centred in fertile basins or river valleys (**Table 8.4**). The earliest complex societies first emerged in regions currently occupied by the desert belt stretching from North Africa, through the Middle East into South Asia and China and also in northern South America (Brooks 2006). This occurred during the late 6th and 5th millennia BP and was characterized by a shift towards drier, more arid conditions. It has been suggested that the earliest known settlements in Mesopotamia emerged after the 8200 cal yr BP drought event which may have lasted up to four centuries across the region (Fagan 2004) (see **Section 5.4** for further details on the nature and causes of this 8200 cal yr BP event). This may have led to a concentration of peoples along the Tigris and Euphrates valleys rather than the desert interior, which had been covered in grassland and used for grazing and herding.

It is interesting to note that these regions were warm environments where high crop yields were possible provided that there was a suitable supply of water.

Table 8.3 Proxy sources for gaining insight into the timing and nature of agricultural practices

Proxy type	Application
1. Botanical remains	Plant macroscopic (e.g. tubers, seeds, seed impressions on mud bricks and pottery, plant fragments, charcoal) and microscopic remains (e.g. pollen grains, phytoliths, starch) and fragments of plants tell what types of plants were associated with the culture and perhaps being eaten and cultivated.
2. Animal remains	Bones, feathers, shells, or other remnants. How much of the diet was animal-based, which animals.
3. Faecal remains	Coprolites (fossilized faeces) can be analysed to see what types of foods were in the diet (seeds, plant fragments) and how eaten (cooked vs raw). Can be used for humans or animals (e.g. diet of animals both wild and domesticated).
4. Tools	Developments in technology, cultivation, storage, grinding, threshing, and perhaps trade.
5. Art and drawings	Graphic representation of plants and animals important in culture.
6. Written accounts	Insight into plants and animals used. Importance of yields and trade.
7. Dating	Determine dates/ages of archaeological materials e.g. ^{14}C, TL, OSL, amino acid racemization.
8. DNA	Insight into the development and breeding of crops and animals. Important for tracing origins of agriculture.

Table 8.4 Regions in which early civilizations evolved along with the timing of their initiation

Region	Basin/river	Date (years ago)
Mesopotamia	Tigris and Euphrates	c. 6000
Mediterranean	Coastal	c. 5200
Egypt	Nile	c. 5200
Indus	Indus	c. 5000
Peru	Coastal	c. 5000
Northern China	Yangtze and Yellow	c. 4000
Mesoamerica	Mexican basin	c. 3500
Andean	Rivers running from Andes to Pacific	c. 3500

Maize, wheat, and barley all developed as desert edge species of large grasses that flourished under hot conditions. In order to sustain high yields hydraulic technology and irrigation became essential to support many of the early state societies. For example, the Egyptian and Indus civilizations were dependent on the receding flood waters of the Nile and Indus respectively, whilst the development of canal irrigation was central to Peru, central Mexico, and Mesopotamia. While city-states tended to develop independently from one another there are examples where they shared contact through trade. For example, the archaeological record

has shown links between Mesopotamia and Egypt and Mesopotamia and the Indus.

Within states the size of populations living as coordinated units rose dramatically above pre-agricultural levels. Hunter-gatherers lived in egalitarian, autonomous bands of about 20 closely related persons, with at most a tribal level of organization above that. Early agricultural villages had 50 to 200 inhabitants, and early cities 10 000 or more. States are centralized political institutions in which the ruling elite controls the population. The state institutions gather revenues from their population in a variety of forms either as taxes, labour, produce, or raw materials. In return they offer protection in times of conflict or famine (Scarre 2005).

A series of views have been put forward to explain the formation and complexities of cities, states and civilizations. The first, by Childe in the 1930s, suggested that an agricultural surplus and craft specialization were the direct outcome from the transition to farming. In his model, where crop yields were high the surplus permitted some individuals to undertake work other than agriculture. This resulted in the development of specialist trades and skills including metal workers, soldiers, and priests (Possehl 2002). The second notion was put forward by Steward in the late 1940s in which irrigation agriculture in arid environments was seen as the key factor and that communities were dependent on irrigation waters controlled by the state for their survival (Wittfogel 1957). The third theory relates to trade. Whilst the early agricultural communities were abundant in crops they lacked other natural resources. Thus cities were formed as a direct result of the need to trade and procure other resources (Possehl 2002). The final theory relates to the notion that as communities grew in size, so did the scale of conflict between communities (Carneiro 1970; Spencer 2003). This would have led to a dominant group or community that led to the formation of a small state. As cities and states developed, warfare between adjoining areas would have led to the formation of larger political entities, driven largely by a desire for security, economic gain, or personal ambition of the rulers and elites.

It is most likely that multi-causal factors drive social complexity and these are required to explain how cities, states, and civilizations actually developed and formed.

Abrupt climate events may have had serious impacts on the rise and fall of civilizations, a number of which are discussed in the following section. The first complex, urban states are thought to have developed in Ancient Mesopotamia during the 6th millennium BP, with a transition from villages to integrated estates and eventually cities (Matthews 2005). Around 7000 cal yr BP occurred the development of the Ubaid period, which was characterized by the development of monumental architecture, pottery production, centralized storage facilities and long-distance trade, including lapis lazuli from Afghanistan and carnelian from the Indus Valley. Around 6000 cal yr BP marks the transition to the Uruk culture during which the first development of towns and cities occurs. The main centre of administration was in the city of Uruk-Warka in southern Mesopotamia. What the Uruk culture lacked in natural resources it imported from outside the region, including semi-precious stones, metals, timber, and stone for building and sculpture. The management of such a burgeoning system is perhaps responsible for the development of the world's earliest writing system, which was done on clay tablets (Potts 1997). The rise of the Uruk culture is thought to have led to increased conflict and fortification of settlements resulting from the competition for resources (Leick 2001). The Uruk culture was followed by the emergence of two city-states, the Akkadian and Babylonian. These were large-scale political entities often termed empires. The Akkadian Empire was initiated by Sargon (2334–2279 BC) and began with his conquest of the cities of Sumer and into northern Mesopotamia, south-east Anatolia and south-west Iran (Matthews 2005). The empire collapsed c. 4200 cal yr BP (see **Section 8.6** for details). This gave way for the later rise of the Ur III dynasty c. 4000 cal yr BP, centred on the city of Ur in southern Mesopotamia, and saw the construction of the massive ziggurat, large royal tombs, and temples. The Ur III kings governed their empires though the system of provincial governors, who recorded in detail the activities of empire.

In the Nile Valley the emergence of the Egyptian state took place after 5200 cal yr BP (Midant-Reynes 1992). In late Predynastic times there was a gradual movement of humans from the deserts to the Nile Valley, the abandonment of pastoralism, and the adoption of intensified agriculture supported by seasonal flooding and artificial irrigation from around 6000 cal yr BP. The emergence of the Egyptian Dynastic civilization increased population density leading to social stratification and the appearance of monumental architecture, including temples and large, ornate tombs (Rice 2003). A social elite developed, controlling the trade in raw material, with skilled workers rising to elevated status through their association with the early Pharoahs (Midant-Reynes 1992; Brooks 2006).

Around 5000 years ago the Indus or Harappan civilization developed along the valleys of the Indus and Ghaggar-Hakra. The Early Harappan changed over a very short period between 5.2 and 4.6 kyr from villages dependent on farming and pastoral activities into major urban centres such as Harappa and Mohenjo-Daro. The Indus civilization covered a region of approximately 500 000 km² from the Makran coast in the west, to the Ganges in the east, Gujarat in the south and Afghanistan in the north (Coningham 2005).

In China the emergence of states took place c. 4000 cal yr BP. The late Neolithic Longshan culture developed c. 5.0 kyr, during which social complexity evolved, including craft specialization in jade carving, ceramics, and an increase in artefacts associated with armed conflict. The Xia Dynasty arose c. 3.7 kyr ago and is regarded as the first dynasty of China. Its centre was located in the city of Erlitou, to the south of the Yellow River (Higham 2005).

In coastal Peru recent work in the valleys of the Norte Chico region has revealed the existence of a complex state-level society c. 5000 cal yr BP (Mann 2005). Over 20 urban complex sites have been identified, with evidence for monumental architecture, including pyramid structures and irrigation agriculture. Brooks (2006, p. 42) points out that 'these findings place the emergence of complex societies in the Americas within a few centuries of their counterparts in the Afro-Asiatic desert belt'.

It should be noted that a number of interesting observations have been made with respect to the development of civilizations and the types of agricultural practice employed:

1. That major civilizations stemmed from groups which practised cereal, particularly wheat, agriculture (e.g. in south-west Asia, Europe, India, parts of south-east Asia; central and parts of north and south America; Egypt, Ethiopia and parts of tropical and west Africa).

2. The rarer nomadic civilizations were based on dairy farming.

3. Groups which practised vegeculture of fruits and tubers, or practised no agriculture at all, did not become civilized to the same extent (e.g. in tropical and south Africa, north and central Asia, Australia, New Guinea and the Pacific, and much of the Americas).

8.6 Climate impacts on human society

Since the end of the last glacial, climate has not, as previously thought, remained stable (see **Chapter 5**) and the Late Glacial and Holocene record is punctuated with long- and short-term events. In recent years there has been an increasing availability of high resolution palaeoclimate records from terrestrial, marine, and ice records across the globe and a striking number of associations between climate and society have been suggested (deMenocal 2001) (**Table 8.5**).

The rise and fall of human societies has been driven by a complex set of factors including: climate, the availability of potable water, food, economic resources, and trade. Archaeologists have often witnessed key societal changes in their preserved records and often they have speculated on the factors behind gradual and abrupt changes.

Post-glacial shifts in aridity

During the Younger Dryas, Natufian peoples, who had occupied the Euphrates Valley of modern-day Syria, deserted much of the area as the climate became colder and drier. Conditions were no longer conducive to their sedentary way of life based on the harvesting of wild cereals and the hunting of gazelle. However, in

some areas (Levant and Zagros foothills) early settlements survived during this period of environmental stress through cultivation of cereal crops (as described in **Section 8.5**). The world's subtropical deserts were considerably wetter than today during the early to mid Holocene (see **Section 5.10**) between 10 and 6 kyr ago. Stalagmite evidence from southern Oman indicates that **monsoon** precipitation during the early Holocene (10 300 to 8200 cal yr BP) was in phase with temperature fluctuations recorded in the Greenland ice record (see **Fig. 5.7**). The Sahara was reoccupied c. 10 kyr; in central regions this was by hunter-gatherers following the monsoonal rain as they moved northwards into the interior (see **Fig. 5.21**) (Bubenzer et al. 2007).

There was an abrupt climate event at 8200 cal yr BP (see **Section 5.4**), which interrupted an otherwise humid period (Fleitmann et al. 2003; Brooks, 2006). Across the Near and Middle East it has been suggested that a drought lasting at least 200 years occurred at this time, which forced the abandonment of agricultural settlements in northern Mesopotamia and the Levant. In Arabia, after the 8200 cal yr BP event, the monsoon rains returned and led to the development of savannah grassland with interspersed *Acacia* tree cover (Parker et al. 2004). Into this landscape migrated the Neolithic herder-gatherers from the Near East with sheep, goats, and cattle. In south-eastern Arabia it has been suggested that **transhumance** was practised between the mountains in the summer and the Arabian Gulf coast in winter (**Fig. 8.10**). On the coast large Neolithic shell middens developed as part of this cycle (**Plate 8.4**). In the central and western regions of the Sahara cattle herding occurred from c. 7000 years ago, having been encouraged by climatic deterioration (Hassan 2002). The move to cattle herding is thought to be related to having predictability of food availability which hunting and gathering does not guarantee. In addition, as conditions became progressively drier, pastoralism based on sheep and goats, which are more tolerant of aridity, became more established (Brooks 2006).

From around 6.0 kyr a reduction in rainfall occurred across North Africa, the Near East, and Arabia owing to a weakening of the monsoon systems due to a reduction in Northern Hemispheric insolation (deMenocal 2000; Fleitmann et al. 2003; Bubenzer et al. 2007). Steig (1999) describes 6.0 kyr as the time of steepest decline in Northern Hemisphere radiation. This ended the practices of Neolithic herding in the desert interior of Arabia and North Africa and forced these peoples to move to the mountain uplands, coasts, oases, or

Table 8.5 Examples of some studies of climatic influences on human history in the Holocene

Location	Approximate date	Source
Collapse of classic Mayan civilization	AD 1000	Hodell et al. (1995)
		Neff et al. (2006)
Expansion of cattle keepering into the Sahara	Early to mid Holocene	Petit-Maire et al. (1997)
Abandonment of Greenland settlements	AD 1300–1500	Grove (2004)
Abandonment of Highland settlement in SE Scotland	AD 1300–1600	Parry (1978)
Anasazi collapse, SW USA	End of 13th century AD	Grove (2004)
Spread of cultivation into highland Britain	Medieval	Lamb (1982)
Decline of Middle Eastern empires (e.g. the Akkadian)	4200 cal yr BP	Weiss et al. (1993)
Culture change in the Atacama Desert, Chile	Mid Holocene	Grosjean et al. (1997)
Cultural change in coastal Peru	Mid Holocene	Wells and Noller (1999)
Domestication	Early Holocene/Late Pleistocene	Sherratt (1997)
Southwards expansion of Bantu	After c. 4200 cal yr BP	Burroughs (2005)
Neolithic settlement in Harz Mountains	7600–4550 cal yr BP	Voigt (2006)
Mid Neolithic decline In NE Morocco	6500–6000 cal yr BP	Zielhofer and Linstädter (2006)
Collapse of Tiwanaku state	c. 1000 years ago	Ortloff and Kolata (1993)
Emergence of Egyptian dynastic state	from c. 6000 cal yr BP	Brooks (2006)

riverine areas. This may have been a contributing factor for the development of city-states, as discussed above.

Around 5200 cal yr BP a century-scale rapid drying and cooling event took place across the Middle East, resulting in the migration of people in the Anatolian and Iranian plateaux. In southern Mesopotamia, intense drought led to the collapse of the urban centred Uruk culture, which had been dependent on irrigation for agriculture (Brooks 2006). In Arabia, as the rains failed and the grassland vegetation cover disappeared, the Neolithic herders of the Rub' al-Khali abandoned the desert interior in favour of upland and coastal sites (**Fig. 8.10**) (Parker et al. 2006).

Around 4200–4100 cal yr BP evidence for a major shift in climate from across a number of regions of the globe has been recognized. Perhaps the best documented is the collapse of the Akkadian Empire in Mesopotamia (Weiss et al. 1993). This culture was established on the broad, alluvial plain between the Tigris and Euphrates c. 4300–4200 cal yr BP under the rule of Sargon of Akkad. The empire was reliant on rain-fed agriculture in the upper reaches of river catchments along with irrigation agriculture in southern Mesopotamia. Around 4100 cal yr BP the empire collapsed owing to the onset of a prolonged drought. Archaeological evidence suggests that there was a large influx of refugees into southern Mesopotamia from the north, which led to the building of a 180 km long wall to stem the influx of people. At Tell Leilan in north-eastern Syria the occupation layer representing the societal collapse was overlain by 1 m of wind-blown silt that was sterile of any archaeological artefacts (Weiss et al. 1993).

Further west, in the Arabian Gulf region, there was a marked cultural change at 4100 cal yr BP between the Umm an-Nar and Wadi Suq periods (**Fig. 8.10**). Evidence from a palaeolake at Awafi, in the Emirate of Ras al-Khaimah, United Arab Emirates, revealed evidence for a short-lived, intense arid event leading to lake desiccation and infilling of the basin with sand (Parker et al. 2004) (**Plate 8.5**). The evidence for regional aridity and increased **aeolian** activity is supported from a deep-sea core from the Gulf of Oman, where mineralogical and geochemical analyses of the sediments revealed a large increase in wind-blown dust derived from Mesopotamia (**Fig. 8.11**) (Cullen et al. 2000). These indicated a 300-year aridity event commencing at 4100 cal yr BP.

This event has also been traced in sediment from Lake Van in Turkey and has been suggested as the cause of the collapse for the Indus Valley civilization in the north-west of the Indian subcontinent. However, some

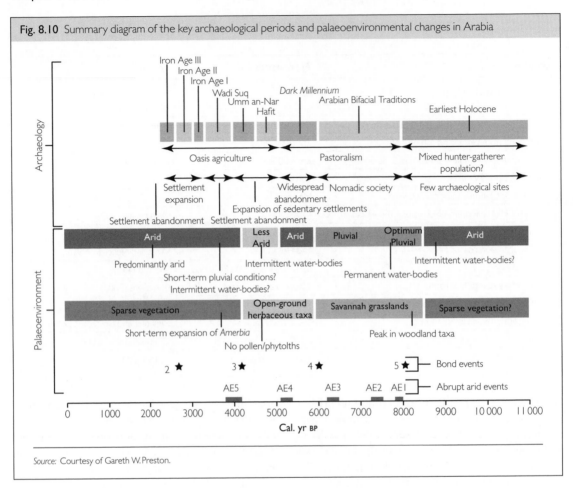

Fig. 8.10 Summary diagram of the key archaeological periods and palaeoenvironmental changes in Arabia

Source: Courtesy of Gareth W. Preston.

workers doubt the validity of a climate-driven cause for the collapse along the Indus and prefer socio-economic reasons, along with invasion, as the key mechanisms.

Outside the Middle East and south-west Asia, evidence for climate change at 4100 cal yr BP is recognized from the Mount Kilimanjaro **ice core**. In Egypt the downfall of the centralized government at 4.2 kyr coincides with diminished Nile flood discharge and the migration of dune sands in the Nile Valley. Booth et al. (2005) have demonstrated how severe drought affected the mid-continent of North America between 4.3 and 4.1 kyr, with large and widespread ecological effects, including dune reactivation, forest fires, and long-term changes in forest composition, highlighting a clear ecological vulnerability of the landscape to abrupt climate change.

Some regions at high latitudes, including northern Europe and Siberia, experienced cooler and/or wetter

conditions at *c.* 4200 cal yr BP. It has been inferred that widespread mid-latitude and subtropical drought, associated with increased moisture at some high latitudes, has been linked in the instrumental record to an unusually steep sea surface temperature (SST) gradient between the tropical eastern and western Pacific Ocean (La Niña of the **ENSO**) and increased warmth in other equatorial oceans. Similar SST patterns may have occurred at 4.2 kyr, possibly associated with external forcing or amplification of these spatial modes by variations in solar irradiance or volcanism. This phase of change also correlates with an episode of glacier advance in western Canada (Menounos et al. 2008) as well as increased ice rafted debris in North Atlantic sediment cores (Bond et al. 1997).

The Mapungubwe agro-pastoral society in the Limpopo Valley, South Africa, persisted for 300 years before disappearing in about AD 1290 as a result of a

Fig. 8.11 Aeolian mineral concentration data for the 6000–2000 calendar yr BP interval in core M5-422 from the Gulf of Oman. The peak in mineral dust concentration c. 4.2–4.0 cal kyr BP corresponds with collapse of the Akkadian Empire in Mesopotamia. The collapse is thought to have been driven by an intense phase of aridity lasting approximately 300 years (from Cullen et al. 2000, Fig. 4). Reprinted with permission of the Geological Society of America

decrease in mean annual rainfall from about 500 mm to the current 340 mm (O'Connor and Kiker 2004).

In Central America the pre-classic Maya culture occupied the lowland and highland regions between the second millennium BC and AD 250. From AD 250, in the Classic Maya periods, there was rapid growth and the development of expansive urban centres and trade networks spanning Mesoamerica. In the Yucatan peninsula at c. 1200 cal yr BP the Maya culture collapsed at its peak (Neff et al. 2006). Sediment composition and **stable isotope** analyses of **ostracods** in lake sediments reveal the region was subjected to a 200-year-long drought. The lowland Maya people were dependent on surface water supplies for agricultural as well as human use. These sources became exhausted during this prolonged drought and the human population of the region declined rapidly. The Quelccaya ice core record from Peru shows an

increase in dust accumulation at the same time as the Mayan collapse, supporting the notion of reduced rainfall across the tropical regions of the Americas (Thompson et al. 1995).

It has been suggested that a number of the key arid events identified above from across the low latitudes are due to weakened monsoonal rainfall. The causal mechanism is likely to be linked with weakening of the North Atlantic conveyor system (**thermohaline circulation**) which resulted in lower sea surface temperatures in the tropical seas of the Northern Hemisphere – a theme discussed further in **Chapter 9**.

The Medieval Warm Period (Little Optimum) and the Little Ice Age

In much of northern Europe, including Britain, the **Little Optimum** (also known as the **Medieval Warm**

Period or MWP, see **Section 5.9**) and some of the preceding centuries were times of extension of settlement to marginal and highland areas. In addition, Scandinavian settlers colonized the islands of the North Atlantic. Between AD 800 and 1000, the Norse settled the Orkneys, Shetland, the Faroes, Iceland, and the coastal region of southern Greenland.

In Greenland the Norse settled three main areas along the coast during the MWP. An estimated 500 plus farms and a population of *c.* 4000 people were present during its peak. In the Eastern Settlement of Greenland, Edwards et al. (2008) have demonstrated that the pre-Norse landscape comprised *Betula pubescens* (birch) woodland. Around AD 950–1020, Norse settlers cleared the tree cover, aided by the use of fire, to provide grassland pasture for domestic stock, mainly sheep and cattle. Analysis of soils and pollen from Norse settlements give insight into the development of agriculture, and in particular hay production, along with the use of irrigation channels between the mid eleventh and late fourteenth centuries (Golding et al. 2011). In addition marine resources supplemented the Norse diet at this time.

Climatic deterioration in the fourteenth century may have decreased productivity of hay meadows and agricultural output with a greater reliance on winter fodder to feed livestock. Overgrazing may have led to accelerated levels of soil erosion leading to the abandonment of some farms (Edwards et al. 2008). The analysis of isotopes from bones has shown that fishing played an increasing economic role towards the end of the Norse settlement in Greenland. Arneborg et al. (1999) showed that the Norse settlers' diet changed from 80% agricultural products and 20% marine food during the initial colonization to 80% marine and 20% agricultural from the fourteenth century.

Climate change and cooling during the **Little Ice Age (LIA)** first led to abandonment of the Western settlement. In addition, harsher winters, fewer visits from trading ships, competition for resources from the Inuit, environmental degradation, and the possibility of inbreeding and disease led to the decline and eventual disappearance of the Norse from Greenland by the mid fifteenth century.

During the Little Ice Age (see **Section 5.9** for a discussion of its climate) the population of Iceland declined from over 70 000 people to less than 40 000 in 400 years. This decline was due to a series of severe cold years and volcanic eruptions. This led to the abandonment of cereal growing and forced the population

to depend solely on fish and shellfish for their survival. The decrease in sea temperatures is thought to have affected fish stocks, and in particular in causing cod to migrate southwards out of range of the fishing ships.

In Finland the impact of climatic deterioration on agrarian populations during the Little Ice Age was analysed from historical weather diaries and rye and barley grain figures (ratio between sown and harvested grain) between the sixteenth to the nineteenth centuries (Holopainen and Helama 2009). It was found that the desertion of farms in the sixteenth and the seventeenth centuries correlates in time with long-term summer temperature cooling resulting in poor crop yields that significantly impeded agriculture as a means of subsistence.

In central Norway, the limits of settlement, forest clearance, and farming were pushed 100 to 200 m farther up the hillsides and valleys in Viking times AD 800–1000, whilst in many parts of Britain there is evidence of medieval tillage on high ground far above anything that would be reasonable now, even in wartime (Lamb 1966). For instance, the thirteenth century limit of tillage in Northumberland appears to have been around 300–350 m above sea level, 120–150 m above the viable limit at the present day. This was also a period when medieval vineyards spread into a number of localities in southern and eastern England (Unwin 1990). After a while, however, these extended settlements underwent fairly considerable decline, and much of this decline started before the Black Death. Of nearly 50 deserted villages in Oxfordshire and 34 in Northamptonshire only about 10 per cent were attributable to the Black Death of AD 1348. All appear to have suffered serious decline in the years of disastrous summers and famines between AD 1314 and AD 1325 (Lamb 1967).

In Scotland, it has been proposed that secular deterioration of climate since the early Middle Ages caused much of the high-lying cultivation and grazing in south-east Scotland to become profoundly sub-marginal in the seventeenth century. During the LIA the growing season was shortened by around a month. Grazing on the summer pastures did not start until May and ended by September. This led to overgrazing and reduced productivity. In addition, the altitudinal range of grazing became reduced due to more extreme weather conditions. The coupled effects of a shorter growing season and the lowering in altitude at which both crops could be grown and animals grazed severely reduced productively (Parry 1975, p. 5–11; Tipping 2004). The pressure, however, was not limited only to the higher-altitude summer grazing but also impacted hay production in low-level meadows, diminishing supplies of winter fodder (Oram and Adderley 2008).

The consecutive harvest failures of the 1690s and the 1780s may have been the immediate stimulus to abandonment of many areas, but the response to these stimuli would neither have been so widespread nor so permanent if the potential for cropping in the upland areas had not been so severely reduced over the preceding three centuries. On the basis of the data presented by Lamb and others for the degree of climatic change, and by a study of the present-day climatic limitations on oat-ripening, Parry (1975) suggested that in the early MWP the chances of crop failure were only about one year in twenty in the Lammermuir area. By the mid fifteenth century this had grown to one year in three, and in all about 4950 ha of land seem to have been abandoned in the Lammermuir-Stow Uplands.

Figure 8.12 plots the progress of the climatic deterioration in that area using two parameters. One measure of the intensity of summer warmth is the accumulated temperature calculated over a base of 4.4 °C . The isopleth of 1150 day-degrees C showed a marked correspondence with the 1860 cultivation limit. PWS (Potential Water Surplus) is a measure of summer wetness, expressed as the excess of a middle and late summer surplus (up to 31 August) over an early summer deficit.

deMenocal (2001, p. 672), in addressing the influence of climate on human-society relationships, states:

The climatic perturbations associated with ... Holocene societal dislocations were extreme in their duration and intensity, far surpassing droughts recorded during the modern instrumental period ... well-dated and detailed palaeoclimate records from climatically sensitive locations bear witness to the occurrence and severity of these multi-decadal-to-multi-century droughts.

These chilling words bear testimony to the impact that climate can have upon societies, especially within marginal regions. It should also be noted that the driving

Fig. 8.12 Climatic deterioration AD 1300–1600 in the Lammermuir-Stow uplands of Scotland represented by the fall of continued isopleths of 1050 day degrees C and 60 mm PWS (modified from Parry 1975, Fig. 5)

mechanisms behind many of the events described above are natural and not anthropogenic in nature.

8.7 Pre-industrial human impacts on the environment

Human impacts on the environment, whether accidental or deliberate, have been profound during the Holocene. Hunter-gathering, the use of fire, the advent of agriculture, development of city-states, trade, economy, and the exploitation of material resources have directly and indirectly changed the landscape.

In Mesopotamia the increase in the scale of irrigation led to an increase in the levels of salt in the soils because of the evaporation of surface water. Sumerian written records outline the salinity problem, which record declining harvest yields from 2500 litres per hectare at 4400 cal yr BP to less than 900 litres some 700 years later. As the soils became more saline there was a shift from emmer wheat, to einkorn wheat, and eventually to the more salt tolerant barley. As a result it is thought that the southern alluvial plain declined in political importance (Jacobsen 1982).

Terraced agriculture is a common feature of mountain landscapes, and is often related to restricted space availability for pasture and the growing of crops. In prehistory, terracing played a major role in the development of the physical landscape, as it was a factor in promoting landscape stability. In addition, terracing promotes soil depth, moisture management, and the modification of microclimate. In the southern Levant, terraced fields are thought to date from the middle to late Bronze Age. In the Yemen highlands of southern Arabia valley floor terraces can be traced back as far as the late Neolithic (6000–5000 cal yr BP), whilst valley-slope terraces are dated from 5000–4000 cal yr BP. By 3000 cal yr BP a significant amount of the Yemeni landscape had been terraced (**Plate 8.6**) (Wilkinson 2005).

The impacts of ancient mining provide an important background to phases of resource exploitation and development of technology, economy, and trade. In the Taurus Mountains of Turkey the remains of over 800 mines have been recorded from the Bolkardağ district (Yener and Özbal 1987). These produced metals and metal ores of gold, silver, lead, zinc, iron, and tin from the early Bronze Age. Such industry is reliant on the availability of wood resources for smelting and it is likely that woodland was managed by coppicing

to conserve the supply of available fuel. While there is evidence for burning and some colluviation there is little evidence to suggest that the mining had exerted a massive stress on the local environment. In contrast to this, extensive erosion resulting from clearance of vegetation and burning associated with the smelting of iron in Swaziland during the nineteenth century has led to the development of deep, eroded gullies known as 'dongas' (**Plate 8.7**).

In the northern Oman Mountains extensive copper deposits have been mined from Bronze Age times. The region is known from early Mesopotamian texts as the land of Magan and it provided copper to the city-states of Sumer and Akkad via trade routes across the Arabian Gulf. Extensive mining continued in the Islamic period, when there is abundant evidence for mines and smelting sites. In Ras al-Khaimah, UAE, the onset of widespread dune reactivation c. 1000 cal yr BP was linked with the large-scale removal of trees in support of copper smelting – an important trade commodity during the Islamic period. This is possibly the earliest evidence yet recorded for **desertification** caused by overgrazing of the landscape to support the local population and clearance of tree cover to support the smelting of copper ore (Stokes et al. 2003).

In lowland Britain alluviation occurred in the river catchments from the mid Holocene onwards. In the River Thames Basin a late Bronze Age and Iron Age rise in the regional water table has been attributed to wetter conditions driven by climate (Parker 2000). Despite this, extensive alluviation did not occur at first. However, widespread forest clearance, the advent of the iron plough and the sowing of winter cereals, especially in the surrounding upland areas such as the Cotswolds, led to widespread soil erosion, large-scale flooding and extensive alluviation during the late Iron Age and Roman periods (Robinson and Lambrick 1984). Although alluviation continued throughout the rest of the Holocene it was at a much reduced scale until the late Saxon and Medieval periods when population pressure and expansion led to an increase in alluviation once again.

In continental Europe a similar pattern of alluviation associated with Bronze Age forest clearance and also early Medieval settlement has been recorded (e.g. in the Sudeten Mountains of Poland). In the Glubczyce **loess** Plateau of the Upper Odra Basin in southwestern Poland, alluvial fans developed as a result of sediments derived from forest clearance, intensive agricultural activity, and slope instability during the

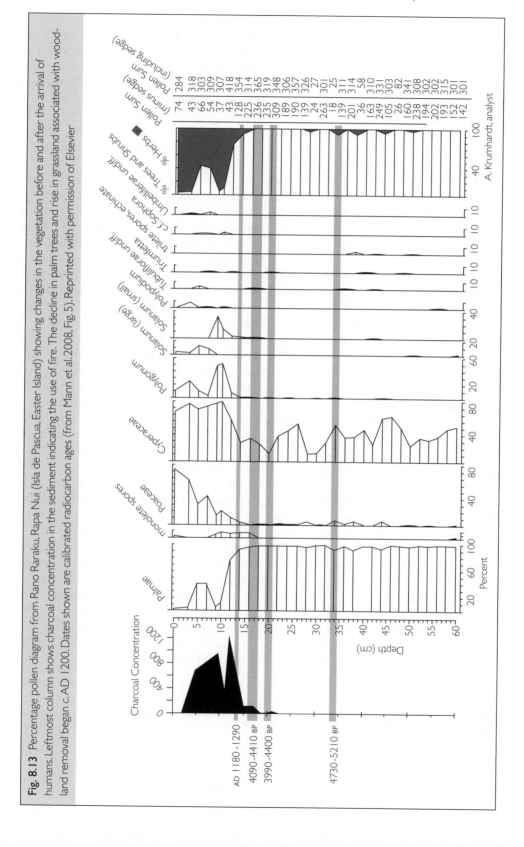

Fig. 8.13 Percentage pollen diagram from Rano Raraku, Rapa Nui (Isla de Pascua, Easter Island) showing changes in the vegetation before and after the arrival of humans. Leftmost column shows charcoal concentration in the sediment indicating the use of fire. The decline in palm trees and rise in grassland associated with woodland removal began c. AD 1200. Dates shown are calibrated radiocarbon ages (from Mann et al. 2008, Fig. 5). Reprinted with permission of Elsevier

Bronze Age and at the beginning of the Iron Age (Zygmunt 2004).

Sediment cores from Lake Patzcuaro, central Mexico, indicate extensive soil erosion events prior to the introduction of livestock-rearing and ploughing by the Spanish in the sixteenth century AD. Soil erosion into the lake, indicated by a predominance of clay and minerals associated with agricultural land, occurred 3600 to 2900 years ago and corresponded to the rise of maize cultivation in central Mexico. A more intensive period of soil erosion took place between 2500 and 1200 cal yr BP. A third phase took place between 850 and 350 years ago, reflecting extensive forest clearing by the Tarascan people, who dominated the region at the time of the Spanish conquest in AD 1521 (O'Hara et al. 1993).

Easter Island occupies an exceptionally isolated position in the Pacific Ocean. It is entirely volcanic, and is famous for its giant statues. Late Quaternary sediments have been investigated in three craters – Rano Raraku, Rano Aroi, and Rano Kao – giving a continuous record over the past 30 000 years. Pollen records indicate that the island was formerly forested.

Although it remains unclear when humans first discovered Rapa Nui (Flenley and Bahn 2007), the onset of deforestation, agricultural settlement, and burning in the landscape started around AD 1200 (**Fig. 8.13**) (Hunt 2007; Mann et al. 2008). Between AD 1200 and AD 1650 huge sculptures were erected across the island. It has been proposed that the natural vegetation was destroyed by a large population explosion, resulting in environmental degradation which led to the societal collapse and near extinction of the human inhabitants. It has been suggested that by felling all the trees, partly to help move large stone sculptures around the island, the people destroyed their ability to make boats and to escape from the ecosystems they had degraded (Flenley and King 1984).

An alternative scenario has been suggested by Hunt (2007) who proposed that deforestation and biological impoverishment occurred rapidly after human arrival but then stabilized. The population maintained itself in much smaller numbers than previously suggested and that the eventual collapse of this society was triggered by new diseases resulting from contact with European explorers. Hunt (2007) also suggested that the initial deforestation of Rapa Nui was due both to the Polynesian settlers and the rats they introduced, which damaged the flora and in particular the seeds.

In addition to the above examples of pre-industrial human impact on landscapes, W.F. Ruddiman has proposed that humans began modifying the global climate during prehistory. It has long been thought that, despite local and regional modifications to the environment, people did not begin to have a significant impact on the global climate until the industrial revolution when the concentration of greenhouse gases in the atmosphere began rising sharply due to the burning of fossil fuels. This view has been challenged by evidence from ice cores showing that from 8000 years ago atmospheric CO_2 began increasing in contrast with the pattern seen during previous interglacials; probably as a result of deforestation across large areas of Europe and southern Asia as agriculture spread. At c. 5500 years ago atmospheric methane also began increasing, in this case because of the dramatic expansion of rice paddy cultivation in southern Asia. It is estimated that the human-caused elevation in the concentration of greenhouse gases from 8000 years ago up to the start of the industrial revolution could have contributed an average global warming of 0.8 °C – a value slightly higher than the 0.6 or 0.7 °C of warming averaged for the Earth's surface over the past 100 years (Ruddiman 2005a,b). From climate modelling studies, Ruddiman and colleagues have also suggested that the global mean temperature today would probably be almost 2 °C cooler than it is now if it were not for the combined effect of pre-industrial and industrial human enhancement of the **greenhouse effect**. The role of greenhouse gases in Earth's climate over different timescales and the 'Ruddiman Hypothesis' are discussed further in **Section 9.6**.

SELECTED READING

A number of key texts have been published which explore the evolution of hominins. This is a controversial area with varying thoughts on the ancestral development of humans. Good examples to explore include: C. Scarre (ed.) (2005) *The Human Past: World Prehistory and the Development of Human Societies*, C. Stringer and P. Andrews (2005) *The Complete World of Human Evolution*, and R. Boyd and J.B. Silk (2011) *How Humans Evolved*, 6th edition.

Although now old, an important book for Late Pleistocene megafaunal extinctions is *Quaternary Extinctions: A Prehistoric Revolution* edited by P.S. Martin and R.G. Klein (1984). Species extinctions during the Holocene are reviewed in *Holocene Extinctions* (2009) edited by S.T. Turvey.

The origins of agriculture are addressed in greater depth by P. Bellwood (2005) *First Farmers: the Origins of Agricultural Societies* and by D. Rindos (1984) *The Origins of Agriculture: an Evolutionary Perspective*.

Details of regional archaeology can be found in the following books: I. Lilley (ed.) (2006) *Archaeology of Oceania: Australia and the Pacific Islands*, P. Mitchell (2002) *The Archaeology of Southern Africa*, T.J. Wilkinson (2003) *Archaeological Landscapes of the Near East*; and the book edited by B. Cunliffe (1994) *The Oxford Illustrated History of Prehistoric Europe*.

W.J. Burroughs (2005) provides an accessible overview linking climatic change with prehistoric events in *Climate Change in Prehistory: The End of the Reign of Chaos*.

For details on the rise of civilization see B. Fagan (2005) *The Long Summer: How Climate Changed Civilization* and B. Trigger (2003) *Understanding Early Civilizations: A Comparative Study*.

REVIEW QUESTIONS

1. Is climate change the key driver in the evolution of hominins?

2. Compare and contrast the various models proposed for the dispersal of hominins. What impact has recent genetic studies had on these theories?

3. What evidence is there to support the view that modern humans left Africa post 60 kyr ago via a coastal route and rapidly reached Australia in a single dispersal?

4. To what extent was the 4.2 kyr event global in nature? Discuss with reference to the palaeoenvironmental and archaeological records.

5. What impacts did the 'Medieval Warm Period' and the 'Little Ice Age' have on human society–environment relationships in Europe?

9

The Causes of Climatic Change

Chapter overview

Why has there been environmental change during the Quaternary? This is a complex and intriguing question. How do we explain the frequency, timing, and abruptness of change? Many hypotheses have been put forward that involve changes in solar activity, in the orbit of the Earth with respect to the Sun, in the amount of greenhouse gases, dust and other particles in the atmosphere, in the geography of the oceans, land masses and mountains, and in the reflectivity (**albedo**) of the Earth's surface. Change may be accentuated and made more abrupt as a result of a range of feedback mechanisms and the crossing of 'thresholds' within the climate system. Unravelling the causes of Quaternary climatic change is made more complex because of the interrelationships between different 'forcing mechanisms' and because different factors become more or less important depending on the timescale that is studied. Finally, particularly since the start of the industrial revolution, humans have themselves had an increasingly potent influence on the nature of our climate.

9.1 Introduction

The climatic changes that have been established and described, and which formed the basis for associated environmental changes such as those of sea level described in **Chapter 6**, have caused a great deal of discussion about their causes. The purpose of this chapter is to introduce and summarize some of the main opinions that have been put forward, to stress the variety of factors involved, and to show the doubts still associated with the major hypotheses.

An indication of the complexity of factors that needs to be considered in any attempt to explain climatic change is given in **Fig. 9.1**. This flow diagram starts with the ways in which the input of solar radiation into the Earth's atmosphere can fluctuate. For reasons such as varying the tidal pull being exerted on the Sun by the planets, the quality and quantity of outputs of solar radiation may change. The receipt of such radiation in the Earth's atmosphere will be affected by the position and configuration of the Earth and by such factors as the presence or absence of interstellar dust. Once the incoming radiation reaches the atmosphere

its passage to the surface of the Earth is controlled by the gases, moisture, and particulate matter that are present. These materials may either be of natural or man-made origin. At the Earth's surface the incoming radiation may be absorbed or reflected according to the nature of the surface (the albedo). The effect of the received radiation on climate also depends on the distribution and altitude of land masses and oceans. These too are subject to change in a wide variety of ways – continents may move to or from areas where ice caps might accumulate, mountain belts may grow or subside to affect world wind belts and local climates, and the arrangement of the climatically highly important ocean currents may be controlled by changes in sill depths and the width of the seas, oceans, and channels. The situation is complicated, as the flow diagram suggests, by the existence of various feedback loops within and between the ocean, atmosphere and land systems.

In addition it needs to be remembered that the potential causative factors in climatic change operate over a very wide range of different timescales, so that some factors will be more appropriate than others to

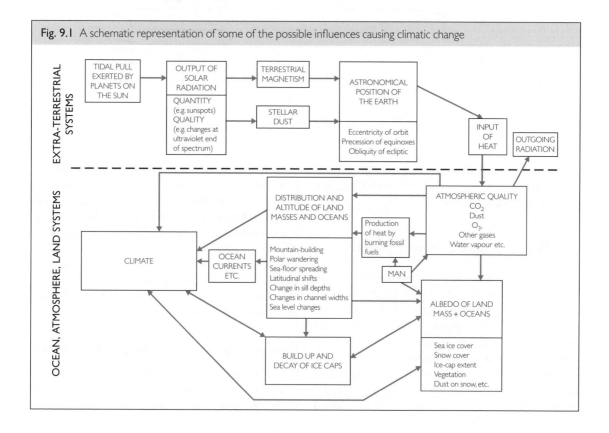

Fig. 9.1 A schematic representation of some of the possible influences causing climatic change

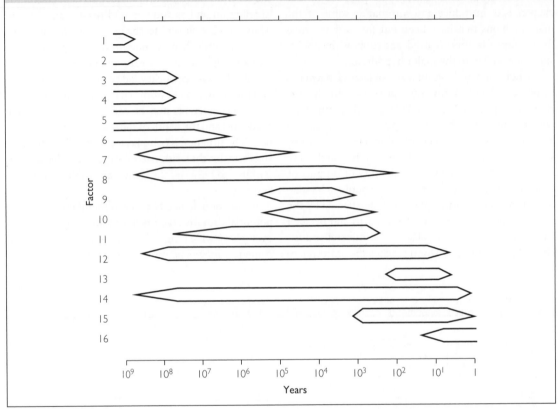

Fig. 9.2 Potential causative factors in climatic change and the probable range of timescales of change attributable to each (after Mitchell 1968, p. 156): (1) evolution of Sun; (2) gravitational waves in the Universe; (3) galactic dust; (4) mass and composition of air (except CO_2, H_2O, and O_3); (5) polar wandering; (6) continental uplift; (7) orogeny and continental uplift; (8) CO_2 in the air; (9) Earth-orbital element; (10) air, sea, ice cap feedback; (11) abyssal ocean circulation; (12) solar variability; (13) CO_2 in the air (fossil fuel combustion); (14) volcanic dust in the stratosphere; (15) ocean-atmosphere auto-variation; (16) atmosphere auto-variation

account for a climatic fluctuation or change of a particular span of time. An attempt to show this diagrammatically is made in **Fig. 9.2**.

9.2 Solar radiation hypotheses

'The history of meteorology is littered with whitened bones of claims to have demonstrated the existence of reliable cycles in the weather' (Burroughs 1992, p. xi). Changes in the output of solar radiation have played a central role in the quest for weather cycles, and have often been invoked to explain decadal to century timescale climatic changes during the Holocene. Indeed, following through the flow diagram (**Fig. 9.1**) it is clear that changes in the receipt of solar radiation

at the Earth's surface should cause changes in climate. Moreover, it has long been recognized that the Sun's radiation changes both in quantity (through association with those enigmatic blemishes on the face of the Sun – sunspots) and in quality (through changes in the ultraviolet range of the solar spectrum). The problem is to establish firm connections between solar changes and changes in the Earth's climate. Support for a possible solar-climate connection has aptly been described as 'flickering' (Crowley and North 1991, p. 106).

Cycles of variation and effects on climate

Cycles of solar activity have been established for the short term by many workers (see Meadows 1975 and Burroughs 1992), with 11- and 22-year cycles being the ones particularly noted and measured instrumentally. Cycles in solar activity of *c.* 90, 210, 2300, and 6000 years have also been postulated. Over the period of instrumental record it has been found by some workers (see, for example, Wood and Lovett 1974) that there has been some correlation between, for example, sunspot activity and East African rainfall and lake levels. Sometimes, however, the correlations may suddenly break down, while other correlations may not necessarily be statistically significant. Nevertheless, some of the more significant associations may have predictive value.

Although the role of changes in solar activity has frequently been attacked, especially with regard to cycles, some striking correlations have been found at certain times between changes in solar activity and certain major characteristics of temperature, general atmospheric circulation, and other climatic variables. There has been an increase in sunspot cycle amplitude and overall solar activity since the end of the **Little Ice Age (LIA)** (**Fig. 9.3**), and it has been suggested that this accounts for some of the warming averaged across the Earth during the late nineteenth and early twentieth centuries. However, in a recent review Benestad and Schmidt (2009) claim that any contribution to twentieth-century warming was unlikely to have amounted to more than 0.1 to 0.2 °C and that since 1980 solar forcing has had a negligible effect on the observed climate trend.

One difficulty with solar radiation hypotheses concerns the relatively low range of solar irradiance change associated with sunspot cycles. For instance, measurements of the solar constant over recent 22-year solar cycles indicate variation of only about 0.1 per cent around the mean (Crowley and Kim 1993) and through the Little Ice Age up to the present, variation has probably not exceeded 0.25 per cent. Modelling studies have suggested that even a 0.5 per cent reduction in the solar constant (at the upper end of estimates for total solar irradiance variability spanning the Holocene) would only directly cause about 0.5 °C of global cooling (Rind and Overpeck 1993). Steinhilber et al. (2009) estimated that total solar irradiance variation over the past 9300 years equates to *c.* 2 W m^{-2}. Therefore advocates of solar forcing have had to invoke a variety of different feedback mechanisms and non-linear responses to explain the correlations that have been identified between solar cycles and climate (Stuiver and Braziunas 1993). For example, Svensmark and Friis-Christensen (1997) proposed that cooling associated with reduced solar radiation is amplified by the associated increase in the cosmic ray flux which has the effect of enhancing cloud formation and thereby increasing planetary albedo. Other proposed feedbacks involve changes in the 'spectral solar irradiance' which varies the per cent of solar energy that arrives as ultraviolet. This is thought to influence the production of ozone in the stratosphere, and through that the atmosphere's overall radiation balance and the latitudinal position of large-scale features such as the jet streams and descending limbs of the Hadley Cells. In their review, Beer and van Geel (2008) suggested that this mechanism was probably more significant during the Holocene than the cosmic ray-cloud formation hypothesis.

Fig. 9.3 Sunspot numbers since AD 1600. The solid line indicates the trend in sunspot cycle amplitude peaks, with dashed lines enclosing the 1-sigma confidence range. The 'Maunder Minimum' correlates with one of the coldest periods of the Little Ice Age (from Hathaway and Wilson 2004, Figure 9). Reprinted with permission by Springer-Verlag

The observation of sunspots as a proxy for solar activity has a long history (Harvey 1978). Telescopic observations on sunspots exist since AD 1610, while naked eye sunspot observations, though less reliable, exist from 527 BC. One very striking feature of the record, as shown in **Fig. 9.3**, is the near absence of sunspots between AD 1645 and AD 1715, a period sometimes called the **Maunder Minimum**. It is perhaps significant that this minimum occurred during some of the more extreme years of the Little Ice Age (Shindell et al. 2001). An earlier sunspot minimum also during the LIA was the Spörer Minimum between AD 1460 and 1550. These and other prolonged periods of low sunspot activity are called 'grand solar minima'.

Over a longer timescale it is more difficult to show that the Sun's output of radiation varied enough to significantly affect the Earth's climate. Nevertheless, this is a hypothesis which has received much support. Some evidence to support it comes from studies of the oscillation in the concentration of atmospheric ^{14}C, which in turn depends partly upon variation in the emission of solar radiation. When solar radiation is reduced, there is less shielding of the atmosphere from the incoming cosmic ray neutrons that transform nitrogen atoms into ^{14}C in the upper atmosphere – hence more ^{14}C is produced. A potential problem for reconstructing variations in solar output from ^{14}C studies is variation in the global carbon cycle (for example changes in ocean circulation have a major influence on fluxes of CO_2). However through analysis of other cosmogenic isotopes, such as beryllium-10 (^{10}Be) concentrations in **ice cores**, it is possible to isolate the solar output signal (Steinhilber et al. 2009).

As introduced in **Section 2.5** in relation to radiocarbon dating, ^{14}C levels have fluctuated during the Holocene, and Denton and Karlén (1973) have argued that the major intervals of high atmospheric ^{14}C activity coincide with periods of **neoglaciation** (at c. 300, 2800, and 5300 cal yr BP), while the intervals of relatively low ^{14}C activity coincide with intervals of glacier contraction. Karlén and Kuylenstierna (1996) have compared a chronology of climatic changes, based on a large number of ^{14}C dates from *Pinus sylvestris* wood samples, proglacial lacustrine sediments, and alpine glacier moraines from Scandinavia, with an index of solar activity obtained from the deviation between dendro-age (the age obtained by counting of annual tree rings) and the ^{14}C age of tree rings. They found that a correlation existed between these records, which suggested to them that solar forcing did indeed

affect the climate during the Holocene. For most of the last 9000 years, they suggest, there has been a good correspondence between the timing of cold events in Scandinavia and the timing of major δ^{14}C anomalies (associated with low solar irradiation).

Magny (1993) presented correlations between lake-level fluctuations in the Jura and French subalpine ranges and varying solar activity: lake **transgressions** at c. 400, 2800, and 5500 cal yr BP correlate well with the pattern of neoglacial expansion discussed by Denton and Karlén. The c. 2800 cal yr BP event identified by Magny is likely to be equivalent to the important climatic deterioration discussed by van Geel et al. (1996, 1999) that they relate to the traditional SubBoreal–SubAtlantic transition, and which they argue was caused by solar forcing. This change, and others, also appear in a reconstruction of past water table heights from British peatlands (Charman et al. 2006). A more recent study of glacier length records and laminated lake sediments from the European Alps shows good agreement with solar forcing – glacier advances are associated with episodes of grand solar minima over the past 1000 years, perhaps due to reductions in summer season ablation related to a combination of solar minima and other factors. Furthermore, broad agreement is seen between Alpine glacier advance and retreat and the long-term 'Hallstatt cycle' of solar activity operating with a 2000 to 2400 year periodicity (Nussbaumer et al. 2011).

Further support for a connection between solar variability and Holocene climate comes from the study by Bond et al. (2001) of Holocene layers of ice rafted debris (IRD) in subpolar North Atlantic deep-sea sediments. This study showed a similarity between the dates of IRD layers (indicating south and eastward extension of drift ice and cooler sea surface temperatures) and phases of lower solar output as inferred from ^{14}C and ^{10}Be reconstructions. Their reconstruction, which extends back 12 000 years, also agrees with the recent **Medieval Warm Period** (MWP)–Little Ice Age pattern. They also proposed that this solar-climate link in the North Atlantic has correlatives in **monsoon** strength and precipitation changes in parts of the low latitudes and have suggested that the response of the North Atlantic region in general may be an important climate amplifier of relatively weak variations in solar irradiance.

In summary, there are grounds for accepting that variation in solar output has a discernible influence on climate; however, its importance continues to be debated, and there remains much uncertainty about precisely how such small changes in total and spectral solar irradiance are amplified to produce the climate effects that have been attributed to them. In addition to the cosmic ray and stratospheric ozone feedbacks mentioned previously, it has also been suggested that the *length* of the time intervals of relatively high or low solar activity might be as important as the magnitude of the changes for influencing climate on centennial to millennial timescales.

The causes of variations in solar energy receipt over much longer timescales are still imperfectly understood, but one possible cause of variability is the presence of clouds of fine interstellar matter through which the Earth might from time to time pass, or which might interpose themselves between the Sun and the Earth. These would tend to lead to a reduction in the receipt of solar radiation. Similarly, the passage of the solar system through a dust lane bordering a spiral arm of the Milky Way galaxy might cause a temporary variation of the radiation output of the Sun, and so contribute to an ice epoch on Earth (McCrea 1975).

9.3 Atmospheric transparency hypotheses

The effects of incoming radiation from the Sun are moderated by changes in the atmospheric composition of the Earth. Particular importance has been attached to the role of volcanic dust emissions (see Chester 1988, for a review). The presence of elevated dust levels in the atmosphere increases the backscattering of incoming radiation (thus encouraging cooling). In addition volcanic dust might reduce sunshine totals further by promoting cloudiness, for dust particles, by acting as nuclei, can promote the formation of ice crystals in sub-freezing air saturated with water vapour. An important climatic effect of volcanic eruptions occurs due to the emission of sulphur dioxide gas: this gas is converted into tiny sulphuric acid droplets which can block incoming radiation and reduce atmospheric temperatures (Bluth et al. 1993). Long-term records of volcanic sulphate deposition are provided in ice cores (e.g. Zielinski et al. 1994, 1996; Castellano et al. 2005). Lamb (1970) has been a major exponent of the idea that climatic deterioration could be produced by the volcanic production of a dust veil in the lower stratosphere.

More specifically, the effect of a volcanic eruption is an increase of the optical thickness of the atmosphere involving a reduction of the insolation reaching the Earth's surface. Such a reduction can have a relatively large impact which lasts about one to three years, and atmospheric cooling can be sustained longer than this when successive large eruptions cluster together in time. In their recent modelling study, Zhong et al. (2011) simulated one way in which heightened volcanic activity can trigger climate cooling which lasts longer than would be expected given the short residence time of volcanic **aerosols** in the atmosphere: given the right spacing between explosive eruptions, multi-year cooling causes Arctic sea ice to expand which in turn exerts a positive feedback on cooling through reducing ocean–atmosphere heat exchange, through the **ice-albedo feedback**, and through suppressing the transport of warm water to higher latitudes.

Hence, the radiative and chemical effects of aerosol clouds can produce non-linear responses in the climate system. By scattering the solar radiation back to space, aerosols lead to a cooling of the surface, but by absorbing both solar and terrestrial radiation the aerosol layer heats the stratosphere. It has been suggested that sulphur aerosol emissions from volcanic eruptions also break down stratospheric ozone. Large volcanic eruptions in tropical latitudes disperse aerosols widely throughout both hemispheres tending to cause a more global climate response than higher latitude eruptions. Northern Hemisphere temperatures tend to respond more strongly to an eruption-induced change in the radiation balance due to the higher percentage of land cover compared with the Southern Hemisphere, and temperature decline is more uniformly felt in summer as opposed to winter (Wanner et al. 2008).

Correlations between volcanism and climatic change

The ash emissions of Krakatoa in the 1880s and Katmai in 1912 produced a global decrease in solar radiation of 10 to 20 per cent lasting 1 to 2 years, and the ash from Krakatoa was injected into the stratosphere, reaching a height of 32 km. It has been shown that many of the coldest and wettest summers in Britain, such as 1695, 1725, the 1760s, 1816 (the 'year without a summer'), the 1840s, 1879, 1903, and 1912 occurred in conjunction with times of high volcanic dust inputs into the stratosphere and upper atmosphere (Lamb 1971). Moreover, the period of warming in the Northern Hemisphere that was experienced in the 1920s, 1930s, and 1940s coincides with a period when there were no major volcanic eruptions in the Northern Hemisphere, suggesting the possibility that the absence of a volcanic-dust pall in those decades was one factor in the warming process. More recent volcanic eruptions have also caused some short-term cooling, with the most notable example being provided by the eruption of Mount Pinatubo in the Philippines in 1991 (McCormick et al. 1995).

Similarly, the MWP and the LIA seem to correspond to periods of low and high volcanic activity respectively, as do the glacial fluctuations of the past 1000 years (Porter 1981, 1986). A comparison of Greenland and Antarctic ice core sulphate concentrations spanning the last 6000 years (Wanner et al. 2008) shows a striking increase in the number of eruptions that affected both hemispheres between the thirteenth century and the present compared with earlier times; and it has been proposed that a series of eruptions in the second half of the thirteenth century may account for the transition from the MWP to the LIA. Notable frost events (determined by the presence of frost rings in bristlecone pines in the western United States) correspond reasonably well with the known history of dust-loadings back to AD 1500 (LaMarche and Hirschboeck 1984). From their recent study of Alpine glacier advances and retreats during the Holocene, Nussbaumer et al. (2011) suggested that the timing of LIA advances was affected by volcanism as well as by solar variation. They noted that large eruptions in 1809 and 1815 (Tambora) caused a decade of cool summers that promoted a strong phase of glacier advance in the later stages of the LIA. The Tambora eruption itself is thought to have caused a temperature decline in the following year of 1 to 2 °C in Europe and perhaps 0.5 °C averaged over the Northern Hemisphere (Turney 2008).

For the Holocene as a whole, Bray (1974) suggested on the basis of examination of ^{14}C dates that the major advances of alpine and polar glaciers were exactly contemporaneous with the major post-glacial volcanic activity phases in New Zealand, Japan, and southern South America. Nesje and Johannessen (1992) came to similar conclusions. They compared a global compilation of Holocene alpine glacier advances with the acidity records of the Crête and Camp Century Greenland ice cores (peaks in acidity are from the fallout of large volcanic eruptions) and found a strong similarity between the two data sets.

A particularly notable post-glacial eruption that changed climate was the eruption of the island of Thera (Santorini) in the Aegean Sea during the Bronze Age. The precise date of this eruption has been a matter of dispute for many years; but **dendrochronological** study of an olive tree buried in the tephra confirms that it occurred shortly before 1600 BC (probably between 1628 and 1600 BC) (Friedrich et al. 2006). Not only did this eruption have a devastating effect upon the Minoan civilization centred on Crete through tsunamis and ash fall, tephra fallout was heavy as far as Turkey (Sullivan 1988); but it also caused significant cooling across the Northern Hemisphere. This has been detected, for instance, in narrow growth rings of subfossil Irish oak dating to the time of the eruption (Baillie and Munro 1988).

Looking beyond the Holocene into the late Pleistocene, Bryson (1989) has found evidence from radiocarbon dates on volcanic deposits for a peak of volcanic activity during the **Late Glacial**. This is supported in a more recent reconstruction of explosive volcanism from Antarctic ice which shows a greater number of tephra layers during the later part of the **Last Glacial Maximum** and Late Glacial compared with the period between 20 000 and 35 000 years ago (Narcisi et al. 2010). It has been argued that the Toba super-eruption in Sumatra, dated as occurring around 74 000 years ago and which released some 2800 km³ of magma, was so large that it created a dense stratospheric dust pall and aerosol clouds. (For comparison, the 1815 eruption of Tambora ejected about 160 km³.) A phase of 'Volcanic Winter' (Rampino and Self 1992) with a decade long global temperature drop of 3 to 5 °C may have accelerated the shift to glacial conditions that was already underway, by inducing perennial snow cover and increased sea ice extent at sensitive northern latitudes.

The effects of Toba on global climate and ecosystems were severe and appear to correlate with a population 'bottleneck' in our genetic ancestry suggesting that the eruption may have pushed *Homo sapiens* to the brink of extinction (Ambrose 1998) (see **Section 8.3** for a discussion of the impact on *H. sapiens*). Williams et al. (2009) argued that the Toba eruption probably triggered, or at least contributed to, the 1800-year cold spell that immediately followed Greenland **interstadial** 20 of the scheme of **Dansgaard–Oeschger cycles** in the Greenland ice core record (see **Fig. 3.18**). Additional support for the role of volcanism generally in Pleistocene climate comes from the analysis of coarse volcanic ash in 320 deep-sea sediment cores. Kennett and Thunell (1975) found that such ash is very frequent in the Quaternary, being about four times the **Neogene** average.

What is intriguing, therefore, is that there is a great deal of prima facie evidence for periods of high volcanic activity being associated with periods of colder climate. Equally intriguing is that these periods cover a wide range of timescales from the decadal fluctuations of the last few centuries, through the neoglacial events of the Holocene, to the climatic oscillations of the Late Glacial, to the Last Glacial Maximum, and to the Pleistocene as a whole. Although correlations of time series do not necessarily imply causality, the implications are certainly strong. There is the possibility, however, that for some of the longer timescales (i.e. 10^3 to 10^6 years) the relationship may be reversed, with climatic changes creating changes in volcanic activity. For example, increased volcanism might result from the stresses imparted by **glacio-isostatic** and hydro-isostatic loading and unloading of the Earth's crust associated with ice cap growth and retreat, and changes in the volume of water in the oceans. Support for this concept is given by Zielinski et al. (1996, 1997) who analysed the sulphate record in the GISP 2 (Greenland) ice core over the last 110 000 years. As Zielinski et al. (1997) wrote in their abstract:

The continuous time series of volcanic SO_4^{2-} for the last 110,000 years show a strong relationship between periods of increased volcanism and periods of climatic change. The greatest number of volcanic SO_4^{2-} signals, many of very high magnitude, occur during and after the final stages of deglaciation (6000–17,000 years ago), possibly reflecting the increased crustal stresses that occur with changing volumes of continental ice sheets and with the subsequent changes in the volume of water in ocean basins (sea level change).

Effects of non-volcanic dust

Changing volcanism is not the only way in which climatically important changes in atmospheric transparency might occur. For example, dust can be emplaced into the atmosphere by deflation of fine-grained surface materials by wind. Dust particles in the atmosphere exert both direct and indirect influences on climate. An example of the former is the effect that dust particles have on radiation budgets. Indirect influences include those brought about by the effects of

dust on biogeochemical cycling (Moreno and Canals 2004), and, for instance, carbon dioxide levels in the atmosphere. Likewise, dust stimulation of phytoplankton production releases DMS (dimethylsulphide) which may increase cloud albedo and so contribute to cooling of the atmosphere. In addition, it needs to be remembered that the relationship between **aeolian** dust and climate is bidirectional, since climate plainly has a major impact on dust generation, transport, and deposition.

Radiative forcing (the perturbation of the radiation balance caused by an externally imposed factor) by dust is complex (Tegen 2003), since it not only scatters but also partly absorbs incoming solar radiation, and also absorbs and emits outgoing long wave radiation. Changes in the amount of dust in the atmosphere would cause changes in the radiation balance and thus also in surface temperatures. However, the magnitude and even the sign of the dust forcing can vary (Arimoto 2001), for it depends on the optical properties of the dust (which relate to its particle size, shape, and mineralogy), on its vertical distribution, on the presence or otherwise of clouds, and on the albedo of the underlying surface. Because of this complexity there is no clear consensus about the extent to which substantially increased dust loadings at the Last Glacial Maximum contributed to cooling, or indeed even a slight warming poleward of 45° N. However, it is believed that the cooling effect of dust at lower latitudes was significantly larger than at higher latitudes (see, for example, Claquin et al. 2003). In addition, it is possible that dust additions to ice caps and glaciers could modify their surface albedo, leading to changes in mass balance. A thin layer of dust across the surface of a glacier may promote ablation through lowering albedo; however, a thick layer could provide insulation during a warming climate, thereby slowing ablation. As in the atmosphere, dust modification of the energy budget at the surface is complex.

Ice cores from Greenland (Taylor et al. 1997), Antarctica (EPICA community members 2004) and tropical mountain glaciers (Thompson et al. 1995), and deep ocean cores show much greater concentrations of mineral dust during colder phases (see **Section 3.2** for dust in Antarctic ice cores). This suggests that there was more dust around in the world's atmosphere during cold periods than during warm phases. As the Earth becomes drier and colder deserts expand, there is reduced global frictional drag owing to the growth of polar ice sheets and loss of vegetation in low latitudes. Among the many other views about the role of dust discussed above, it has also been suggested that this results in higher wind speeds sending more desert dust into the atmosphere where it may reinforce the cold and dryness by forming stable 'inversion' layers that block sunlight and prevent rain-giving convective processes (Adams et al. 1999).

The EPICA Dome C ice core shows a *c.* 25-fold increase in dust flux during each of the last eight glacial cycles compared with the **interglacial** values, and, as shown in **Fig. 9.4**, changes in dust flux are correlated with Antarctic temperature during glacial times but not during interglacial times (Lambert et al. 2008). Lambert et al. interpret the increased dust flux at Dome C as coming primarily from South America, and they suggest correlation with Antarctic temperatures is likely to represent a coupling whereby increased glacial dust flux was both a result of and amplifier for shifts to colder climate during glacials.

Dust and atmospheric CO_2

The presence of carbon dioxide in the atmosphere has been, is, and will continue to be a major influence on the radiation balance of the Earth, and dust loadings in the atmosphere may be interrelated with changes in atmospheric CO_2. Ridgwell (2002), for example, has argued that dust may affect climate by fertilizing ocean biota which in turn draw down CO_2 from the atmosphere, which in turn reduces the **greenhouse effect**. He noted that currently there are some parts of the ocean where a supply of Fe (iron) is a limiting factor in terms of phytoplankton growth. However, during glacials, when global dust production and deposition were greater than today, it is possible that a series of feedbacks could lead to enhanced climatic change. One scenario is that any intensification in glacial state would tend to produce an increase in dust availability and transport efficiency. This in turn could produce a decrease in CO_2 (through Fe fertilization of the Southern Ocean), which would cause further intensification in the glacial state and thus enhanced dust supply, and so on. As he argued (p. 2922):

Operation of this feedback loop would come to an end once the global carbon cycle has reached a second state, one in which biological productivity becomes insensitive to further increases in aeolian Fe supply, perhaps through the onset of limitation by NO_3. If aeolian Fe supply were then to decrease sufficiently to start limiting biological productivity again, the feedback loop operating in the

Fig. 9.4 EPICA Dome C plot of Antarctic dust flux (lower time series) with a time series of the strength of correlation between logarithmic values for dust flux and δD-derived temperatures. The dashed line indicates an r² (coefficient of determination) value of 0.27, above which the correlation is statistically significant with 95% confidence. The numbers beneath the dust flux time series refer to marine oxygen isotope stages (MIS). Notice that correlation between dust flux and temperature is evident during glacial times (LGM and even numbered MIS) but not during interglacials (from Lambert et al. 2008, Figure 2). Reprinted with permission from Macmillan Publishers Ltd: *Nature*

opposite direction would act so as to reverse the original climatic change. That the Earth system might exhibit two distinct states, one of 'high-xCO$_2$ low-dust' and the other 'low-xCO$_2$ high-dust', is consistent with developing views of the climate system as being characterized by the presence of different quasi-steady-states with abrupt transitions between them.

It is also possible, though as yet largely unproven, that dust may have encouraged growth of iron-hungry N$_2$-fixing cyanobacteria, thus alleviating nitrate limitations. This 'iron hypothesis', first advanced by Martin (1990), has been the subject of considerable ongoing research (see, for example, Ridgwell 2003 and Blain et al. 2007), and the extent to which dust stimulated phytoplankton growth leads to CO$_2$ drawdown of the magnitude shown in ice cores is still an uncertainty, though changes in the relative contribution of phytoplankton to total productivity during glacial cycles has been established through analysis of Tasman Sea cores by Calvo et al. (2004).

Dust and clouds

Dust nuclei may modify cloud characteristics. As Toon (2003, pp. 623–624) has explained:

Dust may affect clouds in two ways. All water droplets start off by forming on pre-existing particles. As the number of particles increases, for instance due to a dust storm, the number of cloud droplets may increase. If there are more cloud droplets, the droplets will be smaller because the mass of condensing water is usually fixed by air motions and ambient humidity. Smaller cloud droplets make for a greater surface area, and hence brighter clouds...A less well-studied phenomenon is that smaller droplets are also much less likely to collide with each other and create precipitation...By acting as nuclei for triggering ice formation, dust particles can also affect clouds by causing the water droplets to freeze at higher temperatures than expected...Dust may thus be triggering precipitation in low-altitude clouds that otherwise would be too warm to have produced rain, or be triggering rain at lower levels in convective clouds that otherwise would not have produced rain until reaching much higher altitudes where it is colder...Dust may therefore inhibit precipitation by making more and smaller droplets, or enhance it by adding ice particles to warm clouds.

Rosenfeld et al. (2001) argued that the inhibiting effect on precipitation was most likely and that Saharan dust

provides very large concentrations of cloud condensation nuclei, mostly in the small size range, which mean that clouds are dominated by small droplets so that there is little coalescence. This results in suppressed precipitation, drought enhancement, and more dust emissions, thereby providing a possible **desertification** feedback loop.

Desert dust is also undoubtedly associated with strong ice nucleating behaviour, and high concentrations of dust particles acting as ice nuclei in clouds could lead to changes in cloud microphysical and radiative properties, latent heating and precipitation. Interest has started to build in recent years in the possible role that Saharan dust plays in modifying convective storm activity, affecting anvil cloud development and precipitation.

Another way in which rainfall may be affected is through changes in convective activity brought on by the modification of temperature gradients in the atmosphere created by the presence of dust (Maley 1982). In addition, the radiative effects of dust may lead to the intensification of easterly waves in North Africa (Jones et al. 2003) with consequent effects on numerous climatic parameters, including precipitation. It is probable that dust loadings in the atmosphere were both affected by past climatic changes and had an effect on such changes through complex feedback processes (Harrison et al. 2001).

9.4 Earth geometry theories: the Croll–Milankovitch hypothesis

Following through **Fig. 9.1** it is reasonable to assume that if the position and configuration of the Earth as a planet in relation to the Sun were to change, so might the receipt of insolation from the Sun. Such changes do take place, and there are three main astronomical factors which have been identified as of importance, all three occurring in a cyclic manner (**Fig. 9.5**): changes in the **eccentricity** of the Earth's orbit (a c. 100 000-year cycle), the **precession of the equinoxes** (with cycles of 19 000 and 23 000 years which average out to c. 22 000 years), and changes in the **obliquity** of the ecliptic (the angle between the plane of the Earth's orbit and the plane of its rotational equator). This last has a periodicity of about 41 000 years.

The Earth's orbit round the Sun is not a perfect circle but an ellipse. If the orbit were a perfect circle then

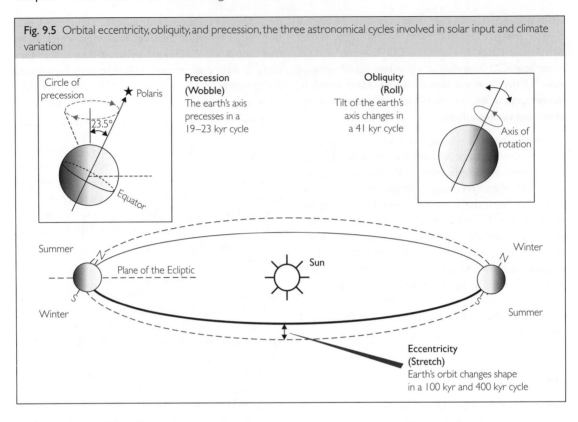

Fig. 9.5 Orbital eccentricity, obliquity, and precession, the three astronomical cycles involved in solar input and climate variation

the summer and winter parts of the year would be equal in their length. With greater eccentricity there will be a greater difference in the length of the seasons. Over a period of about 100 000 years the Earth's orbit can 'stretch' by departing much further from a circle and then revert to almost true circularity.

The precession of the equinoxes simply means that the time of year at which the Earth is nearest the Sun (perihelion) varies. The reason is that the Earth wobbles like a top and swivels round its axis. At the moment the perihelion comes in January. In 11 000 years it will occur in July.

The third cycle perturbation, change in the obliquity of the ecliptic, involves the variability of the tilt of the axis about which the Earth rotates. The values vary from 21°39′ to 24°36′. This movement has been likened to the roll of a ship. The greater the tilt, the more pronounced is the difference between winter and summer (Calder 1974).

Appreciation of the possible significance of these three astronomical fluctuations of the Earth goes back to at least 1842, when J.F. Adhemar made the suggestion that climate might be affected by them. However,

his views were developed by Croll in the 1860s and by Milankovitch in the 1920s (Beckinsale 1965; Mitchell 1965). Milutin Milankovitch postulated that glacials are interrupted by interglacials when the three orbital parameters combine to maximize the receipt of solar energy during the summer at high latitudes in the Northern Hemisphere, causing more snow to melt away in the summer than fell in the winter. Whereas during glacials the orbital parameters are such that summer temperatures at high latitudes are cooler so that some snowfall from winter survives the summer allowing glaciers to grow. Milankovitch was also the first to calculate the periodicities of the three cycles and the resulting changes in insolation at different latitudes. Through the Quaternary, the amount of insolation in the Northern Hemisphere has been more important than in the Southern Hemisphere for pacing global climate, because the north has much more land: the land is heated and cooled more easily than the sea and allows for the growth and expansion of large ice sheets.

Critics of the orbital theory (or **Milankovitch Theory**) have argued that it does not explain the medium- and short-term changes of the Holocene, that it cannot explain the spacing of the major ice ages (the Quaternary being just the most recent **Ice Age** or time of **icehouse climate**), that it does not account for the timing and initiation of an Ice Age, and that the computed variations in insolation resulting from the mechanism are too small by themselves to cause significant change. In one sense many of these arguments are valid – orbital changes do not explain all scales of climatic change, and they may well need to be intensified by other mechanisms (see Elkibbi and Rial 2001 for a review). However, as the compilation of Berger et al. (1984) has shown, the orbital mechanism is one of immense power and significance for explaining some of the major features of the Pleistocene climatic fluctuations. There is substantial evidence to link this mechanism to the longer scales of climatic change (Goreau 1980), and the influence of orbital fluctuations has been traced back in the geological record over millions of years (Kerr 1987).

The major attraction of these ideas is that while the cycles involve a modulation of the insolation received at different latitudes in different seasons, with relatively modest change at a given latitude in terms of W m^{-2} (see **Fig. 3.17** showing changes on the order of 10 per cent at 60° N), the periodicity of these fluctuations is largely comparable with the periodicity of the ice advances and retreats of the Pleistocene. **Radiometric dating** has shown that the record of sea level changes preserved in the coral terraces of Barbados and elsewhere, and the record of heating and cooling in deep-sea cores, correlate well with theoretical insolation curves based on those of Milankovitch (Mesollela et al. 1969; Broecker et al. 1968).

Indeed the variations in the Earth's orbit have been seen as 'the pacemaker' of Pleistocene glacial/interglacial cycles (Imbrie and Imbrie 1979), because detailed statistical analysis of the ocean cores shows that they possess statistically significant wave-like fluctuations with amplitudes closely matching the orbital cycles (**Fig. 9.6**). This has allowed for the age-equivalent method of dating deep-sea sediments known as 'orbital tuning' pioneered by SPECMAP (Imbrie et al. 1984) (see **Section 2.5**). The most important of these cycles for glacial/interglacial cycles over the past 900 000 years is the longest one, corresponding to variations in eccentricity (Hays et al. 1976; Pisias and Moore 1981) (see **Fig. 3.2**). Spectral analysis has shown broadly comparable wavelengths in Chinese **loess** profiles laid down over the last 700 000 years (Lu 1981), with spectral peaks at 41 700 and 25 000 year periodicities seeming to correspond to the obliquity of the ecliptic and the precession of the equinoxes respectively.

They may also help to explain the very marked expansion of lakes that took place in tropical and subtropical areas in the early Holocene. Analyses of theoretical insolation levels by Kutzbach (1981) indicated that Northern Hemisphere radiation receipts in July at that time were larger than now by about 8 per cent (see **Fig. 5.6**) and that this led to an intensification of the monsoonal circulation and associated precipitation. Changes in hydrological conditions, as exemplified by studies of lake-level fluctuations in low latitudes, show a good correspondence with the changes simulated by general circulation models incorporating orbitally-induced changes in insolation receipts (Kutzbach and Street-Perrott 1985; Street-Perrott et al. 1990; Kutzbach et al. 1998).

Although there is now excellent evidence for a very close relationship between orbital forcing and major climatic changes during the Pleistocene, there is still a need to consider the various ways in which the effects of the Milankovitch signal can be magnified to cause the observed degree and speed of climatic changes. One possibility is the role of fluctuations in carbon dioxide levels in the atmosphere as an intensifying mechanism (see **Section 9.6**). Another possibility is that the oceans play a role by responding in a non-linear way to an initial climatic perturbation, and switching from one configuration of circulation characteristics to another. For example, the orbitally-induced melting of an ice cap might liberate large quantities of fresh water into the North Atlantic, which might affect the density and temperature of that water, which might in turn affect the whole pattern of heat-exchange in the oceans of the world, with a whole series of concomitant climatic effects. This is the view put forward by Broecker and Denton (1989, p. 2465):

We propose that Quaternary glacial cycles were dominated by abrupt reorganizations of the ocean-atmosphere system driven by orbitally induced changes in fresh-water transports which impact salt structure in the sea. These reorganizations mark switches between stable modes of operation of the ocean-atmosphere system. Although we think that glacial cycles were driven by orbital change, we see no basis for rejecting the possibility that the mode changes are part of a self sustained internal operation that would operate even in the absence of changes in the Earth's orbital parameters. If so...orbital cycles can merely modulate and pace a self-oscillating climate system.

Fig. 9.6 Orbital and climatic variations over the past 600 000 years (kyr = 1000 years ago). Variance spectra to the right, with their dominant period, indicated in thousands of years are calculated for the time series on the left. From top: factor variations in eccentricity, obliquity, and precession index. Bottom: variations in marine $\delta^{18}O$ (from Imbrie et al. 1984)

One such model of coupled ocean–atmosphere change that has been invoked to account for rapid deglaciation is that put forward by Ruddiman and McIntyre (1981a, b). They proposed that when, as a result of orbital change, there was a larger receipt of insolation at high latitudes in the Northern Hemisphere, this would cause a reduction in the volume of continental ice, and would be accompanied by iceberg calving into the oceans. This would cause sea level to rise (which would make ice-shelves unstable so that accelerated iceberg calving occurred) and the ocean water in the North Atlantic to cool. Ocean cooling would lead to an increased area of winter sea ice and decreased moisture provision to the atmosphere, which would in turn cause less nourishment of ice caps, thereby promoting their demise.

Another coupled ocean–atmosphere model which has been proposed to account for the 'rapidity, near synchroneity, and global extent of the events associated with the termination of the last glacial cycle' is that of Broecker and Denton (1990). They maintain that towards the end of the last glacial cycle large masses of fresh water were delivered to the North Atlantic Ocean by rivers draining from the wasting Northern Hemisphere ice caps. This in turn affected the salinity and density of the oceans which in turn caused a change in the pattern of ocean circulation. They concluded (p. 336):

> The atmosphere and ocean are tied together in a non–linear manner, making the combined system susceptible to abrupt mode switches … We propose that mode switches occur when orbitally induced seasonality changes alter the flow of water vapour from one part of the ocean to another. The Atlantic conveyor belt appears to be the most vulnerable part of the system. Therefore, salinity changes in this part of the sea are most likely to create the instabilities that cause the ocean-atmosphere system to reorganize.

Explaining the 'mid-Pleistocene revolution'

Another feature of the Pleistocene that must be explained in relation to amplifying mechanisms is the 'mid-Pleistocene revolution' about 900 000 years ago (see **Section 3.2**) when the pacing of glacial/interglacial cycles shifted from a periodicity of *c.* 41 000 years (matching the obliquity cycle) to one of *c.* 100 000 years (matching the eccentricity cycle). What is puzzling about this transition is that changes in obliquity have a far bigger effect on seasonal insolation receipt at high latitudes (W m^{-2}) than changes in eccentricity. It has been proposed by some workers that interplanetary dust may be involved (Farley and Paterson 1995; Muller and MacDonald 1997). The suggestion is that gradual changes in the plane of the Earth's orbit cause variations in the quantity of interplanetary dust through which the Earth travels on a *c.* 100 kyr cycle – perhaps this effect became enhanced around a million years ago.

There are also internally-driven feedback mechanisms to account for the mid-Pleistocene revolution (Clark et al. 1999). Over repeated cycles of glaciation during the early Pleistocene, erosion would have removed much of the softer underlying sediments beneath the great ice sheets. As glacier ice flows more slowly over hard material than soft sediments, this would cause ice sheets later in the Pleistocene to develop a progressively thicker profile. Around a million years ago the big Northern Hemisphere ice sheets may have reached such a volume and thickness that they were able to survive the obliquity and precession induced peaks in insolation: it would therefore require the eccentricity enhancement of the other cycles to initiate a glacial **termination**. In addition to this, increased ice thickness would also have led to greater **isostatic** depression of the crust thereby gradually lowering the altitude of the ice surface causing it to become warmer and experience increased ablation. The timescale of this crustal adjustment, along with the slight increase in insolation caused by eccentricity modulation, could account for the crossing of a threshold every *c.* 100 kyr that then led to the rapid end of the glacial phase as the greenhouse gas, albedo, and ocean feedback mechanisms mentioned above took hold. Over the course of the subsequent interglacial and early glacial, the orbital parameters begin to cool summers, the crust rebounds increasing the altitude of the surface, and ice sheet growth is renewed.

It is important to remember that the 23 and 41 kyr cycles remain strongly embedded within the 100 kyr cycle. For instance, over the last glacial cycle relatively warm interstadial times, such as during **marine isotope stages (MIS)** 5 and 3, correlate with peaks in precession. Furthermore, as Ruddiman (2006) explained, glacial terminations have tended to occur after either 82 000 to 92 000 years (two obliquity cycles and four precession cycles) or 115 000 to 123 000 years (three obliquity cycles and five precession cycles) – this averages out to the *c.* 100 kyr cycle. Important for the timing of the termination is when the obliquity and precession cycles are in phase with each other – during a phase of relatively high eccentricity. When the cycles are closely in phase, and of relatively strong amplitude, the corresponding deglaciation is more rapid; whereas if the cycles are relatively subdued and out of phase, such as during Termination V, deglaciation is slower. The relatively long interglacial of MIS 11 referred to in **Chapter 3** (which followed Termination V) is also thought to relate to this orbital configuration. As for the timing of the mid-Pleistocene revolution itself, Ruddiman (2008) suggests that it may simply represent a threshold response to the long-term cooling trend seen through the entire Quaternary Period (see **Figs 3.2** and **3.3**) in which a point was crossed when summer temperatures in the high northern latitudes no longer climbed high enough during each obliquity peak to initiate the melting of all the ice that had formed during the preceding span of lower insolation.

9.5 Hypotheses involving changes in terrestrial geography

While orbital cycles affect climate on a timescale of tens of thousands of years, changes over millions of years, including possibly the initiation of glaciation in certain parts of the world, may have resulted from changes in the positions of the continents, shifts in the position of the polar axis, and uplift of continents by orogeny (Hay 1996). Of these the first two are probably relatively unimportant in terms of the Pleistocene, in that the rates of change involved are very slow and over the time span concerned rather slight (Cox 1968). The rate of continental drift is a little higher, with a mean rate of around 1.0×10^{-7} degrees yr^{-1}, which is equal to a displacement of only $1°$ during the last 10^7 years (only about $0.2°$ since the start of the classic glaciations). Even with a maximum postulated rate of 60×10^{-7} degrees yr^{-1}, the displacement is not of very great significance. However, Ewing (1971) suggested that if sea-floor spreading operated at a rate of about $2\,cm\,yr^{-1}$ the width of a rift such as that between Spitzbergen and Greenland would increase by about $200\,km$ in 10 million years, sufficient to affect the entry of ocean currents into the Arctic and thereby the climate of the surrounding areas. Hence over a much longer timescale than the Pleistocene, changes in the relative positions of the continents and the opening and closing of certain sea connections (the **gateway hypothesis**) played a role in the global **Cainozoic climatic decline** as discussed in **Section 1.4**. On somewhat shorter timescales than continental drift, vertical movement of land, through orogeny, volcanic, and isostatic processes, is thought to have had important climatic effects through the Late Tertiary and Pleistocene.

Looking at how local or regional climate can be affected by uplift, if one assumes a rate of uplift in a

tectonically active area of $10\,m$ per 1000 years it would only require about $10\,000$ years to obtain a mean temperature fall of $0.65\,°C$ with every $100\,m$ rise in altitude. Thus in the course of the Pleistocene it would be quite possible for a mountain to develop sufficiently quickly to cause a marked temperature depression at its summit. Also, total precipitation is known to show a general increase with altitude, at least up to about $3000\,m$, so that the overall increase in height of mountains would serve to produce veritable snow traps.

Some support for this hypothesis is given by the fact that not all areas appear to have undergone multiple glaciation, and in many cases it seems possible or probable that uplift in the mid and Late Pleistocene brought the mountains of some areas into a position where ice could accumulate. Thus, for example, Mauna Kea (Hawaii), Tasmania, and the Pyrenees have been cited as areas which anomalously only suffered one major glacial phase, relatively late into the Pleistocene. However, orogenic effects could also have had a global dimension, and the role of the mountains and plateaux of High Asia and the south-west USA have been accorded particular importance in this respect (Ruddiman et al. 1989). Tibet and the high Himalayas (**Fig. 9.7**) have undergone a net uplift of 2–3 km during the past 3 million years, so that 'As much as 75 per cent of the net elevation of Tibet (an area approximately one third of the size of the contiguous

Fig. 9.7 The uplift history of the Tibetan Plateau and the Himalayas over the last 60 million years. The curve is inferred from palaeobotanical data (based on work by Xu Ren in Molnar and England 1990, Figure 4). Reprinted with permission from Macmillan Publishers Ltd: *Nature*

United States) may have been attained during the time of Northern Hemisphere glacial inception and intensification' (Ruddiman and Raymo 1988, p. 420). Significant Plio-Pleistocene uplift has also occurred in several regions of the south-west United States and parts of the Andes in South America.

Such uplift has had two effects on Northern Hemisphere climate that could be relevant to glacial inception and intensification. One of these is to enhance albedo-temperature feedback on a globally significant scale (Ruddiman and Raymo 1988, pp. 420–421):

The cold air over the large, newly created region of high-standing topography at middle latitudes provides a temperature discontinuity that accelerates the normal southward movement of the snow-line in autumn-winter. This enlarged high-albedo surface in turn leads to global cooling during the transitional and winter seasons.

The second of these mechanisms is a change in the form of the major waves in the general atmospheric circulation. It is proposed that the increased elevation of both the Tibetan-Himalayan region and regions of the western USA would have altered the planetary wave structure in such a way as to cool the North American and European land masses and increase their susceptibility to orbitally driven insolation changes (the Milankovitch mechanism). Ruddiman and Raymo (1988, p. 425) wrote: 'Before late-Pliocene mountain uplift early Pliocene cold air masses were probably more confined to the northernmost parts of North America and Europe . . . We suggest that the degree of southward penetration of the waves over North America and Europe gradually deepened during the late Pliocene and Pleistocene.'

Raymo and Ruddiman (1992) have argued that tectonically driven increases in chemical weathering (particularly the uplift of the Himalayas) may have resulted in a decrease of atmospheric CO_2 concentrations over the last 40 million years contributing to the late Cainozoic cooling trend and the development of icehouse conditions (see also Filippelli 1997). The Eocene to Oligocene transition saw a large increase in the volume of Antarctic ice which is thought to relate to global cooling associated with declining CO_2 (Bowen 2007). Recent studies have improved the long-term reconstruction of CO_2 concentrations during the Cainozoic (see **Fig. 1.7**), and there is an increasing body of evidence from marine sediments that steps towards cooler climate during the Neogene were

related to times of lowered atmospheric CO_2 (Tripati et al. 2009). The mechanism whereby mountain uplift leads to accelerated chemical weathering and increased draw-down of atmospheric CO_2 is explained in **Section 1.4**. Notwithstanding the evidence for such a connection between uplift and enhanced chemical weathering of silicate minerals, it is very difficult to quantify its effect on the carbon cycle (and through that atmospheric temperature) as there are many other factors that also affect weathering rates (see the review in Goudie and Viles 2012).

Another tectonic event of significance for the eventual onset of Northern Hemisphere glaciation was the emergence of the Isthmus of Panama and the closure of the connection between the Atlantic and Pacific that previously existed between South and Central America. By c. 4.6 million years ago the connection between the two oceans had become shallow enough to significantly alter the mode of ocean circulation in the North Atlantic Ocean (Haug and Tiedemann 1998). The effect of this was to intensify the meridional circulation of the Atlantic, and by 3.6 million years ago the **thermohaline circulation** had approached its Pleistocene characteristics (the thermohaline circulation is discussed further in **Section 9.7**). The Gulf Stream and North Atlantic Drift currents had become established, and by transporting warmer waters towards the Arctic, the water vapour content of the atmosphere was increased, increasing the potential for precipitation at high latitudes. This was an important precondition for the growth of Northern Hemisphere ice sheets that eventually occurred around 2.7 million years ago, when the obliquity cycle became stronger and when enhanced seasonality in the North Pacific Ocean caused atmospheric moisture to also be enhanced to the west of North America (Haug et al. 2005).

9.6 The role of the greenhouse gases

In recent years, because of an interest in future global warming, there has been an increasing amount of attention paid to the role of carbon dioxide (CO_2) and various other trace gases in modifying the Earth's radiation budget (Weart 2003).

The CO_2 levels in the atmosphere have an effect on the world's heat balance because CO_2 is virtually transparent to incoming solar radiation but absorbs outgoing terrestrial infra-red radiation – radiation that would otherwise escape to space and result in heat loss from the lower atmosphere. Through the mechanism of this 'greenhouse effect', increases or decreases in the concentration of atmospheric CO_2 would be expected to lead to warming or cooling, and as described above (and in **Chapter 1**) long-term climatic variation over the Cainozoic is believed to relate to changes in the strength of the greenhouse effect.

Until recent decades there was no method by which this and related hypotheses could be tested directly against palaeo-data. However, it has now proved possible to retrieve carbon dioxide from bubbles in layers of ice of known age in deep ice cores. Analyses of changes in carbon dioxide concentrations in these cores have provided truly remarkable results. The early work of Delmas et al. (1980) on the Dome Core in Antarctica showed for the first time that round about the Last Glacial Maximum the level of atmospheric carbon dioxide was only about 50 per cent that of the present. The Vostok ice core (see **Fig. 3.5**) shows CO_2 fluctuating in step with the temperature changes inferred from deuterium data over the last four glacial cycles. There is a seeming coincidence of cold temperatures and low CO_2 levels. Subsequent ice core analyses have confirmed the findings of the Delmas group, and have demonstrated that carbon dioxide changes and climatic changes have progressed in approximate synchrony from recent times well back into the middle Pleistocene (Jouzel et al. 2007). As shown in **Fig. 9.8**, the last interglacial was a time of high CO_2 levels, the Last Glacial Maximum of low CO_2 levels, and the early Holocene a time of very rapid rise in CO_2 levels.

The explanation for this remarkable coincidence is still the matter of intense speculation and debate, and a large number of models have been produced to explain the causes of the fluctuating atmospheric CO_2 levels. Because the oceans are such a large store of CO_2 in comparison with terrestrial sinks, the explanation almost certainly lies in the oceans and changes in their circulation and turnover, or in changes in the productivity of the various organisms that live in them (Sarnthein et al. 1988).

An initial hypothesis was put forward by Broecker (1981). He hypothesized that a possible cause of the high level of CO_2 during the interglacials may be a loss of phosphorus to continental shelf sediments during transgressions of the oceans. This would reduce the amount of plant matter formed in the sea per unit area of upwelled water and would thereby increase the

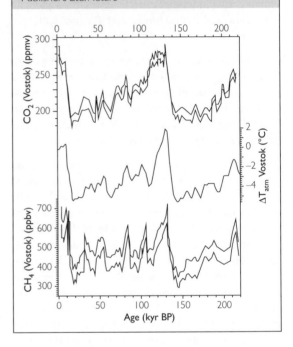

Fig. 9.8 Analysis of air trapped in the Vostok Antarctic ice core showing that methane and carbon dioxide concentrations were closely correlated with the local temperature over the last 200 000 years. The double lines for CO_2 and CH_4 indicate the error range. The temperatures were derived by measuring the proportion of deuterium in the ice (from Jouzel et al. 1993, Fig. 5). Reprinted with permission from Macmillan Publishers Ltd: *Nature*

CO_2 pressure in surface water and in the atmosphere. **Regressions** during glacials would tend to reverse this trend. Given that transgressions and regressions of the oceans are tied up so closely with the decay and growth of ice caps, these CO_2 changes could amplify the effects of orbital variations of the type postulated by the Croll–Milankovitch theory.

Another ingenious explanation for decreased CO_2 levels in glacials, already referred to in **Section 9.3**, was put forward by Martin (1990). He suggested that phytoplankton consume carbon dioxide, and that phytoplankton require iron. With more iron-rich dust in the atmosphere during glacial times, he suggested that the enhanced flux of dust to the oceans enhanced phytoplankton growth and hence the oceanic sink for CO_2. Changes in biological productivity could also be achieved by changes in processes such as low-latitude **upwelling** (Sarnthein et al. 1988) or in the temperatures of high-latitude surface waters (Sarmiento and Toggweiler 1984). The role of changes in ocean circulation patterns are discussed in Siegenthaler and Wenk (1984) and Broecker and Denton (1989). It is also possible that changes in carbon budgets and CO_2 loadings in the atmosphere could be associated with changes in the state of vegetation assemblages and that there could be feedbacks involving cooling and enhancement of the northern soil and peatland formation: this could help explain the initiation of cold phases as peatland progressively expands and increases in importance as a carbon sink during the later stages of interglacials (Klinger et al. 1996).

The role of methane (CH_4) alongside CO_2 must also be taken into account as molecule for molecule it is a highly effective greenhouse gas. As with CO_2, CH_4 levels are greater in their atmospheric concentrations during interglacials than in glacials (Raynaud et al. 1988). As **Fig. 9.8** indicates, CH_4 concentrations have tended to be less than 400 ppbv in glacials probably due to reduced biological activity on the colder, drier land surfaces, and over 600 ppbv in interglacial times. Nisbet (1990) suggested that during the glacials, many high-latitude bogs and wetlands (which are major sources of the gas) would have been covered by ice or rendered less productive by extensive **permafrost** distribution. Moreover, many natural gas-fields which give out methane to the atmosphere, would have been sealed off by ice sheets and permafrost. By contrast, in an interglacial wetlands would produce large amounts of methane, and at the transition between glacial and interglacial conditions the methane that had been trapped in and beneath ice would be released into the atmosphere as the ice cover decreased. The rapid release of the gas would contribute to atmospheric warming which would in turn lead to more melting, and more gas release. Methane concentrations, as revealed from analyses of the GRIP ice core in central Greenland, have also shown significant variations during the Holocene – by up to 15% – and this reflects expansion of tropical lakes and wetlands related to intensified monsoonal circulation as well as changes at higher latitudes (Blunier et al. 1995). The Holocene methane minimum occurred between 5000 and 6000 years ago, which is consistent with a phase of maximum aridity at low latitudes.

It has been suggested that another mechanism, involving sudden and short-lived releases of massive amounts of methane from the ocean floors, could sometimes have resulted in rapid warming phases.

Such a process is believed to be implicated in the **Palaeocene–Eocene Thermal Maximum (PETM)** discussed in **Section 1.4**. However, high resolution ice core records have failed to show any evidence for sudden bursts of methane of this type over the time period that they span (Chappellaz et al. 1993; Taylor et al. 1997). Nonetheless, the role of methane releases, whether as an initiator or response to climate change, is a topic of ongoing research (Nisbet 2002) and changes in methane release as tundra lands warm up in coming decades may have a positive feedback role in anthropogenic climate warming (Bridgham et al. 1995).

In addition to methane, fluctuations in water vapour must not be ignored. Water vapour is a more important greenhouse gas than carbon dioxide and its atmospheric concentration can vary rapidly (Adams et al. 1999). This has led some researchers such as Broecker to suggest that changes in the flux of water vapour into the atmosphere from the oceans, e.g. from changes in sea ice extent or rapid changes in vegetation cover, could possibly amplify climate changes. It has been suggested that ^{18}O changes in tropical ice core records from the Andes (Thompson et al. 1995) may be related to large changes in the water vapour content of the atmosphere. It is important to remember, however, that the residence time of water vapour in the atmosphere is measured over days, as opposed to the hundred or so years for CO_2; and therefore the role of water vapour is seen as a responder and amplifier of climatic change rather than as a driving mechanism.

There is little doubt that fluctuations in greenhouse gases have played an important role as an amplifier of the changes paced by orbital cycles over the Quaternary Period. Research continues into the extent to which changes in greenhouse gases have themselves been a driving force and into quantifying the contribution greenhouse gas fluctuations have made to overall temperature changes – both globally and regionally (e.g. Timmermann et al. 2009). This relies on determining leads and lags with respect to insolation, greenhouse gases, and ice volume changes; and to complicate matters, temperature sensitivity to variation of CO_2 declines as its atmospheric concentration rises. Ruddiman (2006) argued that the 90 ppmv or so lowering of CO_2 during the Last Glacial Maximum compared with the pre-industrial Holocene value, combined with about 320 ppbv lowering for CH_4, directly contributed 0.81 °C of radiative cooling.

Adding in the water vapour feedback, he proposed that lower greenhouse gases concentrations can explain *c*. 40% of the total global temperature depression of *c*. 4.5 °C during the LGM compared with recent times. Ruddiman (2006) also made the case that fluctuations in CO_2 are largely in phase with ice volume changes at the 41 kyr and 100 kyr cycles, playing a key positive feedback role (along with albedo) in amplifying the waxing and waning of the ice sheets and explaining the orbital timescale synchrony between the Northern and Southern Hemispheres. For a detailed discussion of climate sensitivity to CO_2 variation at a variety of timescales see Hansen et al. (2008).

The Holocene shares the characteristic of earlier interglacials in that a peak in CO_2 and CH_4 concentration early in the interglacial was followed by declining values. However, where the Holocene differs is that shortly after this decline set in, the trend was reversed and first CO_2 and then CH_4 began to rise (*c*. 8000 years and 5500 years ago respectively). Ruddiman (2005a, b) explained at length how this relates to the spread of agriculture in prehistory and consequent deforestation, and he argued that the overall *pre-industrial* rise of greenhouse gases due to human activity may have caused the globe to be on average *c*. 0.8 °C warmer than it would otherwise be. The resultant effect at high northern latitudes would have been more than this, perhaps *c*. 2 °C, which Ruddiman considered significant enough to offset some of the late Holocene cooling trend referred to in **Chapter 5** and inhibit the incipient glaciation in northern Canada that should have occurred by now owing to orbitally induced insolation change. Interestingly, Northern Hemisphere summer insolation will gradually rise again in millennia ahead, not returning to levels similar to those at present for another *c*. 50 000 years or so – the suggestion being that humans may be in the process of causing a much longer interglacial period through elevated greenhouse gases than would otherwise have been the case.

9.7 Feedback hypotheses involving land, oceans, atmosphere, and ice

Thus far in our consideration of the possible causes of climatic change we have suggested that the output of solar radiation, the position and configuration of the Earth in relation to other heavenly bodies, the quality of the atmosphere, and the arrangement of land masses, oceans, and mountains have all been important at different timescales. However, there has been a variety of hypotheses in which it is envisaged that interactions between the atmosphere, the oceans and ice sheets possess a degree of internal instability which might furnish a built-in mechanism of change. It is conceivable that some small change might, through positive feedback, have extensive and long-term effects. As Mitchell (1968, p. 158) put it, 'minor environmental disturbances may have sufficed to "flip" the atmospheric circulation and climate from one state to another, and to "flop" it back again'. Some selected examples will give an indication of the importance of the hypotheses involving feedback relations. It is important to remember that such feedbacks can occur as part of the internal variability of the system, as well as acting to amplify changes caused by an outside factor (such as insolation change) as described in **Section 9.4**. Such feedbacks also help to explain abrupt climatic changes of the past.

Albedo-based hypotheses

One factor which controls heating levels in the Earth–atmosphere system is the degree to which incoming solar radiation is reflected or absorbed by the Earth's surface. Changes in the albedo of that surface, brought about perhaps by seemingly minor events, might be able to cause major changes in climate. For example, the deposition of a layer of relatively dark volcanic dust over the ice caps, as a result of a chance volcanic explosion, might lead to the melting of that ice cap which might in turn set a train of events in motion. Similarly, the presence of an unusually widespread and persistent snow cover over northern Canada as a result of a chance association of snowy winters and cool summers, could either help to trigger off climatic change directly, or could play a role as part of a feedback reaction (Williams 1975). Such a snow cover, persisting through all or most of the summer and autumn, would reflect the Sun and further chill the air (Calder 1974). This in itself might increase the likelihood of snow the next winter. A popular name for this mechanism is 'snowblitz', and towards the end of interglacials, as the orbital cycles trend to lowered Northern Hemisphere summer insolation, such a process helps explain incipient glaciation, leading eventually to a great ice sheet. Apart from the higher latitudes, Kuhle (1987) proposed that a globally important region for ice-albedo feedback is the Himalayas and Tibetan Plateau region.

It is possible that the relatively long-lived ice sheets might occasionally help bring about very rapid changes in climate, by rapidly 'surging' outwards into the sea and giving rise to large numbers of icebergs. The intensely cold **Heinrich events** that punctuated the last glacial (see **Fig. 3.21**), thought to be caused by sudden slippage of the **Laurentide ice sheet**, may have had a secondary effect via albedo change: the rapid freshening and cooling of the North Atlantic caused by the discharge of icebergs led to a great expansion of sea ice which, through its much more reflective surface, intensified cooling in the North Atlantic region, especially in winter (Broecker 2006, 2010).

An important factor in rapid regional or global climate changes may be the changes in the albedo of the land surface that result from changes in vegetation or algal cover on desert and polar desert surfaces. Adams et al. (1999) point out that dark-coloured surface algae or lichens, following a particularly warm or moist year, might provide a 'kick' to the climate system by absorbing more sunlight and also reducing the dust flux from the soil surface to the atmosphere. They also suggest that larger vascular plants and mosses might have the same effect on the timescale of years or decades. Forest change at high northern latitudes also plays a longer-term role: boreal forest has a lower albedo than tundra (conifers shed snow more than tundra plants) so that as the northern forest retreats southward towards the end of an interglacial, regional albedo increases thereby reinforcing the cooling, causing continued southward movement of the tree limit, and promoting ice sheet growth. It has been suggested by some that this feedback could now be working in the opposite direction as boreal forest moves north in response to anthropogenic climate warming (Liess et al. 2012).

Causes of abrupt climate changes

One of the most intriguing developments in Quaternary studies over the last two decades has been the growing appreciation that climatic changes of considerable magnitude can be initiated with relative rapidity. In other words, the world climate system may be less stable than was previously thought. While the broad pattern of climatic changes through the Quaternary may be *paced* by Milankovitch orbital cycles, the actual climatic shifts between and during glacial cycles were far swifter than these cycles would predict – having occurred over 'sub-Milankovitch' timescales. Rapid

climatic change events ('RCCEs') such as the **Younger Dryas** (see **Section 5.2**), the initiation of the last glacial episode, and the Dansgaard–Oeschger cycles spanning the last glacial cycle need to be explained. **Table 9.1** summarizes several causes of abrupt climate change.

Ocean–ice–atmosphere connections in the North Atlantic

Changes in ocean currents and deep circulation probably play a major role in either triggering or amplifying rapid climate changes in the historical and recent geological record. The most dramatic changes have been seen in the mid to high latitude regions of the North Atlantic Ocean, and therefore much research during the 1990s on the links between ocean circulation and millennial-scale climatic change was focused here (Broecker 1995; Keigwin et al. 1994; Jones et al. 1996; Rahmstorf et al. 1996). As better dated and higher resolution deep-sea sediment cores became available, it became increasingly clear that changes in the North Atlantic's **thermohaline circulation** (THC) were connected with the abrupt climatic changes observed in the Greenland ice cores. Discussion of the instability of the THC goes back several decades (Stommel 1961) and has been the focus of numerous empirical and numerical modelling studies (Ruddiman and McIntyre 1981a, b; Stocker et al. 1992; Anderson 1998; Manabe and Stouffer 1994; Tziperman et al. 1994).

Variability in the THC was introduced in **Section 3.6** in the context of the abrupt climatic shifts termed Dansgaard-Oeschger cycles. A generalized diagram of the global thermohaline circulation system (the ocean conveyor) is shown in **Fig. 9.9**, while the system in the North Atlantic is illustrated in **Fig. 3.20**. Within the

Table 9.1 Causes of abrupt climate change
Changes in Atlantic Meridional Overturning Circulation (AMOC)
Changes in carbon dioxide and methane concentrations
Surface albedo, snow, and vegetation
Water vapour as feedback
Dust and particles as feedback
Internal modes of atmospheric variability
Non-linear coupling of the climate subsystems

Fig. 9.9 A simplified diagram of the large-scale salt transport system (ocean/global conveyor) operating at present in the oceans (from Broecker and Denton 1990). Note that the diagram shows the pattern in a generalized form, but in fact the system is more complex. In addition to the high-latitude North Atlantic, deepwater formation also occurs around Antarctica. Reprinted with permission from Elsevier

North Atlantic, the Gulf Stream and North Atlantic Drift currents carry warm and relatively salty surface water from the Gulf of Mexico up to the seas between Greenland, Iceland, and Norway. As these waters migrate northwards they cool and become dense enough to sink into the deep ocean. The 'pull' exerted by this dense sinking water is thought to help maintain the strength of the warm Gulf Stream, ensuring a current of warm tropical water moves into the North Atlantic that, along with prevailing westerly winds, causes mild air masses to cross the European continent (Rahmstorf et al. 1996). It has been estimated that the thermohaline circulation system is currently responsible for the northward transport of some 5×10^{21} calories of heat to the North Atlantic region annually. This, for instance, keeps winters in the British Isles far more mild than what would otherwise be expected for its latitude.

Ocean circulation modelling studies suggest that a relatively small increase in freshwater flux to the Arctic Sea and critical regions of the high-latitude North Atlantic could cause deep water production in the North Atlantic to cease thus providing the trigger for a sudden 'switching off' of deepwater formation in the North Atlantic. The drainage of freshwater from the melting of the Laurentide ice sheet, and in particular a drainage event from glacial Lake Agassiz, during the Late Glacial, is thought to have triggered the Younger Dryas (see **Fig. 5.4**). The 8200 cal yr BP event (see **Section 5.4**) is thought to be related to a later, and much larger, sudden drainage of Lake Agassiz which disrupted the THC as well as causing a **eustatic** sea level rise of 1 to 3 metres.

Such massive pulses of freshwater dilute the dense, salty 'Atlantic Conveyor', forming a temporary lid of fresh water that stops the sinking and pulling of water that drives the Gulf Stream. In consequence, the Gulf Stream weakens or switches off altogether, breaking the 'conveyer belt' and allowing a sea ice cap to form, the expansion of sea ice then exerts a further cooling effect through the albedo feedback (described above) and by lowering atmospheric water vapour content.

Evidence from ocean sediments indicates that deepwater formation in the North Atlantic was diminished during the sudden cold Heinrich events and the Younger Dryas phase. Such a relationship is also found for an abrupt cooling (100 to 200 year) of subtropical sea surface temperatures (SSTs) in the Bermuda Rise area at the end of the Eemian interglacial (Lehman et al. 2002). It is also believed that the reverse occurs when climates suddenly warm up in the North Atlantic region, such as at the beginning of interstadials or the beginning of the present interglacial, when the THC switches to its strong mode (Rasmussen et al. 1997; Lehman and Keigwin 1992).

As described in **Section 3.6**, the last glacial cycle contains a series of abrupt warmings and coolings that were most severe in the North Atlantic region but have correlatives elsewhere, and as shown in **Fig. 3.21** the coldest points of **stadials** recorded in Greenland ice cores correlate with layers of ice rafted debris in deep-sea sediment cores – the Heinrich events. At these times there was tremendous discharge of icebergs into the North Atlantic, primarily from the Laurentide ice sheet, and the resultant pulse of freshwater as the icebergs melted caused the THC to collapse from an already weak glacial mode. Following Heinrich events, climate warming in the North Atlantic region was sudden as a strong mode of the THC was re-established. Over time, the contribution of meltwater to the critical regions of North Atlantic Deep Water formation would again weaken the THC to the glacial mode, and over the course of the 'Bond cycle' conditions became progressively colder, culminating in another Heinrich event.

This sequence of events dramatically illustrates the close linkage between ocean, atmosphere and ice in modulating the climate of the North Atlantic region, and through various climate **teleconnections**, many other parts of the world.

The causes of the Bond cycles and Heinrich events have been the subject of much study and are still uncertain. One of the main difficulties is in determining the extent to which the Heinrich events were internally or externally driven. One internally-based explanation is known as the 'binge-purge' model (MacAyeal 1993). According to this model the Heinrich events are the result of internally-generated instability within the Laurentide ice sheet. Over time, and with increased size of the ice sheet, geothermal energy and heat from friction build up along the base of the ice. This makes the ice sheet more 'warm-based' eventually triggering a surge. The surge itself could have had the effect of making other ice sheets with a marine margin unstable through the resulting rise in sea level, thus explaining the apparent similarity in timing between surges from unconnected ice sheets around the North Atlantic. Following the surge it would take several thousand years for the Laurentide ice sheet to again achieve the thickness required to become unstable and generate another Heinrich event.

A related mechanism involves **glacio-isostasy**: as the Laurentide and Fennoscandian ice sheets thickened, they gradually depressed the underlying bedrock, leading to instability along their marine margins due to the relative sea level rise caused by the crustal depression (Ruddiman 2008). While relative sea level change may be a trigger for instability, the resultant eustatic sea level change (over 10 m with each major Heinrich event) would have transmitted the signal to all ice caps in the world with a marine margin.

While the creation of instability in the Laurentide ice sheet is certainly a prerequisite for a Heinrich event, other workers have suggested that the instability and resultant surge is triggered by external factors. One proposed trigger is a warming of subsurface sea temperature (associated with weakening THC) prior to the major ice surges. Subsurface temperature rise would cause melting of ice shelves from beneath, eventually removing the 'buttressing' effect that ice shelves have on grounded ice behind them (Marcott et al. 2011). It has been argued that the influence of an external factor makes it easier to explain the fact that not all of the IRD material represented in Heinrich layers is from the Laurentide ice sheet (debris also came from Greenland, Iceland, and the British Isles). Moreover, IRD from the Laurentide ice sheet usually lags behind IRD sourced from elsewhere. An external factor also helps in explaining the correlations between some Heinrich events and glacial advances in parts of the Southern Hemisphere. Although difficult to prove statistically, it has been suggested that the D–O cycles and Heinrich events are paced by some combination of orbital cycles or solar activity cycles. The precise way in which minor changes in solar energy receipt could alter SSTs and freshwater budgets in the North Atlantic to affect the THC and ultimately, through a complicated set of feedbacks, trigger large-scale ice sheet surges is an area of ongoing research.

Northern and Southern Hemisphere connections

The above discussion has focused on changes within the North Atlantic region, although it is important to remember that the THC is just one part of the global ocean conveyor circulation system; therefore changes in the North Atlantic have far reaching effects. And indeed, changes elsewhere affect the North Atlantic. Some of the Pleistocene correlations between events in the North Atlantic and other parts of the globe were discussed in **Section 3.6**. One of the most important teleconnections is the anti-phase relationship which exists between the Arctic and the Antarctic as shown in **Fig. 3.18**. When the THC is in a strong mode this leads to enhanced North Atlantic Deep Water (NADW) formation at the expense of Antarctic Bottom Water (AABW) – seas around the Arctic warm causing air temperature over Greenland to increase. When the THC collapses and NADW production is diminished, more AABW is produced and seas warm around the Antarctic causing Antarctic temperatures to increase. Between these two extremes is an intermediate mode with less difference between the North and South Atlantic. The timescale over which changes at one end of the Atlantic are felt in the other is remarkably short, being measurable over decades.

As introduced in **Chapter 3**, this oscillating system in the Atlantic has been variously referred to as the **Atlantic Meridional Overturning Circulation** (AMOC), the 'bipolar seesaw' and 'Atlantic heat piracy' (see Seidov and Maslin 2001 and Alley 2007 for reviews). Further consequences of reduced AMOC (and reduced NADW production) are an intensification of the Antarctic Circumpolar Current, southward migration of Southern Hemisphere mid-latitude westerly winds, reduced Southern Ocean sea ice with enhanced vertical mixing of the seas, and a rise in atmospheric CO_2 due to this vertical mixing of the Southern Ocean (Barker et al. 2009). Ironically, the intense cooling of the North Atlantic plays a role in glacial terminations through association with these changes in the seas around Antarctica. Furthermore, ice shelf disintegration and destabilization of the West Antarctic ice sheet during phases of reduced AMOC probably results in iceberg discharge into the Southern Ocean which then feeds back on the Northern Hemisphere ice sheets through southern driven eustatic sea level rise.

As shown in **Fig. 9.10**, this major oscillatory system in meridional heat transport is unique to the Atlantic Ocean because the two important areas for deepwater formation are in the North Atlantic and in the Southern Ocean. These deepwater currents then circulate through the oceans with upwelling in the North Pacific and Indian Oceans. While switches between modes of AMOC would seem to be driven from the North Atlantic, especially due to the catastrophic injections of freshwater caused by the northern ice sheets, there is increased evidence for southern triggering of AMOC switches. For instance, during Termination I, Antarctica warmed earlier than Greenland; and as AABW formation became suppressed due to changes in the Southern Ocean, over time salinity built up in the North Atlantic, eventually crossing a threshold to trigger vigorous NADW formation. This caused a rapid reduction in northern sea ice extent and the abrupt Bølling warming registered in Greenland ice cores (Knorr and Lohmann 2003).

Another important factor behind variation in AMOC is the connection between the Indian Ocean and the South Atlantic. The transfer of warm, saline water into the South Atlantic around southern Africa is referred to as 'Agulhas leakage', and some deep-sea sediment studies have found that this flow has been highly variable in the past and has an important effect on AMOC (e.g. Peeters et al. 2004; Bard and Rickaby 2009). The latitudinal belt across which this transfer takes place is relatively narrow and is sensitive to the position of the subtropical front (STF) boundary between warm and cold water in the Indian Ocean off south-eastern Africa. During glacial times the STF is at a lower latitude and does not extend far enough south to enable warm water transfer from the Indian Ocean into the South Atlantic; whereas during interglacials, the STF is far enough south, allowing Agulhas leakage. Hence the STF in this area has been referred to as a 'gatekeeper' or 'choke point' for the ocean's global heat and salt conveyor.

SST reconstructions off the African coast have shown that southward migration of the STF and the onset of Agulhas leakage *preceded* Northern Hemisphere deglaciation during the last termination. It has been suggested, therefore, that a tropical response to the recent 100 kyr orbitally induced insolation maximum could have played a critical role in switching the AMOC to the strong mode (perhaps in addition to any weakening of AABW formation), which then led to renewed heat transport to the northern North Atlantic. The precise way in which this might work remains uncertain; but would involve

Fig. 9.10 A more detailed, three-dimensional schematic of the ocean conveyor centred on Antarctica showing surface, intermediate, and deepwater flows (from Gordon 1991 and as Fig. 16 in Richardson 2008). Reprinted with permission from Elsevier and A.L. Gordon

SAMW	Subantarctic Mode Water
AAIW	Antarctic Intermediate Water
RSDW	Red Sea Deep Water
NADW	North Atlantic Deep Water
AABW	Antarctic Bottom Water
NPDW	North Pacific Deep Water
ACC	Antarctic Circumpolar Current

a coupled ocean–atmosphere response to additional insolation involving the STF, Hadley Cells, wind systems, and the reach of the monsoons.

Connections between AMOC and low-latitude climate

In addition to influencing air temperatures at high latitudes, variation in AMOC also has an effect on precipitation in the tropical and subtropical belts. For instance, when the THC is weak, SSTs north of the Equator around the African coast tend to be reduced leading to a weaker monsoon in the Sahel zone. Hence, much of northern and eastern Africa dried out during the Younger Dryas event. Reduced THC also tends to lead to warming of the equatorial Pacific.

Another teleconnection concerns the effect that a reduced THC has upon the pressure gradient between the Azores and Iceland. Lower SSTs in the high-latitude North Atlantic would tend to reduce the 'Icelandic Low' and the overall pressure difference between high and low latitudes. This has

a knock-on effect for precipitation across Eurasia because a weaker Icelandic Low tends to reduce the formation of precipitation bearing storm systems and lead to a more meridional atmospheric circulation pattern. A strengthened THC leading to a stronger Icelandic Low would tend to have the opposite effect, strengthening westerly airflow, and transporting more storms across the continent. On shorter timescales, more minor variation in SSTs in the North Atlantic, and the related high/low latitude pressure difference, is referred to as the **North Atlantic Oscillation** (NAO).

Another clear pattern is seen between the location of the **Intertropical Convergence Zone** (ITCZ) and the mode of AMOC. When in the strong, interglacial mode, the ITCZ migrates further north in summer bringing more rain to subtropical latitudes of the Northern Hemisphere; whereas when AMOC collapses, the resultant cooling of high and mid latitudes of the Northern Hemisphere leads to a southward shift in the 'thermal equator' causing tropical rain belts across South America and Asia to move south (Broecker 2010). Marine sediments from the Cariaco

Basin off the coast of Venezuela (referred to in **Section 3.6**) provide important evidence for this. **Speleothem** evidence from Chinese caves indicate that times of AMOC collapse correlate with reduced northerly reach of the East Asian monsoon (e.g. Cheng et al. 2009a, b).

As with the uncertainties surrounding north or south control of AMOC, there is also a similar and related chicken-egg issue surrounding whether the tropical atmosphere is responding to, or driving, changes in AMOC. For example Leduc et al. (2007) proposed that moisture transport from the Atlantic to the Pacific Ocean via easterly trade winds across Central America is an important factor behind the salinity of the North Atlantic Ocean, and hence the strength of the THC. At times when the ITCZ moves south, so do the trade winds. This results in rain bearing systems dropping precipitation on the eastern side of the Andes, over Amazonia, with the moisture finding its way back to the Atlantic Ocean primarily via the Amazon River. The net salinity difference between the Atlantic and Pacific Oceans is thereby reduced and THC reduction is amplified. On the other hand, northward migration of the ITCZ and trade winds causes more rain systems to cross over Central America enhancing the Atlantic/Pacific salinity difference. If during glacial terminations, the gradual orbitally induced accumulation of heat in the tropical atmosphere causes expansion of the Hadley Cells and greater latitudinal reach of the ITCZ, then the mechanism proposed by Leduc et al. could be a way in which change in the tropics contributes to northern warming. This scenario is given support by work in the Venezuelan Andes showing that northern tropical warming preceded high-latitude northern warming at the onset of the Holocene (Stansell et al. 2010). Moreover, in their high resolution analysis of the NGRIP ice core, Steffensen et al. (2008) found that dust flux dropped before both the abrupt Bølling and Holocene warmings. As the predominant dust source was from Asia, they interpret this as representing an increase of precipitation across Asian deserts. They go on to conclude that rapid temperature increase in high northern latitudes must be led by northward migration of the ITCZ and wetting of the Asian deserts.

Holocene variation in ocean circulation

The last major perturbation of AMOC was the 8200 cal yr BP event. However, it is likely that minor variation in AMOC (within its present interglacial mode) has continued to influence climate through the mid to Late Holocene. It is possible, for instance, that some of the oscillations between wetter and drier conditions that affected Near Eastern civilizations referred to in **Chapter 8** could have been linked to such changes in North Atlantic Ocean circulation. Denton and Broecker (2008) argued that there has been an overall weakening of the THC over recent millennia, and that fluctuation superimposed on this trend correlates with neoglacial advances in the European Alps and elsewhere, including the recent Little Ice Age. A strengthened AMOC probably accounts for the persistently positive NAO conditions reconstructed for the Medieval Warm Period of Europe, with weakening of AMOC and shift to negative NAO conditions characterizing the Little Ice Age (Trouet et al. 2009).

There also appears to be a link with **ENSO**: when Atlantic THC was vigorous through much of the early Holocene, the ENSO was weak or non-existent. About 7000 years ago El Niño events became more frequent and generally increased in frequency until about 1200 years ago (Moy et al. 2002). Superimposed on this trend, time spans of weakened or strengthened ENSO seem to be anti-phased with AMOC strength.

On centennial to decadal timescales there are numerous other coupled ocean–atmosphere climate oscillations that have their various regional patterns and effects on climate. The effects of ENSO and other oscillations on timescales relevant to the instrumental record are reviewed in **Section 7.4**.

9.8 Conclusion

Our knowledge of the causes of climatic changes is still highly imperfect, and it is fair to say that the more that is learnt about the climates of the past, the more apparent the complexity of the climate system becomes. It can be extremely difficult to delineate between driving mechanisms and amplifying mechanisms, and factors that are clearly important at one timescale may be not be on a different timescale. There is no single explanation for climatic change, and it is also clear that no one process acting alone can explain all scales of climatic changes. Some coincidence or combination of processes in time is probably required, and at different times in Earth's history there have been unique combinations of factors involved. Moreover, numerous feedback loops exist and quite small triggers may set off a chain of events which lead to substantial change.

This is certainly the case for glacial/interglacial cycles over the Quaternary Period whereby relatively subtle changes in insolation due to orbital cycles pace very significant environmental changes through a complex set of feedbacks involving oceans, ice, atmosphere, and the biosphere. On shorter timescales, changes in solar irradiance, volcanic eruptions, and oscillatory behaviour of the coupled ocean–atmosphere system all have a role to play.

To test certain hypotheses about how the Earth's climate has changed in the past we need precise knowledge of the exact pattern, magnitude, and timing of past fluctuations. While more quantitative, high resolution proxy data sets are becoming available, their global distribution remains patchy and problems arise when trying to compare proxy data sets with differing chronological control. Given the immensely complex series of interrelated systems involved in climate change, it is unlikely that any simple hypothesis or model of climatic change will have very wide applicability.

Nonetheless, much has been learnt from the study of climatic change over the Quaternary Period that is directly applicable to the study of present climate and to the construction of the climate models used to predict future climate given certain scenarios about future levels of greenhouse gases and other changes wrought by people. Various feedback mechanisms, such as those involving oceans, ice, albedo, and the carbon cycle can be incorporated more realistically into climate models when there is palaeo-data to throw light on the strength and effects of such feedbacks in the past; and, as discussed in **Section 1.6**, validation of climate models through 'hindcasting' is increasingly important if there is to be confidence in a model's ability to forecast future change.

As the debate intensifies about the ways in which human activities will influence climate over the coming decades and centuries, it will be important to remember that the increase in the concentration of greenhouse gases in the atmosphere that began with the advent of agriculture and has accelerated since the industrial revolution will continue to be played out against the background of natural climatic forcing at a range of timescales. Humans have also impacted climate in other ways and at various scales, e.g. through changing surface albedos, particulate pollution, deforestation, etc. (see Goudie 2013). In all these cases an appreciation of the natural variations that the Earth has experienced in the past are essential for putting human impacts into context; and more awareness and understanding of the nature of environmental change among the wider public would be beneficial in terms of society's ability to anticipate and respond to future changes (Anderson and Brown 2010).

As yet, because of the complexity of the atmospheric system and the large number of possible causes, it is difficult fully to assess and quantify the role that man has played, though certain mechanisms of anthropogenic climatic changes on a global (as opposed to a microclimatic scale) can be recognized (**Table 9.2**). Perhaps one of the most important insights from recent research into past climates has been the increasing recognition of abrupt, threshold-style climate shifts, and this has focused attention on the possibility that our alteration of the climate system might inadvertently trigger a non-linear climate response. Past climate reconstructions also provide an essential baseline against which the human impact can be measured and give us an indication of the direction climate would likely take were humans out of the equation. Furthermore, the knowledge we have gained about how the global environment has responded to past climatic changes will be invaluable for an understanding of what the future has in store.

Table 9.2 Possible ways in which human activities may cause climatic change.

Gas emissions
CO_2–industrial and agricultural
methane
chlorofluorocarbons (CFCs) (and HCFCs)
nitrous oxide
krypton 85
water vapour
miscellaneous trace gases
Aerosol generation
Thermal pollution
Albedo change
dust addition to ice caps
deforestation
overgrazing
Extension of irrigation
Alteration of ocean currents by constricting straits
Diversion of fresh waters into oceans
Impoundment of large reservoirs

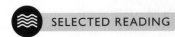 **SELECTED READING**

On the development of ideas on causes of climate change:

There are some excellent surveys of theories on climatic change, which although now old, give a good survey of the development of ideas. In J.B. Whittow and P.D. Wood (eds.), *Essays in Geography for Austin Miller* (1965), R.P. Beckinsale contributes 'Climatic change: a critique of modern theories', pp. 1–38.

Another review, of a similar type but with greater concentration on the auto-variation ideas, is J.M. Mitchell 'Theoretical palaeoclimatology' in H.E. Wright and D.G. Frey (1965) (eds.), *The Quaternary of the U.S.A.*, pp. 881–901.

J.M. Mitchell (1968) has also edited *Causes of climatic change*, Meteorological Monographs, 8.

These sources listed above give lengthy bibliographies and summaries of the major theories, but are now mainly of historical interest.

General overviews:

An excellent overview of causes of climate change across the whole range of timescales is W.F. Ruddiman (2008) *Earth's Climate: Past and Future* (2nd edition).

Another accessible and thorough account is given in W.J. Burroughs (2007) *Climate Change: A Multidisciplinary Approach* (2nd edition).

The NOAA Paleoclimatology website contains a series of explanatory pages on different aspects of climate change in their Paleo Perspectives section: http://www.ncdc.noaa.gov/paleo/perspectives.html

Orbital-scale change and abrupt climatic changes:

The Croll–Milankovitch ideas have been discussed in their historical context by J. Imbrie and K.P. Imbrie (1979), *Ice Ages: solving the mystery*.

They are also discussed in a series of essays in A. Berger et al. (1984) (eds.) *Milankovitch and climate*.

W.S. Broecker (2010) offers a highly readable account of the significance of ocean circulation in *The Great Ocean Conveyor: Discovering the Trigger for Abrupt Climate Change*.

The complex relationship between ice sheets and climate change is discussed in provocative style by T.J. Hughes (1998) *Ice Sheets*.

Decadal to centennial-scale changes:

The book *Natural Climate Variability and Global Warming: A Holocene Perspective* edited by R.W. Battarbee and H.A. Binney (2008) includes chapters about climate forcing mechanisms over short timescales.

A book of essays on solar variability has been edited by E. Nesme-Ribes (1994) *The Solar Engine and its Influence on Terrestrial Atmosphere and Climate*.

On the role of people:

A brief review of the role of man in causing climatic change is given in Goudie (2013) *The Human Impact on the Natural Environment*, 6th edition.

An authoritative account of anthropogenic global warming is given in J. Houghton (2009) *Global Warming: The Complete Briefing*, 4th edition.

The 'Ruddiman hypothesis' on the pre-industrial human impact on Holocene climate is detailed in W.F. Ruddiman (2005) *Plows, Plagues, and Petroleum*.

A diverting history of ideas on the greenhouse effect is S.R. Weart (2003) *The Discovery of Global Warming*.

? REVIEW QUESTIONS

1. (a) Contrast the difficulties of solar forcing and volcanic forcing as explanations for the pattern of mid to late Holocene climate variability.

   (b) What climate feedbacks have been proposed to address these difficulties, and at what times is there good evidence that the two forcing mechanisms acted together?

2. Summarize the factors that have been proposed to account for long-term Cainozoic cooling and the eventual return of the Earth to an 'icehouse' climate. (Discussion of this topic also appears in **Chapter 1**.)

3. Describe the role of orbital variation in millennial-scale climate change and explain why it is only a partial explanation for the changes that have occurred during the Pleistocene.

4. Summarize the phenomenon of the AMOC and describe the role that it plays in abrupt climatic change.

5. Drawing on examples from different timescales, evaluate the usefulness of the study of past climates for predicting future changes.

Glossary

Absolute dating The provision of a relatively precise date in years. Initially absolute dating methods included historical records, dendrochronology and varve analysis, but now they include various radiometric, chemical, and palaeomagnetic methods.

Accelerator Mass Spectrometry (AMS) dating A method of radiocarbon dating that involves direct measurement of the ratio of different isotopes of carbon and can be applied to very small samples.

Aeolian Geomorphological processes involving wind and resultant landforms.

Aerosol A suspension of solid or liquid particles within the atmosphere including dust, sea salt, and soot. They influence the climate directly through scattering and absorbing radiation, and indirectly through their effects on clouds.

Albedo The fraction of incident solar radiation reflected by a surface, whether a cloud, the ground, or vegetation cover.

Altithermal Traditionally, a warm phase of the Holocene between *c.* 8000 and 4000 years ago, also associated with relatively dry conditions in western North America. Also termed the Hypsithermal.

Amino acid geochronology Proteins, composed of amino acids, can survive in bones and shells for long periods but undergo molecular changes after the death of the host organism. Some of these changes are time dependent so that protein residues provide the basis for a relative chronology.

Analogue approach A way of inferring climatic or environmental conditions of the past by looking at the present-day climatic/environmental conditions associated with a modern biological assemblage that is thought to be the best match to an assemblage of organic remains from the time period of interest. Inorganic or landform evidence can be interpreted in a similar way; for instance, observation of processes occurring at contemporary glacier margins can provide an analogue for the processes (and conditions) that must have prevailed to produce relict glacial landforms. The approach is only valid when there is a modern analogue for a fossil assemblage or past process, but this is not always the case.

Antarctic Cold Reversal A phase of colder climate in Antarctica revealed in Antarctic ice cores that correlates with the Late Glacial Interstadial warming recorded in Greenland ice cores.

Atlantic Meridional Overturning Circulation (AMOC) A large-scale pattern of ocean circulation in the Atlantic Ocean involving surface and deepwater currents. It is connected with a global circulation pattern, variously referred to as the 'ocean conveyor', the 'thermohaline circulation' (THC), and the meridional overturning circulation (MOC), which transfers heat and salt through the world's oceans. In the Atlantic, the AMOC causes warm surface water to move towards the higher latitudes of the North Atlantic where strong overturning occurs to produce deepwater currents which flow southwards. Variations in AMOC have been closely linked with climatic changes in the North Atlantic region and elsewhere. See also **thermohaline circulation**.

Biostratigraphy The study of rock and sediment successions, or layers, focusing on biological aspects or the organic remains they contain.

Bond cycles Sawtooth patterns of millennial-scale climatic variation identified from ocean sediment on a roughly 1500-year periodicity. Over the last glacial they are related to the **D–O cycles** seen in Greenland ice core records, with each Bond cycle containing sets of D–O cycles and culminating with a Heinrich event.

Brunhes–Matuyama Boundary The boundary between two 'geomagnetic epochs' of polarity of the Earth's magnetic field the Brunhes (characterized by normal polarity) and the Matuyama (characterized by reversed polarity). The boundary is dated at *c.* 780 000 years ago.

Cainozoic (Cenozoic) climatic decline The long-term cooling of Earth's climate during the Tertiary Period, culminating in the expansion of glacier ice in the high latitudes of the Northern Hemisphere during the Late Pliocene.

Carbonate compensation depth (CCD) The depth in the ocean where the rate of dissolution of calcium carbonate reaches a point equal to the rate of forma-tion. Below this depth dissolution exceeds formation and calcareous microfossils (such as coccoliths and forams) will be absent. The depth varies, but is usually below 4 km. Solubility of calcium carbonate increases with decreasing temperature and increasing pressure.

Chronostratigraphy The study and dating of the suc-cession of time spans over which rocks and sediments are formed. Chronostratigraphic divisions (systems, series, and stages) are assigned precise age bounda-ries based on classic type-sections (**Global Stratotype Section and Points**, or GSSPs) that are used by work-ers as the international standard for correlating strata.

Cirque An amphitheatre-shaped, glacially eroded rock basin with a steep headwall on the side of a mountain. Also termed 'corrie' or 'cwm'.

Coccolithophora One celled marine plants (phyto-plankton) that live in the oceans and have a micro-scopic plating made of calcite. Their calcareous skeletons are found in marine deposits, such as chalks, in vast numbers.

Coral bleaching The loss of zooxanthellae which give coral its pigmentation. Under stress, such as high temperatures, corals may expel the zooxanthellae and so turn white.

Cryosphere Those portions of Earth's surface where water is in a solid form (i.e. ice), including sea, river and lake ice, glaciers, ice sheets and caps, and permafrost.

Dansgaard–Oeschger cycles (D–O cycles) Climatic fluctuations with a duration of the order of a thousand years and with rapid onset and termination. They are identified and numbered in the Greenland ice cores, and correlatives have been found in many other pal-aeoclimate records.

Dendrochronology The study of annual tree rings, based on the measurement of variations in the ring widths caused by variations in climate and the envi-ronment at the time of ring formation.

Denudation The wearing away of the landscape by processes of weathering, erosion, and mass movement.

Desertification The degradation of drylands, and the spread of desert-like conditions, as a result of human activities, climate change, or both.

Diagenesis Any chemical, physical or biological change undergone by a sediment at low tempera-tures and pressures after it is laid down and during and after its lithification, exclusive of weathering and metamorphism.

Diatoms A class of microscopic unicellular algae that live in the surface waters of rivers, lakes and the oceans and form silica-rich shells. They can be used to reconstruct past environmental conditions.

Domestication Human propagation of selected species.

Drumlins Glacial landforms made of till. They form elongated, streamlined hills, often in swarms, and were deposited beneath sediment-rich ice. The steep-est slopes and their summits occur at the up-ice end.

Dust storms Events in which visibility is reduced to less than 1 km as a result of particles entrained by wind being present in the air.

Eccentricity A change in the Earth's orbit around the Sun from near-circular to more elliptical at periods of about 100 000 and 400 000 years.

ENSO The El Niño–Southern Oscillation in the ocean–atmosphere system in the Pacific Ocean. One mode of the oscillation is the El Niño (warm phase) and the other is the La Niña (cool phase).

Epeirogeny Broad upwarping of the land surface with long wavelengths and limited folding. It is a slow process associated with the relatively stable portions of the continents (cratons).

Epoch A 'geochronological' subdivision of the geo-logical timescale that is longer than an age but shorter than a period. The Holocene is an epoch of the Quaternary Period.

Equifinality The concept that an end state may be achieved by many different paths or trajectories. For example, similar landforms might result from a quite different set of processes.

Era A span of geological time that divides up the next-larger division, the eon. The Phanerozoic has three eras of which the last is the Cainozoic (66 million years ago to present).

Erg A large area of desert sand (a sand sea) composed of dunes.

Erratic A rock that lies on contrasting bedrock because of long-distance transport from its origin, usually by glaciers.

Eustasy (eustatic) Vertical displacements of the ocean surface occurring uniformly throughout the world as a result of changes in the volume of the ocean basins or the quantity of water in them.

Flandrian transgression The rise in sea level that followed the Last Glacial Maximum.

Foraminifera Single-celled organisms that possess a hard, calcareous shell (or test). The marine forms have proved useful in climate reconstruction either because of the information that can be gained from particular species or because of isotopic studies of their tests, which accumulate in layers in ocean sediments.

Gateway hypothesis The concept that continental movements in the Cainozoic, especially, for example, the break-up of Gondwanaland, led to changes in oceanic circulation, and thus of climate. One such gateway was the Tasmanian Gateway between Australia and Antarctica. Others were at Panama, the Drake Passage, and off Indonesia.

Gelifluction A type of solifluction (soil flow) that occurs because of seasonally or permanently frozen ground preventing downward percolation of water into underlying rock. The term 'congelifluction' is sometimes used to refer specifically to the process occurring in association with permafrost.

Geochronology The science of dating and determining the time sequence of events in the history of the Earth.

Geomagnetic event An interval of time spanning 10 000 to 100 000 years when the Earth's magnetic field is aligned with either a north or south polarity. Geomagnetic events occur within longer time spans of a predominant polarity termed 'geomagnetic epochs'.

Glacial cycle (glacial) The period covering one successive advance and retreat of ice sheets and glaciers across polar regions and higher latitudes. The global climate spanning the glacial cycle varies, but is on average colder and drier than during interglacials.

Glacial theory A theory developed in the early 1800s that glaciers and ice sheets had once been much more extensive and had covered large tracts of Europe and North America, moulding distinctive landscapes that had formerly been attributed to the Biblical flood.

Glacio-eustasy Worldwide sea level changes brought about in response to the quantity of water stored in ice caps during glaciations and deglaciation.

Glaciofluvial The processes, sediments, and landforms produced by water flowing on, in, and/or under glaciers and away from glacier snouts.

Glacio-isostasy The state of balance that the Earth's crust attempts to achieve when loaded or unloaded by an ice sheet. Loading during a glacial produces subsidence and unloading during an interglacial produces uplift.

Global Stratotype Section and Point (GSSP) An internationally recognized point in a specific stratigraphic sequence used to mark a chronological boundary in the official global stratigraphical column. Establishment of a GSSP (sometimes referred to as a 'golden spike') requires ratification by the International Union of Geological Sciences. See also **chronostratigraphy**.

Greenhouse climate or **Hothouse Earth** A time at which temperatures were high, as were levels of greenhouse gases (e.g. carbon dioxide, methane). Continental ice sheets are absent.

Greenhouse effect The partial trapping by the atmosphere of radiation from the Earth's surface. This is achieved by water vapour and various gases (e.g. methane, carbon dioxide, nitrous oxide, CFCs) and leads to warming.

Guyot A flat-topped submarine volcanic mountain on the sea-floor which does not reach the sea surface and which has been drowned as a result of subsidence associated with sea-floor spreading.

Heinrich events Cold stadials of short duration represented by iceberg debris in North Atlantic ocean sediment cores. The source of most of the ice rafted debris results from destabilization and surging of the Laurentide ice sheet.

Holocene The present interglacial. The start of the Holocene was traditionally placed at 10 000 ^{14}C yr BP but is now dated in calendar years at 11 700 years b2k (before AD 2000) with the GSSP for its onset located in the NGRIP ice core.

Hominid Members of the family Hominidae of primates including all forms of modern and ancestral humans.

Hominin Member of the subfamily Homininae of Hominidae; includes all modern and ancestral humans but excludes the apes. See **Fig. 8.1**.

Hypsithermal See **Altithermal**.

Ice Age A period of widespread glaciation. During glacial phases of the Pleistocene 'Ice Age' mountain glaciers formed on all continents, the ice caps on Antarctica and Greenland were more extensive than today, and vast ice bodies spread across Eurasia and North America. Other Ice Ages (or times of **icehouse climate**) have occurred, including the Permo-Carboniferous Ice Age.

Ice-albedo feedback A positive feedback where the change in the area of snow and ice covered land alters the albedo, making the ground surface more reflective. This reduces the amount of solar energy absorbed, leading to more cooling. This also applies to the expansion of sea ice across an ocean surface.

Ice cores Cores of ice retrieved from ice caps and glaciers from which palaeoclimatic information can be retrieved.

Icehouse climate A time at which continental ice sheets are present because of low temperatures associated with low levels of greenhouse gases within the atmosphere.

Ice-wedge cast The remains of a former ice wedge that formed in a tundra terrain under cold conditions.

Indicator species Any biological species that defines a characteristic of the environment, such as its climate, degree of pollution, etc.

Interglacial A warm phase separating glacials and leading to glacial retreat, and the replacement of tundra by forest over the now temperate lands of the Northern Hemisphere.

Interpluvial A drier climate phase.

Interstadial A phase of slightly greater warmth which leads to an episode of glacial retreat during a glacial cycle.

Intertropical Convergence Zone (ITCZ) This is the low pressure region in low latitudes where the north-east and south-east trade winds converge. The ITCZ migrates seasonally towards the area of most intense solar heating and is associated with the rainy season in the tropics.

Involution Layers of soil and sediment originally laid down horizontally that have been twisted and contorted by differences in rates of freezing and thawing within the active layer above the permafrost table.

Isopoll map A map that produces a picture of regional vegetation cover for selected time periods. Pollen percentages of selected taxa are obtained for all sites in a region for a particular interval of time and are then plotted on maps which are contoured to reflect spatial trends or variations in regional abundance.

Isostasy A condition to which the Earth's crust and mantle tend but which is disturbed by such processes as sedimentation, erosion, changes in water volume (hydro-isostasy) and the waxing and waning of ice sheets (glacio-isostasy).

Lacustral A phase of high lake levels, generally caused by changes in precipitation and evapotranspiration. High lake levels occur during pluvials and low lake levels during interpluvials.

Last Glacial Maximum (LGM) The culmination of the last glacial advance about 21 000 years ago (or c. 18 000 ^{14}C yr BP) when global ice sheets reached their maximum volume. Used as a chronozone it refers to the span between c. 26 000 and 19 000 cal yr BP during which time most ice sheets across the world were at their maximum size. It lies within (but is not equivalent to) MIS 2.

Late Glacial The terminal Pleistocene and final part of the last glacial stage which spans the period following the **Last Glacial Maximum** to the start of the **Holocene**. Traditionally including the 'Oldest Dryas', 'Older Dryas', and 'Younger Dryas' stadials and the 'Allerød' and 'Bølling' interstadials.

Late Glacial Interstadial The episode of warmth during the Late Glacial which in northern Europe is divided into the earlier 'Bølling' and the later 'Allerød' phases. Equivalent to Greenland Interstadial 1 (GI-1) of the Greenland ice core scheme.

Laurentide ice sheet A massive sheet of ice that covered most of Canada and a large portion of the northern USA, down to as far south as St. Louis, Missouri.

Lichenometry A dating technique using lichen measurements to supply relative or absolute dates for rock surface exposures.

Lithostratigraphy A branch of stratigraphy that deals with the study of units of rock, each of which is dominated by a certain lithology, colour, mineral composition, grain size, etc.

Little Ice Age (LIA) A period of cold and highly variable climate and glacial readvance (a neoglaciation) following the Little Optimum and approximately spanning the fourteenth to nineteenth centuries AD.

Little Optimum A warm phase well represented in Europe and parts of North America in early medieval times. Also called the Medieval Warm Period.

Loam A soil with relatively even proportions of sand, silt, and clay.

Loess Fine, wind-blown silt.

Luminescence dating A method of dating employing heat or light to release trapped electrons. The accumulation of electrons within 'traps' in a crystal lattice (and hence the intensity of luminescence when the electrons are released) is time dependent.

Marine isotope stages (MIS) A chronological system based on variation in $\delta^{18}O$ in ocean sediment cores in which even numbered stages represent times of higher $\delta^{18}O$ (colder with more global ice) and odd numbered stages represent times of lower $\delta^{18}O$. Over 100 stages have been identified spanning the Quaternary. Also known as oxygen isotope stages (OIS).

Mass spectrometry A technique whereby the elemental composition and isotope ratios of a substance can be measured by ionizing a sample and measuring the relative deflection of the charged particles through a magnetic field in a mass spectrometer. Heavier ions are deflected more than lighter ones.

Maunder Minimum A state of low sunspot numbers between 1645 and 1715. It coincided with one of the coldest portions of the Little Ice Age.

Medieval Warm Period see **Little Optimum**.

Mesolithic A period of development in human technology between the Palaeolithic and Neolithic periods of the Stone Age. The age given to the start and end of the period varies in different regions, but it is usually associated with the Holocene until the advent of agriculture.

Milankovitch Theory The theory that the variations in the Earth's orbit around the Sun (see **eccentricity, precession**, and **obliquity**) hold the key to the timing of the main climatic events of the Quaternary.

Mitochondrial DNA Genetic code inherited through the maternal line.

Monsoon A term derived from the Arabic word for season that refers to wind patterns that reverse direction seasonally. The term was originally applied to the winds in the Arabian Sea and Indian Ocean. However, other notable examples include the African and Asian monsoons. They can be responsible for the generation of very heavy rainfall.

Neo-catastrophism The concept that over the course of geological history the environment has, at times, changed markedly and rapidly due to infrequent but high magnitude events, e.g. super-eruptions and asteroid impacts.

Neogene A unit of geological time covering c. 23 million years ago up to the base of the Quaternary Period consisting of the Miocene and Pliocene Epochs. It follows the Palaeogene (the Lower Tertiary and the first part of the Cainozoic).

Neoglaciation A Holocene phase of glacial advance. Multiple advances occurred of which the Little Ice Age is the latest example.

Neolithic This follows the Mesolithic or Epipalaeolithic periods of prehistory and corresponds with the transformation from hunting and gathering to agriculture.

Neotectonics The study of horizontal and vertical crustal movements that have occurred in the geologically recent past and which may be ongoing today.

North Atlantic Oscillation (NAO) An ocean-atmosphere oscillation in the North Atlantic that lasts some decades. When in positive mode, high pressure differences are found between the Azores High and the Atlantic (Icelandic) Low, and westerly winds and Gulf Stream advection are strong. When in negative mode there is a weak pressure gradient which is associated with weak westerly winds and Gulf Stream advection.

Obliquity A change in the tilt of the Earth's axis relative to the plane in which the bodies of the solar system lie (the ecliptic), which varies between 21° and 25°, with a periodicity of *c*. 41 000 years.

Oceanic gyre A large system of rotating ocean currents associated with large wind movements and the Coriolis effect. There are five main ones: the Indian Ocean gyre, the North Atlantic gyre, the South Atlantic gyre, the North Pacific gyre, and the South Pacific gyre.

Orogenic eustasy Changes in worldwide sea levels brought about by changes in the volume of the ocean basins associated with orogenesis (mountain building).

Ostracods Small crustaceans found in rivers, lakes and oceans. They have easily preservable carapaces that can be identified to provide palaeoenvironmental information.

Palaeocene–Eocene Thermal Maximum (PETM) A short phase of extreme climate warming at *c*. 55 million years ago, associated with elevated atmospheric concentration of greenhouse gases. Possibly caused by release of methane by breakdown of gas hydrates beneath the oceans and/or by intense volcanic activity associated with rifting and the opening up of the North Atlantic Ocean.

Palaeoecological transfer functions Mathematical models that relate variations in the relative abundance of certain taxa making up an assemblage of organisms to environmental or climatic factors. They are derived from study of modern assemblages ('training sets') used to quantify how assemblages change with respect to a chosen factor, such as temperature, pH, salinity, etc. The resultant mathematical relationship (or function) is then 'transferred' to a past assemblage to arrive at a quantitative reconstruction of the environmental or climatic variable of interest. The technique has been applied to a wide variety of biological proxy data, notably forams, diatoms, and pollen.

Palaeohydrology The study of past occurrences, distribution, and movements of continental waters.

Palaeolithic This represents the first period in the development of human technology, and in particular, stone tools. The period spans from *c*. 3 million years ago to the end of the Pleistocene.

Palaeomagnetism Past changes in the Earth's magnetic field and especially changes of polarity that are recorded by ferromagnetic sediments at the time of deposition. The polarity is described as *normal* where, as today, the north-seeking magnetism gives a Northern Hemisphere pole, or *reversed* where the north-seeking magnetism gives a Southern Hemisphere pole.

Palaeosol A soil that formed in the past and that has been buried by younger deposits. The organic and inorganic characteristics of palaeosols reflect past climatic and environmental conditions.

Palynology Pollen analysis or pollen stratigraphy. It is used as a means of correlating Quaternary stratigraphic units and to reconstruct vegetational history.

Periglacial Refers to non-glacial, cold climate processes and landforms, often associated with the presence of **permafrost**.

Period A 'geochronological' subdivision of an era (e.g. the Quaternary Period of the Cainozoic Era).

Peripheral forebulge A result of glacio-isostasy. The lateral displacement of mantle material from below the centre of the ice results in a compensating area of slight uplift beyond the area that is depressed by the weight of the ice. The forebulge collapses when the ice body disappears.

Permafrost The condition in which the temperature of soil and rock persists below 0 °C for at least two consecutive years.

Phytolith A rigid microscopic body that occurs in many plants. The most common type is composed of silica and is often called an opal phytolith. They are robust enough to be preserved in many sediments and can therefore be used to reconstruct past vegetation conditions.

Pleistocene Epoch The span of time from the beginning of the Quaternary Period up to the onset of the Holocene. It is characterized by repeated glacial/interglacial cycles.

Pliocene The final epoch of the Neogene Period (and Tertiary Period) (*c*. 5.3–2.6 million years ago). It was followed by the Pleistocene Epoch of the Quaternary Period.

Pluvial A wetter climate phase, especially in the low latitudes, in many areas coinciding with interglacial periods.

Pollen zonation scheme A zonation or subdivision of a pollen record to show times characterized by the dominance of certain types of pollen representing certain past vegetational assemblages. Site-specific pollen diagrams are zoned into 'local pollen assemblage zones' and can be compared with regionally-based pollen zonation schemes based on synthesis of many pollen diagrams.

Precession of the equinoxes Changes in the distance between the Earth and the Sun at any given season exhibiting periodicities of 19 000 and 23 000 years.

Provenance The characteristics of the source areas from which sediments are derived.

Proxy data Data gathered from natural recorders of environmental and climatic variability (tree rings, ice cores, ocean sediments, pollen, etc.) that can be used to understand past environments and/or climates before the availability of the instrumental record.

Quaternary The 'period' of geological time (or 'system' of rock succession) extending from 2.588 million years ago up to the present day. It is divided into two epochs (or series) the Pleistocene and the Holocene.

Radiative forcing Changes in the radiation balance, that between absorbed and radiated energy from the Sun. They can be brought about by changes in such factors as the input of energy from the Sun and reflection or absorption by, for example, clouds, gases, aerosols, and land surfaces.

Radiocarbon plateau A period over which materials of different true ages yield the same or similar radiocarbon age. Such periods result from times when atmospheric ^{14}C concentration is decreasing so that a younger sample in calendar years has a similar ratio of ^{14}C to ^{12}C as an older sample in calendar years, even though the older sample has had more time for ^{14}C to decay. An important cause of radiocarbon plateaux are times of rapid reinitiation of deep ocean circulation.

Radiocarbon years The age of a sample in years calculated from radiocarbon dating before any calibration is undertaken to convert the estimate to the calendar year timescale.

Radiometric dating Dating technique based on the principle that radioactive decay of unstable isotopes is time dependent (e.g. radiocarbon dating, U-series dating, etc).

Regression A phase of relative decline of sea level exposing land that was formerly underwater (the term is also used to refer to a phase of lake level decline).

Relative dating The placing of events in their order of occurrence. It does not tell how many years ago an event occurred.

Sapropels Organic rich muds from the remains of phyto- and zooplankton that form layers on the bottom of water bodies including lakes, estuaries, and the sea-floor.

Series A 'chronostratigraphic' division referring to rocks or deposits laid down over a certain time interval. For further information see **System**.

Snowball Earth Times, earlier than 630 million years ago, when glaciers were present near sea level in tropical latitudes and sea ice was globally extensive. Proposed causes include low-latitude continental distribution, lower total solar irradiance, a reduction in greenhouse gases, shifts in Earth's orbit, and various positive feedbacks including a runaway ice-albedo feedback.

Speleothem A secondary mineral deposit formed in caves in limestone regions as dripstone or flowstone. Stalactites and stalagmites are types of speleothem.

Stable isotopes Whereas unstable isotopes are radioactive isotopes that spontaneously disintegrate over time to form other isotopes, stable isotopes do not appear to do this, but may themselves be produced by radioactive isotopes.

Stadial A phase of relative cold and intense glacial activity and advance during a glacial cycle.

Steric effect In the case of seawater, steric effects are driven by changes in the temperature and/or salinity of the ocean, which cause the water to expand or contract, thereby causing a change in sea level.

Strandline A line of erosion running horizontally along a slope indicating the position of a former shoreline. Strandlines are used to reconstruct past lake levels.

Stratigraphic sequence A sequence, and usually vertical section, of material that consists of layers formed or deposited fairly continuously over a time span. Analysis of the stratigraphy (or layers) contained in the sequence enables reconstruction of changing environmental conditions over the time span that the sequence was formed.

Sunspot A dark region of the Sun's surface that is marked by a lower temperature than its surroundings; however, the 'faculae' around the sunspots emit more energy meaning that total solar irradiance is higher when there are more sunspots. Sunspot numbers vary (see, for example, **Maunder Minimum**) and may be implicated in climate change, with low sunspot numbers being associated with colder phases.

System A 'chronostratigraphic' division referring to rocks or deposits laid down over a certain time interval. Systems contain intervals of shorter duration termed **series**. The equivalent 'geochronological' time spans are 'periods' and 'epochs'; for example the Quaternary Period/System contains the Pleistocene and Holocene Epochs/Series.

Taphonomy The study of decaying organisms through time and of the processes by which they become fossilized and preserved with other remains in an assemblage.

Tephrochronology The study of volcanic ash layers (tephra) for dating. Each ash layer provides a time equivalent marker that can be correlated from one site to another.

Teleconnections Long-distance linkages between the weather patterns of widely separated regions of the Earth.

Terminations The boundaries in deep-sea core sediments that separate pronounced oxygen isotope maxima and minima. In effect they are rapid deglaciations marking transitions from glacials to interglacials, and they are conventionally numbered by Roman numerals in order of increasing age.

Thermohaline conveyor or circulation A slow moving, but extensive, current of water linking the oceans and the surface and deep currents globally. The formation of deep or bottom currents is driven by temperature and salinity differences (density differences), with cold, saline water in some locations (high-latitude North Atlantic and margins of Antarctica) causing strong overturning. The nature of the conveyor can change abruptly in response to the pulses of meltwater from decaying ice sheets, having regional and global climatic effects. See also **Atlantic Meridional Overturning Circulation**.

Till Unconsolidated debris deposited by glacier ice. It is typically unsorted and non-stratified in contrast with debris deposited by glacial meltwater.

Time-transgressive Applies to boundaries of 'climatostratigraphic' units that are not time equivalent over large distances. For example, depending on location and other factors geological and biological responses to palaeoclimatic shifts in temperature and/or precipitation exhibit time lags of differing lengths. This means that proxy data used to reconstruct the timing of a palaeoclimatic event in one place may not correlate in time with similar evidence in another place.

Transgression A phase of relative rise of sea level and inundation of land areas (the term is also used to refer to a phase of lake level rise).

Transhumance Seasonal movement of livestock by human groups between mountain and lowland pastures or grazing over considerable distances following set seasonal patterns.

Uniformitarianism The concept, expounded by Charles Lyell in the nineteenth century, that present-day processes have been operating in a similar manner through geological time. In this sense, present-day processes can be used as analogues for understanding past processes that have shaped the environment.

Upwelling The vertical movement of water towards the sea surface often driven by wind-generated currents. This upwelling water is often rich in nutrients and has a high productivity of phytoplankton. Notable examples of where this occurs include the Gulf of Oman and off the coasts of West Africa, Oregon, and Peru.

Varves Laminations in sediments caused by annual variation in sedimentation as, for example, in proglacial lakes where there are summer and winter contrasts in sediment input.

Younger Dryas A cold climatic interval that took place just before the onset of the Holocene. It was traditionally dated to between 11 000 and 10 000 ^{14}C yr BP by radiocarbon, but work on ice cores and varved lake sediments now shows that the event started c. 12 900 and ended c. 11 700 calendar years ago. It caused a return to glacial conditions in many of the highlands of Britain and Europe. It is equivalent to Greenland Stadial 1 (GS-1) of the Greenland ice core scheme, and correlatives of the European Younger Dryas have been identified in many parts of the world.

References

Aaby, B. (1986), 'Palaeoecological studies of mires', in B.E. Berglund (ed.) *Handbook of Holocene Palaeoecology and Palaeohydrology* (Chichester): 145–160.

Abels, H.A., Hilgen, F.J., Krijgsman, W., Kruk, R.W., Raffi, I., Turco, E. and Zachariasse, W.J. (2005), 'Long-period orbital control on middle Miocene global cooling: Integrated stratigraphy and astronomical tuning of the Blue Clay formation on Malta', *Paleoceanography*, 20: 1–17.

Abu-zeid, M. and Abdel-Dayem, S. (1992), 'Variations and trends in agricultural drainage water re-use in Egypt', *Water International*, 16: 247–253.

Adams, J.M. and Faure, H. (eds) QEN members. (1997), *Review and Atlas of Palaeovegetation: Preliminary Land Ecosystem Maps of the World since the Last Glacial Maximum*. Oak Ridge National Laboratory, TN, USA. http://www.esd.ornl.gov/projects/qen/adams1.html.

—— Maslin, M. and Thomas, E. (1999), 'Sudden climate transitions during the Quaternary', *Progress in Physical Geography*, 23: 1–36.

Adams, J.W., Rodriguez, D., Cox, R.A. (2005), 'The uptake of SO_2 on Saharan dust: a flow tube study', *Atmospheric Chemistry and Physics*, 5: 2679–2689.

Agrawal, D.P., Dodia, R. and Seth, M. (1990), 'South Asian climate and environment at *c.* 18 000 BP', in O. Soffer and C. Gamble (eds) *The World at 18 000 BP*, vol. 2 (London).

Ahlmann, H.W. (1948), 'The present climatic fluctuation', *Geographical Journal*, 112: 165–195.

Aitken, M.J. (1989), 'Luminescence dating: a guide for non-specialists', *Archaeometry*, 31: 147–159.

—— (1998), *An Introduction to Optical Dating* (Oxford).

Alley, R.B. (2002), *The Two-Mile Time Machine* (Princeton, NJ).

—— (2004), 'Abrupt climate change', *Scientific American*, November, 40–47.

—— (2007), 'Wally was right: predictive ability of the North Atlantic "conveyor belt" hypothesis for abrupt climate change', *Annual Review of Earth and Planetary Sciences*, 35: 241–263.

—— Meese, D.A., Shuman, C.A., Gow, A.J., Taylor, K.C., Grootes, P.M., White, J.W.C., Ram, M., Waddington, E.D., Mayewski, P.A. and Zielinski, G.A. (1993), 'Abrupt increase in Greenland snow accumulation at the end of the Younger Dryas event', *Nature*, 362: 527–529.

Alonso, R.N., Jordan, T.E., Tubbett, K.T. and Vandervoort, D.S. (1999), 'Giant evaporite belts of the Neogene central Andes', *Geology*, 19: 401–404.

Alvarez, L.W., Alvarez, W., Asaro, F. and Michel, H.V. (1980), 'Extra-terrestrial causes for the Cretaceous-Tertiary boundary extinction', *Science*, 208: 1095–1108.

Ambrose, S.H. (1998), 'Late Pleistocene human population bottlenecks, volcanic winter and differentiation of modern humans', *Journal of Human Evolution*, 34: 623–651.

An, Z., Liu, T.L.Y., Porter, S.C., Kukla, G., Wu, X. and Hua, Y. (1990), 'The long-term paleomonsoon variation recorded by the Loess-Palaeosol sequence in central China', *Quaternary International*, 7/8: 91–95.

Ananova, E.N. (1967), 'Palynological correlation of the flora and vegetation of the Likhvin-Mazovian 1-Holstein-Neede Interglacial', *Review of Palaeobotany and Palynology*, 4: 175–186.

Andersen, B.G. and Borns, H.W. (1994), *The Ice Age World* (Scandinavian University Press).

Anderson, D.E. (1996), 'Abrupt Holocene climatic change recorded in terrestrial peat sequences from Wester Ross, Scotland', unpublished D. Phil. thesis, University of Oxford.

—— (1997), 'Younger Dryas research and its implications for understanding abrupt climatic change', *Progress in Physical Geography*, 21: 230–249.

—— (1998), 'A reconstruction of Holocene climatic changes from peat bogs in north-west Scotland', *Boreas*, 27: 208–224.

—— (2006), 'The Channeled Scablands: tectonics, glaciers and flooding', *Geography Review*, 20: 14–19.

—— and Brown, E.J. (2010), 'Perspectives on Quaternary outreach and aspirations for the future', *Proceedings of the Geologists' Association*, 121: 455–467.

Anderson, J.B. and Thomas, M.A. (1991), 'Marine ice-sheet decoupling as a mechanism, for rapid episodic sea-level change: the record of such events and their influence on sedimentation', *Sedimentary Geology*, 70: 87–104.

Andrews, J.T. (1970), 'A geomorphological study of Post-Glacial uplift with particular reference to Arctic Canada', *Institute of British Geographers Special Publication*, 2.

—— (1975), *Glacial Systems – An Approach to Glaciers and their Environments* (North Scituate, MA).

——Jennings, A.E., Kerwin, M., Kirby, M., Manley, W. and Miller, G.H. (1995), 'A Heinrich-like event, H-0 (DC-0): source(s) for detrital carbonate in the North Atlantic during the Younger Dryas chronozone', *Paleoceanography*, 10: 943–952.

Antevs, E. (1936), 'Dating records of Early Man in the Southwest', *American Naturalist*, 70: 331–336.

—— (1955), 'Geologic-climatic dating in the West', *American Antiquity*, 20: 317–335.

Arbogast, A.F. (1996), 'Stratigraphic evidence for late-Holocene aeolian sand mobilisation and soil formation in south-central Kansas, USA', *Journal of Arid Environments*, 34: 403–414.

Arendt, A.A., Echelmeyer, K.A., Harrison, W.D., Lingle, C.S. and Valentine, V.B. (2002), 'Rapid wastage of Alaska glaciers and their contribution to rising sea level', *Science*, 297: 382–385.

Argue, D., Donlon, D., Groves, C. and Wright, R. (2006) '*Homo floresiensis*: Microcephalic, pygmoid, Australopithecus, or Homo?' *Journal of Human Evolution* 51, 360–374, doi:10.1016/j.jhevol.2006.04.013

Arimoto, M.O. (2001), 'Eolian dust and climate: relationships to sources, tropospheric chemistry, transport and deposition', *Earth-Science Reviews*, 54: 29–42.

Armitage, S.J., Drake, N.A., Stokes, S., El-Hawat, A., Salem, M.J., White, K., Turner, P. and McLaren, S.J. (2007), 'Multiple phases of North African humidity recorded in lacustrine sediments from the Fazzan Basin, Libyan Sahara', *Quaternary Geochronology*, 2: 181–186.

—— Jasim, S.A., Marks, A.E., Parker, A.G., Usik, V.I. and Uerpmann, H.-P. (2011), 'The Southern Route "out of Africa": Evidence for an Early Expansion of Modern Humans into Arabia', *Science*, 331: 453–456.

Arneborg, J., Heinemeier, J., Lynnerup, N., Nielsen, H.L., Rud, N. and Sveinbjörnsdóttir, Á.E. (1999), 'Change of diet of the Greenland Vikings determined from stable carbon isotope analysis and ^{14}C dating of their bones'. *Radiocarbon*, 41: 157–168.

Arz, H.W., Lamy, F., Pätzold, J., Müller, P.J. and Prins, M. (2003), 'Mediterranean moisture source for an Early-Holocene humid period in the northern Red Sea', *Science*, 300: 118–121.

Atkinson, T.C., Briffa, K.R. and Coope, G.R. (1987), 'Seasonal temperatures in Britain during the past 22 000 years reconstructed using beetle remains', *Nature*, 325: 587–592.

Bai, A., Chen, X., Li, J., Yang, L. and Fang, H. (2011), 'Changes in the area of inland lakes of central Asia during the past 30 years', *Environmental Monitoring and Assessment*, 178: 247–256.

Bailey, G.N. and Flemming, N.C. (2008), 'Archaeology of the continental shelf: marine resources, submerged landscapes and underwater archaeology', *Quaternary Science Reviews*, 27: 2153–2165.

Baillie, M.G.L. and Munro, M.A.R. (1988), 'Irish tree rings, Santorini and volcanic dust veils', *Nature*, 322: 344–346.

Baker, A., Smart, P.L., Edwards, R.L. and Richards, D.A. (1993), 'Annual growth banding in a cave stalagmite', *Nature*, 364: 518–520.

Bakke, J., Øyvind, L., Heegaard, E., Dokken, T., Haug, G.H., Birks, H.H., Dulski, P. and Nilsen, T. (2009), 'Rapid oceanic and atmospheric changes during the Younger Dryas cold period', *Nature Geoscience*, doi:10.1038/NGEO439.

Ballantyne, A.P., Lavine, M., Crowley, T.J., Liu, J. and Baker, P.B. (2005), 'Meta-analysis of tropical surface temperatures during the Last Glacial Maximum', *Geophysical Research Letters*, 32: L05712, doi: 1029/2004GL021217.

Ballantyne, C.K. and Harris, C. (1994), *The Periglaciation of Great Britain* (Cambridge).

—— McCarroll, D., Nesje, A. and Dahl, S.O. (1997), 'Periglacial trimlines, former nunataks and the altitude of the last ice sheet in Wester Ross, northwest Scotland', *Journal of Quaternary Science*, 12: 225–238.

Ballard, R.D. and Uchupi, E. (1970), 'Morphology and Quaternary history of the continental shelf of the Gulf Coast of the United States', *Bulletin of Marine Science*, 20: 547–559.

Bandy, O.L. (1968), 'Changes in Neogene paleo-oceanography and eustatic changes', *Palaeogeography, Palaeoclimatology, Palaeoecology*, 5: 63–75.

Barber, K.E. (1981), *Peat Stratigraphy and Climatic Change* (Rotterdam).

—— and Charman, D. (2005), 'Holocene palaeoclimate records from peatlands', in A. Mackay, R. Battarbee, J. Birks and F. Oldfield (eds) *Global Change in the Holocene* (London): 210–226.

Bard, E. and Rickaby, R.E.M. (2009), 'Migration of the subtropical front as a modulator of glacial climate', *Nature*, 460: 380–384.

—— Laberyrue, L.D., Pichon, J.J., Labracherie, M., Arnold, M., Duprat, J., Moyes, J. and Duplessy, J.C. (1990), 'The last deglaciation in the southern and northern hemispheres', in V. Bleil and J. Thiede (eds) *Geological History of the Polar Oceans: Arctic versus Antarctic* (Dordrecht): 405–415.

—— Rostek, F. and Sonzogni, C. (1997), 'Interhemispheric synchroneity of the last deglaciation inferred from alkenone palaeothermometry', *Nature*, 385: 707–710.

—— Hamelin, B. and Delanghe-Sabatier, D. (2010), 'Deglacial meltwater pulse 1B and Younger Dryas sea levels revisited with boreholes at Tahiti', *Science*, 327: 1235–1237.

Barker, G.W.W., Barton, H., Bird, M., Daly, P., Datan, I., Dykes, A., Farr, L., Gilbertson, D., Harrisson, B., Hunt, C., Higham, T., Kealhofer, L., Krigbaum, J., Lewis, H., McLaren, S., Paz, V., Pike, A., Piper, P., Pyatt, B., Rabett, R., Reynolds, T., Rose, J., Rushworth, G., Stephens, M., Stringer, C., Thompson, J. and Turney, C. (2007). 'The human revolution in lowland tropical southeast Asia: The antiquity and behavior of anatomically modern humans at Niah Cave (Sarawak, Borneo)', *Journal of Human Evolution*, 52: 243–261.

Barker, S., Diz, P., Vautravers, M.J., Pike, J., Knorr, G., Hall, I.R. and Broecker, W.S. (2009), 'Interhemispheric Atlantic seesaw response during the last deglaciation', *Nature*, 457: 1097–1102.

Barrett, E.C. (1966), 'Regional variations of rainfall trends in northern England, 1900–1959', *Transactions, Institute of British Geographers*, 38: 41–58.

Barrett, P.J. (1991), 'Antarctica and global climatic change: a geological perspective', in C.M. Harris and B. Stonehouse (eds) *Antarctica and Global Climate Change* (London).

Barron, E.J. (1985), 'Explanations of the Tertiary global cooling trend', *Palaeogeography, Palaeoclimatology, Palaeoecology*, 50: 45–61.

—— van Andel, T.H. and Pollard, D. (2003), 'Glacial environments II. Reconstructing the climate of Europe in the Last Glaciation', in T.H. van Andel and S.W. Davies (eds) *Neanderthals and Modern Humans in the European Landscape during the Last Glaciation* (Cambridge): 57–78.

Barrows, T.T. and Juggins, S. (2005), 'Sea-surface temperatures around the Australian margin and Indian Ocean during the Last Glacial Maximum', *Quaternary Science Reviews*, 24: 1017–1047.

—— Stone, J.O. and Fifield, L.K. (2004), 'Exposure ages for Pleistocene periglacial deposits in Australia', *Quaternary Science Reviews*, 23: 697–708.

Bartholdy, J., Pedersen, J.B.T. and Bartholdy, A.T. (2010), 'Autocompaction of shallow silty salt marsh clay', *Sedimentary Geology*, 223: 310–319.

Bartlein, P.J., Webb III, T. and Fleri, E. (1984), 'Holocene climatic change in the northern Midwest: pollen-derived estimates', *Quaternary Research*, 22: 361–374.

Barton, R.N.E. (1997), *Stone Age Britain* (English Heritage).

Basagic, H.J. and Fountain, A.G. (2011), 'Quantifying 20th century glacier change in the Sierra Nevada, California', *Arctic, Antarctic and Alpine Research*, 43: 317–330.

Battarbee, R.W. (1986), 'Diatom analysis', in B.E. Berglund (ed.) *Handbook of Holocene Palaeoecology and Palaeohydrology* (Chichester): 527–570.

—— and Binney, H.A. (eds) (2008), *Natural Climate Variability and Global Warming: A Holocene Perspective* (Wiley).

Baulig, H. (1928), *Le Plateau Central de la France et sa Bordure Méditerranéenne* (Paris).

Beck, J.W., Recy, J., Taylor, F., Edwards, R.L. and Cabioch, G. (1997), 'Abrupt changes in Holocene tropical sea surface temperatures derived from coral records', *Nature*, 385: 705–707.

Beckinsale, R.P. (1965), 'Climatic change: a critique of modern theories', in J.B. Whittow and P.D. Wood (eds) *Essays in Geography for Austin Miller* (Reading): 1–38.

—— and Beckinsale, R.D. (1989), 'Eustasy to plate tectonics: unifying ideas on the evolution of the major features of the Earth's surface', in K.J. Tinkler (ed.) *History of Geomorphology from Hutton to Hack* (Boston): 205–221.

Bedford, A., Jones, R.T., Lang, B., Brooks, S. and Marshall, J.D. (2004), 'A Late-glacial chironomid record from Hawes Water, northwest England', *Journal of Quaternary Science*, 19: 281–290.

Beer, J. and van Geel, B. (2008), 'Holocene climate change and the evidence for solar and other forcings', in R.W. Battarbee and H.A. Binney (eds) *Natural climate variability and global warming: A Holocene perspective* (Chichester): 138–162.

Beerling, D. (2007), *The Emerald Planet* (Oxford University Press).

Behl, R.J. and Kennett, J.P. (1996), 'Brief interstadial events in the Santa Barbara basin, NE Pacific, during the past 60 kyr', *Nature*, 379: 243–379.

Behre, K-E. (2007), 'A new Holocene sea-level curve for the southern North Sea', *Boreas*, 36: 82–102.

Bell, M. and Walker, M.J.C. (2005), *Late Quaternary Environmental Change, Physical and Human Perspectives*, 2nd edition (Harlow).

Bellwood, P. (2005), *First Farmers: the Origins of Agricultural Societies* (Oxford).

Belperio, A.P., Harvey, N. and Bourman, R.P. (2002), 'Spatial and temporal variability in the Holocene sea-level record of the South Australian coastline', *Sedimentary Geology*, 150: 153–169.

Benazzi, S., Douka, K., Fornai, C., Bauer, C.C., Kullmer, O., Svoboda, J., Pap, I., Mallegni, F., Bayle, P., Coquerelle, M., Condemi, S., Ronchitelli, A., Harvati, K. and Weber, G.W. (2011), 'Early dispersal of modern humans in Europe and implications for Neanderthal behaviour', *Nature*, 479: 525–528.

Benbow, M.C. (1990), 'Tertiary coastal dunes of the Eucla Basin, Australia', *Geomorphology*, 3: 9–29.

Bendix, J., Bendix, A. and Richter, M. (2000), 'El Niño 1997/1998 in Nordperu; anzeichen eines Ökosystem-Wandels?', *Petermanns Geographische Mitteilungen*, 144: 20–31.

Benestad, R.E. and Schmidt, G.A. (2009), 'Solar trends and global warming', *Journal of Geophysical Research*, 114: D14101, doi: 10.1029/2008JD011639.

Bengtsson, L. and Enell, M. (1986), 'Chemical analysis', in B.E. Berglund (ed.) *Handbook of Holocene Palaeoecology and Palaeohydrology* (Chichester): 423–448.

Benjamin, M.T., Johnson, N.M. and Naeser, C.W. (1987), 'Recent rapid uplift in the Bolivian Andes: evidence from fission track dating', *Geology*, 15: 680–683.

Benson, L.V. and 9 others. (2011), 'The rise and fall of Lake Bonneville between 45 and 10.5 ka.', *Quaternary International*, 235: 57–69.

Berger, A. and Loutre, M.F. (1991), 'Insolation values for the climate of the last 10 million years', *Quaternary Science Reviews*, 10: 297–317.

—— Imbrie, J., Hays, J., Kulka, G. and Saltzman, B. (eds) (1984), *Milankovitch and Climate*, 2 vols. (Dordrecht).

Berger, G.W., Pillans, B.J., Bruce, J.G. and McIntosh, P.D. (2002), 'Luminescence chronology of loess-paleosol sequences from southern South Island, New Zealand', *Quaternary Science Reviews*, 21: 1899–1913.

Berger, J.F., Bravard, J-P., Purdue, L., Benoist, A., Mouton, M. and Braemer, F. (2012), 'Rivers of the Hadramawt watershed (Yemen) during the Holocene: Clues of late functioning', *Quaternary International*, 266: 142–161.

Berger, L. R., de Ruiter, D. J., Churchill, S. E., Schmid, P., Carlson, K. J., Dirks, P. H. G. M., Kibii, J. M. (2010), '*Australopithecus sediba*: a new species of Homo-like australopith from South Africa', *Science*, 328: 195–204.

Berglund, B.E. (ed.) (1986), *Handbook of Holocene Paleoecology and Paleohydrology* (Chichester).

—— Birks, H.J.B., Ralska-Jasiewiczowa, M. and Wright, H.E. (eds) (1996), *Palaeoecological events during the last 15 000 years* (Chichester).

Bernabo, J.C. and Webb, III T. (1977), 'Changing patterns in the Holocene pollen record of north-eastern North America: A mapped summary', *Quaternary Research*, 8: 64–96.

Betancourt, J.L., van Devenser, T.R. and Martin, P.S. (eds) (1990), *Packrat middens. The last 40 000 years of biotic change* (Tucson, AZ).

Bettis, E.A., Muhs, D.R., Roberts, H.M. and Wintle, A.G. (2003), 'Last Glacial loess in the conterminous USA', *Quaternary Science Reviews*, 22: 1907–1946.

Betts, R.A. (2000), 'Offset of the potential carbon sink from boreal forestation by decreases in surface albedo', *Nature*, 408: 187–190.

—— (2001), 'Biogeophysical impacts of land use on present-day climate: near surface temperature change and radiative forcing', *Atmospheric Science Letters*, 1: doi:10.1006/asle.2001.0023.

Beverton, R.J.H. and Lee, A.J. (1965), 'Hydrographic fluctuations in the north Atlantic and some biological consequences', in C.G. Johnson and L.P. Smith (eds) *The biological significance of climatic changes in Britain* (London): 79–107.

Bindoff, N.L. and 12 others. (2007), 'Observations: oceanic climate change and sea level', in S. Solomon and 7 others (eds) *Climate Change 2007: the Physical Science Basis* (Cambridge): 385–432.

Biondi, F., Gershunov, A. and Cayan, D.R. (2001), 'North Pacific decadal climate variability since 1661', *Journal of Climate*, 14: 5–10.

Bird, M.I., Austin, W.E.N., Wurster, C.M., Fifield, L.K., Motjahid, M. and Sargeant, C. (2010), 'Punctuated eustatic sea-level rise in the early mid-Holocene', *Geology*, 38: 803–806.

Birks, H.H. (1975), 'Studies in the vegetational history of Scotland IV: Pine stumps in Scottish blanket peats', *Philosophical Transactions of the Royal Society, London*, B, 270: 181–226.

Birks, H.J.B. (1982), 'Holocene (Flandrian) chronostratigraphy of the British Isles; a review', *Striae*, 16: 99–105.

—— (1986), 'Quaternary biotic changes in terrestial and lacustrine environments, with particular reference to north-west Europe', in B.E. Berglund (ed.) *Handbook of Holocene Palaeoecology and Palaeohydrology* (Chichester): 3–65.

—— (1990), 'Changes in vegetation and climate during the Holocene of Europe', in M.M. Boer and R.S. de Groot (eds) *Landscape-Ecological Impact of Climatic Change* (Amsterdam): 133–158.

—— (2005), 'Quantitative palaeoenvironmental reconstructions from Holocene biological data', in A. Mackay, R. Battarbee, J. Birks and F. Oldfield (eds) *Global Change in the Holocene* (London): 107–123.

—— and Birks, H.H. (1980), *Quaternary Palaeoecology* (London); 2nd edn 2004.

—— and Gordon, A.D. (1985), *Numerical Methods in Quaternary Pollen Analysis* (London).

Bishop, P. and Godley, D. (1994), 'Holocene palaeochannels, north central Thailand: Ages, significance and palaeoenvironmental indications', *The Holocene*, 4: 32–41.

Björck, S. (1995), 'A review of the history of the Baltic Sea, 13.0-8.0 ka BP', *Quaternary International*, 27: 19–40.

—— Walker, M.J.C., Cwynar, L.C., Johnsen, S., Knudsen, K.-L., Lowe, J.J., Wohlfarth, B. and Intimate Members (1998), 'An event stratigraphy for the Last Termination in the North Atlantic region based on the Greenland ice-core record: a proposal by the Intimate group', *Journal of Quaternary Science*, 13: 283–292.

Blackford, J.J. and Chambers, F.M. (1993), 'Determining the degree of peat decomposition for peat-based palaeoclimatic studies', *International Peat Journal*, 5: 7–24.

—— Edwards, K.J., Dugmore, A.J., Cook, G.T. and Buckland, P.C. (1992), 'Icelandic volcanic ash and the mid-Holocene Scots pine (*Pinus sylvestris*) pollen decline in northern Scotland', *The Holocene*, 2: 260–265.

Blain, S. and 46 others. (2007), 'Effect of natural iron fertilization on carbon sequestration in the Southern Ocean', *Nature*, 446: 1070–1074.

Bloom, A.L. (1967), 'Pleistocene shorelines: A new test of isostasy', *Bulletin, Geological Society of America*, 78: 1477–1494.

—— (1971), 'Glacial eustatic and isostatic controls of sea-level since the Last Glaciation', in K.K. Turekian (ed.) *The Late Cenozoic Glacial Ages* (New Haven, CT): 355–379.

—— (1983), 'Sea-level and coastal morphology of the United States through the Late Wisconsin glacial maximum', in S.C. Porter (ed.) *Late Quaternary environments of the United States* (London): 215–229.

Blunier, T., Chappellaz, J., Schwander, J., Stauffer, B. and Raynaud, D. (1995), 'Variations in atmospheric methane concentration during the Holocene epoch', *Nature*, 374: 46–49.

—— Chappellaz, J.A., Schwander, J., Barnola, J.-M., Desperts, T., Stauffer, B. and Raynaud, D. (1993),

'Atmospheric methane record from a Greenland ice core over the last 1000 years', *Geophysical Research Letters*, 20: 2219–2222.

—— Chappellaz, J., Schwander, J., Dällenbach, A., Stauffer, B., Stocker, T.F., Raynaud, D., Jouzel, J., Clausen, H.B., Hammer, C.U. and Johnsen, S.J. (1998), 'Asynchrony of Antarctic and Greenland climate change during the last glacial period', *Nature*, 394: 739–743.

—— and Brook, E.J. (2001), 'Timing of millenial-scale climate change in Antarctica and Greenland during the last glacial period', *Science*, 291: 109–112.

Bluth, G.J.S., Schnetzlet, C.C., Frueger, A.J. and Loalter, L.S. (1993), 'The contribution of explosive volcanism to global atmospheric SO_2 concentrations', *Nature*, 373: 399–404.

Blytt, A. (1876), *Essay on the Immigration of the Norwegian Flora during Alternating Rainy and Dry Periods* (Kristiana, Cammermeyer).

—— (1882), 'Die Theori der wechselnden Kontinentalen und insulaten Klimate', *Botanische Jahrbuch*, Leipzig 2: 1–50.

Bobe, R., Behrensmeyer, A.K. and Chapman, R.E. (2002), 'Faunal change, environmental variability and late Pliocene hominin evolution', *Journal of Human Evolution*, 42: 475–497.

Bonatti, E. (1966), 'North Mediterranean climate during the last Würm Glaciation', *Nature*, 209: 984.

Bond, G.C. and Lotti, R. (1995), 'Iceberg discharges into the North Atlantic on millennial time scales during the Last Glaciation', *Science*, 267: 1005–1010.

—— Broecker, W., Johnsen, S., McManus, J., Labeyrie, L., Jouzel, J. and Bonani, G. (1993), 'Correlations between climate records from North Atlantic sediments and Greenland ice', *Nature*, 365: 143–147.

—— Showers, W., Cheseby, M., Lotti, R., Almasi, P., deMenocal, P., Priore, P., Cullen, H., Hajdas, I. and Bonani, G. (1997), 'A pervasive millennial-scale cycle in North Atlantic Holocene and glacial climates', *Science*, 278: 1257–1266.

—— Kromer, B., Beer, J., Muscheler, R., Evans, M.N., Showers, W., Hoffmann, S., Lotti-Bond, R., Hajdas, I. and Bonani, G. (2001), 'Persistent solar influence on North Atlantic climate during the Holocene', *Science*, 294: 2130–2136.

Bonsall, C., Macklin, M.G., Anderson, D.E. and Payton, R.W. (2002), 'Climate change and the adoption of agriculture in north-west Europe', *European Journal of Archaeology*, 5: 9–23.

Booth, R.K., Jackson, S.T., Forman, S.L., Kutzbach, J.E., Bettis, E.A., Kreig, J. and Wright, D.K. (2005), 'A severe centennial-scale drought in midcontinental North America 4200 years ago and apparent global linkages', *The Holocene*, 15: 321–328.

Bosboom, R.E. and 10 others. (2011), 'Late Eocene sea retreat from the Tarim Basin (west China) and concomitant Asian paleoenvironmental change', *Palaeogeography, Palaeoclimatology, Palaeoecology*, 299: 385–398.

Bouzouggar, A., Barton, R.N.E., Vanhaeren, M., D'Errico, F., Collcutt, S., Higham, T.F.G., Hodge, R., Parfitt, S., Rhodes, E., Schwenninger, J-L., Stringer, C., Turner, E., Ward, S., Moutmir, A. and Stambouli, A., (2007), '82,000-year-old shell beads from North Africa and implications for the origins of modern human behaviour', *Proceedings of the National Academy of Sciences*, 104: 9964–9969.

Bowen, A.J. (1972), 'The tidal regime of the River Thames; long-term trends and their possible causes', *Philosophical Transactions of the Royal Society, London*, A, 272: 187–199.

Bowen, D.Q. (1973), 'The Pleistocene history of Wales and the Borderland', *Geological Journal*, 8/2: 207–224.

—— (1978), *Quaternary Geology* (Oxford).

—— Rose, J., McCane, A.M. and Sutherland, D.G. (1986), 'Correlation of Quaternary glaciation in England, Ireland, Scotland and Wales', *Quaternary Science Reviews*, 5: 299–340.

—— Phillips, F.M., McCabe, A.M., Knutz, P.C. and Sykes, G.A. (2002), 'New data for the Last Glacial Maximum in Great Britain and Ireland', *Quaternary Science Reviews*, 21: 89–101.

Bowen, G.J. (2007), 'When the world turned cold', *Nature*, 445: 607–608.

Bowler, J.M. (1976), 'Aridity in Australia: age, origins and expression in aeolian landforms and sediments', *Earth-Science Reviews*, 12: 279–310.

—— and Price, D.M. (1998), 'Luminescence dates and stratigraphic analyses at Lake Mungo: review and new perspectives', *Archaeology in Oceania*, 33: 156–68.

Boyd, R. and Silk, J.B. (2006), *How Humans Evolved*, 4th edn. (New York).

Boyle, E.A. and Keigwin, L.D. (1987), 'North Atlantic thermohaline circulation during the past 20 000 years linked to high-latitude surface temperature', *Nature*, 330, 35–40.

Bozzano, G., Kuhlmann, H. and Alonso, B. (2002), 'Storminess control over African dust input to the Moroccan Atlantic margin (NW Africa) at the time of maximum boreal summer insolation: a record of the last 220 kyr', *Palaeogeography, Palaeoclimatology, Palaeoecology*, 183: 155–168.

Braconnot, P. and 20 others. (2007), 'Results of PMIP2 coupled simulations of the Mid-Holocene and Last Glacial Maximum – Part 1: experiments and large-scale features', *Climate of the Past*, 3: 261–277.

Bradley, R.S. (1985), *Quaternary Paleoclimatology: Methods of Paleoclimatic Reconstruction* (Boston).

—— (1999), *Paleoclimatology: Reconstructing Climates of the Quaternary*, 2nd edn (London).

—— and Jones, P.D. (1995), *Climate since AD 1500* (London).

—— and Miller, G.H. (1972), 'Recent climatic change and increased glacierization in the eastern Canadian Arctic', *Nature*, 237: 385–387.

—— Diaz, H.F., Eischeid, J.K., Jones, P.D., Kelly, P.M. and Goodess, C.M. (1987), 'Precipitation fluctuations over northern hemisphere land areas since the mid-19th century', *Science*, 237: 171–175.

Bradley, S.L., Milne, G.A., Shennan I. and Edwards, R. (2011), 'An improved glacial isostatic adjustment model for the British Isles', *Journal of Quaternary Science*, 26: 541–552.

Bradwell, T., Fabel, D., Stoker, M., Mathers, H., McHargue, L. and Howe, J. (2008), 'Ice caps existed throughout the Lateglacial Interstadial in northern Scotland', *Journal of Quaternary Science*, 23: 401–407.

Branch, N., Canti, M., Clark, P. and Turney, C. (2005), *Environmental Archaeology: Theoretical and Practical Approaches* (London).

Bräuer, G. (1992), 'Africa's place in the evolution of *Homo sapiens*', in G. Bräuer and F. Smith (eds) *Continuity or replacement? Controversies in Homo sapiens evolution* (Rotterdam): 83–98.

Bray, J.R. (1974), 'Volcanism and glaciation during the past 40 millennia', *Nature*, 252: 679–680.

Brenninkmeijer, C.A.M., van Geel, B. and Mook, W.G. (1982), 'Variations in the D/H and $^{18}O/^{16}O$ ratios in cellulose extracted from a peat bog core', *Earth and Planetary Science Letters*, 61: 283–290.

Briat, M., Royer, A., Petit, J.R. and Lorius, C. (1982), 'Late glacial input of eolian continental dust in the Dome C Ice Core: additional evidence from individual microparticle evidence', *Annals of Glaciology*, 3: 27–31.

Bridgham, S.D., Johnston, C.A., Pastor, J. and Updegraff, K. (1995), 'Potential feedbacks of northern wetlands on climate change', *BioScience*, 45: 262–274.

Briffa, K.R. (1994), 'Mid and late Holocene climate change: evidence from tree growth in northern Fennoscandia', in B.M. Funnell and R.L.F. Kay (eds) *Palaeoclimate of the Last Glacial/Interglacial Cycle*: 61–65. Swindon: NERC Earth Sciences Directorate, Special Publication No. 94/2.

—— (2000), 'Annual climate variability in the Holocene: interpreting the message of ancient trees', *Quaternary Science Reviews*, 19: 87–105.

Broecker, W.S. (1981), 'Glacial to interglacial changes in ocean and atmosphere chemistry', in A. Berger (ed.) *Climatic variations and variability: facts and theories* (Dordrecht): 111–121.

—— (1991), 'The Great Ocean conveyor', *Oceanography*, 4: 79–89.

—— (1995), *The glacial world according to Wally* (Eldigo Press).

—— (2001), 'Was the Medieval Warm Period global?', *Science*, 291: 1497–1499.

—— (2002), 'Dust: climate's Rosetta Stone', *Proceedings of the American Philosophical Society*, 146: 77–80.

—— (2006), 'Abrupt climate change revisited', *Global and Planetary Change*, 54: 211–215.

—— (2010), *The Great Ocean Conveyor: Discovering the Trigger for Abrupt Climate Change*. (Princeton, NJ).

—— and Denton, G.H. (1989), 'The role of ocean–atmosphere reorganizations in glacial cycles', *Geochimica et Cosmochimica Acta*, 53: 2465–2501.

—— and Denton, G.H. (1990), 'The role of ocean–atmosphere reorganizations in glacial cycles', *Quaternary Science Reviews*, 9: 305–341.

—— Goddard, J., Ku, T.L., Matthews, R.K. and Mesollela, K.J. (1968), 'Milankovitch hypothesis supported by precise dating of coral reefs and deep-sea sediments', *Science*, 159: 297–300.

—— Kennett, J.P., Flower, B.P., Teller, J.T., Trumbore, S., Bonani, G. and Wolfli, W. (1989), 'Routing of meltwater from the Laurentide Ice Sheet during the Younger Dryas cold episode', *Nature*, 341: 318–321.

—— Bond, G. and Klas, M. (1990), 'A salt oscillator in the glacial Atlantic? 1. The concept', *Paleoceanography*, 5: 469–477.

—— Denton, G.H., Edwards, R.L., Cheng, H., Alley, R.B. and Putnam, A.E. (2010), 'Putting the Younger Dryas cold event into context', *Quaternary Science Reviews*, 29: 1078–1081.

Bronger, A., Winter, R. and Heinkele, T. (1998), 'Pleistocene climatic history of East and Central Asia based on paleopedological indicators in loess-palaeosol sequences', *Catena*, 34: 1–17.

Brook, G.A. and Nickmann, R.J. (1996), 'Evidence of late Quaternary environments in northeast Georgia from sediments preserved in Red Spider Cave', *Physical Geography*, 17: 465–484.

Brook, G.A., Marais, E., Srivastava, P. and Jordan, T. (2007), 'Timing of lake-level changes in Etosha Pan, Namibia, since the middle Holocene from OSL ages of relict shorelines in the Okondeka region', *Quaternary International*, 175: 29–40.

Brooks, K., Scholz, C.A., King, J.W., Peck, J., Overpeck, J.T., Russell, J.M. and Amoako, P.Y.O. (2005), 'Late-Quaternary lowstands of Lake Bosumtwi, Ghana: Evidence from high-resolution seismic-reflection and sediment-core data', *Palaeogeography, Palaeoclimatology, Palaeoecology*, 216: 235–249.

Brooks, N. (2006), Cultural responses to aridity in the Middle Holocene and increased soil complexity. *Quaternary International*, 151: 29–49.

Brooks, S.J., Langdon, P.G. and Heiri, O. (eds) (2007), *The Identification and Use of Palaearctic Chironomidae Larvae in Palaeoecology*. Technical Guide 10, Quaternary Research Association, London.

Brown, K.S., Sheppard, P.M. and Turner, J.R.G. (1974), 'Quaternary refugia in tropical America: evidence from race formation in Heliconius butterflies', *Proceedings, Royal Society*, B: 369–378.

Brown, P., Sutikna, T., Morwood, M.J., Soejono, R.P., Jatmiko, Saptomo, E.W. and Due, R.A. (2004), 'A new small-bodied hominin from the Late Pleistocene of Flores, Indonesia', *Nature*, 431: 1055–1061.

Brown, R.D. and Robinson, D.A. (2011), 'Northern Hemisphere spring snow cover variability and change over 1922-2010 including an assessment of uncertainty', *The Cryosphere*, 5: 219–229.

Brown, T.A., Farwell, G.W., Grootes, P.M. and Schmidt, F.H. (1992), 'Radiocarbon AMS dating of pollen extracted from peat samples', *Radiocarbon*, 34: 550–556.

Brunet, M., Guy, F., Pilbeam, D., Mackay, H.T., Likius, A., Djimboumalbaye, A. et al. (2002), 'A new hominid from the upper Miocene of Chad, central Africa', *Nature*, 418: 145–151.

Bryant, R.G., Bigg, G.R., Mahowald, N.M., Eckardt, F.D. and Ross, S.G. (2007), 'Dust emission response to climate in southern Africa', *Journal of Geophysical Research*, 112: D09207, doi:10.1029/2005JD007025.

Bryson, R.A. (1989), 'Late Quaternary volcanic modulation of Milankovitch climate forcing', *Theoretical and Applied Climatology*, 39: 115–125.

Bubenzer, O., Besler, H. and Hilgers, H. (2007), 'Filling the gap: OSL data expanding ^{14}C Chronologies of Late Quaternary Environmental Change in the Libyan Desert', *Quaternary International*, 175: 41–52.

Buchardt, B. (1978), 'Oxygen isotope palaeotemperatures from the Tertiary period in the North Sea area', *Nature*, 275: 121–123.

Büdel, J. (1982), *Climatic Geomorphology* (Princeton).

Burga, C.A. (1988), 'Swiss vegetation history during the last 18 000 years', *New Phytologist*, 110: 581–602.

Burns, S.J., Fleitmann, D., Matter, A., Neff, U. and Mangini, A. (2001), 'Speleothem evidence from Oman for continental pluvial events during interglacial periods', *Geology*, 29: 623–626.

Burrough, S.L., Thomas, D.S.G. and Bailey, R.M. (2009), 'Mega-lake in the Kalahari: a late Pleistocene record of the Palaeolake Makgadikgadi system', *Quaternary Science Reviews*, 28: 1392–1411.

Burroughs, W.J. (1992), *Weather Cycles, Real or Imaginary?* (Cambridge).

—— (2005), *Climate Change in Prehistory* (Cambridge).

—— (2007), *Climate change. A Multidisciplinary Approach* (Cambridge).

Burt, T.P. and Shahgedanova, M. (1998), 'An historical record of evaporation losses since 1815 calculated using long-term observations from the Radcliffe Meteorological Station, Oxford, England', *Journal of Hydrology*, 205: 101–111.

Bush, M.B., Miller, M.C., de Oliviera, P.E. and Colinvaux, P.A. (2002), 'Orbital forcing signal in sediment of two Amazonian lakes', *Journal of Paleolimnology*, 27: 341–352.

Butzer, K.W. (1971), 'Recent history of an Ethiopian delta', *University of Chicago, Dept. of Geography, Research Paper*, 136: 184.

—— (1972), *Environment and Archaeology: An Ecological Approach to Prehistory* (London): 730ff.

—— and Hansen, C.L. (1968), *Desert and River in Nubia: Geomorphology and Prehistoric Environments at the Aswan Reservoir* (University of Wisconsin Press).

Cabioch, G. and 7 others. (2008), 'Successive reef depositional events along the Marquesas foreslopes (French Polynesia) since 26 ka', *Marine Geology*, 254: 18–34.

Calder, N. (1974), *The Weather Machine* (London).

Calvo, E., Pelejero, C., Logan, G.A. and de Deckker, P. (2004), 'Dust-induced changes in phytoplankton composition in the Tasman Sea during the last four glacial cycles', *Paleoceanography*, 19: PA2020.

Camoin, G.F., Montaggioni, L.F. and Braithwaite, C.J.R. (2004), 'Late glacial to post glacial sea levels in the Western Indian Ocean', *Marine Geology*, 206: 119–146.

Cane, M.A. and Molnar, P. (2001), 'Closing of the Indonesian seaway as a precursor to east African aridification around 3-4 million years ago', *Nature*, 411: 157–162.

Carlson, A.E. (2008), 'Why there was not a Younger Dryas-like event during the Penultimate Deglaciation', *Quaternary Science Reviews*, 27: 882–887.

—— Clark, P.U., Haley, B.A., Klinkhammer, G.P., Simmons, K., Brook, E.J. and Meissner, K.J. (2007), 'Geochemical proxies of North American freshwater routing during the Younger Dryas cold event', *Proceedings of the National Academy of Sciences*, 104: 6556–6561.

Carneiro, R. (1970), 'A Theory of the Origin of the State', *Science*, 169: 733–738.

Carr, S.J., Holmes, R., Van der Meer, J.J.M. and Rose, J. (2006), 'The Last Glacial Maximum in the North Sea Basin: micromorphological evidence of extensive glaciation', *Journal of Quaternary Science*, 21: 131–153.

—— Lukas, S. and Mills, S.C. (2010), 'Glacier reconstruction and mass-balance modelling as a geomorphic and palaeoclimatic tool', *Earth Surface Process and Landforms*, 35: 1103–1115.

Caruso, A., Cosentino, C., Pierre, C. and Sulli, A. (2011), 'Sea-level changes during the last 41,000 years in the outer shelf of the southern Tyrrhenian Sea: evidence from benthic foraminifera and seismostratigraphic analysis', *Quaternary International*, 232: 122–131.

Caspers, G. and Freund, H. (2001), 'Vegetation and climate in the Early- and Pleni-Weichselian in northern central Europe', *Journal of Quaternary Science*, 16: 31–48.

Castellano, E. and 9 others. (2005), 'Holocene volcanic history as recorded in the sulfate stratigraphy of the European Project for Ice Coring in Antarctica Dome C (EDC96) ice core', *Journal of Geophysical Research*, 110: D06114, doi: 10.1029/2004JD005259.

Causse, C., Coque, R., Ch. Fontes, J., Gasse, F., Gibert, E., Ben Ouezdou, H. and Zouari, K. (1989), 'Two high levels of continental waters in the southern Tunisian chotts at about 90 and 159 ka', *Geology*, 17: 922–925.

Cazenave, A. and Llovel, W. (2010), 'Contemporary sea level rise', *Annual Review of Marine Science*, 2: 145–173.

Chamley, H. (1988), 'Contribution éolienne à la sedimentation marine au large du Sahara', *Bulletin Société Géologique de France*, IV, 6: 1091–1100.

Changnon, S.A. (ed.) (2000), *El Niño 1997–1998* (New York).

Chappell, J. (2002), 'Sea-level change as forced ice breakouts in the Last Glacial cycle: new results from coral terraces', *Quaternary Science Reviews*, 21: 1229–1240.

—— and Shackleton, N.J. (1986), 'Oxygen isotopes and sea level', *Nature*, 324: 137–140.

—— Omura, A., Esat, T., McCulloch, M., Pandolfi, J., Ota, Y. and Pillans, B. (1996), 'Reconciliation of Late Quaternary sea-levels derived from coral terraces at Huon Peninsula with deep sea oxygen isotope records', *Earth and Planetary Science Letters*, 141: 227–236.

—— Rhodes, E.G., Thomas, B.G. and Wallensky, E. (1982), 'Hydro-isostasy and the sea-level isobase of 5500 B.P. in north Queensland, Australia', *Marine Geology*, 49: 81–90.

Chappellaz, J.M., Blunier, T., Ratnaud, D., Barnola, J.M., Schwander, J. and Stauffer, B. (1993), 'Synchronous changes in atmospheric CH_4 and Greenland climate between 40 and 8 kyr bp', *Nature*, 366, 443–445.

Charlesworth, J.K. (1957), *The Quaternary Era* (London): 1700ff.

Charman, D.J. (2002), *Peatlands and Environmental Change* (Chichester).

—— Blundell, A., Chiverrell, R.C., Hendon, D. and Langdon, P.G. (2006), 'Compilation of non-annually resolved Holocene proxy climate records: stacked Holocene peatland palaeo-water table reconstructions from northern Britain', *Quaternary Science Reviews*, 25: 336–350.

—— Blundell, A. and ACCROTELM MEMBERS (2007), 'A new European testate amoebae transfer function for palaeohydrological reconstruction on ombrotrophic peatlands', *Journal of Quaternary Science*, 22: 209–221.

Chase, B.M. and Meadows, M.E. (2007), 'Late Quaternary dynamics of southern Africa's winter rainfall zone', *Earth-Science Reviews*, 84: 103–138.

Chemello, R. and Silenzi, S. (2011), 'Vermetid reefs in the Mediterranean Sea as archives of sea-level and surface temperature changes', *Chemistry and Ecology*, 27: 121–127.

Chen, J., Wan, G., Zhang, D.D., Chen, Z., XU, J., Xiao, T. and Huang, R. (2005), 'The "Little Ice Age" recorded by sediment chemistry in Lake Erhai, southwest China', *The Holocene*, 15: 925–931.

Chen, X.Y., Chappell, J. and Murray, A.S. (1995), 'High (ground) water levels and dune development in Central Australia: TL dates from gypsum and quartz dunes around Lake Lewis (Napparby), Northern Terrirory', *Geomorphology*, 11: 311–322.

Cheng, H., Edwards, R.L., Broecker, W.S., Denton, G.H., Kong, X., Wang, Y., Zhang, R. and Wang, X. (2009), 'Ice Age Terminations', *Science*, 326: 248–252.

—— Fleitmann, D., Edwards, R.L., Wang, X., Cruz, F.W., Auler, A.S., Mangini, A., Wang, Y., Kong, X., Burns, S.J. and Matter, A. (2009), 'Timing and structure of the 8.2 kyr B.P. event inferred from $\delta^{18}O$ records of stalagmites from China, Oman, and Brazil', Geology, 37: 1007–1010.

Chepalyga, A.L. (1984), 'Inland sea basins', in A.A. Velichko (ed.) *Late Quaternary Environments of the Soviet Union* (London): 229–247.

Chester, D.K. (1988), 'Volcanoes and climate: Recent volcanological perspectives', *Progress in Physical Geography*, 12: 1–35.

Chiapello, I., Moulin, C., Prospero, J.M. (2005), 'Understanding the long-term variability of African dust transport across the Atlantic as recorded in both Barbados surface concentrations and large-scale Total Ozone Mapping Spectrometer (TOMS) optical thickness', *Journal of Geophysical Research*, 110: doi: 10.1029/2004JD005132.

Chiew, F.H.S. and McMahon, T.A. (1993), 'Detection of trend or change in annual flow of Australian rivers', *International Journal of Climatology*, 13: 643–653.

Chlachula, J. (2003), 'The Siberian loess record and its significance for reconstruction of Pleistocene climate change in north-central Asia', *Quaternary Science Reviews*, 22: 1879–1906.

Chorley, R.J., Dunn, A.J. and Beckinsale, R.P. (1964), *Geomorphology before Davis: the History of the Study of Landforms, or the Development of Geomorphology*, vol. 1 (London).

Chu Ko-Chan (1973), 'A preliminary study on the climate fluctuations during the last 5000 years in China', *Scientia Sinica*, 16: 226–256.

Church, J.A. and 35 others. (2001), 'Changes in Sea-Level', in J.T. Houghton (ed.) *Climate Change 2001: The Scientific Basis* (Cambridge): 639–693.

Clapperton, C.M. (ed.) (1990), 'Quaternary glaciations in the southern hemisphere', *Quaternary Science Reviews*, 9: 121–304.

—— (1993), *Quaternary geology and geomorphology of South America* (Amsterdam and New York).

—— (1997), 'Fluctuations of local glaciers 30–8 ka bp', *Quaternary International*, 38/39: 1–147.

—— (1998), 'Late Quaternary glacier fluctuations in the Andes: testing the synchrony of global change', *Quaternary Proceedings*, 6: 65–73.

Claquin, T. and 11 others. (2003), 'Radiative forcing by ice-age atmospheric dust', *Climate Dynamics*, 20: 193–202.

Clark, D.L. (1982), 'Origin, nature and world climate effect of Arctic Ocean ice cover', *Nature*, 300: 321–325.

Clark, J.A. (1976), 'Greenland's rapid post-glacial emergence: a result of ice-water gravitational attraction', *Geology*, 4: 310–312.

—— and Lingle, C.S. (1979), 'Predicted relative sea-level changes (18 000 years bp to present) caused by late-glacial retreat of the Antarctic ice-sheet', *Quaternary Research*, 11: 279–298.

Clark, P.U. and Mix, A.C. (2002), 'Ice sheets and sea-level of the Last Glacial Maximum', *Quaternary Science Reviews*, 21: 1–7.

—— Alley, R.B. and Pollard, D. (1999), 'Northern Hemisphere ice sheet influences on global climate change', *Science*, 286: 1104–1111.

—— McCabe, M., Mix, A.C. and Weaver, A.J. (2004), 'Rapid rise of sea level 19,000 years ago and its global implications', *Science*, 304: 1141–1144.

—— Marshall, S.J., Clarke, G.K.C., Hostetler, S.W., Licciardi, J.M. and Teller, J.T. (2001), 'Freshwater forcing of abrupt climate change during the last glaciation', *Science*, 293: 283–287.

—— Mitrovica, J.X., Milne, G.A. and Tamisiea, M.E. (2002a), 'Sea-level fingerprinting as a direct test for the source of global meltwater pulse 1A', *Science*, 295: 2438–2441.

—— Pisias, N.G., Stocker, T.F. and Weaver, A.J. (2002b), 'The role of the thermohaline circulation in abrupt climate change', *Nature*, 415: 863–869.

—— Dyke, A.S., Shakun, J.D., Carlson, A.E., Clark, J., Wohlfarth, B., Mitrovica, J.X., Hostetler, S.W. and McCabe, A.M. (2009), 'The Last Glacial Maximum', *Science*, 325: 710–714.

—— and 30 others. (2012), 'Global climate evolution during the last deglaciation', *Proceedings of the National Academy of Sciences*, doi:10.1073/pnas.1116619109/-/DCSupplemental.

Clarke, J. (2006), 'Antiquity of aridity in the Chilean Atacama Desert', *Geomorphology*, 73: 101–114.

Clemens, S.C. and Prell, W.L. (1990), 'Late Pleistocene variability of Arabian Sea summer monsoon winds and continental aridity: eolian records from the lithogenic component of deep sea sediments', *Paleoceanography*, 5(2): 109–145.

Climap Project Members (1981), 'Seasonal reconstructions of the Earth's surface at the last glacial maximum'. *Geological Society of America Map and Chart Series*, MC-36.

Clymo, R.S. (1984), 'The limits to peat bog growth', *Philosophical Transactions of the Royal Society, London*, B, 303: 605–654.

Cohen, A.S. (2003), *Paleolimnology: The History and Evolution of Lake Systems* (Oxford).

Cohen, K.M. and Gibbard, P.L. (2011), *Global chronostratigraphical correlation table for the last 2.7 million years*, Subcommission on Quaternary Stratigraphy (International Commission on Stratigraphy), (Cambridge).

Cohen, M.N. (1989), *Health and the Rise of Civilization*, Yale.

Cohen, T.J. and 10 others. (2011), 'Continental aridification and the vanishing of Australia's megalakes', *Geology*, 39:167–170.

Cohmap Members (1988), 'Climatic changes of the last 18 000 years: Observations and model simulations', *Science*, 241: 1043–1052.

Colhoun, E.A. (1988), 'Recent morphostratigraphic studies of the Australian Quaternary', *Progress in Physical Geography*, 12/2: 264–281.

Colinvaux, P.A. and de Oliveira, P.E. (2000), 'Palaeoecology and climate change of the Amazon basin during the last glacial cycle', *Journal of Quaternary Science*, 15: 347–356.

Collins, J.A. and 13 others. (2011), 'Interhemispheric symmetry of the tropical African rainbelt over the past 23,000 years', *Nature Geoscience*, 4: 42–45.

Collinson, M.E. and Hooker, J.J. (1987), 'Vegetational and mammalian faunal changes in the early Tertiary of southern England', in E.M. Friis, W.G. Chaloner and P.R. Crane (eds) *The origins of angiosperms and their biological consequences* (Cambridge): 259–304.

—— Fowler, K. and Boulter, M.C. (1981), 'Floristic changes indicate a cooling climate in the Eocene of southern England', *Nature*, 291: 315–317.

Comiso, J.C. (2002), 'A rapidly declining perennial sea ice cover in the Arctic', *Geographical Research Letters*, 29: doi: 10.1029/2002 FL01 56 50.

Compton, J.S. (2011), 'Pleistocene sea-level fluctuations and human evolution on the southern coastal plain of South Africa', *Quaternary Science Reviews*, 30: 506–527.

Coningham, R. (2005), 'South Asia: from early villages to Buddhism', in C. Scarre (ed.) *The Human Past: World Prehistory and the Development of Human Societies* (London): 518–551.

Conroy, G.C. (2005), *Reconstructing Human Origins*, 2nd edn. (New York).

Conway, D. (2005), 'From headwater tributaries to international river: observing and adapting to climate variability and change in the Nile basin', *Global Environmental Change*, 15: 99–114.

—— and Hulme, M. (1993), 'Recent fluctuations in precipitation and runoff over the Nile sub-basins and their impact on main Nile discharge', *Climate Change*, 25: 127–151.

Cook, A.J., Fox, A.J., Vaughn, D.G. and Ferrigno, J.G. (2005), 'Retreating glacier fronts on the Antarctic

Peninsula over the past half-century', *Science*, 308: 541–544.

Cook, P.J. and Polach, H.A. (1973), 'A Chenier sequence at Broad Sound, Queensland and evidence against a Holocene high sea-level', *Marine Geology*, 14: 253–268.

Cooke, H.J. (1975), 'The palaeoclimatic significance of caves and adjacent landforms in western Ngamiland, Botswana', *Geographical Journal*, 141: 430–444.

Coope, G.R. (1975a), 'Climatic fluctuations in north-west Europe since the Last Interglacial, indicated by fossil assemblages of Coleoptera', in A.E. Wright and F. Moseley (eds) *Ice Ages: Ancient and Modern* (Liverpool): 153–168.

—— (1975b), 'Mid-Weichselian climatic changes in western Europe reinterpreted from Coleopteran assemblages', *Bulletin, Royal Society of New Zealand*, 13: 101–108.

—— (2001), 'Biostratigraphical distinction of interglacial coleopteran assemblages from southern Britain attributed to Oxygen Isotope Stages 5e and 7', *Quaternary Science Reviews*, 20: 1717–1722.

—— and Lemdahl, G. (1995), 'Regional differences in the Lateglacial climate of northern Europe based on coleopteran analysis', *Journal of Quaternary Science*, 10: 391–395.

—— and Rose, J. (2008), 'Palaeotemperatures and palaeoenvironments during the Younger Dryas: arthropod evidence from Croftamie at the type area of the Loch Lomond Readvance, and significance for the timing of glacier expansion during the Lateglacial period in Scotland', *Scottish Journal of Geology*, 44: 43–49.

—— Morgan, A. and Osborne, P.J. (1971), 'Fossil Coleoptera as indicators of climatic fluctuations during the Last Glaciation in Britain', *Palaeogeography, Palaeoclimatology, Palaeoecology*, 10: 87–101.

Cooper, E.J. and Leduc, S.K. (1995), 'Recent frost date trends in the north-eastern USA', *International Journal of Climatology*, 15: 65–75.

Coque-Delhuille, B. and Veyret, Y. (1989), 'Limite d'englacement et évolution periglaciaire des Îles Scilly: l'interêt des arènes in situ et remainié', *Zeitschrift für Geomorphologie, Supplementband*, 72: 79–96.

Coulthard, T.J. and Macklin, M.G. (2001), 'How sensitive are river systems to climate and land-use

changes? A model-based evaluation', *Journal of Quaternary Science*, 16: 347–351.

Cowling, S.A., Betts, R.A., Cox, P.M., Ettwein, V., Jones, C.D., Maslin, M.A. and Spall, S.A. (2004), 'Contrasting simulated past and future responses of the Amazonian forest to atmospheric change', *Philosophical Transactions of the Royal Society, London*, B, 359: 539–547.

Cox, A. (1968), 'Polar wandering, continental drift and the onset of Quaternary glaciation', *Meteorological Monographs*, 8: 112–125.

—— Doell, R.R. and Dalrymple, G.B. (1968), 'Radiometric time-scale for geomagnetic reversals', *Quarterly Journal of the Geological Society*, 124: 53–66.

Creber, G.T. and Chaloner, W.G. (1985), 'Tree growth in the Mesozoic and early Tertiary and the reconstruction of palaeoclimates', *Palaeogeography, Palaeoclimatology, Palaeoecology*, 52: 35–60.

Crisp, D.J. (1959), 'The influence of climatic changes on animals and plants', *Geographical Journal*, 125: 1–19.

Cronin, T.M. (1999), *Principles of Paleoclimatology* (New York).

—— Szabo, B.J., Ager, T.A., Hazel, J.E. and Owens, J.P. (1981), 'Quaternary climates and sea levels of the U.S. Atlantic coastal plain', *Science*, 211: 233–40.

Cross, S.L., Baker, P.A., Seltzer, G.O., Fritz, S.C. and Dunbar, R.B. (2001), 'Late Quaternary Climate and Hydrology of Tropical South America Inferred from an Isotopic and Chemical Model of Lake Titicaca, Bolivia and Peru', *Quaternary Research*, 56: 1–9.

Crowley, T.J. and North, C.R. (1991), *Palaeoclimatology* (Oxford).

—— and Kim K.Y. (1993), 'Towards development of a strategy for determining the origin of decadal-centennial scale climate variability', *Quaternary Science Reviews*, 12: 375–385.

Crucifix, M. (2008), 'Modeling the climate of the Holocene', in R.W. Battarbee and H.A. Binney (eds) *Natural climate variability and global warming: A Holocene perspective* (Chichester): 98–122.

Cullen, H.M., deMenocal, P.B., Hemkming, S., Hemming, G., Brown, F.H., Guilderson, T. and Sirocko, F. (2000), 'Climate change and the collapse of the Akkadian empire: evidence from the deep sea', *Geology*, 28: 379–382.

Culver, S. and 11 others. (2011). 'Micropaleontologic record of Quaternary paleoenvironments in the

Central Albermarle Embayment, North Carolina, U.S.A.', *Palaeogeography, Palaeoclimatology, Palaeoecology*, 305: 227–249.

Cunliffe, B. (1994), *The Oxford Illustrated History of Prehistoric Europe* (Oxford).

Curran, M.A.J., van Omman, T.D., Morgan, V.I., Phillips, K.L. and Palmer, A.S. (2003), 'Ice core evidence for Antarctic sea ice decline since the 1950s', *Science*, 302: 1203–1206.

Cutler, K.B., Edwards, R.L., Taylor, F.W., Cheng, H., Adkins, J., Gallup, C.D., Cutler, P.H., Burr, G.S. and Bloom, A.L. (2003), 'Rapid sea-level fall and deep ocean temperature change since the last interglacial period', *Earth and Planetary Science Letters*, 206: 253–271.

Cwynar, L.C., Levesque, A.J., Mayle, F.E. and Walker, I. (1994), 'Wisconsinan Late-glacial environmental change in New Brunswick: a regional synthesis', *Journal of Quaternary Science*, 9: 161–164.

Dahl-Jensen, D., Gundestrup, N.S., Miller, H., Watanabe, O., Johnsen, S.J., Steffensen, J.P., Clausen, H.B., Svensson, A. and Larsen, L.B. (2002), 'The North-GRIP deep drilling programme', *Annals of Glaciology*, 35: 1–4.

Dai, A. (2011a), 'Drought under global warming: a review', *Interdisciplinary Reviews: Climate Change*, 2: 45–65.

—— (2011b), 'Characteristics and trends in various forms of the Palmer Drought Severity Index during 1900-2008', *Journal of Geophysical Research*, 116: D12115, doi:10.1029/2010JD015541.

Daley, B. (1972), 'Some problems concerning the Early Tertiary climate of southern Britain', *Palaeogeography, Palaeoclimatology, Palaeoecology*, 11: 177–190.

Dalfes, H.N., Kukla, G. and Weiss, H. (eds) (1997), *Third Millennium bc Climate Change and Old World Collapse* (Berlin).

Daly, R.A. (1934), *The Changing World of the Ice Age* (New Haven, CT).

Dansgaard, W., White, J.W.C. and Johnsen, S.J. (1989), 'The abrupt termination of the Younger Dryas climate event', *Nature*, 339: 532–533.

—— Johnsen, S.J., Clausen, H.B., Dahl-Jensen, D., Gundestrup, N.S., Hammer, C.U., Hvidberg, C.S., Steffensen, J.P., Sveinbjörnsdottir, A.E., Jouzel, J. and Bond, G. (1993), 'Evidence for general instability of past climate from a 250-kyr ice-core record', *Nature*, 364: 218–220.

D'Arrigo, R.D. and Jacoby, G.C. (1991), 'A 1000-year record of winter precipitation from northwestern New Mexico, USA: a reconstruction from tree-rings and its relation to El Niño and the Southern Oscillation', *The Holocene*, 1: 95–101.

Darwin, C. (1936 edn.), *The Origin of Species* (London).

Daveau, S. (1965), 'Dunes ravinées et dépôts du Quaternaire récent dans le Sahel mauretanien', *Revue de Géographie de l'Afrique Occidental*, 1–2: 7–48.

Davies, S.M., Branch, N.P., Lowe, J.J. and Turney, C.S.M. (2002), 'Towards a European tephrochronological framework for Termination 1 and the Early Holocene', *Philosophical Transactions of the Royal Society, London*, A, 360: 767–802.

—— Wastegård, S., Rasmussen, T.L., Svensson, A., Johnsen, J., Steffensen, J.P. and Andersen, K.K. (2008), 'Identification of the Fugloyarbanki tephra in the NGRIP ice core: a key tie-point for marine and ice-core sequences during the last glacial period', *Journal of Quaternary Science*, 23: 409–414.

Davis, B.A.S., Brewer, S., Stevenson, A.C., Guiot, J. and Data Contributors (2003), 'The temperature of Europe during the Holocene reconstructed from pollen data', *Quaternary Science Reviews*, 22: 1701–1716.

Davis, N.E. (1972), 'The variability of the onset of spring in Britain', *Quarterly Journal of the Royal Meteorological Society*, 98: 763–777.

Davis, R.S., Ranov, V.A. and Dodonov, A.E. (1980), 'Early man in Soviet Central Asia', *Scientific American*, 243: 91–102.

Dawson, A.G. (1992), *Ice Age Earth: Late Quaternary Geology and Climate* (London).

Dean, J.S. (1994), 'The medieval warm period in the southern Colorado Plateau', *Climatic Change*, 26: 229–241.

Dean, W.E., Ahlbrandt, T.S., Anderson, R.Y. and Bradbury, J.P. (1996), 'Regional aridity in North America during the middle Holocene', *The Holocene*, 6: 145–155.

De Angelis, M. and Gaudichet, A. (1991), 'Saharan dust deposition over Mont Blanc (French Alps) during the last 30 years', *Tellus*, 43B: 61–75.

DeConto, R.M., Pollard, D., Wilson, P.A., Palike, H., Lear, C.H. and Pagani, M. (2008), 'Thresholds for Cenozoic bipolar glaciation', *Nature*, 455: 652–656.

Deevey, E.S. and Flint, R.F. (1957), 'Post-Glacial hypsithermal interval', *Science*, 125: 182–184.

De Freitas, H.A., Pessenda, L.C.R., Aravena, R., Gouveia, S.E.M., de Souza Ribeiro, A. and Boulet, R. (2001), 'Late Quaternary vegetation dynamics in the Southern Amazon Basin inferred from carbon isotopes in soil organic matter', *Quaternary Research*, 55: 39–46.

De Geer, G. (1890), 'Om Skandinaviens nivåförandringar under kvartärperioden', *Foreningens I Stockholm Förhandlingar*, 12: 61–110.

Degens, E.T. and Hecky, R.E. (1974), 'Palaeoclimatic reconstruction of Late Pleistocene and Holocene based on biogenic sediments from the Black Sea and a tropical African lake', *Colloques Internationaux du CNRS*, 219: 13–23.

d'Errico, F., Henshilwood, C., Vanhaeren, M. and van Niekerk, K. (2005), '*Nassarius kraussianus* shell beads from Blombos Cave: evidence for symbolic behaviour in the Middle Stone Age', *Journal of Human Evolution*, 48: 3–24.

—— and Sanchez Goñi, M.F. (2003), 'Neanderthal extinction and the millennial scale climatic variability of OIS 3', *Quaternary Science Reviews*, 22: 769–788.

Delcourt, H.R. and Delcourt, P.A. (1991), *Quaternary Ecology – A Palaeoecological Perspective* (London).

Delmas, R.J., Ascencio, J.M. and Legrand, M. (1980), 'Polar ice evidence that atmospheric CO_2 20 000 yr bp was 50% of present', *Nature*, 284: 155–157.

Delmonte, B., Basile-Doelsch, I., Petit, J-R., Maggi, V., Revel-Rolland, M., Michard, A., Jagoutz, E. and Grousset, F. (2004a), 'Comparing the Epica and Vostok dust records during the last 220 000 years: stratigraphical correlation and provenance in glacial periods', *Earth-Science Reviews*, 66: 63–87.

—— Petit, J.R., Andresen, K.K., Basile-Doelsch, I., Maggi, V. and Ya Lipenkov, V. (2004b), 'Dust size evidence for opposite regional atmospheric circulation changes over East Antarctica during the last climatic transition', *Climate Dynamics*, 23: 427–438.

Delvaux, D., Kervyn, F., Vittori, E., Kajara, R.S.A. and Kilembe, E. (1998), 'Late Quaternary tectonic activity and lake level change in the Rukwa Rift basin', *Journal of African Earth Science*, 26: 397–421.

DeMenocal, P.B. (1995), 'Plio-Pleistocene African climate', *Science*, 270: 53–59.

—— (2001), 'Cultural responses to climate change during the late Holocene', *Science*, 292: 667–673.

—— (2004), 'African climate change and faunal evolution during the Pliocene-Pleistocene', *Earth and Planetary Science Letters*, 220: 3–24.

—— (2011), 'Climate and human evolution', *Science*, 311: 540–541.

—— and Rind, D. (1993), 'Sensitivity of Asian and African climate to variations in seasonal insolation, glacial ice cover, sea surface temperature and Asian orography', *Journal of Geophysical Research*, 98 (D4): 7265–7287.

—— Ortiz, J., Guilderson, T., Adkins, J., Sarnthein, M., Baker, L. and Yarusinski, M. (2000a), 'Abrupt onset and termination of the African Humid Period: Rapid climate responses to gradual insolation forcing', *Quaternary Science Reviews*, 19: 347–361.

—— Ortiz, J., Guilderson, T. and Sarnthein, M. (2000b), 'Coherent high- and low-latitude climate variability during the Holocene warm period', *Science*, 288: 2198–2202.

Denton, G.H. and Hughes, T.J. (1981), *The Last Great Ice-sheets* (New York).

—— and Karlén, W. (1973), 'Holocene climatic variations: their pattern and possible cause', *Quaternary Research*, 3: 155–205.

—— and Broecker, W.S. (2008), 'Wobbly ocean conveyor circulation during the Holocene?', *Quaternary Science Reviews*, 27: 1939–1950.

Deo, R.C., Syktus, J.I., McAlpine, C.A., Lawrence, P.J., McGowan, H.A. and Phinn, S.R. (2009), 'Impact of historical land cover change on daily indices of climate extremes including droughts in eastern Australia', *Geophysical Research Letters*, 36: L08705, doi:10.1029/2009GL037666.

Depéret, C. (1918–22), 'Essai de coordination chronologique générale de temps quaternaries', *Comptes rendus de l'Académie des Sciences*, 166: 480, 636, 884.

De Ploey, J. (1965), 'Position Géomorphologique, Genèse et Chronologie de Certains Dépôts Superficiels au Congo Occidental', *Quaternaria*, 7: 131–154.

Deschamps, P., Durand, N., Bard, E., Hamelin, B., Camoin, G., Thomas, A.L., Henderson, G.M., Okuno, J. and Yokoyama, Y. (2012), 'Ice-sheet collapse and sea-level rise at the Bølling warming 14,600 years ago', *Nature*, 483: 559–564.

Desruelles, S. and 7 others. (2009), 'Beachrocks and sea level changes since Middle Holocene: comparison between the insular group of Mykonos-Delos-Rhenia (Cyclades, Greece) and the southern coast of Turkey', *Global and Planetary Change*, 66: 19–33.

Dessens, J. and Dinh, P.V. (1990), 'Frequent Saharan dust outbreaks north of the Pyrenees: a sign of climatic change?', *Weather*, 45: 327–333.

Devoy, R.J.N. (1987) (ed.), *Sea Surface Studies* (London).

Diamond, M. (1958), 'Precipitation trends in Greenland during the past 30 years', *Journal of Glaciology*, 3: 177–80.

Dickinson, W.R. (2004), 'Impacts of eustasy and hydro-isostasy on the evolution and landforms of Pacific atolls', *Palaeogeography, Palaeoclimatology, Palaeoecology*, 213: 251–269.

Dickson, R.R., Osborn, T.J., Hurrell, J.W., Meincke, J., Blindheim, J., Adlandsvik, B., Vinje, T., Alekseev, G. and Maslowski, W. (2000), 'The Arctic Ocean response to the North Atlantic Oscillation', *Journal of Climate*, 13: 2671–2696.

Diester-Haass, L. and Schrader, H.J. (1979), 'Neogene coastal upwelling history off north-west and south-west Africa', *Marine Geology*, 29: 39–53.

Dietrich, R., Ivins, E.R., Casassa, G., Lange, H., Wendt, J. and Fritsche, M. (2010), 'Rapid crustal uplift in Patagonia due to enhanced ice loss', *Earth and Planetary Science Letters*, 289: 22–29.

Dillehay, T. (1997), *Monte Verde: a Late Pleistocene Settlement in Chile*, 2 vols. (Washington DC).

Dilli, K. and Pant, R.K. (1994), 'Clay minerals as indicators of the provenance and palaeoclimatic record of the Kashmir loess', *Journal of the Geological Society of India*, 44: 563–574.

Ding, Z., Rutter, N., Han, J. and Liu, T. (1992), 'A coupled environmental system formed at about 2.5 Ma in East Asia', *Palaeogeography, Palaeoclimatology, Palaeoecology*, 94: 223–242.

Ding, Z.L., Ravov, V., Yang, S.L., Finaev, A., Han, J.M. and Wang, G.A. (2002), 'The loess record in southern Tajikistan and correlation with Chinese loess', *Earth and Planetary Science Letters*, 200: 387–400.

Ding, R.Q., Wang, S.G. and Ren, F.M. (2005), 'Decadal change of the spring dust storm in northwest China and the associated atmospheric circulation', *Geophysical Research Letters*, 32, article L02808.

Dodonov, A.E. and Baiguzina, L.L. (1995), 'Loess stratigraphy of Central Asia: palaeoclimatic and

palaeoenvironmental aspects', *Quaternary Science Reviews*, 14: 707–720.

Doerner, J.P. and Carrara, P.E. (2001), 'Late Quaternary vegetation and climatic history of the Long Valley area, west-central Idaho, U.S.A.', *Quaternary Research*, 56: 103–111.

Dominey-Howes, D., Dawson, A. and Smith, D. (1999), 'Late Holocene coastal tectonics at Falasarna, western Crete: a sedimentary study', *Geological Society of London Special Publication*, 140: 343–352.

Donn, W.L., Farrand, W.R. and Ewing, M. (1962), 'Pleistocene ice volumes and sea-level changes', *Journal of Geology*, 70: 206–214.

Doornkamp, J.C. and Temple, P.H. (1966), 'Surface drainage and tectonic instability in part of Southern Uganda', *Geographical Journal*, 132: 238–252.

Dracup, J.A. and Kahya, E. (1994), 'The relationships between U.S. streamflow and La Niña events', *Water Resources Research*, 30: 2133–2141.

Drake, N.A. and Bristow, C. (2006), 'Shorelines in the Sahara: geomorphological evidence for an enhanced monsoon from palaeolake Megachad', *The Holocene*, 16: 901–112

—— Blench, R.M., Armitage, S.J., Bristow, C.S. and White K.H. (2011), 'Ancient watercourses and biogeography of the Sahara explain the peopling of the desert', *Proceedings of the National Academy of Sciences of America*, 108: 458–462.

Drewry, D.J. (1975), 'Initiation and growth of the east Antarctic ice-sheet', *Journal, Geological Society of London*, 131: 255–273.

Dugmore, A., Newton, A., Larsen, G. and Cook, G. (2000), 'Tephrochronology, environmental change and the Norse settlement of Iceland', *Environmental Archaeology*, 5: 21–34.

Dumas, B., Hoang, C.T. and Raffy, J. (2006), 'Record of MIS5 sea-level hillstands based on U/Th dated coral terraces of Haiti', *Quaternary International*, 145/6:106–118.

Dupont, L.M. (1986), 'Temperature and rainfall variation in the Holocene based on comparative palaeoecology and isotope geology of a hummock and a hollow (Bourtangerveen, The Netherlands)', *Review of Palaeobotany and Palynology*, 48: 71–159.

Dupont-Nivet, G., Krijgsman, W., Langereis, C.G., Abels, H.A., Dai, S. and Fang, X. (2007), 'Tibetan plateau aridification linked to global cooling at the Eocene-Oligocene transition', *Nature*, 445: 635–638.

Durant, A.J., Harrison, S.P., Watson, I.M. and Balkanski, Y. (2009), 'Sensitivity of direct radiative forcing by mineral dust to particle characteristics', *Progress in Physical Geography*, 33: 80–102.

Dury, G.H. (1965), 'Theoretical implications of underfit streams', *United States Geological Survey Professional Paper*, 452C: C1–C43.

—— (1967), 'Climatic change as a geographical backdrop', *Australian Geographer*, 10: 231–242.

—— (1973), 'Palaeo-hydrologic implications of some pluvial lakes in north-western New South Wales, Australia', *Bulletin of the Geological Society of America*, 84: 3663–3676.

Dutton, C.G. (1889), 'On some of the greater problems of physical geology', *Bulletin of the Philosophical Society of Washington*, II: 51–64.

Dykoski, C.A., Edwards, R.L., Cheng, H., Yuan, D., Cai, Y., Zhang, M., Lin, Y., Qing, J., An, Z. and Revenaugh, J. (2005), 'A high-resolution, absolute-dated Holocene and deglacial Asian monsoon record from Dongge Cave, China', *Earth and Planetary Science Letters*, 233: 71–86.

Eberl, B. (1930), *Die Eiszeitenfolge im nördlichen Alpenvorlande* (Augsberg).

Eden, D.N. and Hammond, A.P. (2003), 'Dust accumulation in the New Zealand region since the last glacial maximum', *Quaternary Science Reviews*, 22: 2037–2052.

Edwards, K.J., Hirons, K.R. and Newell, P.J. (1991), 'The palaeoecological and prehistoric context of minerogenic layers in blanket peat: a study from Loch Dee, southwest Scotland', *The Holocene*, 1: 29–39.

—— Schofield, J.E. and Mauquoy, D. (2008), 'High resolution paleoenvironmental and chronological investigations of Norse landnám at Tasiusaq, Eastern Settlement, Greenland', *Quaternary Research*, 69: 1–15.

Edwards, T.W.D., Wolfe, B.B. and Macdonald, G.M. (1996), 'Influence of changing atmospheric circulation on precipitation $\delta^{18}O$-temperature relations in Canada during the Holocene', *Quaternary Research*, 46: 211–218.

Ehlers, J. (1996), *Quaternary and Glacial Geology* (New York).

—— and Gibbard, P.L. (eds) (2004), *Quaternary Glaciations – Extent and Chronology, Part I Europe, Part II North America, Part III South America, Asia, Africa, Australia, Antarctica.* (Amsterdam).

—— and Gibbard, P.L (2007), 'The extent and chronology of Cenozoic global glaciation', *Quaternary International*, 164-165: 6–20.

—— Gibbard, P.L. and Hughes, P.D. (eds) (2011), *Quaternary Glaciations – Extent and Chronology: A Closer Look.* (Amsterdam).

Eisma, D. (ed.) (1995), *Climate Change Impact on Coastal Habitation* (Boca Raton, Florida).

Elias, S.A. (2000), 'Late Pleistocene climates of Beringia, based on analysis of fossil beetles', *Quaternary Research*, 53: 229–235.

Elkibbi, M. and Rial, J.A. (2001), 'An outsider's review of the astronomical theory of the climate: is the eccentricity-driven insolation the main driver of the ice ages?', *Earth-Science Reviews*, 56: 161–177.

Elsner, J.B., Kossin, J.P. and Jagger, T.H. (2008), 'The increasing intensity of the strongest tropical cyclones', *Nature*, 455: 92–95.

Emanuel, K.A. (2005), 'Increasing destructiveness of tropical cyclones over the past 30 years', *Nature*, 436: 686–888.

Emeis, K-C., Sakamoto, T., Wehansen, R. and Bournsack, J-J. (2000), 'The sapropel record of the eastern Mediterranean Sea – results of Ocean Drilling Programme Log 160', *Palaeogeography, Palaeoclimatology, Palaeoecology*, 158: 371–395.

Emery, K.O. (1969), 'The continental shelves', *Scientific American*, 221: 107–122.

—— and Uchupi, E. (1972), 'Western North Atlantic Ocean', *Memoir, Association of American Petroleum Geologists*, 17: 1–532.

—— Niino, H. and Sullivan, B. (1971), 'Post Pleistocene levels of the East China Sea', in K.K. Turekian (ed.) *The Late Cenozoic Glacial Ages* (New Haven, CT): 381–390.

Emiliani, C. (1961), 'Cenozoic climatic changes as indicated by the stratigraphy and chronology of deep-sea cores of globigerina-ooze facies', *Annals, New York Academy of Science*, 95: 521–536.

—— and Flint, R.F. (1963), 'The Pleistocene record', in M.N. Hill (ed.) *The Sea*, vol. 3 (New York): 888–927.

Endfield, G.H., Tejedo, I.F. and O'Hara, S.L. (2004), 'Drought and disputes, deluge and dearth: climatic variability and human response in colonial Oaxaca, Mexico', *Journal of Historical Geography*, 30: 249–276.

Enfield, D.B., Mestas-Nunez, A.M. and Trimble, P.J. (2001), 'The Atlantic multidecadal oscillation and its relation to rainfall and river flows in the continental US', *Geophysical Research Letters*, 28: 2077–2080.

Engel, M., Brückner, H., Pint, A., Wellbrock, K., Ginau, A., Voss, P., Grottker, M., Klasen, N. and Frenzel, P. (2012), 'The early Holocene humid period in NW Saudi Arabia–evidence from sediments, microfossils and palaeo-hydrological modeling', *Quaternary International*, 266: 131–141, doi: 10.1016/j.quaint.2011.04.028.

Engelstaedter, S., Tegen, I. and Washington, R. (2006), 'North African dust emissions and transport', *Earth-Science Reviews*, 79: 73–100.

Engelhart, S.E., Horton, B.P., Douglas, B.C., Peltier, W.R. and Törnqvist, T.E. (2009), 'Spatial variability of late Holocene and 20th century sea-level rise along the Atlantic coast of the United States', *Geology*, 37: 1115–8.

—— Horton, B.P., Roberts, D.H., Bryant, C.L. and Corbett, D.R. (2007), 'Mangrove pollen of Indonesia and its suitability as a sea-level indicator', *Marine Geology*, 242: 65–81.

Engstrom, D.R. and Wright, H.E. (1984), 'Chemical stratigraphy of lake sediments as a record of environmental change', in E.Y. Haworth and J.W.G. Lund (eds) *Lake Sediments and Environmental History*, Leicester University Press (Bath): 11–67.

Enzel, Y., Wells, S.G. and Lancaster, N. (2003), 'Late Pleistocene lakes along the Mojave River, southeast California' in Y. Enzel, S.G. Wells and N. Lancaster (eds) *Palaeoenvironments and palaeohydrology of the Mojave and southern Great Basin Deserts* (Boulder, Colorado). *Geological Society of America Special Paper*, 368: 61–77.

Epica community members (2004), 'Eight glacial cycles from an Antarctic ice core', *Nature*, 429: 623–628.

Epica Community Members (2006), 'One-to-one coupling of glacial climate variability in Greenland and Antarctica', *Nature*, 444: 195–198.

Epstein, S., Sharp, R.P. and Gow, A.J. (1970), 'Antarctic ice-sheet: stable isotope analysis of Byrd Station cores and interhemisphere climatic implications', *Science*, 168: 1570–1572.

Esper, J., Cook, E.R. and Schweingruber, F.H. (2002), 'Low-frequency signals in long tree-ring chronolo-

gies for reconstructing past temperature variability', *Science*, 295: 2250–2253.

—— and Frank, D. (2009), 'The IPCC on a heterogeneous Medieval Warm Period', *Climatic Change*, 94: 267–273.

Evans, D.J.A. and Benn, D. (2005), *A practical guide to the study of glacial sediments* (London).

—— Harrison, S., Vieli, A. and Anderson, E. (2012), 'The glaciation of Dartmoor: the southernmost independent Pleistocene ice cap in the British Isles', *Quaternary Science Reviews*, 45: 31–53.

Evans, G. (1979), 'Quaternary transgressions and regressions', *Journal of the Geological Society*, 136: 125–132.

Evans, J.P. (2010), 'Global warming impact on the dominant precipitation processes in the Middle East', *Theoretical and Applied Climatology*, 99: 389–402.

Ewing, M. (1971), 'The Late Cenozoic history of the Atlantic Basin, and its bearing on the cause of the Ice Ages', in K.K. Turekian (ed.) *The Late Cenozoic Glacial Ages* (New Haven, CT): 565–573.

Faegri, K. and Iversen, J. (1989), *Textbook of Pollen Analysis* (Chichester).

Fagan, B. (2004), *The Long Summer: How Climate Changed Civilization* (London).

Fairbanks, R.G. (1989), 'A 17 000-year glacio-eustatic sea level record: Influence of glacial melting rates on the Younger Dryas event and deep ocean circulation', *Nature*, 342: 637–642.

Fairbridge, R.W. (1958), 'Dating the latest movements of the Quaternary sea-level', *Transactions, New York Academy of Sciences*, 20: 471–482.

—— (1981), 'The concept of neotectonics: An introduction', *Zeitschrift für Geomorphologie, Supplementband*, 40: VII–XII.

Farley, K. and Patterson, D.B. (1995), 'A 100 kyr periodicity in the flux of extraterrestrial ^{3}He to the sea floor', *Nature*, 378: 600–603.

Fedje, D.W. and Josenhans, H. (2000), 'Drowned forests and archaeology on the continental shelf of British Columbia, Canada', *Geology*, 28: 99–102.

Feng, Z-D. (2001), 'Gobi dynamics in the Northern Mongolian Plateau during the past 20 000+ yr: preliminary results', *Quaternary International*, 76/77: 77–83.

—— AN, C.B. and Wang, H.B. (2006), 'Holocene climatic and environmental changes in the arid and semi-arid areas of China: a review', *The Holocene*, 16: 119–130.

Fernandes, C.A., Rohling E.J. and Siddall, M. (2006), 'Absence of post-Miocene Red Sea land bridges: bio-geographic implications', *Journal of Biogeography*, 33: 961–966.

Filho, A.C., Schwartz, D., Tatumi, S.H. and Rosique, T. (2002), 'Amazonian paleodunes provide evidence for drier climate phases during the Late Pleistocene-Holocene', *Quaternary Research*, 58: 205–209.

Filippelli, G.M. (1997), 'Intensification of the Asian monsoon and a chemical weathering event in the late Miocene-early Pliocene: implications for late Neogene climate change', *Geology*, 25: 27–30.

Fiol, L.A., Fornós, J.J., Gelabert, B. and Guijarro, J.A. (2005), 'Dust rains in Mallorca (Western Mediterranean): their occurrence and role in some recent geological processes', *Catena*, 63: 64–84.

Firestone, R.B. and 25 others. (2007), 'Evidence for an extraterrestrial impact 12,900 years ago that contributed to the megafaunal extinctions and the Younger Dryas cooling', *Proceedings of the National Academy of Sciences*, 104: 16016–16021.

Fitch, S., Thomason, K. and Gaffney, V. (2005), 'Late Pleistocene and Holocene depositional systems and the palaeogeography of the Dogger Bank, North Sea', *Quaternary Research*, 64: 185–196.

Fitzharris, B. (1996), 'The cryosphere: changes and their impacts' in R.T. Watson, M.C. Zinyowera, R.H. Moss and D.J. Dokken (eds) *Climate change 1995. Impacts, adaptation and mitigation of climate change: scientific and technical analyses.* (Cambridge).

Fitzsimmons, K.E., Rhodes, E.J., Magee, J.W. and Barrows, T.T. (2007), 'The timing of linear dune activity in the Strzelecki and Tirari Deserts, Australia', *Quaternary Science Reviews*, 26: 2598–2616.

Fleitmann, D., Burns, S., Mudelsee, M., Neff, U., Kramers, J., Mangini, A. and Matter, A. (2003), 'Holocene forcing of the Indian Monsoon recorded in a stalagmite from Southern Oman', *Science*, 300: 1737–1739.

—— Burns, S., Mangini, A., Mudelsee, M., Kramer, J., Villa, I., Neff, U., Al- Subbary, A., Buettner, A., Hippler, D. and Matter, A. (2007), 'Holocene ITCZ and Indian monsoon dynamics recorded in stalagmites from Oman and Yemen (Socotra)', *Quaternary Science Reviews*, 26: 170–188.

—— Burns, S.J., Pekala, M., Mangini, A., Al-Subbary, A., Al-Aowah, M., Kramers, J. and Matter, A. (2011), 'Holocene and Pleistocene pluvial periods in Yemen, southern Arabia', *Quaternary Science Reviews*, 30: 783–787.

Flemming, N.C. (1969), 'Archaeological evidence for eustatic change of sea-level and earth movements in the western Mediterranean during the last 2000 years', *Geological Society of America Special Paper*, 109.

—— (1972), 'Relative chronology of submerged Pleistocene marine erosion features in the western Mediterranean', *Journal of Geology*, 80: 633–662.

Flenley, J.R. (1979), 'The late Quaternary vegetational history of the equatorial mountains', *Progress in Physical Geography*, 3: 488–509.

—— (1998), 'Tropical forests under the climates of the last 30 000 years', *Climatic Change*, 39: 177–197.

—— and King, S.M. (1984), 'Late Quaternary Pollen Records from Easter Island', *Nature*, 307: 47–50.

Flint, R.F. (1971), *Glacial and Quaternary Geology* (New York).

—— and Bond, G. (1968), 'Pleistocene sand ridges and pans in western Rhodesia', *Bulletin, Geological Society of America*, 79: 299–313.

Flood, P.G. and Frankel, E. (1989), 'Late Holocene higher sea-level indicators from eastern Australia', *Marine Geology*, 90: 193–195.

Florschutz, F., Menendez Amor, J. and Wijmstra, T.A. (1974), 'Palynology of a thick Quaternary succession in southern Spain', *Palaeogeography, Palaeoclimatology, Palaeoecology*, 10: 233–264.

Flower, B.P. and Kennett, J.P. (1995), 'Middle Miocene deepwater paleoceanography in the southwest Pacific: Relations with East Antarctic Ice Sheet development', *Paleoceanography*, 10: 1095–1112.

Folland, C.K., Karl, T.R. and Vinnikov, K.Y.A. (1990), 'Observed climate variations and change', in J.T. Houghton, G.J. Jenkins and J.J. Ephraums (eds) *Climate Change: the IPCC Scientific Assessment* (Cambridge).

Foltz, G.R. and McPhaden, M.J. (2008), 'Trends in Saharan dust and tropical Atlantic climate during 1980-2006', *Geophysical Research Letters*, 35: L20706, doi:10.1029/2008GL035042.

Forland, E.J., Alexandersson, H., Drebs, A., Hamssen-Bauer, L., Vedin, H. and Tveito, O.E. (1998), 'Trends in Maximum 1 day precipitation in the Nordic region', Dnmi report 14/98, *Klima* (Oslo): 1–55.

Forsström, L. (2001), 'Duration of interglacials: a controversial question', *Quaternary Science Reviews*, 20: 1577–1586.

Francis, D. and Hengeveld, H. (1998), *Extreme Weather and Climate Change* (Downsview, Ontario).

Frechen, M. and Dodonov, A.E. (1998), 'Loess chronology of the Middle and Upper Pleistocene in Tajikistan', *Geologische Rundschau*, 87: 2–20.

—— Oches, E.A. and Kohfeld, K.E. (2003), 'Loess in Europe – mass accumulation rates during the Last Glacial Period', *Quaternary Science Reviews*, 22: 1835–1857.

French, H.M. (1996), *The Periglacial Environment*, 2nd edition (Harlow).

French, J. R., Spencer, T. and Reed, D. J. (eds) (1995), 'Geomorphic responses to sea level rise: existing evidence and future impacts', *Earth Surface Processes and Landforms*, 20: 1–6.

Frenzel, B. (1973), *Climatic Fluctuations of the Ice Age* (Cleveland, Ohio): 300 ff.

—— and Bludau, W. (1987), 'On the duration of the interglacial to glacial transition at the end of the Eemian Interglacial (Deep Sea Stage 5E): Botanical and sedimentological evidence', in W.H. Berger and L.D. Labeyrie (eds) *Abrupt climatic change*: 151–162.

Frey, D.G. (1986), 'Cladocera analysis', in B.E. Berglund (ed.) *Handbook of Holocene Palaeoecology and Palaeohydrology* (Chichester): 667–692.

Freydier, R., Michard, A., De Lange, G. and Thomson, T. (2001), 'Nd isotopic compositions of Eastern Mediterranean sediments: traces of the Nile influence during sapropel SI formation?', *Marine Geology*, 177: 45–62.

Friedrich, M., Kromer, B., Spurk, M., Hofmann, J. and Kaiser, K.F. (1999), 'Palaeo-environment and radiocarbon calibration as derived from Late glacial/Early Holocene tree-ring chronologies', *Quaternary International*, 61: 27–39.

—— Remmele, S., Kromer, B., Hofmann, J., Spurk, M., Kaiser, K.F., Orcel, C. and Küppers, M. (2004), 'The 12,460 year Hohenheim oak and pine tree-ring chronology from Central Europe – a unique annual record for radiocarbon calibration and paleo-environment reconstructions', *Radiocarbon*, 46: 1111–1122.

Friedrich, W.L., Kromer, B., Friedrich, M., Heinemeier, J., Pfeiffer, T. and Talamo, S. (2006), 'Santorini eruption radiocarbon dated to 1627–1600 BC', *Science*, 312: 548.

Friele, P.A. and Clague, J.J. (2002), 'Younger Dryas readvance in Squamish river valley, southern Coast mountains, British Columbia', *Quaternary Science Reviews*, 21: 1925–1933.

Fritts, M.C. (1976), *Tree Rings and Climate* (London).

Fronval, T., Jansen, E., Bloemendal, J. and Johnsen, S.J. (1995), 'Oceanic evidence for coherent fluctuations in Fennoscandian and Laurentide ice sheets on millennium timescales', *Nature*, 374: 443–446.

Fuji, N. (1988), 'Palaeovegetation and palaeoclimate changes around Lake Biwa, Japan during the last *ca.* 3 million years', *Quaternary Science Reviews*, 7: 21–28.

Fullagar, R.L.K., Price, D.M. and Head, L.M. (1996), 'Early human occupation of northern Australia: archaeology and thermoluminescence dating of Jinmium rock-shelter, Northern Territory', *Antiquity*, 70: 751–773.

Galloway, R.W. (1970), 'The full glacial climate in south-western U.S.A.', *Annals of the Association of American Geographers*, 60: 245–256.

Ganor, E. (1994), 'The frequency of Saharan dust episodes over Tel Aviv, Israel', *Atmospheric Environment*, 28: 2867–2871.

—— and Foner, H.A. (1996), 'The mineralogical and chemical properties and the behaviour of aeolian Saharan dust over Israel', in S. Guerzoni and R. Chester (eds) *The Impact of desert dust across the Mediterranean* (Dordrecht): 163–172.

—— Osetinsky, I., Stupp, A. and Alpert, P. (2010), 'Increasing trend of African dust, over 49 years, in the eastern Mediterranean', *Journal of Geophysical Research*, 115: D070201, doi:10.1029/2009JD012500.

Gao, H. and Washington, R. (2010), 'Arctic oscillation and the interannual variability of dust emissions from the Tarim Basin: a TOMS AI based study', *Climate Dynamics*, 35: 511–522.

Gao, T., Su, L.J., Ma, Q.X., Li, H.Y., Li, X. and Yu, X. (2003), 'Climatic analyses on increasing dust storm frequency in the springs of 2000 and 2001 in Inner Mongolia', *International Journal of Climatology*, 23: 1743–1755.

Garcia-Garcia, A., Garcia-Gil, S. and Vilas, F. (2005), 'Quaternary evolution of the Ria de Vigo, Spain', *Marine Geology*, 220: 153–179.

Garreaud, R.D., Molina, A. and Farias, M. (2010), 'Andean uplift, ocean cooling and Atacama hyperaridity', *Earth and Planetary Science Letters*, 292: 39–50.

Gasse, F. (2000), 'Hydrological changes in the African tropics since the Last Glacial Maximium', *Quaternary Science Reviews*, 19: 189–211.

—— and Van Campo, E. (1994), 'Abrupt post-glacial climate events in West Asia and North Africa monsoon domains', *Earth and Planetary Science Letters*, 126: 435–456.

Gavish, E. (1974), 'Geochemistry and mineralogy of a recent sabkha along the coast of Sinai, Gulf of Suez', *Sedimentology*, 21: 397–414.

Gaylord, D.R. (1990), 'Holocene palaeoclimatic fluctuations revealed from dune and interdune strata in Wyoming', *Journal of Arid Environments*, 18: 123–138.

Gear, A.J. and Huntley, B. (1991), 'Rapid changes in the range limits of Scots Pine 4000 years ago', *Science*, 251: 544–547.

Gehrels, W.R. (2010a), 'Sea-level changes since the Last Glacial Maximum: an appraisal of the IPCC Fourth Assessment Report', *Journal of Quaternary Science*, 25: 26–38.

—— (2010b), 'Late Holocene land- and sea-level changes in the British Isles: implications for future sea-level predictions', *Quaternary Science Reviews*, 29: 1648–1660.

—— and Newman, S.W.G. (2004), 'Salt-marsh foraminifera in Ho Bught, western Denmark, and their use as sea-level indicators', *Geografisk Tidsskrift*, 104: 97–106.

—— Kirby, J.R., Prokph, A., Newnham, R.H., Achterbery, E.P., Evans, H., Black, S. and Scott, D.B. (2005), 'Onset of recent rapid sea-level rise in the western Atlantic Ocean', *Quaternary Science Reviews*, 24: 2083–2100.

Geikie, J. (1874), *The Great Ice Age and its Relation to the Antiquity of Man* (London).

Geitzenauer, K.R., Margolis, S.V. and Edwards, D.S. (1968), 'Evidence consistent with Eocene glaciation in a south Pacific deep-sea sedimentary core', *Earth and Planetary Science Letters*, 4: 173–177.

Gellert, J.F. (1991), 'Pleistozän-Kalfzeitliche Vergletscherungen im Hochland von Tibet und im Sudari Kanishen Kapgebirge', *Eiszeitalter und Gegenwart*, 41: 141–145.

Gentilli, J. (1961), 'Quaternary climates of the Australian region', *Annals, New York Academy of Sciences*, 95: 465–501.

Gerasimov, I.B. (1969), 'Degradation of the last European ice-sheet', in H.E. Wright (ed.) *Quaternary Geology and Climate* (Washington, DC): 72–78.

Gerson, R. and Amit, R. (1987), 'Rates and modes of dust accretion and deposition in an arid region: the Negev, Israel', in L.K. Frostick and I. Reid (eds) *Desert Sediments: Ancient and Modern*, Geological Society of London, Special Publication no. 35 (London): 157–169.

Gervais, B.R., MacDonald, G.M., Snyder, J.A. and Kremenetski, C.V. (2002), '*Pinus sylvestris* treeline development and movement on the Kola Peninsula of Russia: pollen and stomate evidence', *Journal of Ecology*, 90: doi:10.1046/J.1365-2745.2002.00697.x

Ghazanfar, S.A. (1999), 'Present flora as an example of palaeoclimate: examples from the Arabian Peninsula', in A.K. Singhvi and E. Derbyshire (eds) *Paleoenvironmental Reconstruction in Arid Lands* (Rotterdam): 263–275.

—— and Fisher, M. (eds) (1998), *Vegetation of the Arabian Peninsula*. (Dordrecht).

Giagante, D.A., Aliotta, S., Ginsberg, S.S., Vecchi, L.G. and Spagnuolo, J.O. (2011), 'Evolution of coastal alluvial deposit in response to the last Quaternary marine transgression, Bahia Blanca estuary, Argentina', *Quaternary Research*, 75: 614–623.

Giannini, A., Savavanan, R. and Chang, P. (2003), 'Oceanic forcing of Sahel rainfall on interannual to interdecadal time scales', *Science*, 302: 1027–1030.

Gibbard, P.L. and Lautridou, J.P. (2003), 'The Quaternary history of the English Channel; an introduction', *Journal of Quaternary Science*, 18: 195–199.

—— and Clark, C.D. (2011), 'Pleistocene glaciation limits in Great Britain', in J. Ehlers, P.L. Gibbard and P.D. Hughes (eds) *Developments in Quaternary Science Vol 15, Quaternary Glaciations – Extent and Chronology: A Closer Look* (London): 75–93.

—— Smith, A.G., Zalasiewicz, J.A., Barry, T.L., Cantrill, D., Coe, A.L., Cope, J.C.W., Gale, A.S., Gregory, F.J., Powell, J.H., Rawson, P.F., Stone, P. and Waters, C.N. (2005), 'What status for the Quaternary?', *Boreas*, 34: 1–6.

—— Head, M.J., Walker, M.J.C. and The Subcommission on Quaternary Stratigraphy. (2010), 'Formal ratification of the Quaternary System/Period and the Pleistocene Series/Epoch with a base at 2.58 Ma', *Journal of Quaternary Science*, 25: 96–102.

Gignoux, M. (1930), 'Charles Déperet (1865–1929)', *Bulletin de la Société Géologique de la France*, 30: 1043–1073.

Gilbert, G.K. (1890), 'Lake Bonneville', *US Geological Survey Monograph* 1.

Gill, E.D. (1961), 'Cainozoic climates of Australia', *Annals, New York Academy of Sciences*, 95: 461–464.

Gillespie, A.R., Porter, S.C. and Atwater, B.F. (eds) (2004), *The Quaternary Period in the United States* (Amsterdam).

Gillespie, R. (2002), 'Dating the first Australians', *Radiocarbon*, 44: 455–472.

Gillette, D.A. and Hanson, K.J. (1989), 'Spatial and temporal variability of dust production caused by wind erosion in the United States', *Journal of Geophysical Research*, 94D: 2197–2206.

Gingele, F.X. and de Deckker, P. (2005), 'Late Quaternary fluctuations of palaeoproductivity in the Murray Canyons area, South Australian continental margin', *Palaeogeography, Palaeoclimatology, Palaeoecology*, 220: 361–373.

Giosan, L., Filip, F. and Constatinescu, S. (2009), 'Was the Black Sea catastrophically flooded in the early Holocene?', *Quaternary Science Reviews*, 28: 1–6.

Gjaeveroll, O. (1963), 'Survival of plants on nunataks in Norway during the Pleistocene glaciation', in A. Love and D. Love (eds) *North Atlantic biota and their history* (Oxford): 261–283.

Glass, B., Ericson, D.B., Heezen, B.C., Opdyke, N.D. and Glass, J.A. (1967), 'Geomagnetic reversals and Pleistocene chronology', *Nature*, 216: 437–442.

Glassford, D.K. and Killigrew, L.P. (1976), 'Evidence for Quaternary westward extension of the Australian Desert into south-western Australia', *Search*, 7: 394–396.

Godwin, H. (1940), 'Pollen analysis and forest history of England and Wales', *New Phytologist*, 39(4): 370.

—— (1956), *The History of the British flora* (Cambridge).

—— (1975), *History of the British Flora: A Factual Basis for Phytogeography* (Cambridge).

—— Suggate, R.P. and Willis, E.H. (1958), 'Radiocarbon dating of the eustatic rise in ocean level', *Nature*, 181: 1518–1519.

Goldenberg, S.B., Landsea, C.W., Mestas-Nunez, A.M. and Gray, W.M. (2001), 'The recent increase in Atlantic hurricane activity: causes and implications', *Science*, 293: 474–479.

Golding. K.A., Simpson, I.A., Schofield, J.E. and Edwards, K.J. (2011), 'Norse–Inuit interaction and landscape change in southern Greenland? A geochronological, pedological, and palynological investigation'. *Geoarchaeology*, 26: 315–345.

Gómez, E.A., Martínez, D.E., Borel, C.M., Guerstein, G.R. and Cusminsky, G.C. (2005), 'Submarine evidence of Holocene sea-level fluctuations in the Bahia Blanca estuary, Argentina', *Journal of South American Earth Sciences*, 20: 139–155.

Gordon, A.L. (1991), 'The role of thermohaline circulation in global climate change', in Lamont-Doherty Geological Observatory 1990 & 1991 Report, Columbia University, Palisades (New York): 44–51.

Goreau, T. (1980), 'Frequency sensitivity of the deep-sea climatic record', *Nature*, 287: 620–622.

—— and Hayes, R.L. (1994), 'Coral bleaching and ocean "hot spots"', *Ambio*, 23: 176–180.

Gornitz, V., Rosenzweig, C. and Hillel, D. (1997), 'Effects of anthropogenic intervention in the land hydrologic cycle on global sea level rise', *Global and Planetary Change*, 14: 147–161.

Görür, N., Çağatay, M.N., Emre, O., Alpar, B., Sakinç, M., Islamoğlu, Y., Algan, O., Erkal, T., KeÇer, M., Akkok, R. and Karlik, G. (2001), 'Is the abrupt drowning of the Black Sea shelf at 7150 yr BP a myth?' *Marine Geology*, 176: 65–73.

Goslar, T., Kuc, T., Ralska-Jasiewiczowa, M., Rozanski, K., Arnold, M., Bard, E., van Geel, B., Pazdur, M.F., Szeroczynska, K., Wicik, B., Wieckowski, K. and Walanus, A. (1993), 'High-resolution lacustrine record of the Late Glacial/Holocene transition in central Europe', *Quaternary Science Reviews*, 12: 287–294.

Gosse, J.C. and Phillips, F.M. (2001), 'Terrestrial *in situ* cosmogenic nuclides: theory and applications', *Quaternary Science Reviews* 20, 1475–1560.

Goudie, A.S. (1972), 'The concept of Post-Glacial progressive desiccation', *Research Paper*, No. 4, School of Geography, University of Oxford.

—— (2002), *Great Warm Deserts of the World: Landscapes and Evolution* (Oxford).

—— (2013), *The Human Impact on the Natural Environment* (Oxford).

—— and Middleton, N.J. (1992), 'The changing frequency of dust storms through time', *Climatic Change*, 20: 197–225.

—— and Middleton, N.J. (2006), *Desert Dust in the Global System* (Berlin).

—— and Viles, H.A. (2012), 'Weathering and the global carbon cycle: Geomorphological perspectives', *Earth-Science Reviews*, 113: 59–71.

—— Allchin, B. and Hegde, K.T.M. (1973), 'The former extensions of the Great Indian Sand Desert', *Geographical Journal*, 139: 243–257.

—— Rendell, H. and Bull, P.A. (1983), 'The loess of Tajik SSR', *Proceedings of the International Karakoram Project*, 1: 399–412.

Graaf van der, W.J.E., Crowe, R.W.A., Bunting, J.A. and Jackson, M.J. (1977), 'Relict early Cainozoic drainages in arid Western Australia', *Zeitschrift für Geomorphologie*, 21: 379–400.

Grasty, R.L. (1967), 'Orogeny, a cause of world-wide regression of the seas', *Nature*, 216: 779.

Gray, S.T., Graumlich, L.J., Betancourt, J.L. and Pederson, G.T. (2004), 'A tree-ring based reconstruction of the Atlantic Multidecadal Oscillation since 1567AD', *Geophysical Research Letters*, 31: L12205, doi:10.1029/2004GL019932.

Green, R,E. and 53 others. (2010), 'A draft sequence of the Neanderthal genome', *Science*, 328: 710–722.

Gregory, K.J. and Benito, G. (2003), *Palaeohydrology: Understanding Global Change* (Chichester).

Gregory, J.M. and Oerlemans, J. (1998), 'Simulated future sea-level rise due to glacier melt based on regionally and seasonally resolved temperature changes', *Nature*, 391: 474–476.

—— and 12 others. (2001), 'Comparisons of results from several AOGCMs for global and regional sea level change 1900-2100', *Climate Dynamics*, 18: 225–240.

Gregory, S. (1956), 'Regional variations in the trend of annual rainfall over the British Isles', *Geographical Journal*, 122: 346–353.

—— (ed.) (1988), *Recent Climatic Change* (Lymington).

Gribbin J. and Gribbin, M. (2001), *Ice Age* (London).

Grootes, P.M. (1978), 'Carbon-14 time scale extended', *Science*, 200: 11–15.

—— Stuiver, M., White, J.W.C., Johnsen, S.J. and Jouzel, J. (1993), 'Comparison of oxygen isotope records from the GISP2 and GRIP Greenland ice cores', *Nature*, 366: 552–554.

Grosjean, M., Nunez, L., Cartajena, I. and Messerli, B. (1997), 'Mid-Holocene climate and culture change in the Atacama Desert, northern Chile', *Quaternary Research*, 48: 239–246.

Grousset, F.E., Parra, M., Bory, A., Martinez, P., Bertrand, P., Shimmield, G. and Ellam, R.M. (1998), 'Saharan wind regimes traced by the Sr-Nd isotopic composition of subtropical Atlantic sediments: last glacial maximum vs today', *Quaternary Science Reviews*, 17: 395–409.

Grove, A.T. (1958), 'The ancient Erg of Hausaland, and similar formations on the south side of the Sahara', *Geographical Journal*, 124: 528–533.

—— (1969), 'Landforms and climatic change in the Kalahari and Ngamiland', *Geographical Journal*, 135: 192–212.

—— and Warren, A. (1968), 'Quaternary landforms and climate on the south side of the Sahara', *Geographical Journal*, 134: 194–208.

—— Street, F.A. and Goudie, A.S. (1975), 'Former lake levels and climatic change in the rift valley of southern Ethiopia', *Geographical Journal*, 141: 177–94.

Grove, J.M. (1966), 'The Little Ice Age in the Massif of Mont Blanc', *Transactions, Institute of British Geographers*, 40: 129–143.

—— (1972), 'The incidence of landslides, avalanches and floods in western Norway during the Little Ice Age', *Arctic and Alpine Research*, 4: 131–138.

—— (1988), *The Little Ice Age* (London).

—— (1996), 'The century time-scale', in T.S. Driver and G.P. Chapman (eds) *Time-scales and Environmental Changes* (Routledge): 39–87.

—— (2004), *Little Ice Ages Ancient and Modern*, 2nd edition (London).

Grundstein, A. and Mote, T.L. (2010), 'Trends in average snow depth across the western United States', *Physical Geography*, 31: 172–85.

Guidion, N. and Delibrias, G. (1986), 'Carbon-14 dates point to man in the Americas 32 000 years ago', *Nature*, 321: 769–771.

Guilcher, A. (1969), 'Pleistocene and Holocene sea-level changes', *Earth-Science Reviews*, 5: 69–98.

Guiot, J., de Beaulieu, J.L., Cheddadi, R., David, F., Ponel, P. and Reille, M. (1993), 'The climate in Western Europe during the last Glacial/Interglacial cycle derived from pollen and insect remains', *Palaeogeography, Palaeoclimatology, Palaeoecology*, 103: 73–93.

Gulliksen, S., Birks, H.H., Possnert, G. and Mangerud, J. (1998), 'A calendar age estimate of the Younger Dryas-Holocene boundary at Kråkenes, western Norway', *The Holocene*, 8: 249–259.

Günster, N., Eck, P., Skowronek, A. and Zöller, L. (2001), 'Late Pleistocene loess and their paleosols in the Granada Basin, southern Spain', *Quaternary International*, 76/77: 241–245.

Gupta, S., Collier, J.S., Palmer-Felgate, A. and Potter, G. (2007), 'Catastrophic flooding origin of shelf valley systems in the English Channel', *Nature*, 448: 342–345.

Gutenberg, B. (1941), 'Changes in sea-level, post-glacial uplift, and mobility of the Earth's interior', *Bulletin, Geological Society of America*, 52: 721–772.

Haberle, S.G. (2005), 'A 23,000-yr pollen record from Lake Euramoo, wet tropics of NE Queensland, Australia', *Quaternary Research* 64, 343–356.

Hack, J.T.N. (1941), 'Dunes of the western Navajo Country', *Geographical Review*, 31: 240–263.

Haffer, J. (1969), 'Speciation in Amazonian forest birds', *Science*, 165: 131–137.

Hagos, S. and Cook, K. (2008), 'Ocean warming and late twentieth-century Sahel drought and recovery', *Journal of Climate*, 21: 3797–3815.

Hahne, J. and Melles, M. (1997), 'Late- and post-glacial vegetation and climate history of the south-western Taymyr Peninsula, central Siberia, as revealed by pollen analysis of a core from Lake Lama', *Vegetation History and Archeobotany*, 6: 1–8.

Haile-Selassie, Y., Suwa, G. and White, T.D. (2004), 'Late Miocene Teeth from Middle Awash, Ethiopia, and Early Hominid Dental Evolution', *Science*, 303: 1503–1505.

Hamilton, R.A. (ed.) (1958), *Venture to the Arctic* (Harmondsworth).

Hamilton, S., Shennan, I., Combellick, R., Mulholland, J. and Noble, C. (2005), 'Evidence for two great earthquakes at Anchorage, Alaska and implications for multiple, great earthquakes through the Holocene', *Quaternary Science Reviews*, 24: 2050–2068.

Hanebuth, T., Stattegger, K. and Grootes, P.H. (2000), 'Rapid flooding of the Sunda Shelf: a Late-glacial sea-level record', *Science*, 288: 1033–1035.

Hanebuth, T.J.J., Stattegger, K. and Bojanowski, A. (2009), 'Termination of the Last Glacial Maximum sea-level lowstand: the Sunda-Shelf data revisited', *Global and Planetary Change*, 66: 76–84.

—— Voris, H.K., Yokoyama, Y., Saito, Y. and Okuno, J. (2011), 'Formation and fate of sedimentary depocentres on Southeast Asia's Sunda Shelf over the past sea-level cycle and biogeographic implications', *Earth-Science Reviews*, 104: 92–110.

Hansen, J. and Renfrew, C. (1978), 'Palaeolithic–Neolithic seed remains at Franchthi Cave, Greece', *Nature*, 271: 349–352, doi:10.1038/271349a0.

Hansen, J. and 9 others. (2008), 'Target atmospheric CO_2: where should humanity aim?', *Open Atmospheric Science Journal*, 2: 217–231.

Harris, G. (1964), 'Climatic changes since 1860 affecting European birds', *Weather*, 19: 70–79.

Harrison, C.G.A. (1990), 'Long-term eustasy and epeirogeny in continents', in National Research Council, *Sea-Level Change*, National Academy Press (Washington D.C): 141–158.

Harrison, S.P. and Dodson, J. (1993), 'Late Quaternary lake-level changes and climates of Australia', *Quaternary Science Reviews*, 12: 211–231.

—— Kohfeld, K.E., Roelandt, C. and Claquin, T. (2001), 'The role of dust in climate changes today, at the last glacial maximum and in the future', *Earth-Science Reviews*, 54: 43–80.

Harsch, M.A., Hulme, P.E., McGlone, M.S. and Duncan, R.P. (2009), 'Are tree-lines advancing? A global meta-analysis of treeline response to climate warming', *Ecology Letters*, 12: 1040–1049.

Harvey, L.D.D. (1978), 'Solar variability as a contributing factor to Holocene climatic change', *Progress in Physical Geography*, 4: 487–530.

Hassan, F.A. (2002), *Droughts, Food and Culture* (New York).

Hastenrath, S. (1984), *The Glaciers of Equatorial East Africa* (Dordrecht).

—— Ming-Chin, W. and Pao-Shin, C. (1984), 'Toward the monitoring and prediction of north-east Brazil droughts', *Quarterly Journal of the Royal Meteorological Society*, 118: 411–425.

Hathaway, D.H. and Wilson, R.M. (2004), 'What the sunspot record tells us about space climate', *Solar Physics*, 224: 5–19.

Haug, G.H. and Tiedemann, R. (1998), 'Effect of the formation of the Isthmus of Panama on Atlantic Ocean thermohaline circulation', *Nature*, 393: 673–676.

—— Ganopolski, A., Sigman, D.M., Rosell-Mele, A., Swann, G.E.A., Tiedemann, R., Jaccard, S.L., Bollmann, J., Maslin, M.A., Leng, M.J. and Eglinton, G. (2005), 'North Pacific seasonality and the glaciation of North America 2.7 million years ago', *Nature*, 433: 821–825.

Hay, W.W. (1996), 'Tectonics and climate', *Geologische Rundschau*, 85: 409–437.

Hays, J.D., Imbrie, J. and Shackleton, N.J. (1976), 'Variations in the Earth's orbit: Pacemaker of the Ice Ages', *Science*, 194: 1121–1132.

Hayward, B.W., Wilson, K., Morley, M.S., Cochran, U. and Grenfell, H.R. (2010), 'Microfossil record of Holocene evolution of coastal wetlands in a tectonically active region of New Zealand', *The Holocene*, 20: 405–421.

Hearty, P.J., Hollin, J.T., Neumann, A.C., O'Leary, M.J. and McCulloch, M. (2007), 'Global sea-level fluctuations during the Last Interglaciation (MIS 5e)', *Quaternary Science Reviews*, 26: 2090–2112.

Heathwaite, A.L., Gottlich, K., Burmeister, E.G., Kaule, G. and Grospietsch, T. (1993), 'Mires: Definitions and Form', in A.L. Heathwaite and K.H. Gottlich (eds) *Mires: Process, Exploitation and Conservation* (Chichester): 1–75.

Heikkinen, O. (1984), 'Forest expansion in the subalpine zone during the past hundred years, Mount Baker, Washington, USA', *Erdkunde*, 38: 194–202.

Heinrich, H. (1988), 'Origin and consequences of cyclic ice-rafting in the Northeast Atlantic Ocean during the past 130 000 years', *Quaternary Research*, 29: 142–152.

Heslop, D., Shaw, J., Bloemendal, J., Chen, F., Wang, J. and Parker, E. (1999), 'Sub-millennial scale variations in East Asian Monsoon Systems recorded by dust deposits from the North-Western Chinese Loess Plateau', *Physics and Chemistry of the Earth*, A, 24: 785–792.

Hesse, P.P. and McTainsh, G.H. (1999), 'Last glacial maximum to early Holocene wind strength in the mid-latitudes of the southern hemisphere from aeolian dust in the Tasman Sea', *Quaternary Research*, 52: 343–349.

Heusser, C.J. (2003), *Ice Age Southern Andes – A Chronicle of Palaeoecological Events* (Amsterdam).

Hewitt, G.M. (1999), 'Post-glacial re-colonization of European biota', *Biological Journal of the Linnean Society* 68: 87–112.

Hewitt, K. (2005), 'The Karakoram Anomaly? Glacier expansion and the 'elevation effect', Karakoram Himalaya', *Mountain Research and Development*, 25: 332–340.

Hey, R.W. (1963), 'Pleistocene screes in Cyrenaica (Libya)', *Eiszeitalter und Gegenwart*, 14: 77–84.

Higgins, C.G. (1965), 'Causes of relative sea-level changes', *American Scientist*, 53: 464–476.

—— (1969), 'Isostatic effects of sea-level changes', in H.E. Wright (ed.) *Quaternary Geology and Climate* (Washington, DC): 141–145.

Higham, C. (2005), 'East Asian agriculture and its impact', in C. Scarre (ed.) *The Human Past: World Prehistory and the Development of Human Societies* (London): 234–263.

Higham, T.F.G., Compton, T., Stringer, C., Jacobi, R., Shapiro, B., Trinkaus, E., Chandler, B., Gröning, F., Collins, C., Hillson, S., O'Higgins, P., FitzGerald, C. and Fagan, M. (2011), 'The earliest evidence for anatomically modern humans in northwestern Europe', *Nature*, 479: 521–524.

Hijma, M.P. and Cohen, K.M. (2010), 'Timing and magnitude of the sea-level jump preluding the 8200 yr event', *Geology*, 38: 275–278.

Hiscott, R.N. and 9 others. (2007), 'A gradual drowning of the southwestern Black Sea Shelf: evidence for a progressive rather than abrupt Holocene reconnection with the eastern Mediterranean through the Marmara Sea Gateway', *Quaternary International*, 167/8: 19–34.

Hobbs, J.E. (1988), 'Recent climatic change in Australasia', in S. Gregory (ed.) *Recent Climatic Change* (London): 285–297.

Hodell, D.A., Curtis, J.H. and Brenner, M. (1995), 'Possible role of climate in the collapse of classic Maya civilisation', *Nature*, 375: 341–347.

Hoelzmann, P., Gasse, F., Dupont, L., Salzmann, U., Staubwasser, M., Leuschner, D.C. and Sirocko, F. (2004), 'Palaeoenvironmental changes in the arid and subarid-belt (Sahara-Sahel-Arabian Peninsula) from 150 ka to present', in R.W. Battarbee, F. Gasse and C.E. Stickley (eds.) *Past Climate Variability Through Europe and Africa, Developments in Paleoenvironmental Research Vol. 6* (Springer): 219–256.

—— Keding, B., Berke, H., Kröpelin, S. and Kruse, H.-J. (2001), 'Environmental change and archaeology: lake evolution and human occupation in the Eastern Sahara during the Holocene', *Palaeogeography, Palaeoclimatology, Palaeoecology*, 169: 193–217.

Hoffman, J.S., Keyes, D. and Titus, J.G. (1983), *Projecting Future Sea-level Rise – Methodology, Estimates in the year 2100 and Research Needs*. US Environmental Protection Agency Report, EPA 230-09-007.

—— Wells, J.B. and Titus, J.G., (1986). 'Future global warming and sea-level rise', in G. Sigbjanarson (ed.) *Proceedings, Iceland Coastal and River Symposium*, 245–266.

Hoffman, P.F., Kaufman, A.J., Halverson, G.P. and Schrag, D.P. (1998), 'A Neoproterozoic Snowball Earth', *Science*, 281: 1342–1346.

Hoinkes, H.C. (1968), 'Glacier variation and weather', *Journal of Glaciology*, 7: 3–20.

Holl, A.F.C. (2004), *Saharan Rock Art. Archaeology of Tassilian Pastoralist Iconography* (New York).

Holliday, V.T. (1989), 'The Blackwater Draw Formation (Quaternary): A 1.4-plus m.y. record of eolian sedimentation and soil formation on the Southern High Plains', *Bulletin, Geological Society of America*, 101: 1598–1607.

Holloway, G. and Sou, T. (2002), 'Has Arctic sea ice rapidly thinned?', *Journal of Climate*, 15: 1691–1701.

Holmes, J.A. and Street-Perrott, F.A. (1989), 'The Quaternary glacial history of Kashmir, north-west Himalaya: a revision of de Terra and Paterson's sequence', *Zeitschrift für Geomorphologie, Supplementband*, 76: 195–212.

—— and Engstrom, D.R. (2005), 'Non-marine ostracod records of Holocene environmental change', in A. Mackay, R. Battarbee, J. Birks and F. Oldfield (eds) *Global Change in the Holocene* (London): 310–327.

Holopainen, J. and Helama, S. (2009), 'Little Ice Age farming in Finland: preindustrial agriculture on the edge of the Grim Reaper's scythe', *Human Ecology*, 37: 213–225.

Holzhauser, H., Magny, M. and Zumbühl, H.J. (2005), 'Glacier and lake-level variations in west-central Europe over the last 3500 years', *The Holocene*, 15: 799–901.

Hone, R., Anderson, D.E., Parker, A.G. and Morecroft, M.D. (2001), 'Holocene vegetation change at Wytham Woods, Oxfordshire', *Quaternary Newsletter*, 94: 1–15.

Hong, S., Candelone, J.-P., Patterson, C.C. and Boutron, C.F. (1994), 'Greenland ice evidence of hemispheric lead pollution two millennia ago by Greek and Roman civilizations', *Science*, 265: 1841–1843.

Hooghiemstra, H. (1989), 'Quaternary and upper-Pliocene glaciations and forest development in the tropical Andes: Evidence from a long high-resolution pollen record from the sedimentary basin of Bogotá, Colombia', *Palaeogeography, Palaeoclimatology, Palaeoecology*, 72: 11–26.

Hopf, F.V.L., Colhoun, E.A. and Barton, C.E. (2000), 'Late-glacial and Holocene record of vegetation and climate from Cynthia Bay, Lake St Clair, Tasmania', *Journal of Quaternary Science*, 15: 725–732.

Horner, R.W. (1972), 'Current proposals for the Thames Barrier and the organization of the investigations', *Philosophical Transactions of the Royal Society, London*, A, 272: 179–185.

Horowitz, A. (1989), 'Continuous pollen diagram for the last 3.5 m.y. from Israel: Vegetation, climate and correlation with the oxygen isotope record', *Palaeogeography, Palaeoclimatology, Palaeoecology*, 72: 63–78.

Horton, B.P. and Shennan, I. (2009), 'Compaction of Holocene strata and the implications for relative sea-level change on the east coast of England', *Geology*, 37: 1083–1086.

Houghton, J.T. (2009), *Global Warming: The Complete Briefing*, 4th edition (Cambridge).

—— Jenkins, G.J. and Ephraums, J.J. (eds) (1990), *Climate Change: The IPCC scientific assessment* (Cambridge).

—— Meira Filho, L.G., Callander, B.A., Harris, N., Kattenberg, A. and Maskell, K. (1996), *Climate Change 1995: The Science of Climate Change* (Cambridge).

—— Ding, Y., Griggs, D.J., Noguer, M., van der Linden, P.J., Dai, X., Maskell, K. and Johnson, C.A. (eds) (2001), *Climate Change 2001: The Scientific Basis* (Cambridge).

Hoyos, C.D., Agudelo, P.A., Webster, P.J. and Curry, J.A. (2006), 'Deconvolution of the factors contributing to the increase in global hurricane intensity', *Science*, 312: 94–97.

Hsieh, Chiao-Min (1976), 'Chu K'O-Chen and China's climatic change', *Geographical Journal*, 142: 248–256.

Huang, C.C., Pang, J. and Zhao, J. (2000), 'Chinese loess and the evolution of the East Asian monsoon', *Progress in Physical Geography*, 24: 75–96.

Hubbard, A., Bradwell, T., Golledge, N., Hall, A., Patton, H., Sugden, D., Cooper, R. and Stoker, M. (2009), 'Dynamic cycles, ice streams and their impact on the extent, chronology and deglaciation of the British-Irish ice sheet', *Quaternary Science Reviews*, 28: 758–776.

Huggett, R.J. (1997), *Environmental Change: The Evolving Biosphere* (London).

Hughen, K.A., Overpeck, J.T., Pewterson, L.C. and Trumbore, S. (1996), 'Rapid climate changes in the tropical Atlantic region during the last deglaciation', *Nature*, 380: 51–54.

Hughes, M.K. and Diaz, H.F. (1994), 'Was there a "Medieval Warm Period" and if so, where and when?', *Climatic Change*, 26: 109–142.

—— Kelly, P.M., Pilcher, J.R. and LaMarche, V.C. Jr. (1982), *Climate from tree rings* (Cambridge).

Hughes, P.D., Woodward, J.C. and Gibbard, P.L. (2006), 'Quaternary glacial history of the Mediterranean mountains', *Progress in Physical Geography*, 30: 334–364.

Hughes, T.J. (1998), *Ice Sheets* (New York).

Hui, W., Cook, B., Ravi, S., Fuentes, J. and D'Odorico, P. (2008), 'Dust-rainfall feedback in the West African Sahel', *Water Resources Research*, 44: 1–6.

Hulme, M. (1992), 'Rainfall changes in Africa: 1931–1960 to 1961–1990', *International Journal of Climatology*, 12: 685–699.

Hunt, T.L. (2007), 'Rethinking Easter Island's ecological catastrophe', *Journal of Archaeological Science*, 34: 485–502.

Huntley, B. and Birks, H.J.B. (1983), *An Atlas of Past and Present Pollen Maps for Europe: 0–13 000 years ago* (Cambridge).

—— and Prentice, I.C. (1988), 'July temperatures in Europe from pollen data, 6000 years before present', *Science*, 241: 687–690.

—— and Prentice, I.C. (1993), 'Holocene Vegetation and Climates of Europe', in H.E. Wright, J.E. Kutzbach,

T. Webb et al. (eds) *Global climates since the last Glacial Maximum* (Minneapolis): 136–168.

Huntley, D.J., Godfrey-Smith, D.I. and Thewalt, M.L. (1985), 'Optical dating of sediments', *Nature*, 313: 105–107.

Hurrell, J.W. (1995), 'Decadal trends in the North Atlantic Oscillation regional temperatures and precipitation', *Science*, 269: 676–679.

Ikeya, M. (1985), 'Electron spin resonance', in N.W. Rutter (ed.) *Dating Methods of Pleistocene Deposits and their Problems*, Geoscience Canada Reprint Series, 2: 73–87.

Ilyashuk, B., Gobet, E., Heiri, O., Lotter, A.F., van Leeuwen, J.F.N., van der Knaap, W.O., Ilyashuk, E., Oberli, F. and Ammann, B. (2009), 'Lateglacial environmental and climatic changes at the Maloja Pass, Central Swiss Alps, as recorded by chironomids and pollen', *Quaternary Science Reviews*, 28: 1340–1353.

Imbrie, J. and Imbrie, K.P. (1979), *Ice Ages: Solving the Mystery* (London).

—— and Kipp, N.G. (1971), 'A new micropalaeontological method of quantitative palaeoclimatology: application to a Late Pleistocene Caribbean core', in K.K. Turekian (ed.) *The Late Cenozoic Glacial Ages* (New Haven, CT): 71–181.

—— Hays, J.D., Martinson, D.G., McIntyre, A., Mix, A.C., Morley, J.J., Pisias, N.J., Prell, W.L. and Shackleton, N.J. (1984), 'The orbital theory of Pleistocene climate: support from a revised chronology of the marine $\delta^{18}O$ record', in A. Berger (ed.) *Milankovitch and Climate* (New York): 269–305.

Imhof, P., Nesje, A. and Nussbaumer, S.U. (2012), 'Climate and glacier fluctuations at Jostedalsbreen and Folgefonna, southwestern Norway and in the western Alps from the 'Little Ice Age' until the present: the influence of the North Atlantic Oscillation', *The Holocene*, 22: 235–247.

Innes, J.L. (1985), 'Lichenometry', *Progress in Physical Geography*, 9: 187–254.

Intergovernmental Panel on Climate Change (IPCC) (1995), *Climate Change 1995: The Scientific Basis* (Cambridge).

—— (2001), *Climate Change 2001: The Scientific Basis* (Cambridge).

—— (2007), *Climate Change 2007: The Physical Science Basis* (Cambridge).

Irino, T., Ikehara, K., Katayama, H., Ono, Y. and Tada, R. (2003), 'East Asian monsoon signals recorded in the Japan Sea sediments', *PAGES News*, 9 (2): 13–14.

Iriondo, M. (2000), 'Patagonian dust in Antarctica', *Quaternary International*, 68/71: 83–86.

Issar, A.S. (2003), *Climate Changes during the Holocene and their Impact on Hydrological Systems* (Cambridge).

Iversen, J. (1944), 'Viscum, Hedera, and Ilex as climate indicators', *Geol fören. Stockh. förh.*, 66: 463–483.

—— (1958), 'The bearing of glacial and interglacial epochs on the formation and extinction of plant taxa', *Uppsala Universiteit Arssk*, 6: 210–215.

Iwashima, T. and Yamamoto, R. (1993), 'A statistical analysis of the extreme events: long-term trend of heavy daily precipitation', *Journal of Meteorological Society of Japan*, 71: 637–640.

Jacobs, P.M. and Mason, J.A. (2005), 'Impact of Holocene dust aggradation on A horizon characteristics and carbon storage in loess-derived Mollisols of the Great Plains, USA', *Geoderma*, 125: 95–106.

Jacobsen, T. (1982), *Salinity and Irrigation Agriculture in Antiquity* (Malibu).

Jakobsson, M., Björck, S., Alm, G., Andrén, T., Lindeberg, G. and Svensson, N-O. (2007), 'Reconstructing the Younger Dryas ice dammed lake in the Baltic Basin: bathymetry, area and volume', *Global and Planetary Change*, 57: 355–370.

Jamieson, T.F. (1865), 'On the history of the last geological changes in Scotland', *Quarterly Journal of the Geological Society of London*, 21: 161–203.

Jayant, K., Tripathi, V. and Rajamani, V. (1999), 'Geochemistry of the loessic sediments on Delhi ridge, eastern Thar desert, Rajasthan: implications for exogenic processes', *Chemical Geology*, 155: 265–278.

Jelgersma, S. (1966), 'Sea-level changes in the last 10 000 years', in *Royal Meteorological Society International Symposium on World Climate from 8000–0 BC*: 54–69.

—— and Tooley, M.J. (1995), 'Sea-level changes during the recent geological past', *Journal of Coastal Research Special Issue*, No. 17: *Holocene Cycles: Climate, Sea-levels and Sedimentation*: 123–139.

Jennings, J.N. (1975), 'Desert dunes and estuarine fill in the Fitzroy Basin, Western Australia', *Catena*, 2: 216–262.

Jessen, K. (1935), 'Archaeological dating in the history of North Jutland's vegetation', *Acta Archaeologica*, 5: 165–214.

—— (1949), 'Studies in late-Quaternary deposits and the flora-history of Ireland', *Proceedings of the Royal Irish Academy*, B52: 6–85.

Jiang, H. and Ding, Z. (2005), 'Temporal and spatial changes of vegetation cover on the Chinese Loess Plateau through the last glacial cycle: evidence from spore-pollen records', *Review of Palaeobotany and Palynology*, 133: 23–37.

Jilbert, T., Reichard, G-J., Aeschlimann, B., Günther, D., Boer, W. and de Lange, G. (2010), 'Climate-controlled multidecadal variability in North African dust transport to the Mediterranean', *Geology*, 38: 19–22.

Johansen, D.C. and Edey, M.A. (1981), *Lucy: the Beginnings of Humankind* (New York).

Johnsen, S.J., Clausen, H.B., Dansgaard, W., Fuhrer, K., Gundestrup, N., Hammer, C.U., Iversen, P., Jouzel, J., Stauffer, B. and Steffensen, J.P. (1992), 'Irregular glacial interstadials recorded in a new Greenland ice core', *Nature*, 359: 311–313.

—— Dahl-Jensen, D., Gundestrup, N., Steffensen, J.P., Clausen, H.B., Miller, H., Masson-Delmotte, V., Sveinbjörnsdottir, A.E. and White, J. (2001), 'Oxygen isotope and palaeotemperature records from six Greenland ice-core stations: Camp Century, Dye-3, GRIP, GISP2, Renland and NorthGRIP', *Journal of Quaternary Science*, 16: 299–307.

Johnson, T.C., Kelts, K. and Odada, E. (2000), 'The Holocene history of Lake Victoria', *Ambio*, 29: 2–11.

—— Scholz, C.A., Talbot, M.R., Kelts, K., Ricketts, R.D., Ngobi, G., Beuning, K., Ssemmanda, I. and McGill, J.W. (1996), 'Late Pleistocene desiccation of Lake Victoria and rapid evolution of Cichlid fishes', *Science*, 273: 1091–1093.

Johnson, W.H. (1990), 'Ice-wedge casts and relict patterned ground in Central Illinois and their environmental significance', *Quaternary Research*, 33: 51–72.

Jones, C., Mahowald, N., Luo, C. (2003), 'The role of easterly waves in African dust transport', *Journal of Climate*, 16: 3617–3628.

Jones, D.K.C. (1981), *South-east and Southern England* (London).

Jones, P.D. and Reid, P.A. (2001), 'Assessing future changes in extreme precipitation over Britain using regional climate model integrations', *International Journal of Climatology*, 21: 1337–1356.

—— Wigley, T.M.L., Folland, C.K., Parker, D.E., Angell, J.K., Lebedeff, S. and Hansen, J.E. (1988),

'Evidence for global warming in the past decade', *Nature*, 228: 790.

—— Bradley, R.S. and Jouzel, J. (eds) (1996), *Climatic Variations and Forcing Mechanisms over the Last 2000 years* (Heidelberg).

—— Lister, D.D., Wilby, R.L. and Kostoppoulou, E. (2006), 'Extended river flow reconstructions for England and Wales, 1865-2002', *International Journal of Climatology*, 26: 219–231.

Jones, R.L. and Keen, D.H. (1993), *Pleistocene Environments in the British Isles* (London).

Jones, R.N., Bowler, J.M. and McMahon, T.A. (1998), 'A high resolution Holocene record of P/E ratio from closed lakes in Western Victoria', *Palaeoclimates*, 3: 51–82.

Joosten, J.H.J. (1995), 'Between diluvium and deluge: the origin of the Younger Dryas concept (extended abstract)', *Geologie en Mijnbouw*, 74: 237–240.

Josenhans, H., Fedje, D., Pienitz, R. and Southon, J. (1997), 'Early humans and rapidly changing Holocene sea levels in the Queen Charlotte Islands-Helcate Strait, British Columbia, Canada', *Science*, 277: 71–74.

Jouzel, J. and 16 others. (1993), 'Extending the Vostok ice-core record of palaeoclimate to the penultimate glacial period', *Nature*, 364: 407–412.

—— and 31 others. (2007), 'Orbital and millennial Antarctic climate variability over the past 800,000 years', *Science*, 317: 793–796.

Jung, S.J.A., Davies, G.R., Ganssen, G.M. and Kroon, D. (2004), 'Stepwise Holocene aridification in NE Africa deduced from dust-borne radiogenic isotope records', *Earth and Planetary Science Letters*, 221: 27–37.

Kafri, U. and Yechieli, Y. (2012), 'Palaeogroundwater dynamics and their importance for past human settlements and today's water management', *Quaternary International*, 257: 83–96.

Kaiser, K. (1969), 'The climates of Europe during the Quaternary Ice Age', in H.E. Wright (ed.) *Quaternary Geology and Climate* (Washington, DC): 10–37.

Kaiser, K.F., Friedrich, M., Miramont, C., Kromer, B., Sgier, M., Schaub, M., Boeren, I., Remmele, S., Talamo, S., Guibal, F. and Sivan, O. (2010), 'Challenging process to make the Lateglacial tree-ring chronologies from Europe absolute – an inventory', *Quaternary Science Reviews*, 36: 78–90.

Kalela, O. (1952), 'Changes in the geographic distribution of Finnish birds and mammals in relation to recent changes in climate', *Fennia*, 75: 38–51.

Karl, T.R. and Knight, R.W. (1998), 'Secular trends of precipitation amount, frequency and intensity in the United States', *Bulletin of the American Meteorological Society*, 79: 1413–1449.

Karlén, W. and Kuylenstierna, J. (1996), 'On solar forcing of Holocene climate: evidence from Scandinavia', *The Holocene*, 6: 359–365.

Kaser, G. (1999), 'A review of the modern fluctuations of tropical glaciers', *Global and Planetary Change*, 22: 93–103.

—— and Osmaston, H. (2002), *Tropical Glaciers* (Cambridge).

Kastner, T.P. and Goñi, M.A. (2003), 'Constancy in the vegetation of the Amazon Basin during the Late Pleistocene: evidence from the organic matter composition of Amazon deep sea fan sediments', *Geology*, 31: 291–294.

Kawahata, H., Okamoto, T., Matsumoto, E. and Ujiie, H. (2000), 'Fluctuations of eolian flux and ocean productivity in the mid-latitude North Pacific during the last 200 kyr', *Quaternary Science Reviews*, 19: 1279–1291.

Keigwin, L.D., Curry, W.B., Lehman, S.J. and Johnsen, S. (1994), 'The role of the deep ocean in North Atlantic climate change between 70 and 130 kyr ago', *Nature*, 371: 323–326.

—— Jones, G.A., Lehman, S.J. and Boyle, E.A. (1991), 'Deglacial meltwater discharge, North Atlantic deep circulation and abrupt climate change', *Journal of Geophysical Research*, 96: 16811-16826.

Kelletat, D. (1991), 'The 1550 BP tectonic event in the Eastern Mediterranean as a basis for assessing the intensity of shore processes', *Zeitschift für Geomorphologie, Supplementband*, 81: 181–194.

Kendall, R.L. (1969), 'An ecological history of the Lake Victoria Basin', *Ecological Monographs*, 39: 121–176.

Kennett, J.P. (1970), 'Pleistocene paleoclimates and foraminiferal biostratigraphy in sub-Antarctic deep-sea cores', *Deep-Sea Research*, 17: 125–140.

—— and Thunell, R.C. (1975), 'Global increase in Quaternary explosive volcanism', *Science*, 187: 497–503.

Kerney, M.P., Brown, E.H. and Chandler, T.J. (1964), 'The late-glacial and post-glacial history of the chalk escarpment near Brook, Kent', *Philosophical Transactions of the Royal Society, London*, B, 248: 135–204.

Kerr, R.A. (1987), 'Milankovitch climate cycles through the ages', *Science*, 235: 973–974.

Khalil, M.A.K. and Rasmussen, R.A. (1987), 'Atmospheric methane: trends over the last 10,000 years', *Atmospheric Environment*, 21: 2445–2452.

Kim, J.H., Schneider, R.R., Mulitza, S. and Mueller, P.J. (2003), 'Reconstruction of SE trade-wind intensity based on sea-surface temperatures in the Southeast Atlantic over the last 25 kyr', *Geophysical Research Letters*, 30: doi 10.1029/2003GL017557.

Kleeman, R. and Power, S.B. (2000), 'Modulation of ENSO variability on decadal and larger timescales', in H.F. Diaz and V. Markgaf (eds) *El Niño and the Southern Oscillations* (Cambridge): 413–441.

Klinger, L.F., Taylor, J.A. and Franzen, L.G. (1996), 'The potential role of peatland dynamics in Ice-Age initiation', *Quaternary Research*, 45: 89–92.

Klitgaard-Kristensen, D., Sejrup, H.P., Haflidason, H., Johnsen, S. and Spurk, M. (1998), 'A regional 8200 cal. yr bp cooling event in northwest Europe, induced by final stages of the Laurentide ice-sheet deglaciation?', *Journal of Quaternary Science*, 13: 165–169.

Klotz, S., Müller, U., Mosbrugger, V., de Beaulieu, J.-L. and Reille, M. (2004), 'Eemian to early Würmian climate dynamics: history and pattern of changes in Central Europe', *Palaeogeography, Palaeoclimatology, Palaeoecology*, 211: 107–126.

Knorr, G. and Lohmann, G. (2003), 'Southern Ocean origin for the resumption of Atlantic thermohaline circulation during deglaciation', *Nature*, 424: 532–536.

Kobashi, T., Severinghaus, J.P., Brook, E.J., Barnola, J.-M. and Grachev, A.M. (2007), 'Precise timing and characterization of abrupt climate change 8200 years ago from air trapped in polar ice', *Quaternary Science Reviews*, 26: 1212–1222.

Kofler, W., Krapf, V., Oberhuber, W. and Bortenschlager, S. (2005), 'Vegetation responses to the 8200 cal. bp cold event and to long-term climatic changes in the Eastern Alps: possible influence of solar activity and North Atlantic freshwater pulses', *The Holocene*, 15: 779–788.

Kohfeld, K.E. and Harrison, S.P. (2003), 'Glacial-interglacial changes in dust deposition on the Chinese Loess Plateau', *Quaternary Science Reviews*, 22: 1859–1878.

Kolla, V. and Biscaye, P.E. (1977), 'Distribution and origin of quartz in the sediments of the Indian Ocean', *Journal of Sedimentary Petrology*, 47: 642–649.

Kolla, V.P., Biscaye, P.E. and Hanley, A.F. (1979), 'Distribution of quartz in Late Quaternary Atlantic sediments in relation to climate', *Quaternary Research*, 11: 261–277.

Krause, J., Fu, Q., Good, J.M., Viola, B., Shunkov, M.V., Derevianko, A.P. and Pääbo, S. (2010), 'The complete mitochondrial DNA genome of an unknown hominin from southern Siberia', *Nature* 464: 894–897, doi:10.1038/nature08976.

Kravtsova, V.I. and Tarasenko, T.V. (2010), 'Space monitoring of Aral Sea degradation', *Water Resources*, 37: 285–296.

Kröhling, D.M. (2003), 'A 54 m thick loess profile in North Pampa, Argentina', *Geological Society of America Abstracts with Programs*: 198.

Krom, M.D., Michard, A., Cliff, R.A. and Strohle, K. (1999), 'Sources of sediment to the Ionian Sea and western Levantine basin of the Eastern Mediterranean during S-1 sapropel times', *Marine Geology*, 160: 45–61.

—— Stanley, J.D., Cliff, R.A. and Woodward, J.C. (2002), 'Nile River sediment fluctuations over the past 7000 yr and their key role in sapropel development', *Geology*, 30: 71–74.

Kropotkin, P. (1904), 'The desiccation of Eur-Asia', *Geographical Journal*, 23: 4–23.

Kudrass, H.R., Erlenkeuser, H., Vollbrecht, R. and Weiss, W. (1991), 'Global nature of the Younger Dryas cooling event inferred from oxygen isotope data from Sulu Sea cores', *Nature*, 349: 406–409.

Kuhle, M. (1987), 'Subtropical mountain- and highland-glaciation as ice-age triggers and the waning of the glacial periods in the Pleistocene', *GeoJournal*, 14: 393–421.

Kukal, K. (1990), 'The rate of geological processes', *Earth-Science Reviews*, 28: 1–284.

Kukla, G.J. (1975), 'Loess stratigraphy of central Europe', in K.W. Butzer and G.L. Isaac (eds) *After the Australopithecines* (The Hague): 99–188.

—— (1977), 'Pleistocene land-sea correlations I. Europe', *Earth-Science Reviews*, 13: 307–374.

Kukla, G. and An, Z.S. (1989), 'Loess stratigraphy in Central China', *Palaeogeography, Palaeoclimatology, Palaeoecology*, 72: 203–225.

—— McManus, J.F., Rousseau, D.-D. and Chuine, I. (1997), 'How long and how stable was the last interglacial?', *Quaternary Science Reviews*, 16: 605–612.

—— and 22 others. (2002), 'Last interglacial climates', *Quaternary Research*, 58: 2–13.

Kumar, R.P., Pant, G.B., Parthascarathy, B. and Sontakka, N.A. (1992), 'Spatial and subseasonal pattern of long-term trends of Indian Summer monsoon rainfall', *International Journal of Climatology*, 12: 153–163.

Kuniholm, P.I. (1996), 'The Prehistoric Aegean: dendrochronological progress as of 1995', in K. Randsborg (ed.) *Absolute Chronology: Archaeological Europe 2500–500* BC, *Acta Archaeologica Supplementum*, 1: 327–335.

Kunkel, K.E. (2003), 'North American trends in extreme precipitation', *Natural Hazards*, 29: 291–305.

Kuper, R. and Kröpelin, S. (2006), 'Climate-Controlled Holocene Occupation in the Sahara: Motor of Africa's Evolution', *Science*, 313: 803–807.

Kurosaki, Y., Shinoda, M. and Mikami, M. (2011), 'What caused a recent increase in dust outbreaks over East Asia?', *Geophysical Research Letters*, 38: L11702, doi:10.1029/2011GL047494.

Kutzbach, J.E. (1980), 'Estimates of past climate of paleolake Chad, North Africa, based on a hydrological energy-balance model', *Quaternary Research*, 14: 210–223.

—— (1981), 'Monsoon climate of the early Holocene: climate experiment with the earth's orbital parameters for 9000 years ago', *Science*, 214: 59–61.

—— and Street-Perrott, F.A. (1985), 'Milankovitch forcing of fluctuations in the level of tropical lakes from 18 to 0 kyr bp', *Nature*, 317: 130–134.

—— and Webb, T. (1993), '*Conceptual Basis for Understanding Late-Quaternary Climates*', in H.E. Wright, J.E. Kutzbach, T. Webb et al. (eds) *Global Climates since the Last Glacial Maximum* (Minneapolis): 5–11.

—— Gallimore, R., Harrison, S., Behling, P., Selin, R. and Laarif, F. (1998), 'Climate and biome simulations for the past 21,000 years', *Quaternary Science Reviews*, 17: 473–506.

Kwong, Y.T.J. and Tau, T.Y. (1994), 'Northward migration of permafrost along the Mackenzie highway and climate warming', *Climatic Change*, 41: 249–257.

LaMarche, V.C. Jr. and Hirschboeck, K.K. (1984), 'Frost rings in trees as records of major volcanic eruptions', *Nature*, 307: 121–126.

Lamb, A.L., Wilson, G.P. and Leng, M.J. (2006), 'A review of coastal palaeoclimate and relative sea-level reconstructions using $\sigma^{13}C$ and C/N ratios in organic material', *Earth-Science Reviews*, 75: 29–57.

Lamb, H.F., Bates, C.R., Coombes, C.V., Marshall, M.H., Umer, M., Davies, S.J. and Dejen, E. (2007), 'Late Pleistocene desiccation of Lake Tana, source of the Blue Nile', *Quaternary Science Reviews*, 26: 287–299.

Lamb, H.H. (1966), *The Changing Climate* (London).

—— (1967), 'Britain's changing climate', *Geographical Journal*, 133: 445–468.

—— (1969), 'Climate fluctuations', in H. Flohn (ed.) *World Survey of Climatology*, vol. 2 (Amsterdam): 173–249.

—— (1970), 'Volcanic dust in the atmosphere: with a chronology and assessment of its meteorological significance', *Philosophical Transactions of the Royal Society, London*, A, 266: 425–533.

—— (1971), 'Volcanic activity and climate', *Palaeogeography, Palaeoclimatology, Palaeoecology*, 10: 203–230.

—— (1977), *Climate: Present, Past and Future*; vol. 2, *Climatic History and the Future* (London).

—— (1982), *Climate, History and the Modern World* (London); 2nd edn 1996.

—— and Morth, H.T. (1978), 'Arctic ice, atmosphere circulation and world climate', *Geographical Journal*, 144: 1–22.

Lambeck, K. (1988), *Geological Geodesy* (Oxford).

—— (1993), 'Glacial rebound and sea-level change: an example of a relationship between mantle and surface processes', *Tectonophysics*, 223: 15–37.

—— (1995), 'Late Devensian and Holocene shorelines of the British isles and North Sea from models of glaciohydro-isostatic rebound', *Journal of the Geological Society of London*, 152: 437–448.

—— (1996), 'Shoreline reconstructions for the Persian Gulf since the last Glacial Maximum', *Earth and Planetary Science Letters*, 142: 43–57.

—— Anzidei, M., Antonioli, F., Benini, A. and Esposito, A. (2004), 'Sea level in Roman time in the Central Mediterranean and implications for recent change', *Earth and Planetary Science Letters*, 224: 563–574.

—— Purcell, A., Flemming, N.C., Vita-Finzi, C., Alsharekh, A.M. and Bailey, G.N. (2011), 'Sea level and shoreline reconstructions for the Red Sea: isostatic and tectonic considerations and implications for hominin migration out of Africa', *Quaternary Science Reviews*, 30: 3542–3572.

Lambert, F. and 9 others. (2008), 'Dust-climate couplings over the past 800,000 years from the EPICA Dome C ice core', *Nature*, 452: 616–619.

Lamothe, L. de (1918), 'Les anciennes nappes alluviales et lignes de rivages du basin de la Somme et leurs rapports avec celles de la Mediterranée occidentale', *Bulletin de la Société Géologique de la France*, 30: 1043–1073.

Lancaster, N. (1979), 'Quaternary environments in the arid zone of southern Africa', *Environmental Studies, University of the Witwatersrand, Occasional Paper*, 22.

—— (1989), 'Late Quaternary palaeoenvironments in the south-western Kalahari', *Palaeogeography, Palaeoclimatology, Palaeoecology*, 70: 367–376.

Landwehr, J.M., Coplan, T.B., Ludwig, K.R., Winograd, I.J. and Riggs, A.C. (1997), 'Data for Devils Hole core DH-11', *US Geological Survey Open-File Report*: 97–792.

Lang, B., Bedford, A., Brooks, S.J., Jones, R.T., Richardson, N., Birks, H.J.B. and Marshall, J.D. (2010), 'Early-Holocene temperature variability inferred from chironomid assemblages at Hawes Water, northwest England', *The Holocene*, 20: 943–954.

Langdon P.G. and Barber, K.E. (2005), 'The climate of Scotland over the last 5000 years inferred from multiproxy peatland records: inter-site correlations and regional variability', *Journal of Quaternary Science*, 20: 549–566.

Larrasoaña, J.C., Roberts, A.P., Rohling, E.J., Winklhofer, M. and Wehausen, R. (2003), 'Three million years of monsoon variability over the northern Sahara', *Climate Dynamics*, 21: 689–698.

Larsen, C.F., Motyka, R.J., Freymuller, J.T., Echelmeyer, K.A. and Ivins, E.R. (2005), 'Rapid viscoelastic uplift in southeast Alaska caused by post-Little Ice Age glacial retreat', *Earth and Planetary Science Letters*, 237: 548–560.

Latrubesse, E.M. and Nelson, B.W. (2001), 'Evidence for Late Quaternary aeolian activity in the Roraima-Guyana Region', *Catena*, 43: 63–80.

Lauritzen, S. (2005), 'Reconstructing Holocene climate records from speleothems', in A. Mackay, R.

Battarbee, J. Birks and F. Oldfield (eds) *Global Change in the Holocene* (London): 242–263.

Lauwaet, D., van Lipzig, N.P.M. and de Ridder, K. (2009), 'The effect of vegetation changes on precipitation and mesoscale convective systems in the Sahel', *Climate Dynamics*, 33: 521–534.

—— van Lipzig, N.P.M., Van Weverberg, K., De Ridder, K. and Goyens, C. (2012), 'The precipitation response to the desiccation of Lake Chad', *Quarterly Journal of the Royal Meteorological Society*, 138: 707–19.

Lavenu, A., Fournier, M. and Sebrier, M. (1984), 'Existence de deux nouveaux episodes lacustres quaternaires dans l'Altiplano peruvo-bolivien', *Cahiers ORSTOM, ser. Géologie*, 14: 103–114.

Lawler, D.M. (1987), 'Climate change over the last millenium in Central Britain', in K.J. Gregory, J. Lewin and J.B. Thornes (eds) *Palaeohydrology in Practice* (Chichester): 99–129.

Lazarenko, A.A. (1984), 'The loess of Central Asia', in Velichko A. (ed.) *Late Quaternary Environments of the Soviet Union* (Longman): 125–131.

Lea, D.W., Pak, D.K., Peterson, L.C. and Hughen, K.A. (2003), 'Synchroneity of tropical and high-latitude Atlantic temperatures over the last Glacial Termination', *Science*, 301: 1361–1364.

Leblanc, M.J., Leduc, C., Stagnitti, F., van Oevelen, P.J., Jones, C., Mofor, L.A., Razack, M. and Favreau, G. (2006), 'Evidence for Megalake Chad, north-central Africa, during the late Quaternary from satellite data', *Palaeogeography, Plaeoclimatology, Palaeoecology*, 230: 230–242.

Le Breton, E., Dauteuil, O. and Biessy, G. (2010), 'Rebound of Iceland during the Holocene', *Journal of the Geological Society*, 167: 417-32.

Ledru, M.P., Rousseau, D.D., Cruz Jr., F.W., Riccomini, C., Karmann, I. and Martin, L. (2005), 'Paleoclimate changes during the last 100 000 yr from a record in the Brazilian Atlantic rainforest region and interhemispheric comparison', *Quaternary Research*, 64: 444–450.

Leduc, G., Vidal, L., Tachikawa, K., Rostek, F., Sonzogni, C., Beaufort, L. and Bard, E. (2007), 'Moisture transport across Central America as a positive feedback on abrupt climatic changes', *Nature*, 445: 908–911.

Lee, J.R., Rose, J., Hamblin, R.J.O., Moorlock, B.S.P., Riding, J.B., Phillips, E., Barendregt, R.W. and Candy,

I. (2011), 'The glacial history of the British Isles during the early and middle Pleistocene: implications for the long-term development of the British Ice Sheet', in J. Ehlers, P.L. Gibbard and P.D. Hughes (eds) *Developments in Quaternary Science Vol 15, Quaternary Glaciations – Extent and Chronology: A Closer Look* (London): 59–74.

Lee-Thorp, J.A., Sponheimer, M., Passey, B.H., de Ruiter, D.J. and Cerling, T.E. (2010), 'Stable isotopes in fossil hominin tooth enamel suggest a fundamental dietary shift in the Pliocene', *Philosophical Transactions of the Royal Society, London*, B, 365: 3389–3396.

Lehman, S.J. and Keigwin, L.D. (1992), 'Sudden changes in North Atlantic Circulation during the last deglaciation', *Nature*, 356: 757–762.

—— Sachs, J.P., Crotwell, A.M., Keigwin, L.D. and Boyle, E.A. (2002), 'Relation of subtropical Atlantic temperature, high-latitude ice rafting, deep water formation and European climate 130 000–60 000 years ago', *Quaternary Science Reviews*, 21: 1917–1924.

Lehmkuhl, F. (1997), 'The spatial distribution of loess and loess-like sediments in the mountain areas of central and High Asia', *Zeitschrift für Geomorphologie, Supplementband*, 111: 97–116.

—— Klinge, M., Rees-Jones, J. and Rhodes, E.J. (2000), 'Late Quaternary aeolian sedimentation in central and south-east Tibet', *Quaternary International*, 68/71: 117–132.

Leick, G. (2001), *Mesopotamia: the Invention of the City* (London).

Leinen, M. and Heath, G.R. (1981), 'Sedimentary indicators of atmospheric activity in the northern hemisphere during the Cenozoic', *Palaeogeography, Paleoclimatology, Palaeoecology*, 36: 1–21.

LeMasurier, W.E. (1972), 'Volcanic record of Antarctic glacial history: implications with regard to Cenozoic Sea-Levels', in R.J. Price and D.E. Sugden (eds) *Polar geomorphology*, Institute of British Geographers Special Publication No. 4: 59–74.

Leng, M.J. (2005), 'Stable-isotopes in lakes and lake sediment archives', in A. Mackay, R. Battarbee, J. Birks and F. Oldfield (eds) *Global Change in the Holocene* (London): 124–139.

—— and Marshall, J.D. (2004), 'Palaeoclimatic interpretation of stable isotope data from lake sediment archives', *Quaternary Science Reviews*, 23: 811–831.

Lericolais G., Popescu, I., Guichard, F. and Popescu, S.M. (2007), 'A Black Sea lowstand at 8500 yr B.P. indicated by a relict coastal dune system at a depth of 90 m below sea level', *Special Paper Geological Society of America*, 426: 171–188.

Leuschner, D.C. and Sirocko, F. (2000), 'The low-latitude monsoon climate during Dansgaard-Oeschger cycles and Heinrich Events', *Quaternary Science Reviews*, 19: 243–254.

Lever, A. and McCave, I.N. (1983), 'Eolian components in Cretaceous and Tertiary North Atlantic sediments', *Journal of Sedimentary Petrology*, 53: 811–832.

Leverett, F. and Taylor, F.B. (1915), 'The Pleistocene of Indiana and Michigan and the Great Lakes', *US Geological Survey Monograph*, 53.

Lewis, C.A. (2008), 'Late Quaternary climatic changes, and associated human responses, during the last 45 000 yr in the Eastern and adjoining Western Cape, South Africa', *Earth Science Reviews*, 88: 167–187.

Lewis, S.E., Wust, R.A.J., Webster, J.M. and Shields, G.A. (2008), 'Mid-late Holocene sea-level variability in eastern Australia', *Terra Nova*, 20: 74–81.

Li, B.Y. and Zhu, L.P. (2001), '"Greatest lake period" and its palaeo-environment on the Tibetan Plateau', *Journal of Geographical Sciences (Acta Geographica Sinica)*, 11: 34–42.

Li, J., Xie, S-P., Cook, E.R., Huang, G., D'Arrigo, R., Liu, F., Ma, J. and Zheng, X-T. (2011), 'Interdecadal modulation of El Niño amplitude during the past millennium', *Nature Climate Change*, 1: 114–118.

Libby, W.F., Anderson, E.C. and Arnold, J.R. (1949), 'Age determination by radiocarbon content: worldwide assay of natural radiocarbon', *Science*, 109: 227–228.

Liess, S., Snyder, P.K. and Harding, K.J. (2012), 'The effects of boreal forest expansion on the summer Arctic frontal zone', *Climate Dynamics*, 38: 1805–1827.

Liljequist, G.H. (1943), 'The severity of the winters at Stockholm, 1757–1942', *Geografiska Annaler*, 25: 81–97.

Lilley, I. (ed.) (2006), *Archaeology of Oceania: Australia and the Pacific Islands* (Oxford).

Lin, C. and Fu, G. (1996), 'The impact of climatic warming on Hydrological Regimes in China: an overview', in J.A.A. Jones (ed.) *Regional Hydrological Response to Climate Change* (Dordrecht).

Lisiecki, L.E. and Raymo, M.E. (2005), 'A Pliocene-Pleistocene stack of 57 globally distributed benthic $\delta^{18}O$ records', Paleoceanography, 20: 1–17.

Lisitzin, E. (1964), 'Land uplift as sea-level problem', Fennia, 89: 7–10.

Liu, C. and Fu, G. (1996), 'The impact of climatic warming on hydrological regimes in China: an overview' in J.A.A. Jones, C. Liu, M.-K. Woo and H.-T. Kung (eds) Regional Hydrological Response to Climate Change (Dordrecht): 133–151.

Liu, J., Hayakawa, N., Lu, M., Dong, S. and Yuan, J. (2003), 'Hydrological and geocryological response of winter streamflow to climate warming in Northeast China', Cold Regions Science and Technology, 37: 15–24.

Liu, T. (1988), Loess in China, 2nd edition (Beijing).

—— and Ding, Z. (1998), 'Chinese loess and the paleomonsoon', Annual Review of Earth and Planetary Science, 26: 111–145.

Liu, W., Yang, H., Cao, Y., Ning, Y., Li, L., Zhou, J. and An, Z. (2005), 'Did an extensive forest ever develop on the Chinese Loess Plateau during the last 130 ka?; a test using soil carbon isotopic signatures', Applied Geochemistry, 20: 519–527.

—— Liu, Z., An, Z., Wang, X. and Chang, H. (2011), 'Wet climate during the 'Little Ice Age' in the arid Tarim Basin, northwestern China', The Holocene, 21: 409–416.

Liu, X.M., Rolph, T., An, Z. and Hesse, P. (2003), 'Paleoclimatic significance of magnetic properties on the Red Clay underlying the loess and paleosols in China', Palaeogeography, Palaeoclimatology, Palaeoecology, 199: 153–166.

Livingstone, I. and Warren, A. (1996), Aeolian Geomorphology: An Introduction (London).

Locker, S.D., Hine, A.C., Tedesco, L.P. and Shinn, E.A. (1996), 'Magnitude and timing of episodic sea-level rise during the last deglaciation', Geology, 24: 827–830.

Loffler, H. (1986), 'Ostracod analysis', in B.E. Berglund (ed.) Handbook of Holocene Palaeoecology and Palaeohydrology (Chichester): 693–702.

Lomax, J., Hilgers, A. and Radtke, U. (2011), 'Palaeoenvironmental change recorded in the palaeodunefields of the western Murray Basin, South Australia – new data from single grain OSL-dating', Quaternary Science Reviews, 30: 723–736.

Long, A.J., Murray-Wallace, C. and Morhange, C. (2002), 'Sea-level changes and neotectonics: an introduction', Journal of Quaternary Science, 17: 385–386.

Loubere, P. (2000), 'Marine control of biological production in the eastern equatorial Pacific Ocean', Nature, 406: 497–500.

Lovell, B. (2010), Challenged by Carbon (Cambridge University Press).

Lowe, J.J. (1993), 'Isolating the climatic factors in early– and mid – Holocene palaeobotanical records from Scotland', in F.M. Chambers (ed.) Climate Change and Human Impact on the Landscape (London): 67–80.

—— and Walker, M.J.C. (1997), Reconstructing Quaternary Environments, 2nd edition (Harlow).

—— Ammann, B., Birks, H.H., Björck, S., Coope, G.R., Cwynar, L., de Beaulieu, J.L., Mott, R.J., Peteet, D.M. and Walker, M.J.C. (1994), 'Climatic changes in areas adjacent to the North Atlantic during the last glacial-interglacial transition (14–9 ka bp): a contribution to IGCP-253', Journal of Quaternary Science, 9: 185–198.

—— and NASP Members (1995), 'Palaeoclimate of the North Atlantic seaboards during the last Glacial / Interglacial transition', Quaternary International, 28: 51–61.

—— Hoek, W. and Intimate Group (2001), 'Interregional correlation of palaeoenvironmental changes during the last glacial-interglacial transition: a protocol for improved precision recommended by the Intimate project group', Quaternary Science Reviews, 20: 1175–1187.

—— Rasmussen, S.O., Björck, S., Hoek, W.Z., Steffensen, J.P., Walker, M.J.C., Yu, Z.C., the INTIMATE group (2008), 'Synchronisation of palaeoenvironmental events in the North Atlantic region during the Last Termination: a revised protocol recommended by the INTIMATE group', Quaternary Science Reviews, 27: 6–17.

Lu, H. and Sun, D. (2000), 'Pathways of dust input to the Chinese Loess Plateau during the last glacial and interglacial periods', Catena, 40: 251–261.

—— Zhang, F., Liu, X. and Duce, R.A. (2004), 'Periodicities of palaeoclimatic variations recorded by loess-palaeosol sequences in China', Quaternary Science Reviews, 23: 1891–1900.

Lu, H.Y., Liu, Z.X., Wu, N.Q., Berné, S., Saito, Y., Liu, B.Z. and Wang, L. (2002), 'Rice domestication and climatic change: phytolith evidence from East China', *Boreas*, 31: 378–385.

Lu Yanchou (1981), *Pleistocene climatic cycles and variations of CaCO₃ contents in a loess profile* (in Chinese).

Lyell, C. (1834), *Principles of Geology* (London).

Mabbutt, J.A. (1971), 'The Australian arid zone as a prehistoric environment', in D.J. Mulvaney and J. Golson (eds) *Aboriginal Man and Environment in Australia* (Canberra): 66–79.

—— (1977), *Desert Landforms* (Cambridge, MA).

MacAyeal, D.A. (1993), 'Binge/purge oscillations of the Laurentide ice sheet as a cause of the North Atlantic's Heinrich events', *Paleoceanography*, 8: 775–784.

Macdonald, F.A., Schmitz, M.D., Crowley, J.L., Roots, C.F., Jones, D.S., Maloof, A.C., Strauss, J.V., Cohen, P.A., Johnston, D.T. and Schrag, D.P. (2010), 'Calibrating the Cryogenian', *Science*, 327: 1241–1243.

Macdonald, G.M. and Case, R.A. (2005), 'Variations in the Pacific decadal oscillation over the past millennium', *Geophysical Research Letters*, 32: L08703, doi:10.1029/2005GL022478.

Mackay, A.W., Battarbee, R.W., Birks, H.J.B. and Oldfield, F. (eds) (2005a), *Global Change in the Holocene.* (London).

—— Jones, V.J. and Battarbee, R.W. (2005b), 'Approaches to Holocene climate reconstruction using diatoms', in A. Mackay et al. (eds) *Global Change in the Holocene* (London): 294–309.

Mackereth, F.J.H. (1966), 'Some chemical observations on post-glacial lake sediments', *Philosophical Transactions of the Royal Society, London*, B, 250: 165–213.

Maclennan, J. and Jones, S.M. (2006), 'Regional uplift, gas hydrate dissociation and the origins of the Paleocene-Eocene Thermal Maximum', *Earth and Planetary Science Letters*, 245: 65–80.

Madole, R.F. (1995), 'Spatial and temporal patterns of late quaternary Eolian deposition, eastern Colorado, U.S.A.', *Quaternary Science Reviews*, 14: 155–177.

Magilligan, F.J. and Goldstein, P.S. (2001), 'El Niño floods and culture change: a late Holocene flood history for the Rio Moquegua, Southern Peru', *Geology*, 29: 431–434.

Magny, M. (1993), 'Solar influences on Holocene climatic cycles illustrated by correlations between past lake-level fluctuations and the atmospheric ¹⁴C record', *Quaternary Research*, 40: 1–9.

—— Bégeot, C., Guiot, J. and Peyron, O. (2003), 'Contrasting patterns of hydrological changes in Europe in response to Holocene climate cooling phases', *Quaternary Science Reviews*, 22: 1589–1596.

Mahaney, W.C. (1989), *Quaternary and Environmental Research on East African Mountains* (Rotterdam).

—— Harmson, R. and Spence, J.R. (1991), 'Glacial and interglacial cycles and development of the Afroalpine ecosystem on East African mountains. Glacial and postglacial geological record and paleoclimate of Mount Kenya', *Journal of African Earth Science*, 12: 505–12.

Maheras, P. (1988), 'Changes in precipitation conditions in the western Mediterranean over the last century', *Journal of Climatology*, 8: 179–189.

—— and Kolyva-Machera, F. (1990), 'Temporal and spatial characteristics of annual precipitation over the Balkans in the twentieth century', *International Journal of Climatology*, 10: 495–504.

Mahowald, N., Kohfeld, K., Hansson, M., Balkanski, Y., Harrison, S.P., Prentice, I.C., Schulz, M. and Rodhe, H. (1999), 'Dust sources and deposition during the last glacial maximum and current climate: A comparison of model results with paleodata from ice cores and marine sediments', *Journal of Geophysical Research*, 104(D): 15895–15916.

—— and 19 others. (2010), 'Observed 20ᵗʰ century desert dust variability: impact on climate and biogeochemistry', *Atmospheric Chemistry and Physics*, 10: 10875–10893.

Maley, J. (1982), 'Dust, clouds, rain types and climatic variations in tropical North Atlantic', *Quaternary Research*, 18: 1–16.

—— (1989), 'Late Quaternary climatic changes in the African rainforest: Forest refugia and the major role of sea surface temperature variations', in M. Leinen and M. Sarnthein (eds) *Paleoclimatology and Paleometeorology: Modern and Past Patterns of Global Atmospheric Transport* (Dordrecht): 585–616.

Malik, J.N., Khadikikar, A.S. and Merh, S.S. (1999), 'Allogenic control on late Quaternary continental sedimentation in the Mahi river basin, Western India', *Journal of the Geological Society of India*, 53: 299–314

Malmer, N. (1992), 'Peat Accumulation and the Global Carbon Cycle', *Catena Supplement*, 22: 97–110.

Mamedov, A.V. (1997), 'The Late Pleistocene-Holocene history of the Caspian Sea', *Quaternary International*, 41/42: 161–166.

Manabe, S. and Stouffer, R.J. (1994), 'Multiple-century response of a coupled ocean-atmosphere model to an increase of atmospheric carbon dioxide', *Journal of Climate*, 7: 5–23.

—— and Stouffer, R.J. (1995), 'Simulation of abrupt climate change induced by freshwater input to the North Atlantic Ocean', *Nature*, 378: 165–167.

Mandryk, C.A.S. and Rutter, N. (eds) (1996), 'The Ice-free corridor revisited', *Quaternary International*, 32.

—— Josenhans, H., Fedje, D.W. and Mathewes, R.W. (2001), 'Late Quaternary palaeoenvironments of northwestern North America: implications for inland versus coastal migration routes', *Quaternary Science Reviews*, 20: 301–314.

Mangerud, J. (1982), 'The chronostratigraphic subdivision of the Holocene in Norden; a review', *Striae* 16: 65–70.

—— Andersen, S.T., Berglund, B.E. and Donner, J.J. (1974), 'Quaternary stratigraphy of Norden, a proposal for terminology and classification', *Boreas*, 3: 109–128.

Manley, G. (1964), 'The evolution of the climatic environment', in W. Watson and J.B. Sissons (eds) *The British Isles: a Systematic Geography* (London).

—— (1966), 'Problems of the climatic optimum: The contribution of glaciology', in *Royal Meteorological Society Symposium on World Climate 8000–0 bc*: 34–39.

Mann, C.C. (2005), 'Oldest civilisation in the Americas revealed', *Science*, 307: 34–35.

Mann, D., Edwards, J., Chase, J., Beck, W., Reanier, R., Mass, M., Finney, B. and Loret, J. (2008), 'Drought, vegetation change, and human history on Rapa Nui (Isla de Pascua, Easter Island)', *Quaternary Research*, 69: 16–28.

Mann, M.E., Bradley, R.S. and Hughes, S.K. (1999), 'Northern hemisphere temperatures during the past millennium: inferences, uncertainties, and limitations', *Geophysical Research Letters*, 26: 759–762.

—— Zhang, Z., Rutherford, S., Bradley, R.S., Hughes, M.K., Shindell, D., Ammann, C., Faluvegi, G. and Fenbiao, N. (2009), 'Global signatures and dynamical origins of the Little Ice Age and Medieval Climate Anomaly', *Science*, 326: 1256–1260.

Mannion, A.M. (1998), *Natural Environmental Change* (Routledge).

Mao, R., Ho, C-H., Shao, Y., Gong, D-Y. and Kim, J. (2011), 'Influence of Arctic Oscillation on dust activity over northeast Asia', *Atmospheric Environment*, 45: 326–337.

Marcott, S.A. and 12 others. (2011), 'Ice-shelf collapse from subsurface warming as a trigger for Heinrich events', *Proceedings of the National Academy of Sciences*, 108: 13415–13419.

Marengo, J.A. (1992), 'Interannual variability of surface climate in the Amazon Basin', *International Journal of Climatology*, 12: 853–863.

—— (1995), 'Variations and change in South American streamflow', *Climatic Change*, 31: 99–117.

Marlow, J.R., Lange, C.B., Wefer, G. and Rosell-Melé, A. (2000), 'Upwelling intensification as part of the Pliocene-Pleistocene climate transition', *Science*, 290: 2288–2291.

Martin, J.H. (1990), 'Glacial-Interglacial CO_2 change: The iron hypothesis', *Paleoceanography*, 5: 1–13.

Martin, R.D., MacLarnon, A.M., Phillips, J.L., Dussebieux, L., Williams, P.R. and Dobyns, W.B. (2006), 'Comment on The Brain of LB1, *Homo floresiensis*', *Science,* 312: 999.

Martiner, N. and Morhange, C. (2007), 'Geoscience of ancient Mediterranean harbours', *Earth-Science Reviews*, 80: 137–194.

Maslin, M. (2004), *Global Warming: a Very Short Introduction* (Oxford).

Mason, J.A. (2001), 'Transport direction of Peoria Loess in Nebraska and implications for loess sources on the Central Great Plains', *Quaternary Research*, 56: 79–86.

—— and Kuzila, M.S. (2000), 'Episodic Holocene loess deposition in central Nebraska', *Quaternary International*, 67: 119–131.

—— Jacobs, P.M., Hanson, P.R., Miao, X. and Goble, R.J. (2003), 'Sources and paleoclimatic significance of Holocene Bignell loess, central Great Plains, USA', *Quaternary Research*, 60: 330–339.

Mason, S.J., Waylen, P.R., Mimmack, G.M., Rajaratnam, B. and Harrison, J.M. (1999), 'Changes in extreme rainfall events in South Africa', *Climatic Change*, 41: 249–257.

Masson, V., Bracconot, P., Jouzel, J., de Noblet, N., Cheddadi, R. and Marchal, O. (2000), 'Simulation of intense monsoons under glacial conditions', *Geophysical Research Letters*, 27, 1747–1750.

Masters, P.M. and Flemming, N.C. (eds) (1983), *Quaternary Coastlines and Marine Archaeology* (London).

Matthes, F.E. (1939), 'Report of Committee on Glaciers', *American Geophysical Union Transactions*, 20: 518–523.

Matthews, J.A. and Briffa, K.R. (2005), 'The "Little Ice Age": re-evaluation of an evolving concept', *Geografiska Annaler* 87A: 17–36.

Matthews, R. (2005), 'The rise of civilisation in southwest Asia', in C. Scarre (ed.) *The Human Past: World Prehistory and the Development of Human Societies* (London): 432–471.

Matthews, R.K. (1990), 'Sea-level change' in National Research Council, *Sea-level Change* (Washington): 88–103.

Mayewski, P.A. and White, F. (2002), *The Ice Chronicles: The Quest to Understand Global Climate Change* (University Press of New England: Hanover, NH).

McBurney, C.B.M. and Hey, R.W. (1955), *Prehistory and Pleistocene Geology in Cyrenaican Libya* (Cambridge).

McCabe, G.J. and Dettinger, M.D. (1999), 'Decadal variations in the strength of ENSO teleconnections with precipitation in the western United States', *International Journal of Climatology*, 19: 1399–1410.

McCarroll, D. (1994), 'The Schmidt hammer as a measure of degree of rock surface weathering and terrain age', in C. Beck (ed.) *Dating in Exposed and Surface Contexts* (Albuquerque): 29–45.

McCave, I.N., Manighetti, B. and Beveridge, N.A.S. (1995), 'Circulation in the glacial North Atlantic inferred from grain size measurements', *Nature*, 374: 149–152.

McCormick, P.M., Thomason, L.W. and Trepte, C.R. (1995), 'Atmospheric effects of Mt. Pinatubo eruption', *Nature*, 373: 399–404.

McCrea, W.H. (1975), 'Ice ages and the galaxy', *Nature*, 255: 607–609.

McGuffie, K. and Henderson-Sellers, A. (2005), *A Climate Modelling Primer* (3rd edition) (New York).

McHenry, H.M. and Coffing, K. (2000), '*Australopithecus* to *Homo*: transformations in body and mind', *Annual Review of Anthropology*, 29:125–46.

McManus, J.F., Francois, R., Gherardi, J.-M., Keigwin, L.D. and Brown-Leger, S. (2004), 'Collapse and rapid resumption of Atlantic meridional circulation linked to deglacial climate changes', *Nature*, 428: 834–837.

McNutt, M. and Menard, H.W. (1978), 'Lithospheric flexure and uplifted atolls', *Journal of Geophysical Research*, 83 (B3): 1206–1212.

McTainsh, G.H., Chan, Y-C., McGowan, H., Leys, J. and TEWS, K. (2005), 'The 23rd October 2002 dust storm in eastern Australia: characteristics and meteorological conditions', *Atmospheric Environment*, 39:1227-1236.

Meadows, A.J. (1975), 'A hundred years of controversy over sunspots and weather', *Nature*, 256: 95–97.

Mengel, R.M. (1970), 'The North American central plans as an isolating agent in bird speciation', in W. Dort and J.K. Jones (eds) *Pleistocene and Recent environments of the central Great Plains* (Kansas University Press): 279–340.

Menounos, B. et al. (2008), 'Western Canadian glaciers advance in concert with climate change circa 4.2 ka', *Geophysical Research Letters*, 35: L07501.2008GeoRL.. 3507501M.

Menzies, J. (2002), 'The Pleistocene legacy: glaciation', in A.R. Orme (ed.) *The Physical Geography of North America* (New York): 36–54.

Mesollela, K.J., Matthews, R.K., Broecker, W.S. and Thurber, D.L. (1969), 'The astronomical theory of climatic change: Barbados data', *Journal of Geology*, 77: 250–274.

Messerli, B., Winniger, M. and Rognon, P. (1980), 'The Saharan and East African Uplards during the Quaternary', in M.A.J. Williams and H. Faure (eds) *The Sahara and The Nile* (Rotterdam), 87–118.

Mestdagh, H., Haesaert, P., Dodonov, A. and Hus, J. (1999), 'Pedosedimentary and climatic reconstruction of the last interglacial and early glacial loess-paleosol sequence in South Tadzhikistan', *Catena*, 35: 197–218.

Meyer, M. and Harff, J. (2005), 'Modelling palaeo coastline changes of the Baltic Sea', *Journal of Coastal Research*, 21: 598–609.

Miao, X., Mason J.A., Goble, R.J. and Hanson, P.R. (2005), 'Loess record of dry climate and Aeolian activity in the early- to mid-Holocene central Great Plains, North America', *The Holocene*, 15: 339–346.

Micklin, P. (2010), 'The past, present, and future of the Aral Sea', *Lakes and Reservoirs: Research and Management*, 15: 193–213.

Midant-Reynes, B. (1992), *Préhistoire de L'Egypte* (Paris).

Miller, G.H., Hollin, J.T. and Andrews, J.T. (1979), 'Aminostratigraphy of U.K. Pleistocene deposits', *Nature*, 281: 539–543.

—— and 12 others. (2012), 'Abrupt onset of the Little Ice Age triggered by volcanism and sustained by sea-ice/ocean feedbacks', *Geophysical Research Letters*, 39: L02708, doi: 10.1029/2011GL050168.

Miller, L. and Douglas, B.C. (2004), 'Mass and volume contribution to twentieth-century global sea level rise', *Nature*, 428: 406–409.

Mills, S.C., Grab, S.W. and Carr, S.J. (2009), 'Recognition and palaeoclimatic implications of late Quaternary niche glaciation in eastern Lesotho', *Journal of Quaternary Science*, 24: 647–663.

Milly, P.C.D., Wetherald, R.T., Dunne, K.A. and Delworth, T.L. (2002), 'Increasing risk of great floods in a changing climate', *Nature*, 415: 514–517.

Min, S-K., Zhang, X., Zwiers, F.W. and Hegerl, G.C. (2011), 'Human contribution to more-intense precipitation extremes', *Nature*, 470, 378–381.

Miri, A., Moghaddamnia, A., Pahlavanravi, A. and Panjehkek, N. (2010), 'Dust storm frequency after the 1999 drought in the Sistan region, Iran', *Climate Research*, 41: 83–90.

Mitchell, G.F. (1972), 'Soil deterioration associated with prehistoric agriculture in Ireland', *20th International Geological Congress Symposium*, 1: 59–68.

—— and Orme, A.R. (1967), 'The Pleistocene deposits of the Isles of Scilly', *Quarterly Journal, Geological Society of London*, 123: 59–92.

Mitchell, J.M. (1963), 'On the world-wide pattern of secular temperature change', *Arid Zone Research*, 20: 161–181.

—— (1965), 'Theoretical palaeoclimatology', in H.E. Wright and D.G. Frey (eds) *The Quaternary of the U.S.A.*: 881–901.

—— (ed.) (1968), *Causes of Climatic Change*, Meteorological Monographs, 8.

Mitchell, P. (2002), *The Archaeology of Southern Africa* (Cambridge).

Mitchell, R.M., Campbell, S.K. and Qin, Y. (2010), 'Recent increase in aerosol loading over the Australian arid zone', *Atmospheric Chemistry and Physics*, 10: 1689–1699.

Mithen, S. (1994), 'The Mesolithic Age', in B. Cunliffe (ed.) *The Oxford Illustrated History of Prehistoric Europe* (Oxford): 79–135.

Mix, A.C., Bard, E. and Schneider, R. (2001), 'Environmental processes of the ice age: land, oceans, glaciers (EPILOG)', *Quaternary Science Reviews*, 20: 627–657.

Mölg, T., Georges, C. and Kaser, G. (2003), 'The contribution of increased incoming shortwave radiation to the retreat of the Rwenzori Glaciers, East Africa, during the 20th Century', *International Journal of Climatology*, 23: 291–303.

Molnar, P. and England, P. (1990), 'Late Cenozoic uplift of mountain ranges and global climate change: chicken or egg?', *Nature*, 346: 29–34.

Montford, H.M. (1970), 'The terrestrial environment during Upper Cretaceous and Tertiary Times', *Proceedings of the Geologists' Association*, 81: 181–204.

Moore, P.D. (1975), 'Origin of blanket mires', *Nature*, 256: 267–269.

—— Webb J.A. and Collinson, M.E. (1991), *Pollen Analysis* (Oxford).

—— Chaloner, B. and Stott, P. (1996), *Global Environmental Change* (Oxford).

Moore, T.C. (2005), 'The Younger Dryas: From whence the fresh water?', *Paleoceanography*, 20: PA4021 pp.1–8.

Moreau, R.E. (1963), 'Vicissitudes of the African biomes in the Late Pleistocene', *Proceedings, Zoological Society of London*, 141: 395–421.

Moreno, A. and Canals, M. (2004), 'The role of dust in abrupt climate change: insights from offshore Northwest Africa and Alboran Sea sediment records', *Contributions to Science (Barcelona)*, 2: 485–498.

—— Cacho, I., Canals, M., Prins, M.A., Sanchéz-Goñi, M.-F., Grimalt, J.O. and Weltje, G.J. (2002), 'Saharan dust transport and high-latitude glacial climatic variability: the Alboran Sea record', *Quaternary Research*, 58: 318–328.

—— Taragona, J., Henderiks, J., Canals, M., Freudenthal, T. and Meggers, H. (2001), 'Orbital forcing of dust supply to the North Canary Basin over the last 250 kyr', *Quaternary Science Reviews*, 20: 1327–1339.

Moreno, P.I., Kaplan, M.R., François, J.P., Villa-Martinez, R., Moy, C.M., Stern, C.R. and Kubik, P.W. (2009), 'Renewed glacial activity during the Antarctic cold reversal and persistence of cold conditions until 11.5 ka in southwestern Patagonia', *Geology*, 37: 375–378.

Morhange C., Bourcier M., Laborel, J., Giallanella, C., Goiran, J-P., Crimaco, L. and Vecchi, L. (1999), 'New data on historical sea level movements in Pozzuoli, Phlaegrean Fields, southern Italy', *Physics and Chemistry of the Earth*, A, 24: 349–354.

Mori, F., Ponti, R., Messina, A., Flieger, M., Havlicek, V. and Sinibaldi, M. (2006), 'Chemical characterization and AMS radiocarbon dating of the binder of a prehistoric rock pictograph at Tadrart Acacus, southern west Libya', *Journal of Cultural Heritage*, 7: 344–349.

Mörner, N.A. (1971), 'Eustatic and climatic oscillations', *Arctic and Alpine Research*, 3: 167–171.

—— (ed.) (1980), *Earth Rheology, Isostasy and Eustasy* (New York).

Morwood, M.J., Brown, P., Jatmiko Sutikna, T., Wahyu Saptomo, E., Westaway, K.E., Rokus Awe Due, Roberts, R.G., Maeda, T., Wasisto, S. and Djubiantono, T. (2005), 'Further evidence for small-bodied hominins from the Late Pleistocene of Flores, Indonesia', *Nature* 437: 1012–1017.

—— Soejono, R.P., Roberts, R.G., Sutikna, T., Turney, C.S.M., Westaway, K.E., Rink, W.J., Zhao, J.X., Van-Den Bergh, G.D., Due, R.A., Hobbs, D.R., Moore, M.W., Bird, M.I. and Fifield, L.K. (2004), 'Archaeology and age of a new hominin from Flores in eastern Indonesia', *Nature*, 431: 1087–1091.

Moss, J.H. (1951), 'Late glacial advances in the Southern Wind River Mountains, Wyoming', *American Journal of Science*, 249: 865–883.

Motyka, R.J., O'Neal, S., Connor, C.L. and Echelmeyer, K.A. (2002), 'Twentieth century thinning of Mendenhall Glacier, Alaska, and its relationship to climate, lake calving and glacier run-off', *Global and Planetary Change*, 35: 93–112.

Moulin, C., Guillard, F., Dulac, F. and Lambert, C.E. (1997), 'Long-term daily monitoring of Saharan dust load over ocean using Meteosat ISCCP-B2 data: 1: methodology and preliminary results for 1983-1994 in the Mediterranean', *Journal of Geophysical Research*, 102:16947-16958.

Moy, C.M., Seltzer, G.O., Rodbell, D.T. and Anderson, D.M. (2002), 'Variability of El Niño/Southern Oscillation activity at millennial timescales during the Holocene epoch', *Nature*, 420: 162–165.

Mrani-Alaoui, M. and Anthony, E.J. (2011), 'New data and a morphodynamic perspective on mid- to late-Holocene palaeoenvironmental changes in the French Flanders coastal plain, southern North Sea', *The Holocene*, 21: 445–453.

Muhs, D.R. and Bettis, E.A. (2000), 'Geochemical variations in Peoria loess of western Iowa indicate paleowinds of midcontinental North America during last glaciation', *Quaternary Research*, 53: 49–61.

—— Swinehart, J.B., Loope, D.B., Aleinikoff, J.N. and Been, J. (1999), '200 000 years of climate change recorded in eolian sediments of the High Plains of eastern Colorado and western Nebraska', *Geological Society of America Field Guide* 1: 71–91.

—— McGeehin, J.P., Beann, J. and Fisher, E. (2004), 'Holocene loess deposition and soil formation as competing processes, Matanuska Valley, southern Alaska', *Quaternary Research*, 61: 265–276.

—— Stafford, T.W., Jr., Swinehart, J.B., Cowherd, S.D., Mahan, S.A., Bush, C.A., Madole, R.F. and Maat, P.B. (1997), 'Late Holocene eolian activity in the mineralogically mature Nebraska Sand Hills', *Quaternary Research,* 48: 162–176.

Muller, R.A. and MacDonald, G.J. (1997), 'Glacial cycles and astronomical forcing', *Science*, 277: 215–218.

Munson, S., Belnap, J. and Okin, G.S. (2011), 'Responses of wind erosion to climate-induced vegetation changes on the Colorado Plateau', *Proceedings of the National Academy of Sciences*, 108: 3854–3859.

Murton, D.K. and Murton, J.B. (2012), 'Middle and Late Pleistocene glacial lakes of lowland Britain and the southern North Sea Basin', *Quaternary International*, 260: 115–142.

Murton, J.B., Bateman, M.D., Dallimore, S.R., Teller, J.T. and Yang, Z. (2010), 'Identification of Younger Dryas outburst flood path from Lake Agassiz to the Arctic Ocean', *Nature*, 464: 740–743.

Mylroie J.E., Carew, J.L. and Moore, A.I. (1995), 'Blue holes: definition and genesis', *Carbonates and Evaporites*, 10: 225–233.

Naeser, C.W. and Naeser, N.D. (1988), 'Fission-track dating of Quaternary events', in D.J. Easterbrook (ed.) *Dating Quaternary sediments, Geological Society of America Special Paper*, No. 227.

Nanson, G.C., Chen, X.Y. and Price, D.M. (1995), 'Aeolian and fluvial evidence of changing climate and wind patterns during the past 100 ka in the western Simpson Desert, Australia', *Palaeogeography, Palaeoclimatology, Palaeoecology*, 113: 87–102.

Narcisi, B., Petit, J.R. and Chappellaz, J. (2010), 'A 70 ka record of explosive eruptions from the TALDICE ice core (Talos Dome, East Antarctic Plateau)', *Journal of Quaternary Science*, 25: 844–849.

Nash, D.J. and Endfield, G.H. (2002), 'A 19th century climate chronology for the Kalahari region of central southern Africa derived from missionary correspondence.' *International Journal of Climatology*, 22: 821–841.

National Institute of Sciences (1952), *Proceedings of the Symposium on the Rajputana Desert* (Delhi).

Natsagdorj, D.L., Jugder, D. and Chung, Y.S. (2003), 'Analysis of dust storms observed in Mongolia during 1937–1999', *Atmospheric Environment*, 37: 1401–1411.

Neff, H., Pearsall, D.M., Jones, J.G., Arroyo de Pieters, B. and Freidel, D.E. (2006), 'Climate change and population history in the Pacific lowlands of Southern Mesoamerica', *Quaternary Research*, 65: 390–400.

Neff, U., Burns, S.J., Mangini, A., Mudelsee, M., Fleitmann, D. and Matter, A. (2001), 'Strong coherence between solar variability and the monsoon in Oman between 9 and 6 kyr ago', *Nature*, 411: 290–293.

Nelson, D.A. (1954), 'The carbon dioxide exchange between the North Atlantic Ocean and the atmosphere', *Tellus*, 6: 342–350.

Nesje, A. and Dahl, S.O. (2000), *Glaciers and Environmental Change* (London).

—— and Johannessen, T. (1992), 'What were the primary forcing mechanisms of high-frequency Holocene climate and glacier variations?', *The Holocene*, 2: 79–84.

—— Lie, O. and Dahl, S.O. (2000), 'Is the North Atlantic Oscillation reflected in glacier mass balance records?', *Journal of Quaternary Science*, 15: 587–601.

—— Bakke, J., Dahl, S.O., Lie, O. and Matthews, J.A. (2008), 'Norwegian mountain glaciers in the past, present and future', *Global and Planetary Change*, 60: 10–27.

Nesme-Ribes, E. (ed.) (1994), *The Solar Engine and its Influence on Terrestial Atmosphere and Climate* (Berlin).

New, M., Todd, M., Hulme, M. and Jones, P. (2001), 'Precipitation measurements and trends in the twentieth century', *International Journal of Climatology*, 21: 1899–1922.

—— Liverman, D., Schroeder, H. and Anderson, K. (2011), 'Four degrees and beyond: the potential for a global temperature increase of four degrees and its implications', *Philosophical Transactions of the Royal Society, London*, A, 369: 6–19.

Newell, N.D. and Bloom, A.L. (1970), 'The reef flat and "two-meter eustatic terrace" of some Pacific atolls', *Bulletin, Geological Society of America*, 81: 1881–1893.

Newman, W.S., Fairbridge, R.W. and March, S. (1971), 'Marginal subsidence of glaciated areas: United States, Baltic and North Seas', *Quaternaria*, 14: 39–40.

Newnham, R.M., Vandergoes, M.J., Sikes, E., Carter, L., Wilmshurst, J.M., Lowe, D.J., McGlone, M.S. and Sandiford, A. (2012), 'Does the bipolar seesaw extend to the terrestrial southern mid-latitudes?' *Quaternary Science Reviews*, 36: 214–222.

New Zealand Geological Survey (1973), *Quaternary Geology: South Island 1:1 000 000*, New Zealand Geological Series Miscellaneous Publications (Lower Hut).

Nichol, J.E. (1991), 'The extent of desert dunes in northern Nigeria as shown by image enhancement', *Geographical Journal*, 157: 13–24.

Nichol, S.L. and Brooke, B.P. (2011), 'Shelf habitat distribution as a legacy of Late Quaternary marine transgressions: a case study from a tropical carbonate province', *Continental Shelf Research*, 31: 1845–1857.

Nicholls, N., Gruza, G.,V., Jouzel, J., Karl, L.A. and Parker, D.E. (1996), 'Observed climatic variability and change', in J.T.Houghton, L.G. Meira Filho, B.A. Callendar, N. Harris, A. Kattenberg and K. Maskell (eds) *Climate Change 1995: The Science of Climate Change* (Cambridge): 133–192.

—— and Lavery, B. (1992), 'Australian rainfall trends during the twentieth century', *International Journal of Climatology*, 12: 153–163.

Nicholls, R.J., Marinova, N., Lowe, J.A., Brown, S., Vellinga, P., de Gusmão, D., Hinkel, J. and Tol, R.S.J. (2011), 'Sea-level rise and its possible impacts given a 'beyond 4°C world' in the twenty-first century', *Philosophical Transactions of the Royal Society, London*, A, 369: 161–181.

Nicholson, S. (2011), *Dryland Climatology* (Cambridge).

Nickl E., Willmott, C.J., Matsuura K. and Robeson, S.M. (2010), 'Changes in annual land-surface precipitation over the twentieth and early twenty-first century', *Annals of the Association of American Geographers*, 100: 729–739.

Nigam, S., Guan, B. and Ruiz-Barradas, A. (2011), 'Key role of the Atlantic Multidecadal Oscillation in 20th century drought and wet periods over the Great Plains', *Geophysical Research Letters*, 38: L16713, doi:10.1029/2011GL048650.

Nilson, E. and Lehmkuhl, F. (2001), 'Interpreting temporal patterns in the late Quaternary dust flux from Asia to the North Pacific', *Quaternary International*, 76/77: 67–76.

Nilsson, E. (1940), 'Ancient changes in climate in British East Africa and Abyssinia', *Geografiska Annaler*, 1–2: 1–79.

Nisbet, E.G. (1990), 'The end of the ice age', *Canadian Journal of Earth Sciences*, 27: 148–157.

—— (2002), 'Have sudden large releases of methane from geological reservoirs occurred since the Last Glacial Maximum and could such releases occur again?' *Philosophical Transactions of the Royal Society, London*, A, 360: 581–607.

Nobre, C.A., Dias, M.A.S., Culf, A.D., Polcher, J., Gash, J.H.C., Marengo, J.A. and Avissar, R. (2004), 'The Amazonian climate', in P. Kabat, M. Claussen, P.A.Dirmeyer, J.H.C.Gash, L.B. de Guenni, M. Meybeck, R.A.Pielke, C.S. Vörösmarty, R.W.A. Hutjes and S. Lütkemeier (eds) *Vegetation, Water, Humans and the Climate (Berlin)*: 79–92.

North Greenland Ice Core Project Members (NorthGrip) (2004), 'High-resolution record of Northern Hemisphere climate extending into the last interglacial period', *Nature*, 431: 147–151.

Nott, J., Bryant, E. and Price, D. (1999), 'Early-Holocene aridity in tropical northern Australia', *The Holocene*, 9: 231–236.

N'tchayi, G.M., Bertrand, J., Legrand, M. and Baudet, J. (1994),Temporal and spatial variations of the atmospheric dust loading throughout West Africa over the last thirty years, *Annales Geophysicae*, 12:265-273.

Nunn, P.D. (1986), 'Implications of migrating geoid anomalies for the interpretation of high-level fossil coral reefs', *Bulletin, Geological Society of America*, 97: 946–952.

Nussbaumer, S.U. and 10 others. (2011), 'Alpine climate during the Holocene: a comparison between records of glaciers, lake sediments and solar activity', *Journal of Quaternary Science*, 26: 703–713.

Nyamweru, C. (1989), 'New evidence for the former extent of the Nile drainage system', *Geographical Journal*, 155: 179–188.

O'Brien, S.R., Mayewski, P.A., Meeker, L.D., Meese, D.A., Twickler, M.S. and Whitlow, S.I. (1995), 'Complexity of Holocene climate as reconstructed from a Greenland ice core', *Science*, 270: 1962–1964.

O'Connor, S. and Fankhauser, B. (2001), 'Art at 40 000 bp? One step closer, an ochre covered rock from Carpenter's Gap Shelter 1, Kimberley Region, W.A.', in A. Anderson, I. Lilley and S. O'Conner (eds) *Histories of Old Ages: Essays of Rhys Jones* (Canberra): 287–300.

—— and Veth, P. (2006), 'Revisiting the past: changing interpretations of Pleistocene settlement, subsistence and demography in Northern Australia', in I. Lilley (ed.) *Archaeology of Oceania* (Oxford): 31–47.

O'Connor, T.G. and Kiker, G.A. (2004), 'Collapse of the Mapungubwe society: vulnerability of pastoralism to increasing aridity', *Climatic Change*, 18: 49–66.

Oerlemans, J. (1994), 'Quantifying global warming from the retreat of glaciers', *Science*, 264: 243–250.

—— (2005), 'Extracting a climate signal from 169 glaciers', *Science*, 308: 675–677.

O'Hara, S.L., Street-Perrott, F.A. and Burt, T.P. (1993), 'Accelerated soil erosion around a Mexican highland lake caused by prehispanic agriculture', *Nature*, 362: 48–51.

O'Hare, G. (2011), 'Updating our understanding of climate change in the North Atlantic: the role of global warming and the Gulf Stream', *Geography*, 96: 5–15.

Ohmura, A., Kasser, P. and Funk, M. (1992), 'Climate at the equilibrium line of glaciers', *Journal of Glaciology*, 38: 397–411.

Olago, D.O. (2001), 'Vegetation changes over palaeo-time scales in Africa', *Climate Research*, 17: 105–121.

Olausson, E. and Olsson, I.V. (1969), 'Varve stratigraphy in a core from the Gulf of Aden', *Palaeogeography, Palaeoclimatology, Palaeoecology*, 6: 87–103.

Oldfield, F. (2005), 'Introduction: the Holocene, a special time', in A. Mackay, R. Battarbee, J. Birks and F. Oldfield (eds) *Global Change in the Holocene* (London): 1–9.

Onac, B.P., Viehmann, I., Lundberg, J., Lauritzen, S.E., Stringer, C. and Popita, V. (2005), 'U-Th ages constraining the Neanderthal footprint at Vartop Cave, Romania', *Quaternary Science Reviews*, 24: 1151–1157.

Oram, R.D. and Adderley, W.P. (2008), 'Lordship and Environmental Change in Central Highland Scotland, c.1300-c.1400', *Journal of the North Atlantic*, 1: 74–84.

Orme, A.R. (2002), 'The Pleistocene legacy: beyond the ice-front', in Orme, A.R. (ed.). *The Physical Geography of North America* (Oxford): 55–85.

Ortloff, C.R. and Kolata, A.L. (1993), 'Climate and collapse: agro-ecological perspectives on the decline of the Tiwanaku state', *Journal of Archaeological Science*, 20: 195–221.

Osborne, A.H., Vance, D., Rohling, E.J., Barton, N., Rogerson, M. and Fello, N. (2008), 'A humid corridor across the Sahara for the migration of early modern humans out of Africa 120,000 years ago', *Proceedings of the National Academy of Sciences*, 105: 16444–16447.

Osborne, P.J. (1974), 'An insect of early Flandrian Age from Lea Marston, Warwickshire, and its bearing on the contemporary climate and ecology', *Quaternary Research*, 4: 471–486.

Osborn, T.J. (2011), 'Winter 2009/2010 temperatures and a record-breaking North Atlantic Oscillation index', *Weather*, doi:10.1002/wea.660.

—— Hulme, M., Jones, P.D. and Basnett, T.A. (2000), 'Observed trends in the daily intensity of United Kingdom precipitation', *International Journal of Climatology*, 20: 347–364.

Oswald, W.W., Foster, D.R., Doughty, E.D. and Faison, E.K. (2009), 'A record of Lateglacial and early Holocene environmental and ecological change from southwestern Connecticut, USA', *Journal of Quaternary Science*, 24: 553–556.

Ottersen, G., Planque, B., Belgrano, A., Post, E., Reid, P.C. and Stenseth, N.C. (2001), 'Ecological effects of the North Atlantic Oscillation', *Oecologia*, 128: 1–14.

Overpeck, J., Rind, D., Lacis, A. and Healy, R. (1996), 'Possible role of dust-induced regional warming in abrupt climate change during the last glacial period', *Nature*, 384: 447–449.

Oxnard, C., Obendorf, P.J. and Kefford, B.J. (2010), 'Post–Cranial Skeletons of Hypothyroid Cretins Show a Similar Anatomical Mosaic as *Homo floresiensis*', PLoS ONE 5(9): e13018. doi:10.1371/journal.pone.0013018.

Pachur, H.J. and Kröpelin, S. (1987), 'Wadi Howar: palaeoclimatic evidence from an extinct river system in the southeastern Sahara', *Science*, 237: 298–300.

Paduan J.B., Clague, D.A. and Davis, A.S. (2009), 'Evidence that three seamounts off Southern California were ancient islands', *Marine Geology*, 265: 146–156.

Pagani, M., Freeman, K.H. and Arthur, M.A. (1999), 'Late Miocene atmospheric CO_2 concentrations and the expansion of C4 grasses', *Science*, 285: 876–879.

—— Zachos, J.C., Freeman, K.H., Tipple, B. and Bohaty, S. (2005), 'Marked decline in atmospheric carbon dioxide concentrations during the Paleogene', *Science*, 309: 600–603.

Palmer, A.S. and Pillans, B.J. (1996), 'Record of climatic fluctuations from ca. 500 ka loess deposits and paleosols near Wanganui, New Zealand', *Quaternary International*, 34–36: 155–162.

Pant, G.B. and Hingane, L.S. (1988), 'Climatic change in and around the Rajasthan Desert during the twentieth century', *Journal of Climatology*, 8: 391–401.

Pant, R.K., Basavaiah, N., Juyal, N., Saini, N.K., Yadava, M.G., Appel, E. and Singhvi, A.K. (2005), 'A 20-ka climate record from Central Himalayn loess deposits', *Journal of Quaternary Science*, 20: 485–492.

Parker, A.G. (2000), 'Biotic response to Late Quaternary global change – the pollen record: a case study from the Upper Thames Valley, England', in S.J. Culver and P. Rawson (eds) *Biotic Response to Global Change: the Last 145 Million Years* (Cambridge): 265–287.

—— and Goudie, A.S. (2008), 'Geomorphological and palaeoenvironmental investigations in the southeastern Arabian Gulf region and the implication for the archaeology of the region', *Geomorphology*, 101, 458–470.

—— Goudie, A.S., Anderson, D.E., Robinson, M.A. and Bonsall, C. (2002), 'A review of the mid-Holocene elm decline in the British Isles', *Progress in Physical Geography*, 26: 1–45.

—— Eckersley, L., Smith, M.M., Goudie, A.S., Stokes, S., White, K. and Hodson, M.J. (2004), 'Holocene vegetation dynamics in the northeastern Rub' al-Khali desert, Arabian Peninsula: a pollen, phytolith and carbon isotope study', *Journal of Quaternary Science*, 19: 665–676.

—— Wilkinson, T.J. and Davies, C. (2006), 'The early-mid Holocene period in Arabia: some recent evidence from lacustrine sequences in eastern and southwestern Arabia', *Proceedings of the Seminar for Arabian Studies*, 36: 243–255.

Parker, D.E. and Folland, C.K. (1988), 'The nature of climatic variability', *Meteorological Magazine*, 117: 201–210.

Parkin, D.W. (1974), 'Trade-winds during the Glacial Cycles', *Proceedings of the Royal Society of London*, A, 337: 73–100.

Parmenter, C. and Folger, D.W. (1974), 'Eolian biogenic detritus in deep-sea sediments: a possible index of equatorial ice-age aridity', *Science*, 185: 695–698.

Parrish, J.T. (1987), 'Global palaeogeography and palaeoclimate of the Late Creaceous and Early Tertiary', in E.M. Friis, W.G. Chaloner and P.R. Crane (eds) *The Origins of Angiosperms and their Biological Consequences* (Cambridge).

Parry, M.L. (1975), 'Secular climatic change and marginal agriculture', *Transactions, Institute of British Geographers*, 64: 1–13.

—— (1978), *Climate Change, Agriculture and Settlement* (Folkestone).

Parungo, F., Li, Z., Li, X., Yang, D. and Harris, J. (1994), 'Gobi dust storms and The Great Green Wall', *Geophysical Research Letters*, 21: 999–1002.

Patrickson, S.J., Sack, D., Brunelle, A.R. and Moser, K.A. (2010), 'Late Pleistocene to early Holocene lake level and paleoclimate insights from Stansbury Island, Bonneville basin, Utah', *Quaternary Research*, 73: 237–246.

Pearce, R.H. and Barbetti, M. (1981), 'A 38 000 year-old site at Upper Swan, W.A.', *Archaeology in Oceania*, 16: 173–178.

Pearsall, W.H. (1964), 'After the ice retreated', *New Scientist*, 383: 757–759.

Peck, J.A., Green, R.R., Shanahan, T., King, J.W., Overpeck, J.T. and Scholz, C.A. (2004), 'A magnetic mineral record of Late Quaternary tropical climate variability from Lake Bosumtwi, Ghana', *Palaeogeography, Palaeoclimatology, Palaeoecology*, 215: 37–57.

Pedoja, K. and 11 others. (2011), 'Relative sea-level fall since the last interglacial stage: are coasts uplifting worldwide?', *Earth-Science Reviews*, 108: 1–15.

Peeters, F.J.C., Acheson, R., Brummer, G.-J.A., de Ruijter, W.P.M., Schneider, R.R., Ganssen, G.M., Ufkes, E. and Kroon, D. (2004), 'Vigorous exchange between the Indian and Atlantic oceans at the end of the past five glacial periods', *Nature*, 430: 661–665.

Peltier, W.R. (2005), 'On the hemispheric origins of meltwater pulse 1a', *Quaternary Science Reviews*, 24: 1655–1671.

—— and Fairbanks, R.G. (2006), 'Global glacial ice volume and Last Glacial Maximum duration from an extended Barbados sea level record', *Quaternary Science Reviews*, 25: 3322–3337.

Penck, A. and Brückner, E. (1909), *Die Alpen in Eiszeitalten* (Leipzig).

Pennington, W. (1947), 'Lake sediments: pollen diagrams from the bottom deposits of the north basin of Windermere', *Philosophical Transactions of the Royal Society, London*, B, 233: 137–175.

—— (1969), *The History of British Vegetation* (London).

—— Haworth, E.Y., Bonny, A.P. and Lishman, J.P. (1972), 'Lake sediments in Northern Scotland', *Philosophical Transactions of the Royal Society, London*, B, 264: 191–294.

Perissoratis, C. and Conispoliatis, N. (2003), 'The impacts of sea-level changes during latest Pleistocene and Holocene times on the morphology of Ionian and Aegean seas (SE Alpine Europe)', *Marine Geology*, 196: 145–146.

Perkins, S. (2002), 'Once upon a lake: The life, times, and demise of the world's largest lake', *Science News*, 162: 283–284.

Perry, A. (2000), 'North Atlantic Oscillation. An enigmatic see-saw', *Progress in Physical Geography*, 24: 289–294.

Peterson, B.J., Homes, R.M., McClelland, J.W., Vörösmarty, C.J., Lammers, R.B., Shiklomanov, A. I., Shiklomanov, I. A. and Rahmstorf, S. (2002), 'Increasing river discharge to the Arctic Ocean', *Science*, 298, 2171–1273.

Petit, J.R., Jouzel, J., Raynaud, D., Barkov, N.I., Barnola, J.-M., Basile, I., Bender, M., Chappellaz, J., Davis, M., Delaygue, G., Delmotte, M., Kotlyakov, V.M., Legrand, M., Lipenkov, V.Y., Lorius, C., Pépin, L., Ritz, C., Saltzman, E. and Stievenard, M. (1999), 'Climate and atmospheric history of the past 420 000 years from the Vostok ice core, Antarctica', *Nature*, 399: 429–436.

Petit-Maire, N. (1989), 'Interglacial environments in presently hyperarid Sahara: Palaeoclimatic implications', in M. Leinen and M. Sarnthein (eds) *Palaeoclimatology and Palaeometeorology: Modern and Past Patterns of Global Atmospheric Transport* (Dordrecht): 637–661.

—— Beaufort, L. and Page, N. (1997), 'Holocene climatic change and man in the present-day Sahara desert', in N. Dalfes, G. Kukla and H. Weiss (eds) *Third Millennium BC Climate Change and Old World Collapse* (Heidelberg): 297–308.

—— Burollet, P.F., Ballais, J.-L., Fontugne, M., Rosso, J.-C. and Lazaar, A. (1999), 'Paléoclimats Holocènes du Sahara septentionale. Dépôts lacustres et terrasses alluviales en bordure du Grand Erg Oriental à l'extrême–Sud de la Tunisie', *Comptes Rendus Académie des Sciences*, Series 2, 312: 1661–1666.

Petraglia, M.D. (2003), 'The Lower Paleolithic of the Arabian Peninsula: Occupations, adaptations, and dispersals', *Journal of World Prehistory*, 17: 141–180.

Pettke, T., Halliday, A.N., Hall, C.M. and Rea, D.K. (2000), 'Dust production and deposition in Asia and the North Pacific Ocean over the past 12 Myr', *Earth and Planetary Science Letters*, 178: 397–413.

Pezza, A.B. and Simmonds, I. (2005), 'The first South Atlantic hurricane: unprecedented blocking, low shear and climate change', *Geophysical Research Letters*, 32: L15712, doi: 10.1029/2005 GL023390.

Philander, S.G. (1990), *El Niño, La Niña and the Southern Oscillation* (London).

Pichevin, L., Cremer, M., Giraudeau, J. and Bertand, P. (2005), 'A 190 ky record of lithogenic grain-size on the Namibian slope: forging a tight link between past wind-strength and coastal upwelling dynamics', *Marine Geology*, 218: 81–96.

Pirazzoli, P.A. (1989), 'Present and near-future global sea-level changes', *Palaeogeography, Palaeoclimatology, Palaeoecology*, 75: 241–258.

—— (1991), *World Atlas of Holocene Sea-level Changes* (Amsterdam).

—— (1994), 'Tectonic shorelines', in R.W.G. Carter and C.D. Woodroffe (eds) *Coastal Evolution* (Cambridge): 451–476.

—— (1996), *Sea Level Changes: The Last 20,000 Years* (Chichester).

—— (2001), *World Atlas of Holocene Sea-level Changes* (Amsterdam).

—— (2005), 'A review of possible eustatic, isostatic and tectonic contribution in eight late-Holocene sea-level histories from the Mediterranean area', *Quaternary Science Reviews*, 24: 1989–2001.

Pisias, N.G. and Moore, T.C. (1981), 'The evolution of Pleistocene climate: A time series approach', *Earth and Planetary Science Letters*, 52: 450–458.

Placzek, C., Quade, J. and Betancourt, J.L. (2001), 'Holocene lake-level fluctuations of Lake Aricota, southern Peru', *Quaternary Research*, 56: 181–190.

Plag, H.-P. and 11 others. (1996), 'Late Quaternary relative sea–level changes and the role of glaciation upon continental shelves', *Terra Nova*, 8: 213-222.

Pokras, E.M. (1989), 'Pliocene history of South Saharan/Sahelian aridity: record of freshwater diatoms (Genus Melosira) and opal phytoliths, ODP sites 662 and 664', in M. Leinen and M. Sarnthein (eds) *Palaeoclimatology and Palaeometeorology: Modern and Past Patterns of Global Atmospheric Transport.* (Dordrecht): 795–804.

Porter, S.C. (1981), 'Recent glacier variations and volcanic eruptions', *Nature*, 291: 139–141.

—— (1986), 'Pattern and forcing of northern hemisphere glacier variations during the last millennium', *Quaternary Research*, 26: 27–48.

—— (2001), 'Chinese loess record of monsoon climate during the last glacial-interglacial cycle', *Earth-Science Reviews*, 54: 115–128.

—— and Denton, G.H. (1967), 'Chronology of neoglaciation in the North American Cordillera', *American Journal of Science*, 265: 177–210.

—— and An, Z.S. (1995), 'Correlation between climate events in the North Atlantic and China during the last glaciation', *Nature*, 375: 305–308.

Possehl, G.L. (2002), *The Indus Civilization: A Contemporary Perspective* (Oxford).

Potts, D.T. (1997), *Mesopotamian Civilisation: the Material Foundations* (Ithaca, NY).

Potts, R. (1998), 'Environmental hypotheses of hominin evolution', *Yearbook of Physical Anthropology* 41, 93–136.

Pourmand, A., Marcantonio, F. and Schulz, H. (2004), 'Variations in productivity and eolian fluxes in the northeastern Arabian Sea during the past 110 ka', *Earth and Planetary Science Letters*, 221: 39–54.

Praeger, R.Ll. (1892), 'Report on the estuarine clays of the north-east of Ireland', *Proceedings of the Royal Irish Academy*, 2: 212.

Prahl, F.G. and Wakeham, S.G. (1987), 'Calibration of unsaturation patterns in long-chain ketone compositions for palaeotemperature assessment', *Nature*, 330: 367–369.

Prance, G.T. (1973), 'Phytogeographic support for the theory of Pleistocene forest refuges in the Amazon basin, based on evidence from distribution patterns in Caryocaraceae, Chrysobalanaceae,

Dichapetalaceae and Lecythidaceae', *Acta Amazonica*, 3: 5–28.

Price, W.A. (1958), 'Sedimentology and Quaternary geomorphology of South Texas', *Transactions, Gulf Coast Association of Geological Societies*, 8: 41–75.

Price-Williams, D., Watson, A. and Goudie, A.S. (1982), 'Quaternary colluvial stratigraphy, archaeological sequences and palaeoenvironments in Swaziland, southern Africa', *Geographical Journal*, 148: 50–67.

Pritchard, H.D. and Vaughan, D.G. (2007), 'Widespread acceleration of tidewater glaciers on the Antarctic Peninsula', *Journal of Geophysical Research*, 112: F03S29, doi:10.1029/2006JF000597.

Probst, J.L. (1989), 'Hydroclimatic fluctuations of some European rivers since 1800', in G.E. Petts (ed.) *Historical change of large alluvial rivers: Western Europe* (Chichester): 41–55.

—— and Tardy, Y. (1987), 'Long range streamflow and world continental runoff fluctuations since the beginning of this century', *Journal of Hydrology*, 94: 289–311.

Proctor, C.J., Baker, A., Barnes, W.L. and Gilmour, M.A. (2000), 'A thousand year speleothem proxy record of North Atlantic climate from Scotland', *Climate Dynamics*, 16: 815–820.

Prospero, J.M. (1996), 'The atmospheric transport of particles to the ocean', in V. Ittekkot, P. Schafer, S. Honjo and P.J. Depetris (eds) *Particle Flux in the Ocean* (Chichester): 19–52.

Pye, K. (1987), *Aeolian dust and dust deposits* (London).

—— and Zhou, L.-P. (1989), 'Late Pleistocene and Holocene aeolian dust deposition in North China and the northwest Pacific Ocean', *Palaeogeography, Palaeoclimatology, Palaeoecology*, 73: 11–23.

—— Winspear, N.R. and Zhou, L.P. (1995), 'Thermoluminescence ages of loess and associated sediments in central Nebraska, USA', *Palaeogeography, Palaeoclimatology, Palaeoecology*, 118: 73–87.

Qian, W., Qian, L. and Shi, S. (2002), 'Variations of the dust storm in China and its climatic control', *Journal of Climate*, 15: 1216–1229.

Qiang, X.K., Li, Z., Li, X., Powell, C. and Zheng, H.B. (2001), 'Magnetostratigraphic record of the Late Miocene onset of the east Asian monsoon and Pliocene uplift of northern Tibet', *Earth and Planetary Science Letters*, 187: 83–93.

—— and 15 others. (2011), 'New eolian red clay sequence on the western Chinese Loess Plateau linked to onset of Asian desertification about 25 Ma ago', *Science China Earth Sciences*, 54: 136–144.

Quade, J., Cerling, T.E. and Bowman, J.R. (1989), 'Development of Asian monsoon revealed by marked ecological shift during the latest Miocene in northern Pakistan', *Nature*, 342: 163–166.

Quigley, M.C., Horton, T., Hellstrom, J.C., Cupper, M.L. and Sandiford, M. (2010), 'Holocene climate change in arid Australia from speleothem and alluvial records', *The Holocene*, 20: 1093–1104.

Rackham, O. (1980), *Ancient Woodland* (London).

Radić, V. and Hock, R. (2011), 'Regionally differentiated contribution of mountain glaciers and ice caps to future sea-level rise', *Nature Geoscience*, 4: 91–94.

Rahmstorf, S. (1995), 'Bifurcations of the Atlantic thermohaline circulation in response to changes in the hydrological cycle', *Nature*, 378: 145–149.

—— (2007), 'A semi-empirical approach to projecting future sea-level rise', *Science*, 315: 368–370.

—— Marotzke, J. and Willebrand, J. (1996), '*Stability of the Thermohaline Circulation*', in W. Kraus (ed.) *The Warmwatersphere of the North Atlantic Ocean* (Berlin): 129–157.

Raikes, R. (1967), *Water, Weather and Prehistory* (London).

Ramirez, E. and 8 others. (2001), 'Small glaciers disappearing in the tropical Andes: a case study of Bolivia: Glacier Chacaltaya (16°S)', *Journal of Glaciology*, 47: 187–194.

Rampino, M.E. and Self, W.F. (1992), 'Volcanic winter and accelerated glaciation following the Toba supereruption', *Nature*, 359: 50–52.

Rasmussen, S.O. and 15 others. (2006), 'A new Greenland ice core chronology for the last glacial termination', *Journal of Geophysical Research*, 111: D06102, doi:10.1029/2005JD006079.

—— Vinther, B.M., Clausen, H.B. and Andersen, K.K. (2007), 'Early Holocene climate oscillations recorded in three Greenland ice cores', *Quaternary Science Reviews*, 26: 1907–1914.

Rasmussen, T.L., van Weering, T.C.E. and Labeyrie, L. (1997), 'Climatic instability, ice sheets and ocean dynamics at high northern latitudes during the last glacial period (58–10 ka bp)', *Quaternary Science Reviews*, 16: 71–80.

Rasmusson, E.M. (1987), 'Global climate change and variability: effects on drought and desertification in Africa', in M.H. Glantz (ed.) *Drought and Hunger in Africa* (Cambridge): 3–22.

Raymo, M.E. and Ruddiman, W.F. (1992), 'Tectonic forcing of late Cenozoic climate', *Nature*, 359: 117–122.

Raynaud, D., Chappellaz, J., Barnola, J.M., Korotkevich, Y.S. and Lorius, C. (1988), 'Climatic and CH_4 cycle implications of glacial-interglacial CH_4 change in the Vostok ice-core', *Nature*, 33: 655–659.

Rea, D.K. (1994), 'The paleoclimatic record provided by eolian deposition in the deep sea: the geologic history of wind', *Reviews of Geophysics*, 32: 159–195.

Rech, J.A. and 8 others. (2010), 'Evidence for the development of the Andean rain shadow from a Neogene isotopic record in the Atacama desert, Chile', *Earth and Planetary Science Letters*, 292: 371–382.

Reich, D., Patterson, N., Kircher, M., Delfin, F., Nandineni, M.R., Pugach, I., Ko, A. M.-S., Ko, Y.-C., Jinam, T.A., Phipps, M.E., Saitou, N., Wollstein, A., Kayser, M., Paabo, S. and Stoneking, M. (2011), 'Denisova admixture and the first modern human dispersals into Southeast Asia and Oceania', *The American Journal of Human Genetics*, 89: 1–13.

Reichert, B.K., Bengston, L. and Oerlemans, J. (2001), 'Midlatitude forcing mechanism for glacier mass balance investigated using general circulation models', *Journal of Climate*, 14: 3767–3784.

Reid, E.M. and Chandler, M.E.J. (1933), *The London Clay Flora* (London).

Reille, M., De Beaulieu, J.-L., Svobodova, H., Andrieu-Ponel, V. and Goeury, C. (2000), 'Pollen analytical biostratigraphy of the last five climatic cycles from a long continental sequence from the Velay region (Massif Central, France)', *Journal of Quaternary Science*, 15: 665–685.

Reimer, P.J. and 27 others. (2009), 'INTCAL09 and MARINE09 Radiocarbon age calibration curves, 0-50,000 years cal BP', *Radiocarbon*, 51: 1111–1150.

Ren, G. (2007), 'Changes in forest cover in China during the Holocene', *Vegetation History and Archaeobotany*, 16: 119–126.

—— and Beug, H-J. (2002), 'Mapping Holocene pollen data and vegetation of China', *Quaternary Science Reviews*, 21: 1395–1422.

Rendell, H.M. (1989), 'Loess deposition during the Late Pleistocene in northern Pakistan', *Zeitschrift für Geomorphologie, Supplementband*, 76: 247–255.

Rennie, A.F. and Hansom, J.D. (2011), 'Sea level trend reversal: land uplift outpaced by sea level rise on Scotland's coast', *Geomorphology*, 125: 193–202.

Rhodes, E.J. (1988), 'Methodological considerations in the optical dating of quartz', *Quaternary Science Reviews*, 7: 395–400.

Rice, M. (2003), *Egypt's Making. The Origins of Ancient Egypt 5000–2000 BC,* 2nd edition) (London).

Richards, K.S., Arnett, R.R. and Ellis, S. (eds) (1985), *Geomorphology and Soils* (Allen & Unwin).

Richardson, P.L. (2008), 'On the history of meridional overturning circulation schematic diagrams', *Progress in Oceanography*, 76: 466–486.

Richter, D., Moser, J., Nami, M., Eiwanger, J. and Mikdad, A. (2010), 'New chronometric data from Ifri n'Ammar (Morocco) and the chronostratigraphy of the Middle Paleolithic in the Western Maghreb', *Journal of Human Evolution*, 59: 672–679.

Richthofen, F. von (1882), 'On the mode of origin of the loess', *Geological Magazine*, 9, series 2: 293–305.

Ridgwell, A. (2002), 'Dust in the earth system: the biogeochemical linking of land, air and sea', *Philosophical Transactions of the Royal Society, London*, A, 360: 2905–2924.

Ridgwell, A.J. (2003), 'Implications of the glacial CO_2 "iron hypothesis" for Quaternary climate change', *Geochemistry, Geophysics, Geosystems*, 4: 1076.

Rightmire, G.P., Deacon, H.J., Schwartz, J.H. and Tattersall, I. (2006), 'Human foot bones from Klasies River main site, South Africa', *Journal of Human Evolution*, 50: 96–103.

Rignot, E., Velicogna, I., Van Den Broecke, M.R., Monaghan, A. and Lenaerts, J. (2011), 'Acceleration of the contribution of the Greenland and Antarctica ice sheets to sea level rise', *Geophysical Research Letters*, 38: L05503, doi:10.1029/2011GL046583.

Rind, D. and Overpeck, J. (1993), 'Hypothesized causes of decade-to-century-scale climate variability: climate model results', *Quaternary Science Reviews*, 12: 357–374.

Rindos, D. (1984), *The Origins of Agriculture: an Evolutionary Perspective* (New York).

Ritchie, J.C. and Haynes, C.V. (1987), 'Holocene vegetation zonation in the eastern Sahara', *Nature*, 330: 645–647.

—— Eyles, C.H. and Haynes, C.V. (1985), 'Sediment and pollen evidence for an early to mid Holocene humid period in the eastern Sahara', *Nature*, 314: 352–355.

Roberts, H.M., Muhs, D.R., Wintle, A.G., Duller, G.A.T. and Bettis, E.A. (2003), 'Unprecedented last-glacial mass accumulation rates determined by luminescence dating of loess from western Nebraska', *Quaternary Research*, 59: 411–419.

Roberts, N. (1983), 'Age, palaeoenvironments and climatic significance of Late Pleistocene Konya Lake, Turkey', *Quaternary Research*, 19: 154–171.

—— (1998), *The Holocene: an Environmental History*, 2nd edition (Oxford).

—— Taieb, M., Barker, P., Damnati, P., Icole, M. and Williamson, D. (1993), 'Timing of the Younger Dryas event in East Africa from lake-level changes', *Nature*, 366: 146–148.

Roberts, R.G., Jones, R. and Smith, M.A. (1990), 'Thermoluminesence dating of a 50 000 year-old human occupation site in northern Australia', *Nature*, 345: 153–156.

Robinson, M.A. and Lambrick, G.H. (1984), 'Holocene alluviation and hydrology in the upper Thames basin', *Nature*, 308: 809–814.

Robinson, S.A., Black, S., Sellwood, B.W. and Valdes, P.J. (2006), 'A review of the palaeoclimates in the Levant and Eastern Mediterranean from 25 000 to 5000 years bp: setting the environmental background for the evolution of human civilisation', *Quaternary Science Reviews*, 25: 1517–1541.

Rodbell, D.T. (1992), 'Late Pleistocene equilibrium-line reconstructions in the northern Peruvian Andes', *Boreas*, 21: 43–52.

Rodda, J.C. (1970), 'Rainfall excesses in the United Kingdom', *Transactions, Institute of British Geographers*, 49: 49–60.

—— Coudé-Gaussen, G., Revel, M., Grousset, F.E. and Pédemay, P. (1996), 'Holocene Saharan dust deposition on the Cape Verde islands: sedimentological and Nd-Sr isotopic arguments', *Sedimentology*, 43: 359–366.

Rogora, M., Mosello, R. and Marchetto, A. (2004), 'Long-term trends in the chemistry of atmospheric deposition in Northwestern Italy: the role of increasing Saharan dust depostition', *Tellus*, 56B: 426–434.

Röhl, U., Bralower, T.J., Norris, R.D. and Wefer, G. (2000), 'New chronology for the late Paleocene thermal maximum and its environmental implications', *Geology*, 28: 927–930.

Rohling, E.J., Fenton, M., Jorissen, F.J., Bertrand, P., Ganssen, G. and Caulet, J.P. (1998a), 'Magnitudes of sea-level lowstands of the past 500 000 years', *Nature*, 394: 162–165.

—— Hayes, A., de Rijk, S., Kroon, D., Zacharasse, W.J. and Eisma, D. (1998b), 'Abrupt cold spells in the northwest Mediterranean', *Paleoceanography*, 13: 316–322.

—— and 10 others. (2002), 'African monsoon variability during the previous interglacial maximum', *Earth and Planetary Science Letters*, 202: 61–75.

—— Liu, Q.S., Roberts, A.P., Stanford, J.D., Rasmussen, S.O., Langen, P.L. and Siddall, M. (2009), 'Controls on the East Asian monsoon during the last glacial cycle, based on comparison between Hulu Cave and polar ice-core records', *Quaternary Science Reviews*, 28: 3291–3302.

Rokosh, D., Bush, A.B.G., Rutter, N.W., Ding, Z. and Sun, J. (2003), 'Hydrologic and geologic factors that influenced spatial variations in loess deposition in China during the last interglacial-glacial cycle: results from proxy climate and GCM analyses', *Palaeogeography, Palaeoclimatology, Palaeoecology*, 193: 249–260.

Rose, J. (1990), 'Raised shorelines', in A.S. Goudie (ed.) *Geomorphological Techniques* (London): 456–475.

Rose, J.I. (2004), 'The Question of Upper Pleistocene connections between East Africa and South Arabia', *Current Anthropology*, 45: 551–555.

—— Usik, V.I., Marks, A.E., Hilbert, Y.H., Galletti, C.S. et al. (2011), 'The Nubian Complex of Dhofar, Oman: An African Middle Stone Age Industry in Southern Arabia', *PLoS ONE* 6(11): e28239. doi:10.1371/journal.pone.0028239.

Rosenberg, T.M., Preusser, F., Fleitmann, D., Schwalb, A., Penkman, K., Schmid, T.W., Al-Shanti, M.A., Kadi, K. and Matter, A. (2011), 'Humid periods in southern Arabia: Windows of opportunity for modern human dispersal', *Geology* 39, 1115–1118.

Rosenfeld, D., Rudich, Y. and Lahav, R. (2001), 'Desert dust suppressing precipitation: a possible desertification feedback loop', *Proceedings of the National Academy of Sciences*, 98: 5975–5980.

Rossetti, D. de F., Toledo, P.M. de and Góes, A.M. (2005), 'New geological framework for western Amazonia (Brazil) and implications for biogeography and evolution', *Quaternary Research*, 63: 78–89.

Rossignol-Strick, M. (1985), 'Mediterranean Quaternary sapropels, an immediate response of the African monsoon to variation of insolation', *Palaeogeography, Palaeoclimatology, Palaeoecology*, 49: 237–263.

—— (1995), 'Sea-land correlation of pollen records in the Eastern Mediterranean for the glacial-interglacial transition: biostratigraphy versus radiometric time-scale', *Quaternary Science Reviews*, 14: 893–915.

—— and Duzer, D. (1980), 'Late Quaternary West African climate inferred from palynology of Atlantic deep-sea cores', *Palaeoecology of Africa*, 12: 227–228.

Rost, K.T. (1997), 'Observations on the distribution and age of loess-like sediments in the high-mountain ranges of central Asia', *Zeitschrift für Geomorphologie, Supplementband*, 111: 117–129.

Rouchy, J.M., Servant, M., Fournier, M. and Causse, C. (1996), 'Extensive carbonate algal bioherms in upper Pleistocene saline lakes of the central Altiplano of Bolivia', *Sedimentology*, 43: 973–993.

Rousseau, D.D. and 11 others. (2002), 'Abrupt millennial climatic changes from Nussloch (Germany) Upper Weichselian eolian records during the Last Glaciation', *Quaternary Science Reviews*, 21: 1577–1582.

Rowley-Conwy, P. (2011), 'Westward Ho! The spread of agriculture from central Europe to the Atlantic', *Current Anthropology*, 52: 431–451.

Ruddiman, W.F. (1997), 'Tropical Atlantic terrigenous fluxes since 25 000 years bp', *Marine Geology*, 136: 189–207.

—— (2001), *Earth's Climate: Past and Future* (San Francisco).

—— (2003), 'The anthropogenic greenhouse era began thousands of years ago', *Climatic Change*, 61: 261–93.

—— (2005a), 'How did humans first alter global climate?', *Scientific American*, 292: 46–53.

—— (2005b), *Plows, Plagues, and Petroleum: How Humans Took Control of Climate*. (Princeton, NJ).

—— (2006), 'Ice-driven CO_2 feedback on ice volume', *Climate of the Past*, 2: 43–55.

—— (2008), *Earth's Climate: Past and Future*, 2nd edition (New York).

—— and McIntyre, A. (1981a), 'The North Atlantic ocean during the last deglaciation', *Palaeogeography, Palaeoclimatology, Palaeoecology*, 35: 145–214.

—— and Mcintyre, A. (1981b), 'Oceanic mechanisms for amplification of the 23 000-year ice-volume cycle', *Science*, 212: 617–627.

—— and Raymo, M.E. (1988), 'The past three million years: Evolution of climatic variability in the North Atlantic region', *Philosophical Transactions of the Royal Society, London*, B, 318: 411–429.

—— and Thomsen, J. S. (2001), 'The case for human causes of increased atmospheric CH_4 over the last 5000 years', *Quaternary Science Reviews*, 20: 1769–1777.

—— Prell, W.L. and Raymo, M.E. (1989), 'Late Cenozoic uplift in southern Asia and the American west: Rationale for general circulation modeling experiments', *Journal of Geophysical Research*, 94, D: 18379–18391.

Rutter, N.W., Rokosh, D., Evans, M.E., Little, E.C., Chlachula, J. and Velichko, A. (2003), 'Correlation and interpretation of paleosols and loess across European Russia and Asia over the last interglacial-glacial cycle', *Quaternary Research*, 60: 101–109.

Ryan, W.B.F., Pitman, W.C., Major, C.O., Shimkus, K., Moskalenko, V., Jones, G.A., Dimitrov, P., Gorür, N., Sakinç, M. and Yüce, H. (1997), 'An abrupt drowning of the Black Sea shelf', *Marine Geology*, 138: 119–126.

Saarinen, T. (1998), 'High-resolution palaeosecular variation in northern Europe during the last 3200 years', *Physics of the Earth and Planetary Interiors*, 106: 301–311.

Saarnisto, M. (1986), 'Annually laminated lake sediments', in B.E. Berglund, (ed.) *Handbook of Holocene Palaeoecology and Palaeohydrology* (Chichester): 343–370.

Said, R. (1981), *The Geological Evolution of the River Nile* (New York).

Sala, J.Q., Cantos, J.O. and Chiva, E.M. (1996), 'Red dust within the Spanish Mediterranean area', *Climatic Change*, 32: 215–228.

Salinger, M.J. and Gunn, J.M. (1975), 'Recent climatic warming around New Zealand', *Nature*, 256: 396–398.

—— and McGlone, M.S. (1990), 'New Zealand Climate – the past two million years', in Royal Society of New Zealand, *New Zealand Climate Report 1990* (Wellington): 13–17.

Salzer, M.W. and Hughes, M.K. (2006), 'Bristlecone pine tree rings and volcanic eruptions over the last 5000 yr', *Quaternary Research*, 67: 57–68.

—— Hughes, M.K., Bunn, A.G. and Kipfmueller, K.F. (2009), 'Recent unprecedented tree-ring growth in bristlecone pine at the highest elevations and possible causes', *Proceedings of the National Academy of Sciences*, 106: 20348–20353.

Samartin, S., Heiri, O., Vescovi, E., Brooks, S.J. and Tinner, W. (2012), 'Lateglacial and early Holocene summer temperatures in the southern Swiss Alps reconstructed using fossil chironomids', *Journal of Quaternary Science*, 27: 279–289.

Sampaio, G., Nobre, C., Costa, M.H., Satyamurty, P., Soares-Filho, B.S. and Cardoso, M. (2007), 'Regional climate change over eastern Amazonia caused by pasture and soybean cropland expansion', *Geophysical Research Letters*, 34: L17709, doi:10.1029/2007GL030612.

Santer, B.D. and 19 others. (2006), 'Forced and unforced ocean temperature changes in Atlantic and Pacific tropical cyclogenesis regions', *Proceedings of the National Academy of Sciences*, 103: 13905–13910.

Sapozhnikov, V.V., Mordasova, N.V. and Metreveli, M.P. (2010), 'Transformations in the Caspian Sea ecosystem under the fall and rise of the sea level', *Oceanology*, 50: 488–497.

Sarmiento, J.L. and Toggweiler, J.R. (1984), 'A new model for the role of the oceans in determining atmospheric PCO$_2$', *Nature*, 308: 621–624.

Sarnthein, M. (1972), 'Sediments and history of the Postglacial transgression in the Persian Gulf and northwest Gulf of Oman', *Marine Geology*, 12: 245–266.

—— (1978), 'Sand deserts during Glacial Maximum and climatic optimum', *Nature*, 272: 43–46.

—— and Diester-Hass, L. (1977), 'Eolian-sand turbidites', *Journal of Sedimentary Petrology*, 47: 868–890.

—— and Koopman, B. (1980), 'Late Quaternary deep-sea record of north-west African dust supply and wind circulation', *Palaeoecology of Africa*, 12: 238–253.

—— Thiede, J., Pflaumann, U., Erlenkeuser, H., Futterer, D., Koopman, B., Lange, H. and Seibold, E. (1982), 'Atmospheric and oceanic circulation patterns off northwest Africa during the past 25 million years', in U. von Rad, K. Hinz, M. Sarnthein and E. Seibold (eds) *Geology of the Northwest African Continental Margin* (Berlin): 545–604.

—— Winn, K., Duplessy, J.C. and Fontugne, M.R. (1988), 'Global variations of surface ocean productivity in low and mid latitudes: Influence of CO$_2$ reservoirs on the deep ocean and atmosphere during the last 21 000 years', *Paleoceanography*, 3: 361–399.

Sartori, M., Evans, M.E., Heller, F., Tsatskin, A. and Han, J.M. (2005), 'The last glacial/interglacial cycle at two sites in the Chinese Loess Plateau: mineral magnetic, grain-size and ^{10}Be measurements and estimates of palaeoprecipitation', *Palaeogeography, Palaeoclimatology, Palaeoecology*, 222: 145–160.

Sauer, C.O. (1948), 'Environment and culture during the last deglaciation', *Proceedings, American Philosophical Society*, 92: 65–77.

Saunders, M.A. and Lee, A.S. (2008), 'Large contribution of sea surface warming to recent increase in Atlantic hurricane activity', *Nature*, 451: 557–560.

Scarre C. (ed.) (2005), *The Human Past: World Prehistory and the Development of Human Societies* (London).

Schaub, M., Kaiser, K.F., Frank, D.C., Buentgen, U., Kromer, B. and Talamo, S. (2008), 'Environmental change during the Allerod and Younger Dryas reconstructed from tree-ring data', *Boreas*, 37: 74–86.

Schell, I.I. (1962), 'On the iceberg severity off Newfoundland and its prediction', *Journal of Glaciology*, 4: 161–172.

Scherler, D., Bookhagen, B. and Strecker, M.R. (2011), 'Spatially variable response of Himalayan glaciers to climate change affected by debris cover', *Nature Geoscience*, 4: 156–159.

Schmidt, G.A. and LeGrande, A.N. (2005), 'The Goldilocks abrupt climate change event', *Quaternary Science Reviews*, 24: 1109–1110.

Scholz, C.A., Cohen, A.S., Johnson, T.C., King, J., Talbot, M.R. and Brown, E.T. (2011), 'Scientific drilling in the Great Rift Valley: the 2005 Lake Malawi Scientific Drilling Project – an overview of the past 145,000 years of climate variability in Southern Hemisphere East Africa', *Palaeogeography, Palaeoclimatology, Palaeoecology*, 303: 3–19.

Schramm, C.T. (1989), 'Cenozoic climatic variation recorded by quartz and clay minerals in North Pacific sediments', in M. Leinen and M. Sarnthein (eds) *Palaeoclimatology and Palaeometeorology: Modern and Past Patterns of Global Atmospheric Transport* (Dordrecht): 805–839.

Schubert, C. (1984), 'The Pleistocene and recent extent of the glaciers of the Sierra Nevada de Merida, Venezuela', *Erdwissenschaftliche Forschung*, 18: 269–278.

Schulz, E. and Whitney, J.W. (1986), 'Upper Pleistocene and Holocene lakes in the An Nafud, Saudi Arabia', *Hydrobiologica*, 143: 175–190.

Schuster, M., Duringer, P., Ghienne, J-F., Vignaud, P., Mackaye, H.T., Likius, A. and Brunet, M. (2006), 'The age of the Sahara Desert', *Science*, 311: 821.

Scicchitano, G. and 7 others. (2011), 'Millstones as indicators of relative sea-level changes in northern Sicily and southern Calabria coastlines, Italy', *Quaternary International*, 232: 92–104.

Scrivner, A.E., Vance, D. and Rohling, E.J. (2004), 'New neodymium isotope data quantify Nile involvement in Mediterranean anoxic episodes', *Geology*, 32: 565–568.

Scuderi, L.A. (2002), 'The Holocene environment', in A.R. Orme (ed.) *The Physical Geography of North America* (New York): 86–97.

Seager, R. and Vecchi, G.A. (2010), 'Greenhouse warming and the 21st century hydroclimate of southwestern North America', *Proceedings of the National Academy of Sciences*, 107: 21277–21282.

Seidov, D. and Maslin, M. (2001), 'Atlantic Ocean heat piracy and the bipolar climate see-saw during Heinrich and Dansgaard-Oeschger events', *Journal of Quaternary Science*, 16: 321–328.

Senut, B., Pickford, M., Gommery, D, Mein, P., Cheboi, K. and Coppens, Y. (2001), 'First hominid from the Miocene (Lukeino Formation, Kenya)', *Comptes Rendus de l'Académie de Sciences*, 332: 137–144.

Seppä, H. and Bennett, K.D. (2003), 'Quaternary pollen analysis: recent progress in palaeoecology and palaeoclimatology', *Progress in Physical Geography*, 27: 548–579.

—— Nyman, M., Korhola, A. and Weckström, J. (2002), 'Changes of treelines and alpine vegetation in relation to post-glacial climate dynamics in northern Fennoscandia based on pollen and chironomid records', *Journal of Quaternary Science*, 17: 287–301.

Seppälä, M. (2004), *Wind as a Geomorphic Agent in Cold Climates* (Cambridge).

Sexton, P.F., Norris, R.D., Wilson, P.A., Pälike, H., Westerhold, T., Röhl, U., Bolton, C.T. and Gibbs, S. (2011), 'Eocene global warming events driven by ventilation of oceanic dissolved organic carbon', *Nature*, 471: 349–352.

Shackleton, N.J. (1975), 'The stratigraphic record of deep-sea cores and its implications for the assessment of glacials, interglacials, stadials and interstadials in the mid-Pleistocene', in K.W. Butzer and G.L. Isaac (eds) *After the Australopithecines* (The Hague): 1–24.

—— (1987), 'Oxygen isotopes, ice volume and sea-level', *Quaternary Science Reviews*, 6: 183–190.

—— Berger, A. and Peltier, W.R. (1990), 'An alternative astronomical calibration of the lower Pleistocene timescale based on ODP Site 677', *Transactions of the Royal Society of Edinburgh: Earth Sciences*, 81: 251–261.

—— and Crowhurst, S. (1996), Timescale Calibration, ODP 677. *IGBP PAGES/World Data Center-A for Paleoclimatology Data Contribution Series # 96-018*. NOAA/NGDC Paleoclimatology Program, Boulder CO, USA.

—— and Opdyke, N.D. (1973), 'Oxygen isotope and palaeomagnetic stratigraphy of equatorial Pacific core V28–238: oxygen isotope temperatures and ice volume on a 10^5 and 10^6 year scale', *Quaternary Research*, 3: 39–55.

Shakun, J.D., Burns, S.J., Clark, P.U., Cheng, H. and Edwards, R.L. (2011), 'Milankovitch-paced Termination II in a Nevada speleothem?', *Geophysical Research Letters*, 38: L18701, doi:10.1029/2011GL048560.

Shaler, N.S. (1874), 'Preliminary report on the recent changes of level on the coast of Maine', *Memoir Boston Society of Natural History*, 2: 320–340.

Shanahan, T.M. and 8 others. (2006), 'Paleoclimatic variations in West Africa from a record of late Pleistocene and Holocene lake level stands of Lake Bosumtwi, Ghana', *Palaeogeography, Palaeoclimatology, Palaeoecology*, 242: 287–302.

Shaw, A. and Goudie, A.S. (2002), 'Geomorphological evidence for the extension of the mega-Kalahari into south-central Angola', *South African Geographical Journal*, 84: 182–194.

Shea, J.J. (2008), 'Transitions or turnovers? Climatically-forced extinctions of Homo sapiens and Neanderthals in the east Mediterranean Levant', *Quaternary Science Reviews*, 27: 2253–2270.

Shennan, I. (1983), 'Flandrian and late Devensian sea-level changes and crustal movements in England and Wales', in D.E. Smith and A.G. Dawson (eds) *Shorelines and Isostasy* (London): 255–283.

—— (1986a), 'Flandrian sea-level changes in the Fenland I: The geographical setting and evidence of relative sea-level changes', *Journal of Quaternary Science*, 1/2: 119–153.

—— (1986b), 'Flandrian sea-level changes in the Fenland II: Tendencies of sea-level movement, altitudinal changes and local and regional factors', *Journal of Quaternary Science*, 1: 155–179.

—— (1992), 'Late Quaternary sea-level changes and coastal movements in eastern England and eastern Scotland: an assessment of models of coastal evolution', *Quaternary International*, 15/16: 161–173.

—— Lambeck, K., Flather, R., Horton, B., McArthur, J., Innes, J., Lloyd, J., Rutherford, M. and Wingfield, R. (2000), 'Modelling western North Sea palaeogeographies and tidal changes during the Holocene', *Geographical Society, London, Special Publications*, 166: 299–319.

—— Bradley, S., Milne, G., Brooks, A., Bassett, S. and Hamilton, S. (2006), 'Relative sea-level changes, glacial isostatic modelling and ice-sheet reconstructions from the British Isles since the Last Glacial maximum', *Journal of Quaternary Science*, 21: 585–599.

—— Milne, G. and Bradley, S. (2011), 'Late Holocene vertical land motion and relative sea-level changes: lessons from the British Isles', *Journal of Quaternary Science*, 26: 541–552.

Shepard, F.P. (1963), 'Thirty-five thousand years of sea-level', in T. Clements (ed.) *Essays in Marine Geology in Honor of K.O. Emery* (Los Angeles).

Sherratt, A. (1997), 'Climatic cycles and behaviour revolutions: the emergence of modern humans and the beginning of farming', *Antiquity*, 71: 271–287.

Shevenell, A.E., Kennett, J.P. and Lea, D.W. (2004), 'Middle Miocene Southern Ocean cooling and Antarctic cryosphere expansion', *Science*, 305: 1766–1770.

Shi, P., Yan, P., Yuan, Y. and Nearing, M.A. (2004), 'Wind erosion research in China: past, present and future', *Progress in Physical Geography*, 28: 366–386.

Shichang, K., Dahe, Q., Mayewski, P.A., Wake, C.P. and Jiawen, R. (2001), 'Climatic and environmental records from the Far East Rongbuk ice core, Mt. Qomolangma (Mt. Everest)', *Episodes*, 24: 176–181.

Shindell, D.T., Schmidt, G.A., Mann, M.E., Rind, D. and Waple, A. (2001), 'Solar forcing of regional climate change during the Maunder Minimum', *Science*, 294: 2149–2152.

Shotyk, W. (1992), 'Organic soils', in I.P. Martin and W. Chesworth (eds) *Weathering, Soils and Paleosols*. (Amsterdam): 203–224.

—— (1996), 'Natural and anthropogenic enrichments of As, Cu, Pb, Sb and Zn in ombrotrophic versus minerotrophic peat bog profiles, Jura mountains, Switzerland', *Water, Air, & Soil Pollution*, 90: 375–405.

Shulmeister, J. (1999), 'Australasian evidence for mid-Holocene climate change implies precessional control of Walker Circulation in the Pacific', *Quaternary International*, 57/58: 81–91.

Shuman, B. (2012), 'Patterns, processes, and impacts of abrupt climate change in a warm world: the past 11,700 years', *Wiley Interdisciplinary Reviews: Climate Change*, 3: 19–43, doi: 10.1002/wcc.152.

Siegenthaler, U. and Wenk, Th. (1984), 'Rapid atmospheric CO_2 variations and ocean circulation', *Nature*, 308: 624–626.

Simmons, I.G. (1989), *Changing the Face of the Earth* (Oxford).

—— (1993), 'Vegetation change during the Mesolithic in the British Isles: some amplifications', in F.M. Chambers (ed.) *Climate Change and Human Impact on the Landscape* (London): 109–118.

Singarayer, J.S., Richards, D.A., Ridgwell, A., Valdes, P.J., Austin, W.E.N. and Beck, J.W. (2008), 'An oceanic origin for the increase of atmospheric radiocarbon during the Younger Dryas', *Geophysical Research Letters*, 35: L14707, doi:10.1029/2008GL034074.

Singh, G. (1971), 'The Indus valley culture seen in context of Post-Glacial climate and ecological studies in north-west India', *Archaeology and Anthropology in Oceania*, 6: 177–189.

Singhvi, A.K. and 9 others. (2010), 'A ~200 ka record of climatic change and dune activity in the Thar Desert, India', *Quaternary Science Reviews*, 29: 3095–3105.

Sirocko, F., Sarnthein, M., Large, H. and Erlenkeuser, H. (1991), 'Atmospheric summer circulation and coastal upwelling in the Arabian Sea during the Holocene and the Last Glaciation', *Quaternary Research*, 36: 72–93.

Sissons, J.B. (1962), 'A re-interpretation of the literature on late-glacial shorelines in Scotland with particular reference to the Forth area', *Transactions of the Edinburgh Geological Society*, 19: 83–99.

—— (1976), *Scotland* (London).

—— (1980a), 'The Loch Lomond Advance in the Lake District, northern England', *Transactions, Royal Society of Edinburgh, Earth Sciences*, 71: 13–27.

—— (1980b), 'Palaeoclimatic inferences from Lomond advance glaciers', in J.J. Lowe, J.M. and J.E. Robinson (eds) *Studies in the La[...] North-west Europe* (Oxford): 31–43.

—— and Sutherland, D.G. (1976), 'Climatic inferences from former glaciers in the south-east Grampian Highlands', *Journal of Glaciology*, 17: 325–346.

Six, D., Reynaud, L. and Letreguilly, A. (2001), 'Bilans de masse des glaciers alpines et scandinaves, leurs relations avec l'oscillation due climat de l'Atlantique nord', *Comptes Rendus Academie des Sciences, Sciences de la Terre et des Planètes*, 333: 693–698.

Sluijs, A., Schouten, S., Pagani, M., Woltering, M., Brinkhuis, H., Sinninghe Damsté, J.S., Dickens, G.R., Huber, M., Reichart, G.-J., Stein, R., Matthiessen, J., Lourens, L.J., Pedentchouk, N., Backman, J., Moran, K. and The Expedition 302 Scientists. (2006), 'Subtropical Arctic Ocean temperatures during the Palaeocene/Eocene thermal maximum', *Nature*, 441: 610–613.

—— Schouten, S., Donders, T.H., Schoon, P.L., Röhl, U., Reichart, G.-J., Sangiorgi, F., Kim, J.-H., Sinninghe Damsté, J.S. and Brinkhuis, H. (2009), 'Warm and wet conditions in the Arctic region during Eocene Thermal Maximum 2', *Nature Geoscience*, 2: 777–780.

Smalley, I.J. and Vita-Finzi, C. (1968), 'The formation of fine particles in sandy deserts and the nature of "desert" loess', *Journal of Sedimentary Petrology*, 38: 766–774.

Smith, A.G. (1970), 'The influence of Mesolithic and Neolithic man on British vegetation: a discussion', in D. Walker and R.G. West (eds) *The Vegetational History of the British Isles* (Cambridge): 81–96.

—— and Pickering, K.T. (2003), 'Oceanic gateways as a critical factor to initiate icehouse Earth', *Journal of the Geological Society*, 160: 337–340.

Smith, D.E. and Dawson, A.G. (eds) (1983), *Shorelines and Isostasy* (London).

—— Davies, M.H., Brooks, C.L., Mighall, T.M., Dawson, S., Rea, B.R., Jordan, J.T. and Holloway, L.K. (2010), 'Holocene relative sea levels and related prehistoric activity in the Forth lowland, Scotland, United Kingdom', *Quaternary Science Reviews*, 29: 2382–2410.

—— Harrison, S., Firth, C.R. and Jordan, J.T. (2011), 'The early Holocene sea level rise', *Quaternary Science Reviews*, 30: 1846–1860.

Smith, F. (1992), 'The role of continuity in modern human origins', in G. Bräuer and F. Smith (eds.) *Continuity or Replacement? Controversies in Homo sapiens Evolution* (Rotterdam): 145–156.

Smith, G.I. (2009), 'Late Cenozoic geology and lacustrine history of Searles Valley, Inyo and San Bernadino Counties, California', *US Geological Survey Professional Paper*: 1727.

Smith, T.M., Tafforeau, P., Reid, D.J., Grün, R., Eggins, S., Boutakiout, M. and Hublin J.-J. (2007), 'Earliest evidence of modern human life history in North African early *Homo sapiens*', *Proceedings of the National Academy of Sciences*, 104: 6128–6133.

Smithers, S.G. and Woodroffe, C.D. (2000), 'Micro-atolls as sea-level indicators on a mid-ocean swell', *Marine Geology*, 168: 61–78.

Smithson, P., Addison, K. and Atkinson, K. (2002), *Fundamentals of the physical environment*, 3rd edition (London).

Snyder, C.T. and Langbein, W.B. (1962), 'The Pleistocene lake in Spring Valley, Nevada and its climatic implications', *Journal of Geophysical Research*, 67: 2385–2394.

Sombroek, A.G., Mbuvi, J.P. and Okwaro, H.W. (1976), 'Soils of the semi-arid savanna zone of north-eastern Kenya', *Kenya Soil Survey, Miscellaneous Soil Paper*, No. M2.

Soreghan, G.S., Soreghan, M.J. and Hamilton, M.A. (2008), 'Origin and significance of loess in late Paleozoic western Pangaea: A record of tropical cold?', *Palaeogeography, Palaeoclimatology, Palaeoecology*, 268: 234-259.

Soulet, G., Ménot, G., Lericolais, G. and Bard, E. (2011), 'A revised calendar age for the last reconnection of the Black Sea to the global ocean', *Quaternary Science Reviews*, 30: 1019–1029.

Southwood, T.R.E. (2003), *The Story of Life* (Oxford).

Spaulding, W.G. (1991), 'Pluvial climatic episodes in North America and North Africa: types and correlation with global climate', *Palaeogeography, Palaeoclimatology, Palaeoecology*, 84: 217–229.

Spencer, C.S. (2003), 'War and early state formation in Oaxaca, Mexico', *Proceedings of the National Academy of Sciences*, 100: 11185–11187.

Stager, J.C., Ryves, D.B., Chase, B.M. and Pausata, F.S.R. (2011), 'Catastrophic drought in the Afro-Asian monsoon region during Heinrich Event 1', *Science*, 331: 1299–1302.

Stanford, J.D., Hemingway, R., Rohling, E.J., Challenor, P.G., Medina-Elizalde, M. and Lester, A.J. (2011), 'Sea-level probability for the last deglaciation:

A statistical analysis of far-field records', *Global and Planetary Change*, 79: 193–203.

Stanley, D.J. and Chen, Z. (1993), 'Yangtze delta, eastern China: I Geometry and subsidence of Holocene depocenter', *Marine Geology*, 112: 1–11.

—— Krom, M.D., Cliff, R.A. and Woodward, J.C. (2003), 'Nile flow failure at the end of the old kingdom, Egypt: strontium isotopic and petrologic evidence', *Geoarchaeology*, 18: 395–402.

Stansell, N.D., Abbott, M.B., Rull, V., Rodbell, D.T., Bezada, M. and Montoya, E. (2010), 'Abrupt Younger Dryas cooling in the northern tropics recorded in lake sediments from the Venezuelan Andes', *Earth and Planetary Letters*, 293: 154–163.

Starkel, L. (1966), 'Post-glacial climate and the moulding of European relief', in J.S. Sawyer (ed.) *World Climate from 8000 to 0 BC* (London): 15–33.

Stauffer, B., Flückiger, J., Wolff, E. and Barnes, P. (2004), 'The Epica deep ice cores: first results and perspectives', *Annals of Glaciology*, 39: 93–100.

Steffen, H. and Wu, P. (2011), 'Glacio-isostatic adjustment in Fennoscandia – a review of data and modelling', *Journal of Geodynamics*, 52: 169–204.

Steffensen, J.P. and 19 others. (2008), 'High-resolution Greenland ice core data show abrupt climate change happens in few years', *Science*, 321: 680–684.

Stein, R. (1985), 'The post-Eocene sediment record of DSDP site 366: implications for African climate and plate tectonic drift', *Geological Society of America, Memoir*, 163: 305–315.

Steiner, G.M. (1997), 'The bogs of Europe', in L. Parkyn, R.E. Stoneman and H.A.P. Ingram (eds) *Conserving Peatlands* (Wallingford): 4–24.

Steinhilber, F., Beer, J. and Fröhlich, C. (2009), 'Total solar irradiance during the Holocene', *Geophysical Research Letters*, 36: L19704, doi: 10.1029/2009GL040142.

Stewart, I.S., Sauber, J. and Rose, J. (2000), 'Glacio-seismotectonics: ice sheets, crustal deformation and seismicity', *Quaternary Science Reviews*, 19: 1367–1389.

Stocker, T.F. (2001), 'Physical climate processes and feedbacks', in J.T. Houghton (ed.) *Climate Change 2001: The Scientific Basis* (Cambridge): 417–470.

—— Wright D.G. and Broecker, W.S. (1992), 'The influence of high-latitude surface forcing on the global thermohaline circulation', *Paleoceanography*, 7: 529–541.

Stockton, E.W. (1990), 'Climatic variability on the scale of decades to centuries', *Climatic Change*, 16: 173–183.

Stoddart, D.R. (1973). 'Coral reefs: the last two million years', *Geography*, 58: 313-333.

Stoll, H.M. (2006), 'The Arctic tells its story', *Nature*, 441: 579–581.

Stommel, H. (1961), 'Thermohaline convection with two stable regimes of flow', *Tellus*, 13: 224–228.

Stokes, S. (1999), 'Luminescence dating applications in geomorphological research', *Geomorphology*, 29: 153–171.

—— and Bray, H.E. (2005), 'Late Pleistocene eolian history of the Liwa region, Arabian Peninsula', *Bulletin of the Geological Society of America*, 117: 1466–1480.

Storey, M., Duncan, R.A. and Swisher, C.C. III (2007), 'Paleocene-Eocene Thermal Maximum and the opening of the Northeast Atlantic', *Science*, 316: 587–589.

Stout, J.E. and Lee, J.A. (2003), 'Indirect evidence of wind erosion trends on the southern High Plains of North America', *Journal of Arid Environments*, 55:43-61.

Street-Perrott, F.A., Mitchell, J.F.B., Marchand, D.S. and Brunner, J.S. (1990), 'Milankovitch and albedo forcing of the tropical monsoons: a comparison of geological evidence and numerical simulations for 9000 YBP', *Transactions, Royal Society of Edinburgh: Earth Sciences*, 81: 407–427.

—— Huang, Y., Perrott, R.A., Eglinton, G., Barker, P., Khelifa, L.B., Harkness, D.D. and Olago, D.O. (1997), 'Impact of lower atmospheric carbon dioxide on tropical mountain ecosystems', *Science*, 278: 1422–1426.

Stride, A.H. (1959), 'On the origin of the Dogger Bank in the North Sea', *Geological Magazine*, 96: 33–44.

Stringer, C. (2002), 'Modern human origins: progress and prospects', *Philosophical Transactions of the Royal Society, London, B*, 357: 563–579.

—— and Gamble, C. (1993), *In Search of the Neanderthals* (London).

Stuiver, M. and Braziunas, T.F. (1993), 'Sun, ocean, climate and atmospheric $^{14}CO_2$: An evaluation of causal and spectral relationships', *The Holocene*, 3: 289–305.

—— Braziunas, T.F., Becker, B. and Kromer, B. (1991) 'Climatic, solar, oceanic and geomagnetic influen on Late-Glacial and Holocene atmospheric $^{14}C/^{12}C$ change', *Quaternary Research*, 35: 1–24.

Stuut, J-B., Zabel, M., Ratmeter, V., Helmke, P., Schefuss, E., Lavik, G. and Schneider, R. (2005), 'Provenance of present-day eolian dust collected off NW Africa', *Journal of Geophysical Research*, 110: doi: 10.1029/2004JD005161.

Subbramayya, I. and Naidu, C.V. (1992), 'Spatial variations and trends in the Indian monsoon rainfall', *International Journal of Climatology*, 12: 597–609.

Suess, E. (1888), *Das Antlitz der Erde* (Vienna).

Sugden, D.E. (1996), 'The East Antarctic Ice Sheet: unstable ice or unstable ideas?', *Transactions of the Institute of British Geographers*, 21: 443–454.

—— and John, B.S. (1976), *Glaciers and Landscape* (London).

Sukumar, R., Ramesh, R., Pant, R.K. and Rajagopalan, G. (1993), 'A $\delta^{13}C$ record of late Quaternary climate change from tropical peats in southern India', *Nature*, 364: 703–706.

Sullivan, D.G. (1988), 'The discovery of Santorini Minoan tephra in western Turkey', *Nature*, 333: 552–554.

Summerfield, M.A. (1991), *Global Geomorphology* (Harlow).

Sun, J. (2002), 'Source regions and formation of the loess sediments on the high mountain regions of northwestern China', *Quaternary Research*, 58: 341–351.

Suppiah, R. and Hennessy, K.J. (1998), 'Trends in total rainfall, heavy rain events and number of dry days in Australia', *International Journal of Climatology*, 18: 1141–1164.

Surić, I. M., Juračić, M., Horvatinčić, N. and Krajcarbronić I. (2005), 'Late Pleistocene-Holocene sea level rise and the pattern of coastal karst inundation: records from submerged speleothems along the Eastern Adriatic Coast (Croatia)', *Marine Geology*, 214: 163–175.

Svensmark, H. and Friis-Christensen, E. (1997), 'Variation of cosmic ray flux and global cloud coverage – a missing link in solar-climate relationships', *Journal of Atmospheric and Solar-Terrestrial Physics*, 59: 1225–1232.

Svensson, A., Biscaye, P.E. and Grousset, F.E. (2000), 'Characterization of late glacial continental dust in the Greenland Ice Core Project ice core', *Journal of Geophysical Research*, 105 (D4): 4637–4656.

Swann, G.E.A. and Leng, M.J. (2009), 'A review of diatom $\delta^{18}O$ in palaeoceanography', *Quaternary Science Reviews*, 28: 384–398.

Sylla, M.B., Gaye, A.T., Jenkins, G.S., Pal, J.S. and Giorgi, F. (2010), 'Consistency of projected drought over the Sahel with changes in the monsoon circulation and extremes in a regional climate model projections', *Journal of Geophysical Research*, 115: D16108, doi:10.1029/2009JD012983.

Sylvestre, F., Servant, M., Servant-Vildary, S., Causse, C., Fournier, M. and Ybert, J.P. (1999), 'Lake-level chronology on the southern Bolivian Altiplano (18–23°S) during late-glacial time and the early Holocene', *Quaternary Research*, 51: 54–66.

Talbot, M.R. (1981), 'Holocene changes in tropical wind intensity and rainfall: Evidence from southeast Ghana', *Quaternary Research*, 16: 201–220.

—— Williams, M.A.J. and Adamson, D.A. (2000), 'Strontium isotope evidence for Late Pleistocene reestablishment of an integrated Nile drainage network', *Geology*, 28: 343–346.

Tallis, J.H. (1975), 'Tree remains in southern Pennine peats', *Nature*, 256: 483–484.

Talon, B. (2010), 'Reconstruction of Holocene high-altitude vegetation cover in the French southern Alps: evidence from soil charcoal', *The Holocene*, 20: 35–44.

Tan, A., Adam, J.C. and Lettenmaier, P. (2011), 'Change in spring snowmelt timing in Eurasian Arctic snowmelt rivers', *Journal of Geophysical Research*, 116: D03101, doi:10.1029/2010JD014337.

Tarasov, L. and Peltier, W.R. (2005), 'Arctic freshwater forcing of the Younger Dryas cold reversal', *Nature*, 435: 662–665.

Taschetto, A.S. and England, M.H. (2007), 'An analysis of late twentieth century trends in Australian rainfall', *International Journal of Climatology*, 29: 791–807.

Taylor, K.C., Hammer, C.U., Alley, R.B., Clausen, H.B., Dahl-Jensen, D., Gow, A.J., Gundestrup, N.S., Kipfstuhl, J., Moore, J.C. and Waddington, E.D. (1993), 'Electrical conductivity measurements from the GISP2 and GRIP Greenland ice cores', *Nature*, 366: 549–552.

—— and 12 others. (1997), 'The Holocene: Younger Dryas transition recorded at Summit, Greenland', *Science*, 278: 825–827.

Tchakerian, V.P. and Lancaster, N. (2001), 'Late Quaternary arid/humid cycles in the Mojave Desert and

Western Great Basin of North America', *Quaternary Science Reviews*, 21: 799–810.

Tegen, I. (2003), 'Modelling the mineral dust aerosol in the climate system', *Quaternary Science Reviews*, 22: 1821–1834.

Tetzlaff, G. and Peters, M. (1986), 'Deep-sea sediments in the eastern equatorial Atlantic off the African coast and meteorological flow patterns over the Sahel', *Geologische Rundschau*, 75: 71–79.

—— Peters, M., Janssen, W. and Adams, L.J. (1989), 'Aeolian dust transport in West Africa', in M. Leinen and M. Sarnthein (eds) *Palaeoclimatology and Palaeometeorology: Modern and Past Patterns of Global Atmospheric Transport* (Dordrecht): 1985–2203.

Thom, B.G., Hails, J.R. and Martin, A.R.H. (1969), 'Radiocarbon evidence against Post-Glacial higher sea-levels in eastern Australia', *Marine Geology*, 7: 161–168.

Thomas, D.S.G. and Shaw, P.A. (1991), *The Kalahari Environment* (Cambridge).

—— Burrough, S.L. and Parker, A.G. (2012), 'Extreme events as drivers of early human behaviour in Africa? The case for variability, not catastrophic drought', *Journal of Quaternary Science*, 27: 7–12.

Thomas, M.F. (1994), *Geomorphology in the Tropics* (Chichester).

Thomas, R.H. (1986), 'Future sea level rise and its early detection by satellite remote sensing', in J.G. Titus (ed.) *Effects of Changing Stratospheric Ozone and Climate Change*, volume 4, (Washington, DC): 19-36.

Thompson, L.G. (2000), 'Ice core evidence for climate change in the Tropics: implications for our future', *Quaternary Science Reviews*, 19: 19–35.

—— and Mosley-Thompson, E. (1981), 'Microparticle concentration variations linked with climatic changes: evidence from polar ice cores', *Science*, 212: 812–815.

—— Mosley-Thompson, E., Bolzan, J.F. and Koci, B.R. (1985), 'A 1500-Year Record of Tropical Precipitation in Ice Cores from the Quelccaya Ice Cap, Peru', *Science*, 229: 971–973.

—— Mosley-Thompson, E., Dansgaard, W. and Grootes, P.M. (1986), 'The Little Ice Age as recorded in the stratigraphy of the tropical Quelccaya Ice Cap', *Science*, 234: 361–364.

—— Mosley-Thompson, E., Davis, J.F., Bolzan, J.F., Dai, J. and Klei, L. (1990), 'Glacial stage ice core records from the subtropical Dunde Ice Cap, China', *Annals of Glaciology*, 14: 288–297.

—— Mosley-Thompson, E., Davis, M.E., Lin, P.-N., Henderson, K.A., Cole-Dai, J., Bolzan, J.F. and Liu, K.-B. (1995), 'Late glacial stage and Holocene tropical ice core records from Huascarán, Peru', *Science*, 269: 46–50.

Thompson, R. and Oldfield, F. (1986), *Environmental Magnetism* (London).

Thompson, W.G., Curran, H.A., Wilson, M.A. and White, B. (2011), 'Sea-level oscillations during the last interglacial highstand recorded by Bahamas corals', *Nature Geoscience*, 4: 684–687.

Thorne, A.G. (1980), 'The longest link: human evolution in Southeast Asia and the settlement of Australia', in J.J. Fox, R.G. Garnaut, P.T. McCawley and J.A.C. Maukie (eds) *Indonesia: Australian Perspectives* (Canberra), 35–43.

Tibby, J. (2012), 'The Younger Dryas: Relevant in the Australian region?', *Quaternary International*, 253: 47–54.

Timmermann, A., Timm, O., Stott, L. and Menviel, L. (2009), 'The roles of CO_2 and orbital forcing in driving Southern Hemispheric temperature variations during the last 21 000 Yr', *Journal of Climate*, 22: 1626–1640.

Tipping, R. (2004), 'Palaeoecology and political history: evaluating driving forces in historic landscape change in southern Scotland', in I.D. Whyte and A.J.L. Winchester (eds) *Society, Landscape and Environment in Upland Britain* (Exeter): 11–20.

Todhunter, P.E. and Cihacek, L.J. (1999), 'Historical reduction of airborne dust in the Red River Valley of the North', *Journal of Soil and Water Conservation*, 54:543-551.

Tolonen, K. (1986), 'Charred particle analysis', in B.E. Berglund (ed.) *Handbook of Holocene Palaeoecology and Palaeohydrology* (Chichester): 485–496.

Tooley, M.J. (1978), *Sea-level Changes: North-west England during the Flandrian stage* (Oxford).

Toon, O.B. (2003), 'African dust in Florida clouds', *Nature*, 424: 623–624.

Torii, M., Lee, T.-Q., Fukuma, K., Mishima, T., Yamazaki, T., Oda, H. and Ishikawa, N. (2001), 'Mineral magnetic study of the Taklimakan desert sands and its relevance to the Chinese loess', *Geophysical Research International*, 146: 416–424.

Törnqvist, T.E., Bick, S.J., González, J.L.,van der Borg, K. and de Jong, A.F.M. (2004), 'Tracking the sea-level signature of the 8.2 ka cooling event: new constraints from the Mississippi Delta', *Geophysical Research Letters*, 31: 1–4.

—— González, J.L., Newsom, L.A., van der Borg, K., de Jong, A.f.m. and Kurnik, C.W. (2004), 'Deciphering Holocene sea-level history on the U.S. Gulf Coast: high resolution record from Mississippi Delta', *Bulletin, Geological Society of America*, 116: 1026-39.

—— and 9 others. (2008), 'Mississippi Delta subsidence primarily caused by compaction of Holocene strata', *Nature Geoscience*, 1: 173–176.

Toscano, M.A. and Macintyre, I.G. (2003), 'Corrected western Atlantic sea-level curve for the last 11,000 years based on calibrated ^{14}C dates from *Acropora palmata* framework and intertidal mangrove peat', *Coral Reefs*, 22: 257–270.

Toth, N. and Schick, K. (2005), 'African origins', in C. Scarre (ed.) *The Human Past: World Prehistory and the Development of Human Societies* (London): 46–83.

Toucanne, S. and 8 others. (2010), 'The first estimation of Fleuve Manche palaeoriver discharge during the last deglaciation: evidence for Fennoscandian ice sheet meltflow in the English Channel ca 20-18 ka ago', *Earth and Planetary Science Letters*, 290: 459–473.

Trauth, M.H., Maslin, M., Deino, M.A., Deino, A. and Strecker, M.R. (2005), 'Late Cenozoic Moisture History of East Africa', *Science*, 203: 2051–2053.

Trembour, F. and Friedman, I. (1984), 'The present status of obsidian hydration dating', in W.C. Mahaney (ed.) *Quaternary Dating Methods* (Amsterdam): 141–151.

Trenberth, K. (2005), 'Uncertainty in hurricanes and global warming', *Science*, 308: 1753–1754.

—— and 11 others. (2007), 'Observations: surface and atmospheric climate change', in S. Solomon and 7 others (eds) *Climate Change 2007: The Physical Science Basis* (Cambridge): 235–336.

Tricart, J. (1974), 'Existence de Periodes Sèches au Quaternaire en Amazonie et dans les Regions Voisines', *Révue de Géomorphologie Dynamique*, 23: 145–158.

Trigger, B. (2003), *Understanding Early Civilizations: A Comparative Study* (New York).

Trimble, S.W. (1981), 'Changes in sediment storage in the Coon Creek basin, driftless area, Wisconsin, 1853 to 1975', *Science*, 214: 181–183.

Tripaldi, A. and Forman, S.L. (2007), 'Geomorphology and chronology of Late Quaternary dune fields of western Argentina', *Palaeogeography, Palaeoclimatology, Palaeoecology*, 251: 300–320.

Tripati, A.K., Roberts, C.D. and Eagle, R.A. (2009), 'Coupling of CO_2 and ice sheet stability over the major climate transitions of the last 20 million years', *Science*, 326: 1394–1397.

Troels-Smith, J. (1956), 'Neolithic period in Switzerland and Denmark', *Science*, 124: 876–879.

Trouet, V., Esper, J., Graham, N.E., Baker, A., Scourse, J.D. and Frank, D.C. (2009), 'Persistent positive North Atlantic Oscillation mode dominated the Medieval Climate Anomaly', *Science*, 324: 78–80.

Tucker, G.B. (1975), 'Climate: is Australia's changing?', *Search*, 6: 323–328.

Tuenter, E., Weber, S.L., Hilgen, F.J. and Lourens, L.J. (2003), 'The response of the African summer monsoon to remote and local forcing due to precession and obliquity', *Global and Planetary Change*, 36: 219–235.

Turner, C. and West, R.G. (1968), 'The subdivision and zonation of interglacial periods', *Eiszeitalter und Gegenwart*, 19: 93–101.

Turner, J. (1962), 'The Tilia decline: an anthropogenic interpretation', *New Phytologist*, 61: 328–341.

Turney, C. (2006), *Bones, Rocks and Stars: The Science of When Things Happened* (London).

—— (2008), *Ice, Mud and Blood: Lessons From Climates Past* (London).

—— Bird, M., Fifield, L., Roberts, R., Smith, M., Dortch, C., Grun, R., Lawson, E., Aycliffe, L., Miller, G., Dortch, J. and Cresswell, R. (2001), 'Early human occupation at Devil's Lair, Southwestern Australia 50 000 years ago', *Quaternary Research*, 55: 3–13.

Tyson, P.D. (1986), *Climatic Change and Variability in Southern Africa* (Cape Town).

—— and Preston-Whyte, R.A. (2000), *The Atmosphere, Weather and Climate of Southern Africa* (Cape Town).

Tzedakis, P.C. (2005), 'Towards an understanding of the response of southern European vegetation to orbital and suborbital climate variability', *Quaternary Science Reviews*, 24: 1585–1599.

—— Hooghiemstra, H. and Palike, H. (2006), 'The last 1.35 million years at Tenaghi Philippon: revised chronostratigraphy and long-term vegetation trends', *Quaternary Science Reviews*, 25: 3416–3430.

Tziperman, E., Toggweiler, J.R., Feliks, Y. and Bryan, K. (1994), 'Instability of the thermohaline circulation with respect to mixed boundary conditions: Is it really a problem for realistic models?', *Journal of Physical Oceanography*, 24: 217–232.

United States Committee for the Global Atmospheric Research Program (1975), *Understanding Climatic Changes: A Program for Action* (Washington, DC).

Unwin, T. (1990), 'Saxon and early Norman viticulture in England', *Journal of Wine Research*, 1: 61–67, doi:10.1080/09571269008717857.

van Andel, T.H. (1989), 'Late Quaternary sea-level changes and archaeology', *Antiquity*, 63: 733–745.

van der Hammen, T. (1974), 'The Pleistocene changes of vegetation and climate in tropical South America', *Journal of Biogeography*, 1: 3–26.

—— and Hooghiemstra, H. (2000), 'Neogene and Quaternary history of vegetation, climate and plant diversity in Amazonia', *Quaternary Science Reviews*, 19: 725–742.

—— Wijmstra, T.A. and Zagwijn, W.H. (1971), 'The floral record of the Late Cenozoic of Europe', in K.K. Turekian (ed.) *The Late Cenozoic Glacial Ages* (New Haven, CT): 391–424.

van der Plassche, O. (ed.) (1986), *Sea-level Research: A Manual for the Collection and Evaluation of Data* (Norwich).

van der Ween, C.J. (2002), 'Polar ice sheets and global sea-level: how well can we predict the future?', *Global and Planetary Change*, 32: 165–194.

van Geel, B., Buurman, J. and Waterbolk, H.T. (1996), 'Archaeological and palaeoecological indications of an abrupt climate change in The Netherlands and evidence for climatological teleconnections around 2650 bp', *Journal of Quaternary Science*, 11: 451–460.

—— and Mook, W.G. (1989), 'High resolution ^{14}C dating of organic deposits using natural atmospheric ^{14}C variations', *Radiocarbon*, 31: 151–155.

—— Raspopov, O.M., Renssen, H., van der Plicht, J., Dergachev, V.A. and Meijer, H.A.J. (1999), 'The role of solar forcing upon climate change', *Quaternary Science Reviews*, 18: 331–338.

van Zinderen Bakker, E.M. (1984), 'Elements for the chronology of late Cainozoic African climate', in W.C. Mahoney (ed.) *Correlation of Quaternary Chronologies* (Norwich): 23–37.

Vaz, J.E. and García-Miragaya, J. (1992), 'Thermoluminescence dating and chemistry of Quaternary sodic alluvial soils in the Venezuelan savanna', *Catena*, 19: 209–220.

Veeh, H.H. and Chappell, J. (1970), 'Astronomical theory of climatic change: Support from New Guinea', *Science*, 167: 862–865.

—— and Veevers, J.J. (1970), 'Sea-level at -175 M off the Great Barrier Reef, 13 600 to 17 000 years ago', *Nature*, 226: 526–527.

Veenstra, H. J. (1970), 'Quaternary North Sea coasts', *Quaternaria*, 12: 169–79.

Velichko, A. and Spasskaya, I. (2003), 'Climate change and the development of landscapes', in M. Shahgedanova (ed.) *The Physical Geography of Eurasia* (Oxford): 43–66.

Venteris, E.R. (1999), 'Rapid tide water glacier retreat: a comparison between Columbia Glacier, Alaska and Patagonian calving glaciers', *Global and Planetary Change*, 22: 131–138.

Vermeer, M. and Rahmstorf, S. (2009), 'Global sea level linked to global temperature', *Proceedings of the National Academy of Sciences*, 106: 21527–21532.

Vermeesch, P., Fenton, C.R., Kober, F., Wiggs, G.F.S., Bristow, C.S. and Xu, S. (2010), 'Sand residence times of one million years in the Namib Sand Sea from cosmogenic nuclides', *Nature Geoscience*, 3: 862–865.

Verosub, K.L. and Roberts, A.P. (1995), 'Environmental magnetism: past, present and future', *Journal of Geophysical Research*, 100 (B2): 2175–2192.

Verstappen, H. Th. (1970), 'Aeolian geomorphology of the Thar Desert and palaeoclimates', *Zeitschrift für Geomorphologie, Supplementband*, 10: 104–120.

Vibe, C. (1967), 'Arctic animals in relation to climatic fluctuations', *Meddelelser om Grønland*, 170: 227ff.

Vidic, N.J. and Montañez, I.P. (2004), 'Climatically driven glacial-interglacial variations in C_3 and C_4 plant proportions on the Chinese Loess Plateau', *Geology*, 32: 337–340.

Viles, H.A. and Goudie, A.S. (2003), 'Interannual, decadal and mutidecadal scale climatic variability and geomorphology', *Earth-Science Reviews*, 61: 105–131.

Vincent, E. and Berger, W.H. (1985), 'Carbon dioxide and polar cooling in the Miocene: The Monterey Hypothesis', in E.T. Sundquist and W.S. Broecker (eds) *The Carbon Cycle and Atmospheric CO$_2$: Natural*

Variations Archean to Present (Washington DC, American Geophysical Union): 455–468.

Vinnikov, K.Y., Robock, A., Stouffer, R.J., Walsh, J.E., Parkinson, C.L., Cavalieri, D.J., Mitchell, J.F.B., Garrett, D. and Zakharov, V.F. (1999), 'Global warming and Northern Hemisphere sea ice extent', *Science*, 286: 1934–1937.

Vita-Finzi, C. (1973), *Recent Earth History* (London).

—— (1986), *Recent Earth Movements: An Introduction to Neotectonics* (London).

Vogel, J.C. (ed.) (1984), *Late Cainozoic Palaeoclimates of the Southern Hemisphere* (Rotterdam).

Voigt, R. (2006), 'Settlement history as reflection of climate: the case study of Lake Jues (Harz Mountains, Germany)', *Geografiska Annaler*, 88A: 97–105.

Völkel, J. and Grunert, J. (1990), 'To the problem of dune formation and dune weathering during the Late Pleistocene and Holocene in the southern Sahara and the Sahel', *Zeitschrift für Geomorphologie*, 34: 1–17.

von Grafenstein, U., Erlenkueser, H., Müller, J., Jouzel, J. and Johnsen, S. (1998), 'The cold event 8200 years ago documented in oxygen isotope records of precipitation in Europe and Greenland', *Climate Dynamics*, 14: 73–81.

—— Erlenkueser, H., Brauer, A., Jouzel, J. and Johnsen, S.J. (1999), 'A Mid-European decadal isotope-climate record from 15,500 to 5000 years B.P.', *Science*, 284: 1654–1657.

Voris, H.K. (2000), 'Maps of Pleistocene sea levels in Southeast Asia: shorelines, river systems and time durations', *Journal of Biogeography*, 27: 1153–1167.

Walden, J., Oldfield, F. and Smith, J. (eds) (1999), *Environmental Magnetism: A Practical Guide*. Technical Guide 6, Quaternary Research Association, London.

Walker, I.R. (2001), 'Midges: Chironomidae and related Diptera', in J.P. Smol, H.J.B. Birks and W.M. Last (eds) *Tracking Environmental Change using Lake Sediments, Vol. 4: Zoological indicators* (Dordrecht).

Walker, M.J.C. (2005), *Quaternary Dating Methods* (Chichester).

—— and 18 others. (2009), 'Formal definition and dating of the GSSP (Global Stratotype Section and Point) for the base of the Holocene using the Greenland NGRIP ice core, and selected auxiliary records', *Journal of Quaternary Science*, 24: 3–17.

—— Lowe, J., Blockley, S.P.E., Bryant, C., Coombes, P., Davies, S., Hardiman, M., Turney, C.S.M. and Watson, J. (2012), 'Lateglacial and early Holocene palaeoenvironmental 'events' in Sluggan Bog, Northern Ireland: comparisons with the Greenland NGRIP GICC05 event stratigraphy', *Quaternary Science Reviews*, 36: 124–138.

Walsh, K. (2004), 'Tropical cyclones and climate change: unresolved issues', *Climate Research*, 27: 77–83.

Walsh, R.P., Hudson, R.N. and Howells, K.A. (1982), 'Changes in the magnitude-frequency of flooding and heavy rainfalls in the Swansea Valley since 1875', *Cambria*, 9: 36–60.

—— Hulme, M. and Campbell, M.D. (1988), 'Recent rainfall changes and their impact on hydrology and water supply in the semi-arid zone of the Sudan', *Geographical Journal*, 154: 181–198.

Wanamaker, A.D., Heinemeier, J., Scourse, J.D., Richardson, C.A., Butler, P.G., Eiriksson, J. and Knudsen, K.L. (2008), 'Very long-lived mollusks confirm 17th century AD tephra-based radiocarbon reservoir ages for North Icelandic shelf waters', *Radiocarbon*, 50: 399–412.

Wang, X., Auler, A.S., Edwards, R.L., Cheng, H., Cristalli, P.S., Smart, P.L., Richards, D.A. and Shen, C.-C. (2004a), 'Wet periods in northeastern Brazil over the past 210 kyr linked to distant climate anomalies', *Nature*, 432: 740–743.

—— Dong, Z., Zhang, J., Liu, L. (2004b), 'Modern dust storms in China: an overview', *Journal of Arid Environments*, 58: 559–574.

Wang, X.L. and Swail, V.R. (2001), 'Changes of Extreme Wave Heights in Northern Hemisphere Oceans and Related Atmospheric Circulation Regimes', *Journal of Climate*, 14: 2204–2221.

—— Eerdun, H., Zhou, Z. and Liu, X. (2007), 'Significance of variations in the wind energy environment over the past 50 years with respect to dune activity and desertification in arid and semiarid northern China', *Geomorphology*, 86: 252–266.

Wang, Y., Chang, H., Edwards, R.L., Kong, X., Shao, X., Chen, S., Wu, J., Jiang, X., Wang, X. and An, Z. (2008), 'Millennial- and orbital-scale changes in the East Asian monsoon over the past 224,000 years', *Nature*, 451: 1090–1093.

Wanner, H. and 17 others. (2008), 'Mid- to Late Holocene climate change: an overview', *Quaternary Science Reviews*, 27: 1791–1828.

Ward, J.D. (1988), 'Eolian, fluvial and pan (playa) facies of the Tsondab Sandstone Formation in the Central Namib Desert, Namibia', *Sedimentary Geology*, 55: 143–162.

Warrick, R.A. and Oerlemans, J. (1990), 'Sea level rise', in J.T. Houghton, G.J. Jenkins, G.J.J. Ephraums (eds), *Climate Change: the IPCC Scientific Assessment*, (Cambridge): 257-281.

Wasson, R.J. (1984), 'Late Quaternary palaeoenvironments in the desert dunefields of Australia', in J.C. Vogel (ed.) *Late Cainozoic Palaeoclimates of the Southern Hemisphere* (Rotterdam): 419–432.

—— Fitchett, K., Mackey, B. and Hyde, R. (1988), 'Large-scale patterns of dune type, spacing and orientation in the Australian continental dunefield', *Australian Geographer*, 19: 80–104.

Waters, M.R., Forman, S.L., Jennings, T.A., Nordt, L.C., Driese, S.G., Feinberg, J.M., Keene, J.L., Halligan, J., Lindquist, A., Pierson, J., Hallmark, C. T., Collins, M. B. and Wiederhold, J. B. (2011a), 'The Buttermilk Creek Complex and the Origins of Clovis at the Debra L. Friedkin Site, Texas', *Science*, 331: 1599–1603.

—— Stafford, T.W., McDonald, H.G., Gustafson, C., Rasmussen, M., Cappellini, E., Olsen, J.V., Szklarczyk, D., Jensen, L.J., Gilbert, M.T.P. and Willerslev, E. (2011b), 'Pre-Clovis mastodon hunting 13,800 years ago at the Manis site, Washington', *Science*, 334: 351–353.

Watts, A.B. (2001), *Isostasy and Flexure of the Lithosphere* (Cambridge).

Weart, S.R. (2003), *The Discovery of Global Warming* (Cambridge, MA).

Webb, R.S., Rind, D.H., Lehman, S.J., Healy, R.J. and Sigman, D. (1997), 'Influence of ocean heat transport on the climate of the Last Glacial Maximum', *Nature*, 385: 695–699.

Webb, S. (2009), 'Palaeotrophic reconstruction and climatic forcing of mega-Lake Eyre in late Quaternary Central Australia: a review', *Boreas*, 39: 312–324.

Webb III, T., Cashing, E.J. and Wright, H.R. (1983), 'Holocene changes in the vegetation of the Midwest', in H.R. Wright, Jr. (ed.) *Late Quaternary Environments of the United States, Vol. 2: The Holocene* (Minneapolis), 142–165.

Weber, M.E., Clark, P.U., Ricken, W., Mitrovica, J.X., Hostetler, S.W. and Kuhn, G. (2011), 'Interhemispheric ice-sheet synchronicity during the Last Glacial Maximum', *Science*, 334: 1265–1269.

Webster, J.M. and 10 others. (2009), 'Coral reef evolution on rapidly subsiding margins', *Global and Planetary Change*, 66: 129–148.

Webster, P.J., Holland, G.J., Curry, J.A. and Chang, H-R. (2005), 'Changes in tropical cyclone number, duration, and intensity in a warming environment', *Science*, 309: 1844–1846.

Wegmann, E. (1969), 'Changing ideas about moving shorelines', in C.J. Scheer (ed.) *Toward a History of Geology* (Cambridge, MA): 386–414.

Wehmiller, J.F. and Miller, G.H. (2000), 'Aminostratigraphic dating methods in Quaternary geology', in J.S. Noller, J.M. Sowers and W.R. Letts (eds) *Quaternary Geochronology, Methods and Applications*, American Geophysical Union: Washington, D.C., Reference Shelf 4: 187–222.

Weiss, H., Courty, M.A., Wellerstrom, W., Guichard, F., Senior, L., Meadow, R. and Currow, A. (1993), 'The genesis and collapse of Third Millennium north Mesopotamian Civilization', *Science*, 291: 995–1088.

Wells, G.L. (1983), 'Late glacial circulation over North America revealed by aeolian features', in F.A. Street-Perrott, M. Bevan and R. Ratcliffe (eds) *Variation in the Global Water Budget* (Dordrecht): 317–330.

Wells, L.E. and Noller, J.Y. (1999), 'Holocene coevolution of the physical landscape and human settlement in northern coastal Peru', *Geomorphology*, 14: 755–789.

Wells, P.V. (1976), 'Macrofossil analysis of wood rat (*Neotoma*) middens as a key to the Quaternary vegetational history of arid America', *Quaternary Research*, 6: 223–248.

Werner, M., Tegen, I., Harrison, S.P., Kohfeld, K.E., Prentice, I.C., Balkanski, Y., Rodhe, H. and Roelandt, C. (2002), 'Seasonal and interannual variability of the mineral dust cycle under present and glacial climate conditions', *Journal of Geophysical Research*, 107: D24, doi: 10.1029/2002JD002365.

Werth, D. and Avissar, R. (2002), 'The local and global effects of Amazon deforestation', *Journal of Geophysical Research – Atmospheres*, 107 (D20): article no. 8087.

West, R.G. (1977), *Pleistocene Geology and Biology*, 2nd edition (London).

Whittow, J. B. (1970) 'Shoreline evolution and the eastern coast of the Irish Sea', *Quaternaria*, 12: 185–96.

White, K.H. and Mattingley, D.J. (2006), 'Ancient lakes of the Sahara', *American Scientist*, 94: 58–65.

White, T.D., Suwa, G. and Asfaw, B. (1994), '*Australopithecus ramidus*, a new species of early hominid from Aramis, Ethiopia', *Nature*, 371: 306–312.

——Asfaw, B., Beyene,Y., Haile-Selassie, Y., Lovejoy, C.O., Suwa, G. and WoldeGabriel, G. (2009), '*Ardipithecus ramidus* and the Paleobiology of Early Hominids', *Science,* 326: 75–86.

Whittlesey, D. (1868), 'Depression of the ocean during the ice period', *American Association for the Advancement of Science, Proceedings*, 16: 92–97.

Wigley, T.M.L. and Atkinson, T.C. (1977), 'Dry years in south-east England since 1698', *Nature*, 265: 431–434.

—— and Jones, P.D. (1987), 'England and Wales precipitation: a discussion of recent changes in variability and an update to 1985', *Journal of Climatology*, 7: 231–246.

—— and Raper, S.C.B. (1992), 'Implications for climate and sea level of revised IPCC emissions scenarios', *Nature*, 357: 293-300.

—— and Raper, S.C.B. (1993), 'Future changes in global mean temperature and sea level', in R.A.Warrick, E.M. Barrow and T.M.L. Wigley (eds) *Climate and Sea Level Change,* (Cambridge): 111-133.

Wild, C., Wells, C., Anderson, D., Boardman, J. and Parker, A. (2001), 'Evidence for Medieval Clearance in the Seathwaite Valley, Cumbria', *Transactions of the Cumberland and Westmorland Antiquarian and Archaeological Society*, I: 53–68.

Wilkinson, T.J. (2003), *Archaeological Landscapes of the Near East* (Chicago).

—— (2005), 'Soil erosion and valley fills in the Yemen Highlands and Southern Turkey: integrating settlement, geoarchaeology and climatic change', *Geoarchaeology*, 20: 169–192.

Willcox, G., Buxo, R. and Herveux, L. (2009), 'Late Pleistocene and early Holocene climate and the beginnings of cultivation in northern Syria', *The Holocene*, 19: 151–158.

Williams, J. (1975), 'The influence of snowcover on the atmospheric circulation and its role in climatic change: an analysis based on results from the near global circulation model', *Journal of Applied Meteorology*, 14: 137–152.

Williams, J.W., Shuman, B.N., Webb III, T., Bartlein, P.J. and Leduc, P.L. (2004), 'Late-Quaternary vegetation dynamics in North America: scaling from taxa to biomes', *Ecological Monographs*, 74: 309–334.

—— Shuman, B. and Bartlein, P.J. (2009), 'Rapid responses of the prairie-forest ecotone to early Holocene aridity in mid-continental North America', *Global and Planetary Change*, 66: 195–207.

Williams, M.A.J. (1985), 'Pleistocene aridity in tropical Africa, Australia and Asia', in I. Douglas and T. Spencer (eds) *Environmental Change and Tropical Geomorphology* (London): 219–233.

—— and Adamson, D.A. (1974), 'Late Pleistocene desiccation along the White Nile', *Nature*, 248: 584–586.

—— and Adamson, D.A. (eds) (1980), *A Land between Two Niles. Quaternary Geology and Biology of the Central Sudan* (Rotterdam).

—— and Balling, R.C. Jr (1996), *Interactions of Desertification and Climate* (London).

—— and Clarke, M.F. (1995), 'Quaternary geology and prehistoric environments in the Son and Belan valleys, north central India', *Memoirs Geological Society of India*, 32: 282–308.

—— Dunkerley, D.L., de Deckker, P., Kershaw, A.P. and Stokes, T. (1993, 1998), *Quaternary Environments* (London).

—— Adamson, D., Cock, B. and McEvedy, R. (2000), 'Late Quaternary environments in the White Nile region, Sudan', *Global and Planetary Change*, 26: 305–316.

—— Ambrose, S.H., van der Kaars, S., Ruehlemann, C., Chattopadhyaya, U., Pal, J. and Chauhan, P.R. (2009), 'Environmental impact of the 73 ka Toba super-eruption in South Asia', *Palaeogeography, Palaeoclimatology, Palaeoecology*, 284: 295–314.

Williams, R.B.G. (1975), 'The British climate during the Last Glaciation: an interpretation based on periglacial phenomena', in A.E. Wright and F. Moseley (eds) *Ice Ages: Ancient and Modern* (Liverpool): 95–127.

Willis, K.J. and van Andel, T.H. (2004), 'Trees or no trees? The environments of central and eastern Europe during the Last Glaciation', *Quaternary Science Reviews*, 23: 2369–2387.

—— and Whittaker, R.J. (2000), 'The refugia debate', *Science*, 287: 1406–1407.

Wilson, J.S. and Pitts, J.P. (2010), 'Illuminating the lack of consensus among descriptions of earth history data in the North American deserts: a resource for biologists', *Progress in Physical Geography*, 34: 419–441.

Wilson, R.C.L., Drury, S.A. and Chapman, J.L. (2000), *The Great Ice Age: Climate Change and Life* (London).

Winograd, I.J., Coplen, T.B., Landwehr, J.M., Riggs, A.C., Ludwig, K.R., Szabo, B.J., Kolesar, P.T. and Revesz, K.M. (1992), 'Continuous 500 000-year climate record from vein calcite in Devils Hole, Nevada', *Science*, 258, 255–260.

Winspear, N.R. and Pye, K. (1995), 'Textural, chemical and mineralogical evidence for the origin of Peoria Loess in central and southern Nebraska, USA', *Earth Surface Processes and Landforms*, 20: 735–745.

Wittfogel, K. (1957), *Oriental Despotism* (New Haven, CT).

Woillard, G. (1978), 'Grande Pile peat bog: a continuous pollen record for the last 140 000 years', *Quaternary Research*, 9: 1–21.

—— (1979), 'Abrupt end of the last interglacial s.s. in north-east France', *Nature*, 281: 558–562.

—— and Mook, W.G. (1982), 'Carbon-14 dates at Grand Pile: correlation of land and sea chronlogies', *Science*, 215: 159–161.

Wolpoff, M.H., Thorne, A.G., Jelínek, J. and Zhang, Y. (1994), 'The case for sinking *Homo erectus*. 100 years of *Pithecanthropus* is enough!', *Courier Forschungsinstitut Senckenberg*, 171: 341–361.

Wong, S., Dessler, A.E., Mahowald, N., Colarco, P.R. and Da Silva, A. (2008), 'Long-term variability in Saharan dust transport and its link to North Atlantic sea surface temperature', *Geophysical Research Letters*, 35: L07812, doi:10.1029/2007GL032297.

Wood, C.A. and Lovett, R.R. (1974), 'Rainfall, drought and the solar cycle', *Nature*, 252: 594–596.

Wood, K.R. and Overland, J.E. (2010), 'Early 20th century warming in retrospect', *International Journal of Climatology*, 30: 1269-1279.

Wood, W. and Imes, J.L. (1995), 'How wet is wet? Precipitation constraints on late Quaternary climate in the southern Arabian peninsula', *Journal of Hydrology*, 164: 263–268.

Woodroffe, C. (2000), 'Deltaic and estuarine environments and their late Quaternary dynamics on the Sunda and Sahul shelves', *Journal of Asian Earth Sciences*, 18: 393–413.

—— (2002), *Coasts: Form, Process and Evolution* (Cambridge).

—— (2005), 'Late Quaternary sea-level highstands in the central and eastern Indian Ocean: a review', *Global and Planetary Change*, 49: 12–138.

Woodroffe, S.A. and Horton, B.P. (2005), 'Holocene sea-level changes in the Indo-Pacific', *Journal of Asian Earth Sciences*, 25: 29–43.

Woodward, J.C., Macklin, M.G. and Welsby, D. (2001), 'The Holocene fluvial sedimentary record and alluvial geoarchaeology in the Nile Valley of northern Sudan', in D. Maddy, M.G. Macklin and J.C. Woodward (eds) *River Basin Sediment Systems: Archives of Environmental Change* (Cambridge): 327–355.

Wooldridge, S.W. (1951), 'The progress of geomorphology', in G. Taylor (ed.) *Geography in the Twentieth Century* (London): ch. 7.

World Glacier Monitoring Service. (2008), 'Global glacier changes: facts and figures', www.grid.unep.ch/glaciers/ (accessed 29th October, 2011).

World Meteorological Organization (1995), *Climate System Review* (Geneva).

Worsley, P. (1990), 'Lichenometry', in A.S. Goudie (ed.) *Geomorphological techniques* (London): 422–428.

Wright, H.E., Kutzbach, J.E., Webb, T., Ruddiman, W.F., Street-Perrott, F.A. and Bartlein, P.J. (eds) (1993), *Global Climates since the Last Glacial Maximum* (Minneapolis).

Wright, H.E., Jr. and Thorpe, J. (2005), 'Climatic change and the origin of agriculture in the Near East', in A. Mackay, R. Battarbee, J. Birks and F. Oldfield (eds) *Global Change in the Holocene* (London): 49–62.

Wright, W.B. (1914), *The Quaternary Ice Age* (London).

—— (1937), *The Quaternary Ice Age*, 2nd edition (London).

Wrozyna, C., Frenzel, P., Steeb, P., Zhu, L. and Schwalb, A. (2009), 'Recent lacustrine Ostracoda and a first transfer function for palaeo-water depth estimation in Nam Co, southern Tibetan Plateau', *Revista Española de Micropalaeontologia*, 41: 1–20.

Wyatt, M.G., Kravtsov, S. and Tsonis, A.A. (2011), 'Atlantic Multidecadal Oscillation and Northern Hemisphere's climate variability', *Climate Dynamics*: doi:10.1007/s00382-011-1071-8.

Xu, C., Song, C., Jiang, W. and Shi, C. (2000), 'GPS measurements of present-day uplift in the Southern Tibet', *Earth Planets Space*, 52: 735–739.

Xu, M., Chang, C-P., Fu, C., Qi, Y., Robock, A., Robinson, D. and Zhang, H-M. (2006), 'Steady decline of east Asian monsoon winds, 1969-2000: evidence from direct ground measurements of wind speed', *Journal of Geophysical Research*, 111: D24111, doi:10.1029/2006JD007337.

Yang, P., Hong, G., Dessler, A.E., Ou, S.S.C., Liou, K.-N., Minnis, P. and Vardhan, H. (2010), 'Contrails and induced cirrus', *Bulletin of the American Meteorological Society*, April, 2007: 473–478.

Yang, S.L. and Ding, Z.L. (2004), 'Comparison of particle size characteristics of the tertiary "red clay" and Pleistocene loess in the Chinese Loess Plateau: implications for origin and sources of the "red clay"', *Sedimentology*, 51: 77–93.

Yang, Y.Q., Hou, Q., Zhou, C.H., Liu, H.L., Wang, Y.Q. and Niu, T. (2008), 'Sand/dust storm processes in Northeast Asia and associated large-scale circulations', *Atmospheric Chemistry and Physics*, 8: 25–33.

Yatagai, S., Takemura, K., Naruse, T., Kitagawa, H., Fukusawa, H., Kim, M.-H. and Yasuda, Y. (2002), 'Monsoon changes and eolian dust deposition over the past 30 000 years in Cheju Island, Korea', *Transactions, Japanese Geomorphological Union*, 23: 821–831.

Yener, K.A. and Özbal, H. (1987), 'Tin in the Turkish Taurus Mountains: the Bolkardag mining district', *Antiquity*, 61: 220–226.

Yesner, D.R. (2001), 'Human dispersal into interior Alaska: antecedent conditions, mode of colonization, and adaptations', *Quaternary Science Reviews*, 20: 315–327.

Yin, X. and Nicholson, S.E. (2002), 'Interpreting annual rainfall from the Levels of Lake Victoria', *Journal of Hydrometeorology*, 3: 406–416.

Yokoyama, Y., Lambeck, K., de Dekker, P., Johnston, P. and Fifield, L.K. (2000), 'Timing of the Last Glacial Maximum from observed sea level minima', *Nature*, 406: 713–716.

—— Esat, T.H. and Lambeck, K. (2001), 'Coupled climate and sea-level change deduced from Huon Peninsula coral terraces of the last ice age', *Earth and Planetary Science Letters*, 193: 579–587.

Young, N.E., Briner, J.P., Stewart, H.A.M., Axford, Y., Csatho, B., Rood, D.H. and Finkel, R.C. (2011), 'Response of Jakobshavn Isbrae, Greenland, to Holocene climate change', *Geology*, 39: 131–134.

Yu, Z. (2007), 'Rapid response of forested vegetation to multiple climatic oscillations during the last deglaciation in the northeastern United States', *Quaternary Research*, 67:29–303.

Zachos, J., Pagani, M., Sloan, L., Thomas, E. and Billups, K. (2001), 'Trends, rhythms and aberrations in global climate 65 Ma to present', *Science*, 292: 686–693.

Zalasiewicz, J., Williams, M., Haywood, A. and Michaels, E. (2011), 'The Anthropocene: a new epoch of geological time?', *Philosophical Transactions of the Royal Society, London*, A, 369: 835–841.

Zarate, M.A. (2003), 'Loess of southern South America', *Quaternary Science Reviews*, 22: 1987–2006.

Zeeberg, J. and Forman, S.L. (2001), 'Changes in glacier extent of north Novaya Zemlya in the twentieth century', *The Holocene*, 11: 161–175.

Zeng, N. and Yoon, J. (2009), 'Expansion of the world's deserts due to vegetation-albedo feedback under global warming', *Geophysical Research Letters*, 36: L17401, doi:10.1029/2009GL039699.

Zeuner, F.E. (1945), *The Pleistocene Period: Its Climate, Chronology and Faunal Successions* (London).

—— (1959), *The Pleistocene Period: Its Climate, Chronology and Faunal Successions*, 2nd edition (London).

Zhang, L. and Zhou, T. (2011), 'An assessment of monsoon precipitation changes during 1901–2001', *Climate Dynamics*: doi 10.1007/s00382-011-0993-5.

Zhang, X. and 7 others. (2007), 'Detection of human influence on twentieth-century precipitation trends', *Nature*, 448: 461–465.

Zhang, Z. and Sun, J. (2011), 'Palynological evidence for Neogene environmental change in the foreland basin of the southern Tianshan range, northwestern China', *Global and Planetary Change*, 75: 56–66.

—— Zhao, M., Eglington, G., Lu, H. and Huang, C.Y. (2006), 'Leaf wax lipids as paleovegetational and paleoenvironmental proxies for the Chinese loess plateau over the last 170 kyr', *Quaternary Science Reviews*, 25: 575–594.

Zhao, J., Cao, Y. and Shi, J. (2010), 'Spatial variation in the Arctic Oscillation and its long-term change', *Tellus*, 62A: 661–672.

Zheng, H., Tada, R., Jia, J., Lawrence, C. and Wang, K. (2010), 'Cenozoic sediments in the southern Tarim Basin: implications for the uplift of northern Tibet and evolution of the Taklimakan Desert', *Geological Society of London Special Publication*, 342: 67–78.

Zhong, S.L., Zhu, Y.H., Wilems, H. and Xu, J.L. (2003), 'On the use of coccoliths to reconstruct dust source areas and transport routes: an example from the Malan loess of the Penglai district, Northwestern Shandong Peninsula, China', *Courier Forschungsinstitut Senckenberg*, 244: 129–135.

Zhong, Y., Miller, G.H., Otto-Bliesner, B.L., Holland, M.M., Bailey, D.A., Schneider, D.P. and Geirsdottir, A. (2011), 'Centennial-scale climate change from decadally-paced explosive volcanism: a coupled sea ice-ocean mechanism', *Climate Dynamics*, 37: 2373–2387.

Zhou, L.P., Dodonov, A.E. and Shackleton, N.J. (1995), 'Thermoluminescence dating of the Orkutsay loess section in Tashkent Region, Uzbekistan, Central Asia', *Quaternary Science Reviews*, 14: 721–730.

Zhou, X.K.K. and Zhai, P.M.M. (2004), 'Relationship between vegetation coverage and spring dust storms over northern China', *Journal of Geophysical Research*, 109D: article D03104.

Zhou, Z. and Zhang, G. (2003), 'Typical severe dust storms in northern China during 1954–2002', *Chinese Science Bulletin*, 48: doi: 10.1360/03wd0029.

Zhu, R.X., Potts, R., Xie, F., Hoffman, K.A., Deng, C.L., Shi, K.B., Pan, Y.X., Wang, H.Q., Shi, R.P., Wang, Y.C., Shi, G.H. and Wu, H.Q. (2004), 'New evidence on the earliest human presence at high northern latitudes in northeast Asia', *Nature*, 431: 559–562.

Zielhofer, C. and Linstädter, J. (2006), 'Short-term mid-Holocene climatic deterioration in the West Mediterranean region: climatic impact on Neolithic settlement pattern?', *Zeitschrift für Geomorphologie, Supplementband*, 142: 1–17.

Zielinski, G.A., Mayewski, P.A., Meeker, L.D., Whitlow, S., Twickler, M.S., Morrison, M., Meese, D.A., Gow, A.J. and Alley, R.B. (1994), 'Record of volcanism since 7000 b.c. from the GISP 2 Greenland Ice Core and implications for the volcano-climate system', *Science*, 264: 948–952.

—— Mayewski, P.A., Meeker, D.L., Whitlow, S. and Twickler, M.S. (1996), 'A 110 000-yr record of explosive volcanism from the GISP2 (Greenland) Ice Core', *Quaternary Research*, 45: 109–118.

—— Mayewski, P.A., Meeker, D.L, Grönvold, K., Germani, M.S., Whitlow, S., Twickler, M.S. and Taylor, K. (1997), 'Volcanic aerosol records and tephrochronology of the Summit, Greenland, ice cores', *Journal of Geophysical Research*, 102: 26625–26640.

Zinck, J.A. and Sayago, J.M. (2001), 'Climatic periodicity during the late Pleistocene from a loess-paleosol sequence in northwest Argentina', *Quaternary International*, 78: 11–16.

Zonneveld, K.A.F., Ganssen, G., Troelrma, S., Versteegh, G.J.M. and Visser, H. (1997), 'Mechanisms forcing abrupt fluctuations of the Indian Ocean Summer Monsoon during the last deglaciation', *Quaternary Science Reviews*, 16: 187–201.

Index